The Disruption of Thought

Homo Technologicus, Social and Ethical Futures

Editors

James Giordano, Niko Kohls, and John Shook

VOLUME 3

The titles published in this series are listed at *brill.com/htse*

The Disruption of Thought

In an Age of Accelerating Technology

By

Patrick Scannell

With foreword by

Vint Cerf

BRILL

ISSN 2666-8769
ISBN 978-90-04-73714-3
ISBN 978-90-04-73916-1 (PDF)
DOI https://doi.org/10.1163/9789004739161

Library of Congress Control Number: 2026939341

Brill and Koninklijke Brill BV are part of De Gruyter Brill.
www.degruyterbrill.com

Questions about General Product Safety Regulation: productsafety@degruyterbrill.com

ISSN 2666-8769
ISBN 978-90-04-73714-3
ISBN 978-90-04-73916-1 (PDF)
DOI https://doi.org/10.1163/9789004739161

Library of Congress Control Number: 2026939341

Brill and Koninklijke Brill BV are part of De Gruyter Brill.
www.degruyterbrill.com

Questions about General Product Safety Regulation: productsafety@degruyterbrill.com

The Disruption of Thought

In an Age of Accelerating Technology

By

Patrick Scannell

With foreword by

Vint Cerf

BRILL

I dedicate this work, and everything I do, to the lights of my life –
my wife, ***Tami****, and our daughters,* ***Mackenzie and Sloane***

You carried me through this long journey with patience, love, and belief.
This book is a testament to the life and meaning we have built together

The journey is the destination

I also dedicate this work to my mother, ***Rindy Scannell****, who somehow raised five of us on her own; to* ***Mary Lou and Bob****, for modeling what loving, grounded parenting looks like; and to* ***Jay, John, Shannon, Matt, my cousin Dee, the Auntie Camp crew****, and the rest of my family*

I love you all deeply. Family, by birth and by choice, is everything

The views expressed herein are those of the author, and do not reflect the opinions or policies of any employer or organization I may work for or with whom I associate.

•••

Cover image evokes the long artistic tradition of the "thinker contemplating mortality," including Hugo Rheinhold's Affe mit Schädel (1893).

∵

Contents

PART 3
Insights from the Model

PART 4
The Path Forward

Foreword

In 2020, the Atlantic Council set out to understand where the world stood at the dawn of a new decade and to develop a roadmap that could lead to a freer, more secure, and more prosperous world a decade later, in 2031. To this end, they convened a commission of leaders from industry, government, academia, and national security, and I was honored to be included.

Our commission's mandate: report on the geopolitical impacts of new technologies and data, to serve the United States and its allies, members of Congress, the new presidential administration, private industry, policy makers, and like-minded nations.

Our work focused on technology – that forcing function that now ubiquitously dominates our world, impacting geopolitics, global competition, and global opportunities for collaboration. It has created a sophisticated but potentially fragile ecosystem of networks and platforms that now connect people and nations. One of our central goals was to find ways to incorporate resiliency into this ecosystem, which had become a necessary foundational pillar of modern life.

While previous generations have consistently relied on technologies and tools to improve societies, we find ourselves in an unprecedented era where new technologies and data are reshaping societies and geopolitics in novel and even unanticipated ways. Advancing speed, scale, and sophistication of new technologies and data capabilities do, in fact, aid us in many ways, but they also contain unprecedented ways of disrupting our interconnected world.

Just before the report was officially released in May 2021, amid the pandemic, several of us commissioners and project staff held a virtual panel titled Geotech Decade Ahead, where we shared many of the report's findings. During that talk, I was discussing the rapid propagation of misinformation and disinformation, which were enabled in large part by social media, when I made this statement:

> I think we need a much deeper understanding of how people think and why they think the way they think in order to defend against the abusive use of these social media companies that make these applications.

A few months later, Pat Scannell reached out to me about this specific statement, saying he was working on a book that addressed this exact topic and wanted to talk to me about it.

I was familiar with Scannell through some organizations we both participated in, but I didn't know him well. I remembered that he had recently co-authored a book on misinformation and disinformation with several industry experts, aimed at facilitating vaccination campaigns during the height of the pandemic, among other things.

But in his email to me, however, Scannell claimed the issue wasn't just with misinformation, disinformation, AI, or social media, but with *all* technology. He claimed that rapidly accumulating and accelerating technological innovation was literally disrupting how we think, and that this was the biggest issue facing humanity, the root cause of many issues in the world today – from an increasingly polarized society to declining emotional well-being and physical health.

Bold statements, to be sure.

We had a few conversations, and then he asked me to read the book and consider writing this foreword. After reading it, I do believe it helps provide "*a much deeper understanding of how people think and why they think the way they think*."

This book is many things but among the most notable is the extremely deep literature survey across a wide range of disciplines undertaken by Pat Scannell, which led him to this thesis – speaking roughly – that our very means and modes of thought are being disrupted by the technology and tools brought on by the semiconductor revolution and its consequences: computing, networking, and software.

One thing this book is not is alarmist.

Rather, it ironically made me think harder about my own daily use of and reliance on programmable technology. Some of this is so well-rehearsed that I don't even think about it until I am reminded to do so, as I have been, in reading this book.

The reliance on networked, programmable technology feels a little like the "muscle memory" we gain when we learn to ride a bicycle or even tie a tie. I do little of the former and a lot of the latter but in both cases, I don't have to think a lot about how to do it.

As we become more dependent on these technologies, Scannell argues that the nature of our thought processes is changed. The information overload has been with us for a long time, well in advance of the Internet and the World Wide Web. We coped with not watching every movie, not reading every book or newspaper, not listening to every TV or radio show, not even reading all the junk mail that ends up in our mailboxes.

We filtered, got recommendations, and turned to sources we trusted to select what we consumed. In the online world, algorithms do some of this for us – however, we may not be in as much control of these as we have been in the past when we ask friends and experts what to read or attend to in the information sphere.

We no longer have to know about how to care for horses. With some exceptions, most of us rely on others to take care of the 275+ horses in our garages. Even that is changing as more and more of the automobile's function is implemented in software. I drive a Tesla model S and it seems like a computer on wheels that boots up when you activate the car.

Scannell helped me see both the upside and downside of trends, drawing on anecdotes and documents including those from Douglas Engelbart and JCR Licklider, both huge proponents of the augmentation of thought through computing, and both of whom I've had the pleasure to work with in my career.

Scannell makes a strong case that this is a sea change, a true "disruption of thought." He makes a case for the idea that this is a change unlike any other because it never quite settles down as has been the case with earlier "paradigm shifts" [referring to Thomas Kuhn]. The rate of change and diversity of applications puts us in a continual state of adaptation.

He makes a positive case for harnessing this insight, but I also came away feeling that our ability to cope with so much change might prove challenging both to individuals and societies. The solution to this challenge may lie in the very technology that is causing it.

Thankfully, Scannell does not try to sell readers on "AI is the solution to everything." But he does make us think about the role these machine learning technologies may have on our ability to cope with substantial change.

Scannell does something I have not seen elsewhere – at the end of each chapter is a short summary to help readers capture the essence of his intent. Even if you happen to come to different conclusions, this capsule summary is extraordinarily helpful – and something one might read later as a reminder.

So, why would I recommend this book to other readers?

My short summary: It made me think!

Vint Cerf
McLean, VA

Preface – How this Book Came About

> Nearly everything is really interesting if you go into it deeply enough.[1] (Richard Feynman)

> There is nothing like looking, if you want to find something. You certainly usually find something, if you look, but it is not always quite the something you were after.[2] (J.R.R. Tolkien)

The questions that drove this book emerged from an unexpected source: a corporate transaction that gave me time to think.

In 2013, the Google-backed startup where I was an early employee was acquired. The generous severance package allowed me to wrestle with puzzling questions: How do our digital technologies affect us, and how we think? And how does how we think, in turn, influence the technologies we create and use?

These questions worried me because they felt important yet unanswered. While everyone celebrated technological progress, I sensed something deeper happening: a fundamental shift in human cognition that few were tracking.

I had been part of leading this transformation. For years, I helped build the technological infrastructure surrounding us, including launching the world's first national 4G cellular network and catalyzing the global connected device ecosystem we now take for granted. Social media became the 4G killer app, and without ubiquitous high-speed connectivity, today's Facebook, YouTube, and TikTok wouldn't exist.

In boardrooms across the country, I brokered meetings between Cisco and GE executives about "smart grids," strategized with Disney's Chief Technology Officer (CTO) about "monetizing the back seat of the car," met with Amazon to envision what 4G could do for the Kindle, and worked with Sony about transforming mobile gaming. Our cable partners were excited about how 4G would enable their customers to watch television, movies, and videos, or play immersive video games, anytime and anywhere.

Always: more, more, more!

As technologists, we celebrated increasing screen time and connectivity without asking a critical question: what was this doing to people using it?

---

1 Feynman, Richard P. *The Pleasure of Finding Things Out: The Best Short Works of Richard P. Feynman.* Edited by Jeffrey Robbins, Basic Books, 1999, p. 9.

2 Tolkien, J. R. R. *The Hobbit.* HarperCollins, 1995, p. 54.

I suspected that we were part of a problem. Driving down the road, I'd see all the people in the cars around me staring at their phones, even as they drove. My industry was intimately complicit in this.

And it's not just phones. Technologists are revolutionizing nearly every part of life, from smart home devices like doorbells and lightbulbs to self-driving cars and AI-powered tools that reshape work and education. Wearables like the Apple Watch track health, while connected fitness gear like Peloton and Lululemon's Mirror redefine workouts. Even our beds can now adapt to our sleep patterns. This constant innovation transforms how we live, work, and play, with phone-distracted drivers being just one visible symptom of this sweeping change.

1 The Search for Answers

When that acquisition closed and my options vested, I set out to find answers to these basic questions. The existing literature fell into two camps. Techno-utopians like Ray Kurzweil championed technology's potential to create post-scarcity abundance, while critics like Nick Carr and Shoshanna Zuboff warned about unseen dangers such as distracted drivers, Internet-addled brains, surveillance capitalism.

Neither perspective satisfied me. Technology wasn't all good or all bad. I wasn't concerned with one technology or another; they changed so quickly that specific concerns became obsolete before research could develop reliable findings. I worried about cumulative effects and underappreciated acceleration rates.

Fast forward a year and I'd consumed hundreds of books and articles. My severance ran out, and I began tapping into savings and retirement to pursue this quest. What I thought were "basic questions" proved so fundamental that they hadn't been clearly asked or answered yet.

Yes, at that point, I was up to speed on the dopamine arguments and the primacy of visual circuits in our brain that worked hand in hand with dopamine to drive us to seek new and exciting things, creating addictive habits that we in the industry monetize by creating innovative new technologies to serve and exploit those needs. And how the striatum, and particularly the ventral striatum, functions in part to evaluate rewards in social contexts, and is targeted and hijacked by modern social media platforms and other tech. This drives us to seek novelty and undergirds "fear of missing out" (FOMO), among other things.

But those answers didn't satisfy me; they felt too simple, or reductionist. It felt like various people were focusing on different problems, but at the expense of seeing the whole picture. But that's what people get rewarded for economically and professionally, so I didn't fault those writers and thinkers; they've greatly added to our understanding in their areas.[3]

My research questions, initially, were more basic, and, as I was to learn, more profound:

- How does technology affect how we think?
- How does how we think affect how we adopt technology?
- How effective is technology at serving us/our well-being?
- What are the cumulative effects of technology on us individually and collectively?

2 Down the Rabbit Hole

Starting in late 2014, I reached out directly to scholars, asking these questions. Many graciously responded, and these conversations and my reading kept coming back to E.O. Wilson's observation:[4]

> We have created a Star Wars civilization, with Stone Age emotions, medieval institutions, and godlike technology.

Harari, for example, begins his now-classic book, Sapiens, by discussing the human cognitive revolution that began approximately 50,000 years ago, pulling us out of the Stone Age. Others, like McBride, Richtel, or Greenfield, argue that we act the way we do because our brains are hardwired in certain ways, and these traits were passed down to us from our ancestors from 50,000–100,000 years ago.

3 Later, I encountered McGilchrist's invaluable book, *The Master and His Emissary*, which examines the brain's two sides, clinically known as hemispheric lateralization. Stepping aside from the pop psychology surrounding this topic, McGilchrist spent over twenty years rigorously exploring the consequences of this aspect of brain neuroanatomy. He concludes that modern society is driven by traits centered in the left brain, which manipulates the world. It is detail-oriented, analytical, sequential, and reductionist. Modern society rewards these traits. The right hemisphere understands the world; it is creative, integrative, and emotional, recognizing patterns and striving toward holistic outcomes. It turns out I was seeking right-side answers in a left-side world.

4 Wilson, Edward O. *The Social Conquest of Earth. Liveright Publishing Corporation,* 2012, p. 7.

This led to deeper questions. What selection pressures drove the physiology of a 200,000 or 300,000-year-old brain to set us up to not only create seminal technologies but also, in turn, be shaped by them at that point in time, maybe 50,000 years ago? Put simply, what happened in the past to lay the foundation for how we think now? If that change was technology, and it lifted people out of the Stone Age, where are our current innovations taking us? What's the difference in selection pressures that shaped our 300,000-year-old brain, our early behaviorally-modern one of 50,000 years ago, and how have those pressures changed now? Crucially, how are our thoughts performing now under today's pressures?

You can see where this is going – VERY Lewis Carroll, and not in a good way. Several years later, I had drafted two massive books examining how cognition co-evolved with technology[5] across tens of millions of years of primate and human evolution. Oxford University Press showed interest and sent them out for peer review.

I received great feedback from those initial reviews, including some concerns about publishing books on these bold, complex topics from someone who is... shall we say...self-taught.

More than one scholar said the project was the most ambitious and comprehensive treatment of the material they had ever seen. (I'm not sure that was intended as a compliment.) Still, the material and the proposal were encouraged by people who spend a lot of time saying "no" to books, and OUP accepted my proposal and gave the books a green light.

To address the credibility concerns, I reached out to two professors who had provided me with valuable insights and feedback. Both are world leaders in their fields – one in primate neuroscience and the other in human prehistory. Would they like to co-author the books with me?

Yes, from both!

3 When the Future Couldn't Wait

Those two manuscripts are now still awaiting markups and feedback (one of those authors had a very serious long-term illness), and collaborating like this can take some time.

Meanwhile, I took consulting jobs to keep the lights on. This included work for VC-backed startups with innovative new technologies, and helping the US

5 Later we'll get into my definitions of technology, which will help you see why I place biological adaptions like tails and color vision in scope.

cable industry plan its national 5G wireless networks, and envision how that industry's "10G" strategy (10 gigabit wired connections to the home) would transform daily life.[6] I even had several opportunities to serve the US Government and military in various ways. Across these consulting projects, I drove technologies such as cloud, AI, big data, new connected devices, mobile apps for education, various 5G network plans, advanced defense and intelligence community technologies, and national security policy work. Again: more, better, faster.

Meanwhile, since those books had stalled, a friend advised me to write the book I truly wanted, exploring how technology impacts us today, rather than in the distant past. Don't wait for foundational books. Act now.

So, in March 2020, at the beginning of the pandemic, I began waking at 3 AM or 4 AM to read, think, and write this book. I developed a model of human cognition that gradually evolved into presentations I reviewed with dozens of scientists and thought leaders worldwide via Zoom calls.

With each conversation, rough ideas gradually crystallized. Sixty-minute calls stretched into two hours or more of engaged discussion. What I felt intuitively was relevant and interesting gradually crystallized into the realization that thought itself was undergoing a complete transformation with dramatic consequences.

This wasn't just interesting or important. This was potentially the most important issue facing humanity. These conclusions were confirmed through conversations with some of the brightest minds in fields like cognitive science, human evolution, technology, and national defense.

What I discovered over this decade-long investigation will challenge how you see your relationship with technology and even your own mind. These findings don't just touch on personal habits and human behavior; they reshape how we understand human potential in an age of constantly accelerating technology.

The story begins with a quirky parable and a couple of profound insights into what thinking is in our hyperconnected world.

6 Cable Labs, "What Is 10G?," *Cable Labs*, https://www.cablelabs.com/10g. Accessed 19 August, 2025.

Acknowledgments

> Ye live not for yourselves; ye cannot live for yourselves; a thousand fibres connect you with your fellow-men, and along those fibres, as along sympathetic threads, run your actions as causes, and return to you as effects[1]. (Reverend Henry Melvill)

I used to skip this section entirely when picking up new books.

Writing this one taught me why that was a mistake. This book isn't really mine – it's the collective wisdom of everyone generous enough to share their time and ideas on this journey. If you find something valuable here, credit them. If you spot errors or hate what you've read, blame me.

We supposedly stand on the shoulders of giants, but the truth is simpler: that we all stand at all is a function of the communities that buttress us in countless ways, large and small. I never grasped this fully until I worked on this project and discovered what I learned about *us* in the process.

To everyone who helped me wrestle with these ideas – you have my deepest gratitude. To those I've inevitably overlooked – you're loved and appreciated anyway.

One unexpected gift from all your feedback: I now read acknowledgments first when I pick up any book. That's where you discover who *really* created the book.

That includes reading one of the best acknowledgment lines anywhere, in George Dyson's *Analogia* – *"Nonconformists are also like cats, making their way through life the way alley cats survive thanks to people who leave out food."* He goes on to list those folks who *"left food out for me along the way."*[2]

Here's to those who left some food out for this nonconformist, sustaining this project.

First and foremost, I want to thank Mark Holland for encouraging me to tackle it and for the countless hours of interesting, curious, patient, and supportive discussion. You made my pandemic time much more productive. I'll send you my Amazon book tab.

1 Melvill, Henry. *Golden Lectures for 1855*. Sermon "Partaking in Other Men's Sins," 12 June 1855, pp. 454. Quoted in *Six Thousand Illustrations of Moral and Religious Truths*, edited by John Bate, London, 1885, pp. 311–12.

2 While Dyson's line is in my top 5 all-time acknowledgements, he still doesn't hold a candle to Hans Rosling, who dedicated *Factfulness*, one of my favorite books, in part to *"the brave barefoot woman, whose name I do not know but whose rational arguments saved me from being sliced by a mob of angry men with machetes."*

This book would not exist without David Bray, whose contributions are incalculable. He graciously and generously brought me into the brilliant think tank communities he has built, led, or taken part in. His leadership on global technology issues is unparalleled and inspirational. Many people I now consider dear friends and valuable allies in the fight to make the world a better place are connected to David's unique, globe-spanning network(s) and activities, either directly or indirectly. Thanks for the hours we've spent talking, wrestling with, and even debating these ideas. Here's hoping someday I overcome the 30-point gap in our IQ to win one of those arguments.

I've learned so much and found such a supportive home within some of those communities David brought me into, including Net-Econ, People-Centered Internet (PCI), and Global Vanguards – three global next-generation think tanks. These folks include technology's OGs and the new guard, as well as world leaders on other global issues, combined with more than a few community builders, radicals, and nonconformists. All banding together to think about and work on hard topics to make the world a better place. If you are dismayed by reading the news or social media, you only need to take part in virtual communities like these to know that people are good, and there is hope for the future.

Carl Pabo has uniquely dedicated himself for over 20 years to a set of research questions very much like mine. He stepped away from the pinnacle of academia because he became concerned that humans were building themselves into a "crisis of complexity," and he wanted to understand that and raise awareness. Carl and I still feel we may be the only two researchers in the world looking at the issues in this book, at this full scope and scale. Thank you, Carl, for not letting me walk it alone. You are a critically important lighthouse for humanity.

Thanks, of course, to Vint, the OG of the Internet, (and an example of a Bray introduction) for pre-reading an incredibly rough draft of this, writing the foreword, and patiently interacting with me across these several years as this project evolved.

To Jim Giordano, for also patiently putting up with me (let's just assume anyone in this section has been pestered by me over several years with dumb questions and many roughly formed ideas, and that they didn't always tell me to go away, shall we?), and giving me a platform first on his podcast, and then, as a book series editor, recommending this volume for publication.

Many of those in this section represent my "chosen family." Here's a few more of those cherished people, always taking my emails, calls, and texts, and being there for me as I wrestled with this: Darren Culbreath, Tyler Sweatt, Melissa Flagg, Katina Michael, Guy Huntington, Chris Preble, Daniella Taveau,

Pablo Breuer, Ed Kirschner, Tony Aghazarian, Jonathan Lovegrove, Divya Chander, Eric Sapp, Alex Ruiz, Dave Ackerman, Eusebio Scornavacca, Zach Levy, Linda Lindquist-Bishop, Brian Bishop, and last but certainly not least, mi hermano, Ernesto Hernandez.

Mike Keene has shown unwavering belief, support, and commitment to these issues and work. He, my cousin Dee, and our dear family friend Foster Bailey, along with many others, quietly stepped in to help my family during its darkest hours, which occurred during this research. I love you all. Mike – I can't wait to see this movie in theaters.

It's not until I've begun pre-reading for other authors that I learned to appreciate the value of the time, effort, insights, "second set of eyes," and – often most importantly – the frank criticism that pre-readers can offer. So, I'm forever grateful for the many pre-readers of various drafts of this manuscript, especially Leif Ulstrup, Rollie Cole, Paul Budde, and David Weinberger, who first read this when it was a bloated manuscript that contained what is now two (in Leif's case, it may have been four) separate titles and was a hot mess. Fortunately, they recognized it and brought it to my attention in constructive and encouraging ways. David joins Jim Rutt in being my two harshest and favorite critics, though their styles and use of profanity wildly diverge.

Thanks to Marci Harris, Newton Howard, Mark Humphries, Jeff Frazier, Paul Quaiser, Roba Abbas, Scott David, Louis Rosenberg, Pierre Dussault, Andreas Widmer, Scott Shumate, Michael Ross, Bob Gourley, and Jim O'Shea (I'm probably overlooking a dozen people here – apologies!) for their reviews of the later versions that were closer to a finished book, and their feedback and encouragement.

To the first group of people who were kind enough to join me in the pioneering days of the first attempt at building the Future of Thought Consortium long before its time, when my ideas were still in their early stages: Anca Selariu, Pierre Dussault, Federico Petrozzi, Ken Harvey, and Chip Colbert (along with others mentioned previously).

Of all the things that keep me motivated recently, the commitment, time, and widespread support from the second wave of the founding members of the Future of Thought Consortium, when my ideas were beginning to gel, stand out. That includes, among others listed elsewhere in this section – Jumana Abu-Ghazaleh, RJ Cordes, Tami DeWeese, Michael Ross, Scott Shumate, Jason Guesman, Cyndi Coon, Michael Robbins, Marc Prenzky, Neil Sumilas, and Erich Izdepski.

Ideas don't spring out of the ground, whole cloth – they are built and polished through trial and error. Thanks to those who provided platforms for that messy work: Nick Savage, the University of Arizona School for the Future of

Innovation in Society (especially my very dear friend and colleague Katina Michael), PCI, IEEE, European Association of Archaeology, Evanna Hu and NATO, Johns Hopkins University Applied Physics Lab (especially Nat Barrett and Daniel Silver), as well as Trond Undheim and other podcasts.

Work like this wouldn't have been possible if folks hadn't provided me with day jobs and the side hustles I did to keep the lights on and food on the table. Thank you, Peter Cannistra, David Waldrop, Kenneth Parker, Maria Bothwell, John Pappajohn, Keith Belcher, and Sean Chi for those opportunities.

There were many academics who gave their time and attention to my often naïve questions and ideas: Cyrus Foroughi, Peter Whybrow, Dennis McBride, Peter Carruthers, Newton Howard, RE Burnette, Michael Arbib, Harry Wingo, Andreas Widmer, Jaco Hamman, Sam Arbesman, George Dyson, Paul Werbos, Rand Waltzman, Michael Nestor, Ilan Chaby, Gerald Scannell, Fred Coolidge, Thomas Wynn, Steven Stanley, Chris Springer, MG Michael, Yehezkel Dror, Ken Younge, Andy Gilbert, Armen Arevian, Meredith Bastian, Jennifer Botting, "Betsy" Herrelko, Peter Robinson, Marc Bekoff, Kenneth Younge, and Adam Thierer, among many others.

Here's to the fellowship and collaboration provided by the co-authors of my various other writing projects – Chet Sherwood, Tim Taylor, Ajit Mann, Sahil Shah, Scott David, RJ Cortez, Alex Ruiz, and Eric Sapp.

Along the way, conversations (and emails/texts, etc.) really helped me to explore parts of these ideas. Thanks for those goes out to Jamie Arbib, Anthony Scriffignano, Jerome Pickett, Bob Gourley, Pete McCrystal, John Wilson, John Hawker, Henning Schulzrinne, Amy Chaput, Louis Rosenberg, Wil Forbis, Dan Kasun, Russ Hayward, Owen Brown, Kevin Binz, Ben Houston, Nick Kesler, Belal Hemzah, and Chris Savage, alongside the others mentioned here.

My work is inspired by the efforts of an encouraging number of my colleagues who tilt at their own audaciously challenging windmills of world change. Those inspirational candles of hope and light in the world include Michael Robbins, Newton Howard, Paul Werbos, Jim Rutt, Jamie Arbib, Paul Quaiser, Jumana Abu-Ghazaleh, Andrew Titus, Cyndi Coon, Guy Huntington, and others already mentioned here.

Nearly all of this work was done during my "office hours" in the wee hours of the morning at my local Starbucks in Leesburg, Virginia, and I'm extremely grateful for the home and community that was built there by all of the staff (especially Justin, Coco, Charlie, and Becca) and the barrels of coffee over the last 10+ years of reading and writing there. Thanks, Paul Fedor and Mike Fitzgerald, for the companionship and discussions there. And leaving me alone when I needed to work. Sometimes.

In this book, we discussed the importance of "scouting the fringes" of idea spaces. I read many kinds of books, but science fiction holds a special place in helping me to sensing what futures may unfold. In its own way, the writing of science fiction manifests its own influence on the future; we see that when we look at the influence of the original Star Trek, for example, on early and even current technology development. While there is no room here to show the depth of my appreciation to the entire field (I flesh that out a bit in the suggested reading appendix), I'm especially indebted to David Brin, Doug E. Richards, and Karl Schroeder for their "hard science" fiction books and our conversations about how their visions for the future intersect mine.

There were thousands of authors of books, articles, and other content that I used as research for this project, and only a small fraction of whom were appropriate to cite in the footnotes or Suggested Reading section. However, I appreciate the effort they put into their work. Thanks, as well to the librarians and bookstores everywhere who made many of them available to me.

Since I'm Irish, I have a large family, and they always have my back, especially Dee Flaherty, Jay, and Dr. Emily Scannell, Jeff Scannell, Annie Mermelstein, the whole Auntie Camp crew, and all the rest of you who are there for us in good times and bad.

Here's to the family we are born with, and those we make/choose along the way, including the best-ever family of neighbors: Monica & David, Linda & Niroth, Shurronne, Bob, Ed & Rhonda, and AJ & Kate.

Figures and Table

Figures

Table

INTRODUCTION

The Water We Swim In

> An old fish runs into two younger fish; he nods, says, "Morning, boys, how's the water?", and swims off. One young fish looks at the other and says, "What the hell is water?"[1] (David Foster Wallace)

– You don't think the way you think you think.
– Technology disrupts your thinking more profoundly than Netflix transformed entertainment, Uber revolutionized transportation, or Amazon redefined retail.

These two insights frame the most important problem and opportunity we face today.

We become so immersed in our environment that we lose awareness of its presence or fail to consider our relationship with it. We sleepwalk through life, and nothing remains as invisible as the obvious.

Thinking forms the essence of our lives, but most of us take it for granted, and never pause to reflect on it. When we do, we'll see it is changing rapidly as we increasingly outsource cognitive processes to surrounding technology. We no longer remember our children's phone numbers or endure the effort of dialing, we simply ask our phone to call them. Our children need not turn their heads when driving; blind-spot monitors, lane-changing sensors, and backup cameras handle spatial awareness. Artificial intelligence agents track our communications as we type, highlighting errors and suggesting words. ChatGPT and Claude can write our papers, complete our work, and serve as trusted advisors and even dear companions.

In many households, stating a desire aloud prompts Alexa to deliver it tomorrow. Your phone's notifications dictate your attention, priorities, and time allocation.

Despite all the changes we've experienced over the last two decades, what lies ahead will prove far more dramatic:

1 "This Is Water by David Foster Wallace (Full Transcript and Audio)", *Farnam Street*, 2012, https://fs.blog/2012/04/david-foster-wallace-this-is-water/. Accessed Aug. 19, 2025.

 | HTTPS://DOI.ORG/10.1163/9789004739161-002

- **AI** – Former Google CEO Eric Schmidt warns that AI *"will disrupt and transform"* all industries and facets of human life through its capacity *"to learn, evolve, and surprise."*[2]
- **Metaverse** – Louis Rosenberg and Alvin Graylin predict that Augmented Reality (AR) and Virtual Reality (VR) will *"fundamentally change our lives, impacting not only how we work and play, but how we perceive our world, engage our surroundings and interact with each other."*[3]
- **Quantum** – Kevin Coleman warns that quantum technology's *"disruptive potential will make the change of the Internet era look like a small bump in the road."*[4]

While many notice specific aspects and wonder about particular impacts, grasping the complete picture requires examining numerous facets of life and specialized fields that seem incompatible.

1 Seeing the Water

This book cultivates awareness and illuminates reality as Wallace suggested: *"It is about simple awareness – awareness of what is so real and essential, so hidden in plain sight all around us, that we have to keep reminding ourselves, over and over: 'This is water, this is water.'"*[5]

Mathematician Andrew Odlyzko observed that *"very often the key element in an innovation is to realize there is a problem to be solved, not the technical solution."*[6]

Consider climate change discussions in the late 1970s. Scientists debated whether climate change was real and whether CO_2 contributed to it. No one discussed sophisticated solutions like carbon trading schemes yet.

This pattern repeats across major public issues. Before tackling smoking, drunken driving, civil rights, opioids, or football concussions, society first

2 Kissinger, Henry A., et al. *The Age of AI: And Our Human Future*. Little, Brown and Company, 2021, p. 4.

3 Graylin, Alvin W., and Louis Rosenberg. *Our Next Reality: How the AI-Powered Metaverse Will Reshape the World.* Foreword by Neal Stephenson, Nicholas Brealey Publishing, 2024, p. 6.

4 Gillian, Duncan. "How Abu Dhabi's New Quantum Computer Could Help Solve the Mysteries of Science." *The National*, https://www.thenationalnews.com/uae/2023/10/30/how-abu-dhabis-new-quantum-computer-could-help-solve-the-mysteries-of-science/. Accessed Nov. 19, 2025.

5 "This Is Water by David Foster Wallace (Full Transcript and Audio)", *Farnam Street*, 2012, https://fs.blog/2012/04/david-foster-wallace-this-is-water/. Accessed Aug. 19, 2025.

6 Private communication Andrew Odlyzko, 2020.

had to recognize and accept that problems existed. People must understand their dilemma before they can address it.

That's where we stand with technology's cognitive impact – and that's this book's focus.

2 Your Stake in This Story

This transformation affects everyone, but certain groups face particularly urgent stakes. I wrote this for thoughtful individuals who want to gain a deeper understanding of our modern times and invest in building a better future.

Parents grapple daily with decisions about their children's screen time and AI exposure, aware that these choices will shape the developing minds of their children. Technologists find themselves at a crossroads – how can innovative products outcompete existing technologies while genuinely improving users' lives rather than exploiting their attention? Leaders across sectors such as national security, government, academia, social justice, and industry, must navigate emerging trends that fundamentally alter their missions. The cognitive shifts explored in this book ripple through every domain of human endeavor.

This book requires no specialized knowledge; experts in one field may be novices in others, so all topics are presented accessibly for readers of any background.

3 The Road Ahead: Danger and Promise

This book presents neither a definitive statement from someone who has figured everything out nor initially uplifting content. We already face catastrophe fatigue from pandemics, wars, and political strife. We don't need someone bringing new problems, especially one that this book argues represents humanity's biggest problem, one that few know about and even fewer work on.

People are working on this, but within silos, atomistically, addressing symptoms like shortened attention spans, screen time, social polarization, surveillance capitalism, and misinformation. They work on facets, not root causes. There's no blueprint for how the pieces fit together or roadmap for moving forward productively.

But here's the crucial point: this isn't just humanity's biggest problem – it's also our biggest opportunity. The phones in our back pockets can match the computing performance of the world's best supercomputer from just 20 years ago,[7] and we can use those phones to have increasing access to most collected human knowledge, often for free.

This book challenges you to confront our current situation while rewarding you with the ability to see your world in new and more productive ways. It serves as the first draft of a blueprint opening new paths to better solutions for current challenges and those we'll likely encounter.

With shared understanding, we can transform this technological transition from a threat to an opportunity that enhances our collective problem-solving. Solutions will arise from all corners of society, but only with shared situational understanding, focused actions, and ability to leverage each other within a coordinated plan.

4 Your Journey through This Book: What to Expect

- **Part 1** delivers the punchline: the very fabric of how we think is being disrupted. It outlines key findings and explains why this represents humanity's most critical issue.
- **Part 2** introduces my model of human cognition and explains why we need it. Despite thousands of years of research into the human mind, no available model adequately describes how modern humans actually think in our tech-immersed culture.
- **Part 3** explores key insights gained from that model. This section goes beyond the headline of "thought is being disrupted" to provide important details that help you understand thought in the modern world.
- **Part 4** examines how well these dynamics serve us, signals where solutions will emerge, and provides practical frameworks for confronting these ideas in your own life and work.

Finally, this book is not the conclusion; it is the beginning of a dialogue. Even within my own work, I have four other books in various stages of the peer-reviewed writing/publication process, each dealing with different dynamics stemming from my initial research questions. And there's a documentary on my work being developed. It's my hope that the books and other works like the film can be the beginning of bringing people together, to collaborate and move us toward an optimistic future I see, and away from the dystopian one that I sense awaits us on our current path.

7 See Chapter 13 for some math around this awkward comparison.

had to recognize and accept that problems existed. People must understand their dilemma before they can address it.

That's where we stand with technology's cognitive impact – and that's this book's focus.

2 Your Stake in This Story

This transformation affects everyone, but certain groups face particularly urgent stakes. I wrote this for thoughtful individuals who want to gain a deeper understanding of our modern times and invest in building a better future.

Parents grapple daily with decisions about their children's screen time and AI exposure, aware that these choices will shape the developing minds of their children. Technologists find themselves at a crossroads – how can innovative products outcompete existing technologies while genuinely improving users' lives rather than exploiting their attention? Leaders across sectors such as national security, government, academia, social justice, and industry, must navigate emerging trends that fundamentally alter their missions. The cognitive shifts explored in this book ripple through every domain of human endeavor.

This book requires no specialized knowledge; experts in one field may be novices in others, so all topics are presented accessibly for readers of any background.

3 The Road Ahead: Danger and Promise

This book presents neither a definitive statement from someone who has figured everything out nor initially uplifting content. We already face catastrophe fatigue from pandemics, wars, and political strife. We don't need someone bringing new problems, especially one that this book argues represents humanity's biggest problem, one that few know about and even fewer work on.

People are working on this, but within silos, atomistically, addressing symptoms like shortened attention spans, screen time, social polarization, surveillance capitalism, and misinformation. They work on facets, not root causes. There's no blueprint for how the pieces fit together or roadmap for moving forward productively.

But here's the crucial point: this isn't just humanity's biggest problem – it's also our biggest opportunity. The phones in our back pockets can match the

Technology offers incredible potential to solve our most pressing problems, but only if we approach it consciously. We can no longer afford to sleepwalk into a future designed by others. The time has come to wake up, choose deliberately, and harness these powerful tools to create the life and world we actually want.

Like any good story, this one is full of danger and existential threats. When you finish this, you will be more afraid of technology than you are right now. But, if you squint, tilt your head, and look at the material differently, you will see that I'm writing a love letter to technology, our modern way of life, and potential human futures. Work simply needs to be done and changes made to get them to play together as well as they could, and achieve the desired outcomes.

Let's get started.

PART 1

Thought Is Being Disrupted

1 Purpose of This Part of the Book

This part contains the main point of the book: thought is undergoing disruption. First, I do a bit of a literature review of what the word "thought" has meant to different experts over the years. Then, I introduce my model of human cognition (very briefly – it is fully developed in chapters 5–11), and then jump into the case that (a) thought is being disrupted and (b) this is a critical issue for humanity. While the rest of the book substantiates these points, I present the main conclusions upfront to help motivate and orient you, the reader, as you navigate the rest of the book.

 | HTTPS://DOI.ORG/10.1163/9789004739161-003

CHAPTER 1

What Is Thought, and Why Do We Need a New Model?

> In order to keep up with the world of 2050, you will need not merely to invent new ideas and products but above all to reinvent yourself again and again.[1] (Yuval Noah Harari)

> What is called thinking? The question sounds definite. It seems unequivocal. But even a slight reflection shows it to have more than one meaning. No sooner do we ask the question than we begin to vacillate. Indeed, the ambiguity of the question foils every attempt to push toward the answer without some further preparation.[2] (Martin Heidegger)

> Did you ever stop to think, and forget to start again?[3] (Winnie the Pooh)

What is thought?

Humans[4] have thought about thought for thousands of years, and in all that time, our view and understanding of how we think continues to evolve. Even the definition of "thought" has been remarkably fluid, and it is still quite a slippery topic.

For millennia, it was the domain of the philosopher: Plato's seminal definition of thought was that it was indistinguishable from speech, except that thought was "*dialogue of the soul with itself.*"[5] In the seventeenth century, informed by advances in physics by people like Newton and Descartes, Hobbes characterized the mind[6] as mechanical – simply the "*motions of certain parts*

---

1 Harari, Yuval Noah. *21 Lessons for the 21st Century*. Spiegel & Grau, 2018, p. 265.

2 Heidegger, Martin. *What Is Called Thinking?* Translated by J. Glenn Gray, Harper & Row, 1968, p. 3.

3 Milne, Alan Alexander. *Winnie-the-Pooh*. Illustrated by E. H. Shepard, Methuen & Co. Ltd., 1926, p. 67.

4 The scope of this book is limited to humans; while I acknowledge that other animals have cognition (refer to my previous work on this, for example), here we are focused exclusively on *Homo sapiens sapiens*.

5 Plato. *Theaetetus*. Translated by Harold North Fowler, in *Plato in Twelve Volumes*, Vol. 12, Harvard University Press/Heinemann, 1921. 189e.

6 In this book, I treat thought and mind as equivalent and fungible, but I take the position that consciousness and thought are distinctly different and not interchangeable (see Chapter 13).

 | HTTPS://DOI.ORG/10.1163/9789004739161-004

of an organic body."[7] Fast forward to today's literature, and one finds the matter is still open.

Dictionary references are rife with circular definitions, where thought is – "*an idea or opinion produced by thinking, or occurring suddenly in the mind,*" and "*the action or process of thinking,*"[8] while Wikipedia states bluntly, "*Although thinking is an activity of an existential value for humans, there is still no consensus as to how it is adequately defined or understood.*"[9]

Evolutionary psychologists John Tooby and Leda Cosmides claimed that much of the 20th-century work on cognition, starting predominantly with the behavioral approaches pioneered by scholars such as Skinner,[10] Mead,[11] Geertz,[12] Gould,[13] Lewontin, and Kamin[14] could be summarized (somewhat derogatively) as the Standard Social Science Model (SSSM)[15] wherein our cognitive processes are shaped almost entirely by culture (a nurture over nature theme).[16]

Tooby and Cosmides propose an integrated model (IM) that applies evolutionary principles to computational theories of mind, treating neural networks as computational circuits.

Harvard psychologist Steven Pinker epitomizes this neurocentric approach with his definition of the mind: "*The mind is a system of organs of computation, designed by natural selection to solve the kinds of problems our ancestors faced*

Since I believe the majority of cognitive processes are unconscious, I necessarily don't focus on consciousness, and so sidestep the "hard problem of consciousness."

7 Hobbes, Thomas. "Third Objections." In René Descartes, *The Philosophical Writings of Descartes*, Vol. 2, translated by John Cottingham, Robert Stoothoff, and Dugald Murdoch, Cambridge University Press, 1984, pp. 122–123 (AT VII, 172–173).

8 Oxford English Dictionary. "Thought, n.," Oxford English Dictionary, https://www.oed.com/dictionary/thought_n? Accessed Jul. 1, 2020.

9 "Thought." *Wikipedia*, Wikimedia Foundation, en.wikipedia.org/wiki/Thought. Accessed Oct. 31, 2020.

10 Skinner, B. F. *Verbal Behavior*. Appleton-Century-Crofts, 1957.

11 Mead, Margaret. *Coming of Age in Samoa: A Psychological Study of Primitive Youth for Western Civilisation*. William Morrow, 1928.

12 Geertz, Clifford. *The Interpretation of Cultures: Selected Essays*. Basic Books, 1973.

13 Gould, Stephen Jay. *The Mismeasure of Man*. W. W. Norton, 1981.

14 Lewontin, Richard C., Steven Rose, and Leon J. Kamin. *Not in Our Genes: Biology, Ideology and Human Nature*. Pantheon Books, 1984.

15 This list of SSSM scholars is itself, like many aspects of science, controversial and highly debated. I'm simplifying the debate to give you a gist of how the topics have roughly evolved. See, for example, Hampton, Simon J. "Can Evolutionary Psychology Learn from the Instinct Debate?" *History of the Human Sciences*, vol. 19, no. 4, 2006, pp. 57–74.

16 Barkow, Jerome H., Leda Cosmides, and John Tooby, editors. *The Adapted Mind: Evolutionary Psychology and the Generation of Culture*. Oxford University Press, 1992.

CHAPTER 1

What Is Thought, and Why Do We Need a New Model?

> In order to keep up with the world of 2050, you will need not merely to invent new ideas and products but above all to reinvent yourself again and again.[1] (Yuval Noah Harari)

> What is called thinking? The question sounds definite. It seems unequivocal. But even a slight reflection shows it to have more than one meaning. No sooner do we ask the question than we begin to vacillate. Indeed, the ambiguity of the question foils every attempt to push toward the answer without some further preparation.[2] (Martin Heidegger)

> Did you ever stop to think, and forget to start again?[3] (Winnie the Pooh)

What is thought?

Humans[4] have thought about thought for thousands of years, and in all that time, our view and understanding of how we think continues to evolve. Even the definition of "thought" has been remarkably fluid, and it is still quite a slippery topic.

For millennia, it was the domain of the philosopher: Plato's seminal definition of thought was that it was indistinguishable from speech, except that thought was "*dialogue of the soul with itself.*"[5] In the seventeenth century, informed by advances in physics by people like Newton and Descartes, Hobbes characterized the mind[6] as mechanical – simply the "*motions of certain parts*

1 Harari, Yuval Noah. *21 Lessons for the 21st Century*. Spiegel & Grau, 2018, p. 265.

2 Heidegger, Martin. *What Is Called Thinking?* Translated by J. Glenn Gray, Harper & Row, 1968, p. 3.

3 Milne, Alan Alexander. *Winnie-the-Pooh*. Illustrated by E. H. Shepard, Methuen & Co. Ltd., 1926, p. 67.

4 The scope of this book is limited to humans; while I acknowledge that other animals have cognition (refer to my previous work on this, for example), here we are focused exclusively on *Homo sapiens sapiens*.

5 Plato. *Theaetetus*. Translated by Harold North Fowler, in *Plato in Twelve Volumes*, Vol. 12, Harvard University Press/Heinemann, 1921. 189e.

6 In this book, I treat thought and mind as equivalent and fungible, but I take the position that consciousness and thought are distinctly different and not interchangeable (see Chapter 13).

 HTTPS://DOI.ORG/10.1163/9789004739161-004

of an organic body."[7] Fast forward to today's literature, and one finds the matter is still open.

Dictionary references are rife with circular definitions, where thought is – "*an idea or opinion produced by thinking, or occurring suddenly in the mind,*" and "*the action or process of thinking,*"[8] while Wikipedia states bluntly, "*Although thinking is an activity of an existential value for humans, there is still no consensus as to how it is adequately defined or understood.*"[9]

Evolutionary psychologists John Tooby and Leda Cosmides claimed that much of the 20th-century work on cognition, starting predominantly with the behavioral approaches pioneered by scholars such as Skinner,[10] Mead,[11] Geertz,[12] Gould,[13] Lewontin, and Kamin[14] could be summarized (somewhat derogatively) as the Standard Social Science Model (SSSM)[15] wherein our cognitive processes are shaped almost entirely by culture (a nurture over nature theme).[16]

Tooby and Cosmides propose an integrated model (IM) that applies evolutionary principles to computational theories of mind, treating neural networks as computational circuits.

Harvard psychologist Steven Pinker epitomizes this neurocentric approach with his definition of the mind: *"The mind is a system of organs of computation, designed by natural selection to solve the kinds of problems our ancestors faced*

Since I believe the majority of cognitive processes are unconscious, I necessarily don't focus on consciousness, and so sidestep the "hard problem of consciousness."

7 Hobbes, Thomas. "Third Objections." In René Descartes, *The Philosophical Writings of Descartes*, Vol. 2, translated by John Cottingham, Robert Stoothoff, and Dugald Murdoch, Cambridge University Press, 1984, pp. 122–123 (AT VII, 172–173).

8 Oxford English Dictionary. "Thought, n.," Oxford English Dictionary, https://www.oed.com/dictionary/thought_n? Accessed Jul. 1, 2020.

9 "Thought." *Wikipedia*, Wikimedia Foundation, en.wikipedia.org/wiki/Thought. Accessed Oct. 31, 2020.

10 Skinner, B. F. *Verbal Behavior*. Appleton-Century-Crofts, 1957.

11 Mead, Margaret. *Coming of Age in Samoa: A Psychological Study of Primitive Youth for Western Civilisation*. William Morrow, 1928.

12 Geertz, Clifford. *The Interpretation of Cultures: Selected Essays*. Basic Books, 1973.

13 Gould, Stephen Jay. *The Mismeasure of Man*. W. W. Norton, 1981.

14 Lewontin, Richard C., Steven Rose, and Leon J. Kamin. *Not in Our Genes: Biology, Ideology and Human Nature*. Pantheon Books, 1984.

15 This list of SSSM scholars is itself, like many aspects of science, controversial and highly debated. I'm simplifying the debate to give you a gist of how the topics have roughly evolved. See, for example, Hampton, Simon J. "Can Evolutionary Psychology Learn from the Instinct Debate?" *History of the Human Sciences*, vol. 19, no. 4, 2006, pp. 57–74.

16 Barkow, Jerome H., Leda Cosmides, and John Tooby, editors. *The Adapted Mind: Evolutionary Psychology and the Generation of Culture*. Oxford University Press, 1992.

in their foraging way of life, in particular, understanding and outmaneuvering objects, animals, plants and other people."[17]

This school of thought is sometimes called the Savannah Principle, as popularized by Satoshi Kanazawa in a 2004 paper, arguing "*The human brain has difficulty comprehending and dealing with entities and situations that did not exist in the ancestral environment.*"[18] The idea has been around for a long time, dating at least as far back as Darwin[19] and Wallace[20] under various names, including the Savannah hypothesis. Renowned anthropologist Raymond Dart, credited with discovering and naming the human ancestor species *Australopithecus*[21] back in 1925, said:

> For the production of man a different apprenticeship was needed to sharpen the wits and quicken the higher manifestations of intellect – a more open veldt country where competition was keener between swiftness and stealth, and where adroitness of thinking and movement played a preponderating role in the preservation of the species. ... Southern Africa, by providing a vast open country with occasional wooded belts and a relative scarcity of water, together with a fierce and bitter mammalian competition, furnished a laboratory such as was essential to this penultimate phase of human evolution.[22]

The idea is not without many critics,[23] especially regarding the areas of environmental change and how and when humans became bipedal, but it still significantly influences evolutionary psychology.

Other modern psychologists, philosophers, and others who study cognitive function avoid blanket definitions of thought altogether, instead focusing on

---

17 Pinker, Steven. *How the Mind Works.* Penguin UK, 2003, p. 21.

18 Kanazawa, Satoshi. "The Savanna Principle." *Managerial And Decision Economics*, vol. 25.1, 2004, pp. 41–54.

19 Darwin, Charles. *The Descent of Man, and Selection in Relation to Sex,* 2nd ed. John Murray, 1874.

20 Wallace, Alfred Russel. *Darwinism: An Exposition of the Theory of Natural Selection with Some of Its Applications.* London: Macmillan and Co., 1889.

21 His Taung Child discovery at the time (now categorized as *Australopithecus africanus*) was widely thought at the time to be the "missing link" in human evolution. See for, example: Madison, Paige. "Why the 'Missing Link' Fossil Was Almost Missed." *National Geographic*, Oct. 18, 2024.

22 Dart, Raymond A. "Australopithecus Africanus: The Man-Ape of South Africa." *Nature*, vol. 115, no. 2884, 1925, p. 199.

23 See for example Marlowe, Frank W. "Hunter-Gatherers and Human Evolution." *Evolutionary Anthropology: Issues, News, and Reviews*, vol. 14, no. 2, 2005, pp. 54–67. But this critique points out that we adapted to multiple archaic environments, not that we are without evolutionary influence.

more specific or discrete sub-mental processes, including the types of memory,[24] attention,[25] inner speech,[26] mental imagery,[27] decision-making,[28] mental agency,[29] autonomy and planning.[30]

Some develop very specific definitions of thought, bespoke to their specific position. Carruthers, for example, says "*thought will be understood to encompass all and only propositional-attitude events that are both episodic (as opposed to persisting) and amodal in nature (having a non-sensory format).*"[31]

This evolution in understanding *how* the brain works also represents a shift in *who* is looking at the question. The 20th century began with social scientists leading the charge, but ended with "hard scientists" having a growing prominence in how the mind works, thanks largely to advances in technology, in areas such as brain imaging and genetic sequencing. Seminal psychologists like William James,[32] John B. Watson,[33] B.F. Skinner,[34] as well as scholars in other fields like psychiatrist Sigmund Freud,[35] gradually gave way to architects of dominant current models like the Cognitive Theory of the Mind, which frames the brain as a computer.[36]

Cognitive Theory of the Mind examines how the brain processes external stimuli, enabling individuals to perceive, understand, and interact with the world. Central to this theory is the debate of "massive modularity" (MM),

24 Tulving, Endel. *Elements of Episodic Memory*. Oxford University Press, 1983.

25 Posner, Michael I., and Mary K. Rothbart. *Attention, Genes, and the Developing Brain: How Genes and Experiences Shape the Brain, Behavior, and Learning*. Oxford University Press, 2007.

26 Alderson-Day, Ben, and Charles Fernyhough. "Inner Speech: Development, Cognitive Functions, Phenomenology, and Neurobiology." *Psychological Bulletin*, vol. 141.5, 2015, p. 931.

27 Kosslyn, Stephen M. *Image and Mind*. Harvard University Press, 1980.

28 Tversky, Amos, and Daniel Kahneman. "Judgment under Uncertainty: Heuristics and Biases." *Science*, vol. 185, no. 4157, 1974, pp. 1124–1131.

29 Metzinger, Thomas K. "Myth Of Cognitive Agency: Subpersonal Thinking as a Cyclically Recurring Loss of Mental Autonomy." *Frontiers in Psychology*, vol. 4, 2013, 68140.

30 McCarthy-Jones, Simon. "The Autonomous Mind: The Right to Freedom of Thought in the 21st Century." *Frontiers in Artificial Intelligence*, vol. 2, 2019, article 19.

31 Carruthers, Peter. "Illusion of Conscious Thought." *Journal of Consciousness Studies*, no. 24.9–10, 2017, pp. 228–252.

32 James, William. *The Principles of Psychology*. Henry Holt and Company, 1890.

33 Watson, John B. "Psychology as the Behaviorist Views It." *Psychological Review*, vol. 20, no. 2, 1913, pp. 158–177.

34 Skinner, B. F. *Science and Human Behavior*. Macmillan, 1953.

35 Freud, Sigmund. "The Unconscious." *The Standard Edition of the Complete Psychological Works of Sigmund Freud*, translated by James Strachey, vol. 14, Hogarth Press, 1957, pp. 166–215.

36 Neisser, Ulric. *Cognitive Psychology*. Appleton-Century-Crofts, 1967.

which models the mind as a system of specialized modules or subsystems that are responsible for specific cognitive functions, such as language processing, face recognition, and spatial navigation. Massive modularity as a theory has been shaped by the ability to examine neuronal activity at an increasingly sophisticated and granular level.

There are two problems with the current hard science approach to understanding how neurons in the brain form the foundation of human thought. First, it fails to adequately explain Harari's "Cognitive Revolution," which may have occurred 70,000–30,000 years ago, or as Mithen describes it, as a "cultural explosion" that started 60,000–30,000 years ago.[37] It was this time in human history when we saw great advances in symbols, tool use, resource extraction, and, overall, a great leap forward for humans. But it would be hard to show changes in the brain that fostered that unique uplift. And maybe it didn't happen – after all, the reason for humanity's prominent role as a leading species on Earth, according to Matt Ridley, *"lies not in our 5 percent of special DNA or in our ability to learn associations,"* but rather *"in our ability to accumulate culture and transmit information, across the seas and generations."*[38]

MM may provide a laboratory explanation of how your brain and body react to seeing a picture of a face,[39] but it doesn't account for what changed in human systems and thinking to give rise to the Cognitive Revolution/cultural explosion, which was a function of rapidly evolving social interactions, thinking that occurs based on the shared understanding between individuals. As Austrian sociologist Gumplowicz said:

> The great error of individualistic psychology is the supposition that man thinks. It leads to the continual search for the source of thought in the individual and the reason why the individual thinks so and not otherwise; and prompts naïve theologians and philosophers to consider and even to advise how man ought to think. A chain of errors, for it is not man himself who thinks but his social community; the source of his thoughts is in the social medium in which he

37 Mithen, Steven J. *The Prehistory of The Mind: A Search For The Origins Of Art, Religion and Science.* Thames and Hudson, 1996.

38 Ridley, Matt. *Nature via Nurture: Genes, Experience, and What Makes Us Human.* New-York, 2003, p. 201.

39 Or may not; see Lisa Feldman Barrett's *How Emotions Are Made*, in which she makes a compelling case for a wide variety of emotional expression of a state like fear, both across individuals and within a single individual: *"a mental event, such as fear, is not created by only one set of neurons. Instead combinations of different neurons can create fear. Neuroscientists call this principle degeneracy."*

> lives, the social atmosphere in which he breathes, and he cannot think ought else than what the influences of his social environment concentrating upon his brain necessitate.[40] (Gumplowicz, 1899)

The second gap in the Massively Modular theory is its focus on the organic basis of thought, and it overlooks the possibility, as Mark Rowlands states, that "*Cognitive processes are not located exclusively inside the skin of cognizing organisms.*"[41] As we'll talk about in chapter 7, we think not only with our brains, or the people and culture around us, but also through stuff in our lives, including technology.

1 Thinking: The Whole Elephant

I opened the introduction chapter with the point that we don't think the way we think we think.

That's what I've learned after looking at the intersection of cognition and technology *very* deeply and broadly over the last decade.

Part of the problem, as I mentioned, is that we are all so specialized, we see only a small part of the overall processes.

This began to dawn on me as I worked on my first book, which focused on the evolution of cognition in primates, and I was researching gorilla cognition. Notably, it was challenging to extract food from the gorillas' environmental niche. It took a lot of mental skill and manual dexterity for them to get food safely out of difficult plants such as nettles and thistles.[42] It was also interesting to learn that gorillas have a significantly larger cerebellum than earlier primates. While the cortex and specifically the prefrontal cortex just behind your forehead gets a lot of attention when we think about humans' ability to think, the smallish cerebellum, or "little brain" at the base of the back of the brain, has about 50% of all the neurons in the whole brain.[43] Wow, this is

---

40 Gumplowicz, Ludwig. "*The Outlines of Sociology.*" Translated by Frederick W. Moore, American Academy of Political and Social Science, 1899, p. 156.

41 Rowlands, Mark. *The Body in Mind: Understanding Cognitive Processes*. Cambridge University Press, 1999.

42 Byrne, Richard. W., Nadia Corp, and Jeannette M. Byrne. "Manual Dexterity in the Gorilla: Bimanual and Digit Role Differentiation in a Natural Task." *Animal Cognition*, vol. 4, no. 3–4, Nov. 2001, pp. 347–361.

43 Knierim, James. "Cerebellum." *Neuroscience Online*, University of Texas Health Science Center at Houston, 2020, https://nba.uth.tmc.edu/neuroscience/m/s3/chapter05.html. Accessed on Nov. 19, 2025.

some interesting stuff. So, I asked one of my neuroscience friends to recommend some cerebellum research and scholars who could help me understand how this part of the brain may fit into my growing picture of cognition.

"Oh, you don't want to do that. Neuroscientists who begin looking at the cerebellum fall down a hole and never come back," he said.

He was making a little inside joke at these specialists, but his comment revealed something deeper: the dangerous allure of hyper-specialization.

Everywhere I turned, I saw researchers disappearing into their own "holes."

One neuroscientist studies the macro networks across the whole brain, while another zooms in on the synaptic microstructures linking individual neurons. Experimental archaeologists piece together ancient cognitive abilities by studying pottery shards from early farming communities, asking how the mental demands of crafting vessels might have shaped developing minds. Paleoanthropologists perform similar detective work with million-year-old stone tools, reverse-engineering the cognitive skills required to knap flint or fashion hand axes. Philosophers weigh in with theories about what thought is for, while psychologists and psychiatrists work in clinical settings, focusing on how it shapes our actions, and particularly our clinically abnormal or undesired behavior (i.e., sick care, not health care). Primatologists, in turn, study how wild chimpanzees use sticks to fish termites from mounds, or how captive capuchin monkeys solve puzzle boxes in controlled experiments. Yet despite studying the same phenomenon – cognition – these researchers from neuroscience to archaeology rarely collaborate on the whole, leaving us with a patchwork picture of cognition that is as fascinating as it is fragmented.

Massive Modularity is not wrong; it is merely insufficient. It interacts with other planes of thinking, and a significant reason for this is the rapid accumulation of technology. Humans have harnessed technology to construct our own niche of an ecosystem, one that is far different[44] from the past environment, which Pinker claims shaped the selection of our current brain.[45]

Why would a highly regarded discipline like neuroscience create a model with these shortcomings? Was it a mistake, an oversight, or just because these

44 Some scholars believe that our built environment is fully synthetic; Baudrillard, for example, claims we are currently divorced from all reality and meaning, having completely replaced it with symbols and signs, that modern human experience could more accurately be seen as a simulation of reality. See Baudrillard, Jean. *Simulacra and Simulation*. University of Michigan Press, 1994.

45 Pinker, Steven. *How the Mind Works*. W. W. Norton & Company, 1997. See Chapter 1: "*The mind is a system of organs of computation, designed by natural selection to solve the kinds of problems our ancestors faced in their foraging way of life, in particular, understanding and outmaneuvering objects, animals, plants, and other people.*"

other elements were unimportant? Howard Gardner pointed out that things like context, culture, and history were not primary inputs to cognitive science because people thought they were unimportant, but rather because people thought they were enormously complicated, and the field needed to simplify and isolate simple parts of cognition to understand those building blocks.[46]

My perspective, like all individuals, is unique. I am a technologist who shares common experiences with many modern people – I have children and a family, work, interact with diverse teams of people, and lead an often hectic and technology-immersed life.

Based on this complicated, messy, and lived experience of thinking in the modern world, and the applied real-world problems I work on empirically, combined with my research, I now see thought as a set of processes differently than what we might see through the specialized lenses of philosophy, social science, clinical therapy, psychology, sociology, anthropology, neurobiology and/or other cognitive sciences.

My unit of analysis is not the brain, consciousness, or even the "mind"; many others focus on those domains, and I will leave that to them.[47]

I'm interested in how we *think*, and the supporting processes that make those thoughts possible. I'm especially interested in the effects of technology on thought.

Which is why I will not focus on locating one thought or another at a molecular level, or within one neuron, synapse, specific neural node, or module. I won't be pointing to specific areas of the brain that are changing, since that is not my focus. Chapter 10 provides more reasoning for this decision and broader purview, but for now, consider, for a minute, the definition of thinking by Magnus Enquist and colleagues: "*By thinking, we mean mental mechanisms that, at the time of decision making, can examine and recombine information that has been gathered previously, possibly in many different situations.*"[48]

If you really reflect on this definition, you can see how humans uniquely outsource these operations to elements outside of our brain. I'm trying to understand how we think, as Hutchins says, in "*the wild…everyday world, where human cognition adapts to its natural surroundings.*"[49] Thought that struggles to apply itself not just to a single complex problem, task, or occupa-

46 Howard, Gardner. *The Mind's New Science: A History of the Cognitive Revolution.* Basic Books, 1985.

47 See Chapter 10 for my take on the molecular manifestation/instantiation of thoughts within the brain.

48 Enquist, Magnus, Stefano Ghirlanda, and Johan Lind. *The Human Evolutionary Transition: From Animal Intelligence to Culture.* Princeton University Press, 2023, p. 20.

49 Hutchins, Edwin. *Cognition in the Wild.* MIT Press, 1995.

tion, but to multiple simultaneous layered problems related to dozens of values, that are respectively set by each of the many selves that exist within each of us. And thought as it assesses the innumerable values and needs we intuit in each of the scores of people in our lives, and as it tries to keep up with the devices we've come to rely on. In our present-day natural surroundings, experiences are increasingly mediated by technology: we regularly use our phones to augment our memory, and there are now already ubiquitous artificial intelligence agents on our phones or watches to coach us through our daily activities, and we increasingly offload complex cognitive processes to the devices around us, such as driving and parking. And thought, which struggles to cope with all of that, as those parts and the priority we assign to each of those many processes vary across time and the ever-changing environments we traverse in our daily lives.

Now is the perfect time to develop a new model of thought that addresses how we think in the modern present and the near-future niches we are constructing for ourselves. After all, as Hutchins added, *"humans create their cognitive powers by creating the environment in which they exercise those powers."*[50]

2 My Definition of Thought

A new model needs a new definition. While I prefer Enquist's definition over those rooted in biological functions, we will later see the need for a different articulation. I define it as:

> **Thought**: the processes and tools a person uses (consciously or unconsciously) to:
> - evaluate their needs and goals,
> - make sense of their environment (including other people and themselves),
> - consider how to use available resources to achieve valued outcomes (goals),
> - control processes used to manage resources to achieve those valued outcomes.

This definition gives us room to consider thought that happens outside the skull, and the modules of thought that exist culturally within the shared understanding of the social-cultural milieu. Once you begin to follow the

50 Hutchins, p. xvi.

threads of thought, you find you *must* leave the skull to understand how they are really occurring on these three "streams":

- "wet-layer" – the embodied brain
- social dimension – we think through people and social constructions
- the physical – the modern Extended Mind Thesis tools and processes used in the technology examples above, and which we develop further in part two of the book, especially in Chapter 7.

Here, I synthesize them into an eclectic model I call the Three Streams Model of human thought.

3 Plain Language Summary of This Chapter

Models of thought have been around for a long time, and interestingly, they tend to change as new technologies evolve that help to understand thinking in new ways. None of those models, though, really helps us to holistically understand how we think today in our modern niche very well. To overcome this, I bring together models from different fields, each of which helps us to understand a dimension of how we think. Yes, we think with our brains, and through complex networks of synapses and modules in the brain, but those networks and modules are tied to modules in the people and things around us. This model is not completely new; it builds on and extends the ideas of other scholars,[51] but as we will see later in the book, especially in Part 3, what is new is how I integrate the three streams, applying the model to our modern world, and especially the insights we find when we do that.

51 In addition to the other scholarship reviewed here, this conclusion approaches adjacent models in the socio-technical approaches. See, for example, Geels, Frank W. "Technological Transitions as Evolutionary Reconfiguration Processes: A Multi-Level Perspective and a Case-Study." *Research Policy*, vol. 31, no. 8–9, 2002, pp. 1257–1274 and Geels, Frank W., and Johan Schot. "Typology of Transition Pathways." *Research Policy*, vol 36.3, 2007, pp. 399–417.

CHAPTER 2

The Disruption of Thought

> It's unbelievable how much you don't know about the game you've been playing all your life.[1] (Mickey Mantle)

What's my case for the central claim that thought is being disrupted?[2]

I identify four areas that may indicate a discontinuity in our thinking; any of these changes has the potential to reach the level of "disruption." However, the most significant factor is that *all four are happening simultaneously.*

Here I introduce them all, in a simplified, headline way. Call it the USA Today summary. In later chapters, I will provide a more detailed, rigorous, and contextual explanation of how I arrived at these conclusions.

Let's first turn to Thomas Kuhn, a founding father of how we use the word "paradigms" today. He points out that paradigms exist for a reason – they provide a common framework in which scientists can collaborate and communicate effectively with one another. It provides scholars with a common operating picture in which to share common beliefs, languages, and worldviews, within which different theories can be explored and evaluated. It is a common set of fundamentals, and assumptions. When you challenge or radically change these fundamentals, you create a paradigm shift – a disruption.

To understand the issues being disrupted, it's helpful to have a basic understanding of how I think we think. This model was mentioned briefly at the end of the introductory chapter and is developed in more detail in Part 2 of the book, but for now, let me provide a basic sketch:

– **We think with our brains**, of course. What we know about brains and how they work advances with every new breakthrough technique, but right now the brain seems to work through an extremely large number of networks that connect various nodes of the brain into modules, which reconfigure rapidly as we use the brain for various tasks and processes. I'll refer to this as massive modularity. It is also important to remember that the brain doesn't work by itself; it is connected to the nervous and other systems throughout the body. Cognition in our bodies (or "wet thought") works as an extended system, so we should frame it not so much as hap-

1 *Moneyball.* Directed by Bennett Miller, performances by Brad Pitt and Jonah Hill, Columbia Pictures, 2011.

2 See Appendix 1 on Disruption for my take on the word "disruption."

HTTPS://DOI.ORG/10.1163/9789004739161-005

pening in an isolated organ, but rather, I'll call it the "embodied brain" – a part of your body's system of systems.

- **However, we also think through people and culture.** People and culture fundamentally shape our cognitive processes. We develop and express thoughts through conscious and unconscious social interactions, while culturally constructed technologies, such as language and writing, physically reshape our neural pathways. These culturally influenced thought patterns transcend individual brains, gaining meaning, context, and efficacy when shared collectively. Our thoughts extend beyond our physical bodies, continuously adapting through social engagement. Values, which serve as critical organizing principles in cognition, emerge primarily from cultural foundations. Indeed, significant portions of our thought processes are transmitted through cultural institutions, including family structures, educational systems, and broader social frameworks.
- **Finally, we think through technology, and the things around us.** This may seem a little odd, but bear with me, and read through the next part of the book before you pass final judgment. For now, let's just say when you outsource thinking to the things around you – like letting your phone remember the phone numbers of your friends and family – you are literally thinking with technology. And that's a simplified tip of the iceberg. There are several reasons to support this as an actual mode of thought, and we'll see significant implications that help us better understand the world when we realize we think this way, including some we'll talk about in the rest of this chapter.

I call this view of cognition the Three Streams Model of thought. The intention is to highlight that our thinking spans not just nodes and modules in our brain, but goes beyond to connect with nodes and modules in culture and technology as well.[3]

Now, let's examine where these disruptions are occurring within that model.

3 This model is distinct from but runs parallel to another multi-tier model – the Macro-Meso-Micro paradigm or multi-perspective thinking in the socio-technical thinking. See, for example, Geels, Frank W. "Micro-Foundations of the Multi-Level Perspective on Socio-Technical Transitions: Developing a Multi-dimensional Model of Agency through Crossovers between Social Constructivism, Evolutionary Economics and Neo-Institutional Theory." *Technological Forecasting and Social Change* 152, 2020, pp. 119–894.

1 Changes in the Fundamentals of Thought

One of the biggest reasons thought is being disrupted is a radical reshuffling of the *fundamentals of thought*, across all three dimensions of the Three Streams Model. While these are discussed in Part 3 of the book, let me briefly summarize the biggest fundamental changes:

- **Thought is a process for achieving certain values, but the values at the heart of each layer of thought are being radically restructured:**
  - **Brain** – evolved to allow us to produce offspring that outcompete other offspring in periods of abundant risk and scarce resources. Now we have radically changed lifespans, life history, and reproduction, and we've transformed our niche to be risk-scarce and resource-abundant. Brains are anachronistic now, sub-optimal, and buggy. This is a relatively recent phenomenon.
  - **Culture** – cultures encode outsourced cognition. But our rapidly evolving niche has even our cultures, which store our basic beliefs, values, and socially constructed tools, hindering us. Culturally held beliefs about race, gender, and other themes are being updated, while culturally pursued values such as wealth may be suboptimal now for the planet and for each of us.
  - **Technology** – Technology increasingly scaffolds our thoughts and often is a substitute for us having to think. But each platform we interact with has values of its own that shape and pull at "our" values. They are specifically designed to achieve these values through rapidly evolving gamification and other strategies that optimize their value to you, while also containing their own biases, and, in particular, with AI, their predictions on how we will react. The accelerating change to these amplifies the cognitive load and dissonance created by rapidly changing values at the first two levels. Your inherited need to be popular and influential in small groups breaks down when interacting with mass media, social media, and other emerging tech. It may have been easy to overlook the increased use of screens in school and the corresponding dramatic effect on teaching and thinking, but one cannot miss that AI will change how we function in most parts of our lives. And AI is just one very obvious technology that surrounds our daily existence. And almost all are changing much faster than we are adapting.
- **Our ability to control our own thoughts is rapidly eroding.** As cognitive complexity increases, we resort to simplifying tactics, such as relying on cultural cues and artificial technologies to help us navigate that complexity. Yet, those platforms are not only not under our control, but they are

run by actors who often have contrary goals and values to yours. Meanwhile, the many individual selves hardwired within you find themselves being subtly manipulated by increasingly sophisticated external actors, who are leveraging increasingly powerful technologies targeted at you. You are losing agency in your own thoughts, wants, and needs.

- **Thought is not working for us.** Kuhn points out that new paradigms arise to address crises. This is perhaps the biggest crisis we could face: in the modern world, thought, as mediated by increasingly powerful technology, rarely leads to healthier, happier, and more productive people. What will happen as we add more technology to the mix, without first learning of and addressing the paradigm shift, the disruption?

We are already thinking differently, but we are failing at it.

> No more thinking tonight, let's watch TV. (John Denver, in the movie Oh, God! (1977))

2 Changes in the Fundamentals of Technology

The second reason thought is being disrupted is that I see a radical reshuffling of the *fundamentals of Technology* we rely on to think. This is different than the values being pursued or optimized by those technologies. To understand this, let's talk about the phrase "General Purpose Technology," a term coined by Bresnahan and Trajtenberg in 1995.[4] General Purpose Technologies (GPTs) transform economies at national and global scales by revolutionizing production methods, reshaping competitive advantages, and fundamentally altering societal values and cultural character.

Scholars classify the Internet as a General Purpose Technology[5] because it permeates virtually every domain of human endeavor, creating positive feedback loops and non-zero-sum emergent outcomes that continuously accelerate its transformative impact. The feedback loops in other technologies are greatly enhanced or amplified by the Internet, and in total, they are an unprecedented General Purpose Technology disrupting human civilization.

In the past, we would experience new technology, like train travel, and after a period of struggle and adaptation, we would return to a period of some mar-

---

4 Bresnahan, Timothy F., and Manuel Trajtenberg. "General Purpose Technologies 'Engines of Growth?" *Journal of Econometrics.* 65.1. 1995, pp. 83–108.

5 Helpman, Elhanan, editor. *General Purpose Technologies and Economic Growth.* MIT Press, 1998.

ketplace equilibrium. Those days are over – now technologies evolve before we complete our adaptations, and we never regain equilibrium.[6] Multiple feedback loops interconnect transformative technologies, not just the Internet, but also 5G networks, artificial intelligence, machine learning, the Internet of Things, edge computing, and quantum computing. As each technology advances, it bootstraps advances in other technologies.

Leaving thought aside for a minute, we are approaching a disruption in the technologies we rely on. It's not that the technologies themselves are disruptive (they are), it's that we've lost the capacity to understand them. More significantly, we have lost the capacity to restore equilibrium and adapt to any of these changes, let alone all of them. Kuhn's return to a normalization of the new paradigm is gone; our common basis for shared beliefs, values, or frameworks for broad-scale human collaboration is shrinking. The uncertainty around this unprecedented dynamic represents an important vector in the disruption of thought, at least as much as other landmark events in human history, such as the evolution of language.

We've come to depend on these technologies for our day-to-day functioning, not only not understanding how they work, but without realizing the unpredictable ways they can fail and leave us stranded. What happens when the memories we have outsourced to technology, like the phone numbers in our contact list, get wiped out due to some user or platform error? How will we call those we love? What if an AI system helping with corporate strategy introduces errors through faulty coding, inherent biases, or fabricated references that support flawed strategic decisions, all without our awareness? When we lose the ability to understand how something works, we are less able to spot its mistakes.

Few are looking at how such innovations across this broad front of human activity will interact with one another. It is literally impossible to test how software and platforms will work in the voluminous combinations that will happen ad hoc in the real world, or to see how those permutations will affect us. *We* are the guinea pigs.

To understand our frayed grasp on Technology, let's spend a couple more minutes on artificial intelligence.

We use the term all the time, as it is a cutting-edge technology that's starting to change the world in new and important ways. But the current definitions of artificial intelligence stem directly from the pioneering work done in

6 See Hagel et al. in the Appendix 1 on Disruption for the case behind this statement.

the 1950s: *"the capability of a machine to imitate intelligent human behavior,"*[7] or *"the science of making machines do things that would require intelligence if done by men."*[8]

These definitions of AI were set down when the highest aspiration of artificial intelligence was to emulate human thought processing. However, they are still in use, even though computing has advanced significantly, allowing these platforms to perform work not only equal to, but far surpassing that of humans, and much quicker.

Modern AI systems transcend human cognitive limitations rather than replicate them. We design these systems to optimize outcomes through capabilities that humans, constrained by biological hardware, simply cannot achieve. Today's AI operates not just far beyond human cognition, but completely differently, optimized against forward-looking requirements, not for 250,000-year-old use cases. This includes real-time translation across dozens of languages, multimodal processing with parallel thread execution, and concurrent control of millions of processors.

Consider the breakthrough at Imperial College London, where scientists spent several years investigating antibiotic-resistant superbugs. Before they published their results, the lead investigator gave a short prompt to the Google AI known as Co-Scientist, and within 48 hours, it arrived at the same conclusion as the investigators, completing a decade's work in two days. And it also identified four additional plausible explanations the human team had overlooked.[9]

These are NOT things humans can do. So, why do experts in our current AI paradigm misunderstand their own technology by describing it as something that aspires to think like a human?

- Kurzweil (one of today's most influential futurists, 2012) – *"we are already on the road to a world in which computers will operate at the level and with the speed of the human mind."*[10]

7 Merriam-Webster. "Artificial Intelligence." https://www.merriam-webster.com/dictionary/artificial%20intelligence. Accessed Aug. 27, 2021.

8 Dates back to 1955; see, for example, this work, which attributes this definition to Minsky, but others, such as McCarthy, Rochester, and Shannon, have used similar: Stonier, Tom. "The Evolution of Machine Intelligence." *Beyond Information*. Springer, 1992.

9 "AI Cracks Superbug Problem in Two Days That Took Scientists Years." *BBC News*, Mar. 30, 2025, http://www.bbc.com/news/articles/clyz6e9edy30. Accessed Mar. 31, 2025.

10 Keen, Andrew. "Keen On… Ray Kurzweil: How Computers Will Reverse Engineer the Human Mind By 2029." *TechCrunch*, Dec. 3, 2012, techcrunch.com/2012/12/03/keen-on-ray-kurzweil-how-computers-will-reverse-engineer-the-human-mind-by-2029/. Accesse d Jul. 27, 2024.

- Eric Schmidt (Former Google CEO; tech billionaire, 2023) – AI is "*human-like, but not human intelligence.*" When asked specifically about what makes AI "*like a human,*" Schmidt referenced GPT-3, noting that it "*sucks in all the text information around the world and, remarkably, appears to understand it*" and can "*generate answers*" in ways that seem to mirror human cognitive processes.[11]
- Masayoshi Son (SoftBank CEO; tech billionaire, 2024) – AI now can "*think like humans.*"[12]

This persistent mismatch between AI's actual capabilities and how we conceptualize them reflects a broader pattern in our relationship with transformative technologies. The assumption that intelligent behavior must mirror human cognition has become so deeply embedded that even technology leaders default to anthropomorphic frameworks when describing fundamentally alien systems. This conceptual lag isn't merely academic, it shapes how we design, regulate, and integrate these technologies into society. A better frame is provided Stuart Russell and Peter Norvig, authors of a very highly regarded AI textbook: *"The quest for 'artificial flight' succeeded when the Wright brothers and others stopped imitating birds and started using wind tunnels and started learning about aerodynamics. Aerodynamic engineering texts do not define the goal of their field as making 'machines that fly so like pigeons that they can fool other pigeons.'"*[13]

Part 2 reveals how this same pattern emerges when examining human learning itself. For millennia, scholars have emphasized that the engine of human progress is social learning – our uniquely collaborative ability to share insights, imitate skills, and pass knowledge across generations and geographies. This collective bootstrapping of culture is what allowed our species to advance so rapidly. AI systems increasingly demonstrate that intelligence can emerge through radically different pathways.

DeepMind's AlphaZero is AI's answer to the jet engine: it achieved chess mastery not by mimicking human grandmasters – the cognitive equivalent of flapping feathers – but through millions of self-played games that generated entirely alien strategies. Given only the basic rules of chess and no human gameplay data, this artificial intelligence system taught itself exclusively through self-play, engaging in millions of games and continuously

11 Schmidt, Eric. Interview by Reid Hoffman. "You should be running toward AI, with Eric Schmidt." *Masters of Scale,* Apr.5, 2023, mastersofscale.com/you-should-be-running-toward-ai-eric-schmidt/. Accessed Jan. 17, 2024.

12 Son, Masayoshi. SoftBank World 2024. *The Sun,* Oct. 3, 2024.

13 Russell, Stuart, and Peter Norvig. *Artificial Intelligence: A Modern Approach, 3rd edition.* Pearson, 2009.

refining strategies via reinforcement learning without human intervention. This approach fundamentally diverges from how human expertise traditionally develops. AlphaZero mastered chess in less than 24 hours, achieving unparalleled proficiency and becoming the world's strongest player. This occurred in 2017, now several computational generations ago in AI development. Rather than analyzing established human strategies, its iterative methodology generated unprecedented moves and tactics. Chess grandmasters characterized its play as *"chess from another dimension,"* highlighting its radical departure from human cognitive approaches.[14]

This was 2017, already several generations of AI progress ago. The lesson is stark: while human mastery has always rested on shared culture, AI systems can generate superhuman insight by bypassing culture altogether. They don't "think" like us.

Earlier, I quoted from the book *The Age of AI*, by Kissinger, Schmidt, and Huttenlocher; here's the rest of that quote: *"AI is not an industry, let alone a single product. In strategic parlance, it is not a 'domain.' It is an enabler of many industries and facets of life; … The outcome will be the alteration of human identity and human experience of reality at all levels not experienced since the dawn of the modern age."*[15]

Artificial intelligence, profoundly misunderstood (or at least, misrepresented) by many developers and tech leaders, and nearly all users, will fundamentally reshape human experience. Its effects will cascade through society, amplifying the potential of all other technologies. These brief examples showcase how technology thinks in *new* ways, which humans won't even be able to begin to fathom. Rather than continuing to force AI into familiar human frameworks, we might benefit from my colleague David Bray's provocative suggestion: consider AI as "Alien Interaction" rather than "Artificial Intelligence."[16]

How those interactions will shape human thinking is beyond the pale of our understanding, especially since the mistaken understanding of AI as something that thinks like humans is continually perpetuated.

---

14 Knight, Will. "Alpha Zero's 'Alien' Chess Shows the Power, and the Peculiarity, of AI." *MIT Technology Review*, Dec. 8, 2017, http://www.technologyreview.com/2017/12/08/147199/alpha-zeros-alien-chess-shows-the-power-and-the-peculiarity-of-ai/. Accessed Aug. 23, 2025.

15 Kissinger, Henry A., et al. *The Age of AI: And Our Human Future*. Little, Brown and Company, 2021.

16 Krigsman, Michael, host. "When AI Goes Off the Rails: An Executive Action Plan." CXO-Talk, episode 888, featuring Dr. David A. Bray, Jul. 25, 2025. CXOTalk, www.cxotalk.com/episode/when-ai-goes-off-the-rails-an-executive-action-plan. Accessed Sep. 1, 2025.

This chapter presents my initial argument that we are seeing profound transformations in technological foundations and the intellectual frameworks they influence. The book will explore many more examples of this phenomenon.

3 The Changing Nature of Our Relationship to Reality

The third change I see is the changing nature of our relationship to reality. By reality I mean the directly observed and unmediated observation or participation in life. This may be the most provocative of the four areas. Baudrillard presents perhaps the best articulation of it in his 1981 book *Simulacra and Simulation,* in which he examines the relationships between reality, symbols, and society. Baudrillard notes that models are useful; you can carry a map of a mountain around more easily than you can carry a mountain, for example.

But he believes we've become so reliant on models that we have no underlying relationship with reality anymore. In his view, modern human experience is a simulation of reality. The cues that we get about the world are increasingly mediated by symbols, media, and technology, and these cues are inaccurate. Meaning becomes meaningless when infinitely mutated. "*It is no longer a question of imitation, nor duplication, nor even parody. It is a question of substituting the signs of the real for the real,*" he says, in a process he called the "*precession of simulacra.*"[17]

This is vividly illustrated, literally, in *The Lost Words,* written by Robert McFarlane and illustrated by Jackie Morris.[18] The book details changes made to the 2007 Oxford University Press (OUP) Junior Dictionary to delete certain words describing nature and countryside – such as acorn, bluebell, dandelion, fern, otter, and willow – to make way for newer terms like analog, block graph, celebrity, blog, cut-and-paste, voicemail, and MP3 player. There was a subsequent campaign to bring the old words back, and *The Lost Words* was part of that backlash, with beautiful illustrations for 20 of the words that had been cut. McFarland explains how the OUP editor, Vinneta Gupta, justified the change, "*the dictionary needed to reflect the consensus experience of modern-day childhood.... Nowadays the environment has changed.*" McFarland responds, "*There is a realism to her response – but also an alarming acceptance of the idea that*

17 Baudrillard, Jean. *Simulacra and Simulation.* University of Michigan Press, 1994.

18 Macfarlane, Robert, and Jackie Morris. *The Lost Words.* Hamish Hamilton, 2017.

children might no longer see the seasons, or that the rural environment might be so unproblematically disposable."[19]

Baudrillard is not saying our culture is fake; his point is much more profound – that we've lost the ability to distinguish between real and artificial. Children today interact with highly mediated electronic portrayals of celebrities, not physical things like acorns. This process occurred in three steps, according to Baudrillard, starting in the pre-modern era, where we knew something was a fake, such as a Shakesperean or Kabuki play; then the industrial revolution where the distinctions began to break down; and finally, in the postmodern era of the late 20th century, we've now entered his precession of simulacra stage. He claims this happened due to:

- **Media culture** – media is motivated to connect to our most private selves, not reality. We don't acquire things for real needs; we do so for needs that have been shaped by commercialization.
- **Exchange-value** – Capitalist culture substitutes "use-value" (the real uses to which an item will be put) for worth-value (what it can be exchanged for; aka its exchange-value). We think of our time in terms of its value, but we often overlook its true nature and potential uses.
- **Multinational capitalism** – Complex industrial processes divorce our understanding of the world – e.g., we know Starbucks, but we don't know coffee plants.
- **Urbanization** – Our built environments cause us to lose touch with the natural world.
- **Language and Ideology** – we are so reliant on language to structure our perceptions that any representation of reality is always already ideological, always already constructed by simulacra (e.g., culturally mediated or constructed).

Consider this recent experience: I switched on the radio to find two radio personalities dissecting a conversation between two politicians, a discussion that had been recorded for a podcast about Governor DeSantis's relationship with President Trump. The recursive nature of this moment struck me immediately: I was in an Escher-like construct: listening to two people analyze two people who were discussing two other people.

This layered mediation/commentary upon commentary upon commentary – created a dizzying hall of mirrors where the original *actual* political relationship had become almost irrelevant. What mattered now was the interpretation of the interpretation of the perceived relationship. As I sat there

19 Linda. "Robert Macfarlane's Landmarks." *Under the Gables*, Mar. 4, 2017, underthegables.blogspot.com/2017/03/robert-macfarlanes-landmarks.html. Accessed Aug. 23, 2025.

following this chain of removed discourse, I wondered if I might be the only listener in the national audience thinking of Baudrillard's concept of simulacra – where copies of copies eventually replace any meaningful connection to reality.

The irony wasn't lost on me: here I was, contemplating French postmodern theory while consuming fourth-hand political analysis through yet another medium. Perhaps this perfectly encapsulated our current media landscape, where we've become so accustomed to engaging with representations of representations that we rarely question how far we've drifted from any authentic political conversation.

The claim that we are losing our ability to relate to reality may seem provocative. Yet, this assertion is not controversial; we simply lack the conceptual framework to recognize this phenomenon and its significant implications. The following section presents three use cases showing how the three dimensions of thinking integrate within the Three Streams Model. As these examples progress from mundane (selecting television content) to complex (navigating a world permeated by AI guidance and virtual and/or mixed reality), technology increasingly mediates our human interactions, even with individuals physically present. This mediation then exposes us to forces that compromise our cognitive autonomy and reinforce the other concerns previously addressed.[20]

Let me put it this way: a friend is diabetic. I asked him how that affected his life, and he answered that once he got used to it, it wasn't too bad. "Let's say you have a glass of juice at breakfast; I can't, but I can have an orange and a glass of water." Our highly mediated world is not like the experience of peeling and eating a fresh orange, or drinking a glass of "fresh" (but pasteurized) orange juice – which aside from lacking the esthetic experience of the smell, sight, sound, and feeling of peeling, also has significantly less fiber than the orange, possibly lower levels of some key nutrients, and has twice the amount of sugar because it has ~2 oranges in it, and has been processed.[21]

Rather, our modern experience is more like drinking a can of thawed-out orange "drink" concentrate. You know – those cans in the frozen foods section at the grocery store, to which you usually add to a pitcher of water to make a

20 Note: It's important that we don't confuse Baudrillard's point that we've lost touch with reality and now live in a simulacrum of it, with Bostrom and others' hypothetical point that we may all be living in a computer simulation. See Bostrom, Nick. "Are We Living in a Computer Simulation?" *The Philosophical Quarterly, vol.* 53.211, 2003, pp. 243–255.

21 Chanson-Rolle, Aurelie, et al. "Nutritional Composition of Orange Juice: A Comparative Study between French Commercial and Home-Made Juices." *Food and Nutrition Sciences* no. 7.04, 2016, p. 252.

batch of drinks? The orange "drink" isn't orange juice – (let's pretend you bought this one to save a dollar over the more expensive orange "juice"). It might have parts that originally came from real oranges, but it also contains many other ingredients, all condensed down and with all the water removed.

That's the metaphor for modern daily living: we are not experiencing the reality of eating an orange and drinking a glass of water. We're drinking a thick, syrupy sludge of something; we don't know what. But it's very different from the rich sensory experience of peeling and eating an orange, and we've forgotten that there's even a difference.

Far from being an esoteric or philosophical discussion, this issue is critical in our everyday real-world lives. A good real-world illustration of this "orange drink concentrate version of reality" (or ODCVoR for short) is *Lord of the Flies*.

It's common for me to discuss with a range of people how technology affects our lives and the current trend of technology eroding our subjective well-being. At least once a month, someone will bring up *Lord of the Flies* (LotF), usually in the context of how genuine/primitive human nature surfaces in various technology forums or interactions. Many Americans had this novel, written in 1954 by William Golding, as required reading in high school and probably remember that it featured a group of schoolboys who survived a plane crash onto an uninhabited island. With no adults around, the boys established their own social structure. Even people who haven't read the book or seen the 1990 movie know that it didn't end well. By the time they are rescued months later, the island is literally in ashes, and three of the children are dead. The first man ashore is a Royal Navy officer who said, "*I should have thought a pack of British boys would have been able to put up a better show than that.*" At this, one of the boys bursts into tears, and the book tells us, "*Ralph wept for the end of innocence,*" and for "*the darkness of man's heart.*"[22]

The plot features an intensification of fear, power struggles, group polarization and division, and the rise of primal violence. It's well accepted in our society that the story is an allegory that modern civilization is just a thin veneer, and only a few days or weeks are needed to strip it away, revealing we are, at our core, just a bunch of selfish, destructive savages with darker urges waiting to take over. Written a few years after the horrors of WWII, the book would sell tens of millions of copies and be hailed as a literary classic. And, in its success, it helped shape and reinforce our understanding of human nature, without civilization's moderating influences.

This is such a powerful, memorable, and viscerally cynical lesson that makes it a highly effective cultural touchstone used regularly by others when

22 Golding, William. *Lord of the Flies*. Faber & Faber, 1954.

I talk about the intersection of tech, individuals, and society. It helps us understand that many of us have an inner Nazi or savage brute checked only by civilization's diaphanous membrane.

We take that fictional lesson to heart, but very few pay attention to another lesson offered to us from the real world, just a few years after *Lord of the Flies* was published.

In 1965, a real group of 6 schoolboys aged 13–16 found themselves stranded on an uninhabited island for 15 months when a small boat they were in drifted out to sea.[23] The island, 'Ata, wasn't some Tom Hanks tropical paradise with sandy beaches and waving palm trees loaded with coconuts. It was a huge rock, looming more than a thousand feet above the ocean. Now considered unhabitable, the boys, armed only with an old knife blade, built out shelters, gardens, badminton courts, and even an improvised weight room. They developed a rotating work roster, an effective dispute resolution process, and, using an improvised guitar they made, ended each day with prayer and song. Doctors examining the boys after rescue found them in good spirits, well-muscled, and healthy, even though one of the boys had experienced a serious fall and a broken leg during their stay. The other boys rescued him from the steep cliff, set his leg, and volunteered to take over his chores while he healed.

One of the technologies central to both stories is fire. In *Lord of the Flies*, fire symbolized hope and destruction, and catalyzed many of the most barbaric exhibitions. Efforts to establish a signal fire early in the story led to a fire that destroyed part of the island and killed one of the youngest boys. Later, efforts to maintain that fire led to power struggles, and when it goes out (due to inattention and frolicking) during a chance for rescue, it triggers a social breaking point. At the end of the novel, with the group splintered into factions, one group uses it to set the forest on fire to smoke out or kill members of the other groups. Unintentionally, this fictional conflagration has the unintentional consequence of leading the boys to be found and rescued.

In real life, the boys started a fire early on and maintained it rigorously. It never went out during the 15 months, in the face of all kinds of weather. It, too, was the source of the non-fictional boys getting rescued.

How is this comparison between art and real life, over seventy years ago, relevant today? Because many of us accept the artistic writing of a depressed alcoholic as fact and, in our entertained fugue, fail to attend to the optimistic

---

23 Bregman, Rutger. "The Real Lord of the Flies: What Happened When Six Boys Were Shipwrecked for 15 Months." *The Guardian*, May 9, 2020, https://www.theguardian.com/books/2020/may/09/the-real-lord-of-the-flies-what-happened-when-six-boys-were-ship wrecked-for-15-months. Accessed Nov. 19, 2025.

but factual lessons that the world presents to us. The mediation of books, movies, news, and other technologies transforms the real world into an artificial (fake) reality, and we've lost the ability to discern or recognize the transformation and artificiality. The businesses behind most media, not just social media, exist to make money. They seek the spectacular and use it to lure and retain our eyes and ears; the useful facts, good news, and the information we *need* rarely make that cut.

Consider the LotF/ ODCVoR effects in your own life.

Take violence, for example. Your life, in the real world, may actually have little violence. But your mediated life has constant violence; because our brains are hardwired to attend to violent activities, we find it interesting, and companies in industries such as entertainment, news, social media, and advertising use it to attract and keep our attention. That's how they pay the bills and get advertising dollars. Sure, it might be in comedy, such as in the rom-com where one person opens the door and hits the head of the person listening at the door. Or it might be presented as "fact" by a news organization talking about a shooting that killed several people, including a pregnant woman and her unborn child.[24] We've become so surrounded by these signals that we think the world is a violent place. We don't consider that the murder we are seeing on the credible evening news broadcast occurred over a thousand miles away. Between us and this remote violent act, tens of millions of people lived peaceful and even mundane lives, amid historic downturns in violent crime. (We'll get back to this in chapter 24, when we discuss Rosling's great book, *Factfulness*). Like the orange drink concentrate, the artificially condensed perception of violence is just one of many things that are represented inaccurately in the artificial niche we've created for ourselves.

This is not an indictment; it is part of the observation of the world. It is becoming aware of the water, as Foster-Wallace asked us.

To paraphrase Harari,[25] several hundred thousand years ago, we were guided by elders with experience in the world. For thousands of years, people took authority from religion. For the last few centuries, we've begun to shift our focus from secular power structures, such as governments, schools, and science, to seeking authority within ourselves. In recent years, change has

---

24 "Mass Murder Leaves 6 Dead, Including Unborn Child, in Indianapolis." *WANE*, Jan. 24, 2021, www.wane.com/news/mass-murder-leaves-6-dead-including-unborn-child-in-indianapolis/. Accessed Aug. 23, 2025.

25 Harari, Yuval N. "Yuval Noah Harari on Big Data, Google and the End of Free Will." *Financial Times*, Aug. 26, 2016, www.ft.com/content/50bb4830-6a4c-11e6-ae5b-a7cc5dd5a28c. Accessed Aug. 23, 2025.

been accelerating so fast that we are less likely to rely on our hearts and experience, but on data. Harari coined the term "Dataism" to describe this.

But Baudrillard would be the first to say that experience built on data is not real. It is artificial and an inaccurate view of reality. We must first come to grips with this if we are to learn how to live with technology. That understanding will directly help with some of the potential solutions I lay out in part 4 of this book.

This inability to discern our artificial and mediated "reality" from the authentic thing is one of the big reasons we have the Great Irony of Technology problem that we'll discuss in Part 4: why many of us are better off than we used to be a generation or two ago, but we feel worse. It's not AI, television, the Internet, social media, cell phones, or any other tech. It's all of it, in aggregate. In media, we interact with heroes who are all more heroic than us, people are sexier, and everyone has better things. And the world is more dangerous and violent. Everything is more. Nothing is reality. And we've lost the ability to see that.

4 The Big Transformational Opportunity

Finally, there's a fourth and bigger reason thought is being disrupted: since we've offloaded so much thinking to technology, we've created an opportunity to free up our thinking and apply it in completely new ways. I call this the Big Transformational Opportunity. To understand this, you'll want to read Part 2, and particularly Chapter 6, about how we think through culture.

There you'll read about scholars arguing that major improvements in human cognition were culturally derived, in the form of language and the use of writing and symbols. But language evolved over circuits in the brain that evolved to give us hand dexterity, so we can manipulate and grasp things. Dexterous grasping abilities evolved in part due to an evolution toward efficient bipedalism. Bipedalism freed up the hands to not only extract resources differently, through tool use, but also allowed us to move resources from one place to another, which caused us to consider and anticipate our future needs more. And we used those nimble hands to develop writing systems, which may have given rise to the pillars of civilization.[26]

26 Here we are skipping over scores of other compelling events in human history, like the move to sleeping on the ground, the importance of fire, our ability to throw accurately, run without overheating, imitate, and the cultural importance of the grandmother; all factors which contribute plausibly to our cognitive features today.

These three examples – the evolution of efficient bipedalism, the emergence of language, and the innovation of external symbol storage – are landmark events in human cognitive history:

- Bipedalism – evolved over millions of years of foraging on the ground, several million years ago. The unintended consequence of this was the development of primal grasping and forward-thinking (mental time travel),[27] our unique and important evolution of highly effective throwing and efficient ability to migrate long distances, all dramatically shaping our cognition. This had the unintended consequence of setting us up for the language revolution.
- Language – evolved gradually over at least tens of thousands of years, maybe sometime around 50,000 years ago. It evolved not by using older communications circuits in the brain, but by co-opting circuits that developed as part of our shift to bipedalism, most particularly those grasping circuits.[28] Language had the practical effect of enabling cooperative hunting, enhanced teaching, but it also had the unintended consequence of giving rise to a new form of thinking: the narrative, where we developed the ideas of abstractness, myth, the future, and the supernatural.
- Writing and the external storage of ideas – evolved over thousands of years, as we started talking more and differently, and then began to store ideas on things around us. It expanded our cognitive capacity, freeing up resources in our heads to do other things. The unintended consequences of offloading cognitive functions like memory into the things around us include freeing up scarce cognitive resources to perform new functions. We'll discuss in more depth in Chapter 7 how we used that surplus, but essentially it let us reflect, to consider the meta, and during that, we developed the capacity to think more deeply about ideas, and developed science, law, reasoning, arts, and other higher-order cognitive skills.

The brain is an expensive organ. It consumes far more of its fair share of your daily energy intake: composing only about 3% of the body's mass, it uses 22x more energy than the same amount of muscle,[29] up to about 20% of your

27 Suddendorf, Thomas, D. R. Addis, and M. C. Corballis. "Mental Time Travel and Shaping of the Human Mind." Bar M., ed. *Predictions in the Brain: Using Our Past to Generate a Future*, 2011, pp. 344–354.

28 See Arbib, Michael A. *How the Brain Got Language: The Mirror System Hypothesis*. Oxford University Press, 2012.

29 Aiello, Leslie C., and Peter Wheeler. "The Expensive Tissue Hypothesis: The Brain and the Digestive System in Human and Primate Evolution." *Current Anthropology*, vol. 36, no. 2, 1995, pp. 199–221.

calories.[30] Unlike computers, and despite our best attempts to multitask, the brain can only process one complex, conscious, and focused task at a time. We only have so much memory and so much capacity. Every time we can effectively shift some of these functions outside of our skull, we score a big win in freeing up this rare and valuable resource. And when we've successfully achieved cognitive offloads in the past, we've gained significant individual and collective benefits as a species.

Each major transformation sets the stage for the precursors of the *next* major transformation that would follow. It was a massive bootstrapping operation conducted over the vast arc of human history. And each further shaped the structure and function of people who adopted them.

What happens as we increasingly offload not just memories, but many of our learned tasks, such as how to drive, or do simple long division, to technology around us? Do we gain a cognitive surplus? If so, what kind of new thinking might that lead to? Do we fill that cognitive space with distractions suited to anachronistic values? Do we leave behind "real" needs and substitute a taste for manufactured ones, as Baudrillard might claim? When the programmable air fryer buys us 15 more minutes in the day, and the self-driving car another 30, do we use that time to think deeply, or meditate? Or do we fill it up with fantasy football, engaging in polarized clashes on social media, becoming enthralled with how everyone is leading a better life than us, or navigating virtual worlds and games?

The big question is: will we use *this* transformation to leave us better off in ways we value individually and collectively? Or, as I said to the audience after my 2017 TEDx talk, "In a few years, Technology will be able to create and present this talk better than I can; what should I be thinking about, then, when I can outsource this work? What should my brain do, then?"

When technology around us develops new forms of thinking, what are the corresponding changes to human thinking? The things around us no longer merely remember or process things on our behalf; they will have new and emergent properties that unlock vast changes to human potential futures.

Unlike the previous revolutions in human thought, all these changes will play out not in millions or thousands of years, but in an unprecedented pace of just decades.

---

30 Raichle, Marcus E., and Debra A. Gusnard. "Appraising the Brain's Energy Budget." *Proceedings of the National Academy of Sciences*, vol. 99, no. 16, 2002, pp. 10237–10239.

CHAPTER 3

How Will We Think in the Future?

> The real problem is not whether machines think but whether men do.[1] (B. F. Skinner)

If thought is being disrupted, it means what comes next won't look like thought did, previously.

What will it look like?

By my definition of disruption, those of us within the existing paradigm won't be able to know with any certainty what the future will hold. The worldviews will be incommensurable, and most of us won't even sense the paradigm shift – the disruption – because it doesn't make sense from our worldview. Until at some point, the new replaces the old. But, since the "future is already here, just not evenly distributed," we can get clues about what comes after a disruption by "scouting the fringes"[2] of our existing paradigms.

Where are the fringes? To address this, I'll first look back, slightly, and in a different direction at how technology and cognition are merging, before looking forward (slightly).

1 The Cyborg Present: We're Already Transformed

Let's look back at October 1, 2001, the day that Michael Chorost became a cyborg.

That was the day he had a cochlear implant surgically inserted into his head.

He had been hard of hearing all his life, but after he became completely deaf as an adult, he made the decision to get the implant. He detailed that transformative experience in his 2005 book *Rebuilt*.[3]

Chorost was always shy and socially awkward but becoming deaf made it worse. He drew a parallel between his situation and the science fiction book

1 Skinner, B. F. "The Real Question Is Not Whether Machines Think But Whether Men Do." *Contingencies of Reinforcement: A Theoretical Analysis*, 1969.

2 Refer to the Appendix 1 on Disruption for more about "scouting the fringes," and other tactics in responding to paradigm change/disruption.

3 Chorost, Michael. *Rebuilt: How Becoming Part Computer Made Me More Human*. Houghton Mifflin Harcourt, 2005.

Cyborg by Martin Caidin, in which a test pilot crashes and faces near death before being resurrected and transformed with new technology. (Yes, if this sounds familiar to some of you of a certain age, this was the basis for the $6 Million Dollar Man).

In *Rebuilt*, Chorost shares, "*I needed a story not of survival, but of transformation.*" He collaborates with clinicians in his transformation and conducts extensive research on the cyborg topic, detailing his personal struggles and journey. He notes, for example, that the process included a "painful transformation and gradual acceptance of his new body." Ultimately, the shy young man begins successfully dating and re-engaging with his communities. Hooray!

Cochlear implants have been around since the 1970s and continue to improve dramatically in performance. The term cyborg was first put forth in a 1960 issue of the journal Astronautics, written by Manfred Clynes and Nathan Kline. Cyborg was a contraction of an earlier term, cybernetics, which developed out of various fields in the early 20th century. Norbert Wiener, who popularized the term with his 1948 book *Cybernetics and Society*, defined cybernetics as "*the scientific study of control and communication in the animal and the machine.*"[4]

Chorost delves into this history as he considers augmenting himself to overcome his deafness. He considers literary theorist and feminist scholar Donna Haraway's Cyborg Manifesto, in which "*A cyborg is a cybernetic organism, a hybrid of machine and organism, a creature of social reality as well as a creature of fiction. Social reality is lived social relations, our most important political construction, changing fiction.*"[5]

In her focus on gender studies, Haraway thinks characteristics like gender and activities like sex may become moot or highly transformed by technological innovations such as advanced reproductive technologies, virtual avatars, artificial insemination, sexual reassignment surgery, and artificial intelligence. Other distinctions, like life and death, human and machine, virtual and real, may also disappear, according to her:

> By the late twentieth century in United States scientific culture, the boundary between human and animal is thoroughly breached. The last beachheads of uniqueness have been polluted if not turned into amusement parks – language,

---

4 Wiener, Norbert. *Cybernetics: Or Control and Communication in the Animal and the Machine.* MIT Press, 1948.

5 Haraway, Donna J. "Cyborg Manifesto: Science, Technology, and Socialist-Feminism in the Late 20th Century." *Social Review*, no. 80, 1985, pp. 65–108.

> tool use, social behavior, mental events, nothing really convincingly settles the separation of human and animal- And many people no longer feel the need for such a separation...the ubiquity and invisibility of cyborgs is precisely why these sunshine belt machines are so deadly. They are as hard to see politically as materially. They are about consciousness – or its simulation. (Haraway, 1985)

People like Chris Hables Gray helped codify this topic into its own intellectual discipline – Cyborg Anthropology, with his publication of *The Cyborg Handbook* in 1995.[6] Anthropologists like Robbie Davis-Floyd and Joseph Dumas extended the field in new directions with their 1998 book Cyborg Babies.[7] In that, they present the writings of several leading scholars that together make the case that modern culture, dominated by technology, is an environment that converts modern children into cyborgs. Cyborg anthropologist Amber Case put it simply in her 2011 TED talk: "*We are all cyborgs now.*"[8] For her, a cyborg is simply a human who interacts with technology; it need not be implanted, it can extend ourselves. The result, according to Case, is that we now have two selves: one digital, one physical.[9]

With thoughts like this about technology literally in his head, Chorost wrote another book in 2011, *World Wide Mind*. In that, Chorost moves away from the cyborg rhetoric and just embraces the new normal of the technology-infused human. For him, the cochlear implant was an upgrade, one that led him to new communities, expertise, and technology, and he used new language to identify his ongoing transformation. He looks beyond the technology of today to envision technologies of the future that will connect humans into larger communities. In a thought experiment, he considers how cochlear implant technologies that could let us read each other's minds might move us beyond a focus on the individual, and humanity could move into new transpersonal models.

This look back into the recent past gives us a foundation to stand on as we look forward. And we need it, because even in the present, things begin to sound very science-fiction-like, quickly.

---

6 Gray, Chris Hables, editor. *The Cyborg Handbook*. Routledge, 1995.

7 Davis-Floyd, Robbie, and Joseph Dumit, editors. *Cyborg Babies: From Techno-Sex to Techno-Tots*. Routledge, 1998.

8 Case, Amber. "We Are All Cyborgs Now." *TED*. https://www.ted.com/talks/amber_case_we_are_all_cyborgs_now. Accessed Aug. 23, 2025.

9 Ottery, Christine "Cyborg anthropologist Amber Case's memo to her future self." *WIRED UK*, May 23, 2016, https://www.wired.com/story/memo-to-future-self/. Accessed Aug. 23, 2025.

As Case points out, even in the present, even without surgeries, we are all cyborgs. That is one of the core streams of my model and has been with humanity for thousands of years. But technology's exponentially increasing power outside of our skin is part of the disruption. As Haraway says, the invisibility of this cyborgness is astounding. Harari, too, makes the case that our world now relies on algorithms, software, and Big Data to make our decisions, a concept we discussed earlier, which he terms Dataism. He uses Angelina Jolie's decision to get a radical double mastectomy in 2013 as a case in point.[10]

Her 2013 DNA test showed that she carried the BRCA1 gene. Statistical databases showed that she had an 87% chance of contracting breast cancer, so she pre-empted the likely problem by having both her breasts removed.

Is Angelina Jolie a cyborg because she relied on DNA tests and Big Data models to direct her decisions? While that's a semantic decision we need not resolve, it poignantly illustrates how closely intertwined our thinking and decision-making are with our technology. It is the David Foster-Wallace water that surrounds us.

Note that I'm not talking here about how technology is eroding our ability to think or making us stupid. This is a well-trodden path. Over 20 years ago Neil Postman wrote *Amusing Ourselves to Death*, an interesting book about how television was turning all information into entertainment and eroding public discourse.[11]

Over 10 years ago, Nick Carr penned the now classic book *The Shallows*, in which he claimed the Internet's constant access to infinite information made us feel connected, but at a price – we were adapting to our computers, instead of the other way around. We lost our deep "literary" form of thinking and exchanged it for a new form that values efficiency and multitasking. He says, *"when we go online, we enter an environment that promotes cursory reading, hurried and distracted thinking, and superficial learning. It's possible to think deeply while surfing the Net, just as it's possible to think shallowly while reading a book, but that's not the type of thinking the technology encourages and rewards."*[12]

In his essay in "The Atlantic" that preceded the book, he shares his personal experience:

---

10 Harari, Yuval N. "Yuval Noah Harari on Big Data, Google and the End of Free Will." *Financial Times*, 26 Aug. 2016, https://www.ft.com/content/50bb4830-6a4c-11e6-ae5b-a7cc5dd5a28c. Accessed Aug. 23, 2025.

11 Postman, Neil. *Amusing Ourselves to Death: Public Discourse in the Age of Show Business.* Penguin Books, 1985.

12 Carr, Nicholas. *The Shallows: What the Internet Is Doing to Our Brains.* W. W. Norton & Company, 2010, p. 116.

> Over the past few years I've had an uncomfortable sense that someone, or something, has been tinkering with my brain, remapping the neural circuitry, reprogramming the memory. My mind isn't going – so far as I can tell – but it's changing. I'm not thinking the way I used to think. I can feel it most strongly when I'm reading. Immersing myself in a book or a lengthy article used to be easy. My mind would get caught up in the narrative or the turns of the argument, and I'd spend hours strolling through long stretches of prose. That's rarely the case anymore. Now my concentration often starts to drift after two or three pages. I get fidgety, lose the thread, begin looking for something else to do. I feel as if I'm always dragging my wayward brain back to the text. The deep reading that used to come naturally has become a struggle.[13] (Carr, 2008)

He adds that if we dreamt up a thing *"that would rewire our mental circuits as quickly and thoroughly as possible,"* we would envision something like the Internet.

Other popular pundits like Susan Greenfield, the former director at the Royal Institution, warned that *"Children's brains could be left damaged, and they could suffer temporary 'dementia' by playing computer games."*[14] In 2020, Tristan Harris and others launched the movie The Social Dilemma, which told us we are being manipulated and rewired by social media and algorithms designed to hold our attention, make us buy things, and give us a distorted view of the world, ourselves, and each other.[15]

But for every Postman, there's a Steven Johnson who argues the complexity of television shows makes us smarter. Nick Carr's concerns are met by people like Clive Thompson, who say the Internet is ushering in a new style of intelligence.[16] Greenfield is countered by academics like Mark Griffiths, who highlight the features that make video games potentially useful for training and improving our cognitive processes.[17]

---

13 Carr, Nicholas. "Is Google Making Us Stupid?" *The Atlantic*, 2008, Jul./Aug. version, https://www.theatlantic.com/magazine/archive/2008/07/is-google-making-us-stupid/306868/. Accessed Nov. 19. 2025.

14 Whitelocks, Sadie. "Computer Games Leave Children with 'Dementia', Warns Top Neurologist." *Daily Mail*, Oct. 14, 2011, www.dailymail.co.uk/health/article-2049040/Computer-games-leave-children-dementia-warns-neurologist.html. Accessed Aug. 23, 2025.

15 *The Social Dilemma*. Directed by Jeff Orlowski, Netflix, 2020.

16 Thompson, Clive. *Smarter Than You Think: How Technology Is Changing Our Minds for the Better*. Penguin Press, 2013.

17 Nuyens, Filip M., et al. "The Empirical Analysis of Non-Problematic Video Gaming and Cognitive Skills: A Systematic Review." *International Journal of Mental Health and Addiction*, vol. 17.2, 2019, pp. 389–414.

No, in this section, I am not discussing the unintentional incidental consequences of certain aspects of technology on our thinking. Rather, I'm focused on the potential aggregate effects of the whole of technology on our thinking. This is more about the opportunity that JZ Young talked about in his 1950 lectures entitled "Doubt and Certainty in Science" – "*We create tools and mould ourselves through our use of them.*"[18]

2 Intelligence Augmentation: The Visionary Lineage

The easiest way to understand intended effects is the idea of intelligence augmentation.

This idea or opportunity was put forward in tantalizing fashion in Vannevar Bush's article "As We May Think," written in 1945 as World War II ended. In it, Bush observed that science had changed the world, but it leaves behind mountains of records. Human experience was expanding prodigiously, but the processes for dealing with it were the same as those used when the Navy employed sailing ships to fight. Specialization was necessary to navigate increasingly sophisticated fields, but it also led to a breakdown in communication, as we struggle to understand one another.

What's needed, in Bush's opinion, is future solutions to these problems.

> Consider a future device … in which an individual stores all his books, records, and communications, and which is mechanized so that it may be consulted with exceeding speed and flexibility. It is an enlarged intimate supplement to his memory…It consists of a desk, and while it can presumably be operated from a distance, it is primarily the piece of furniture at which he works. On the top are slanting translucent screens, on which material can be projected for convenient reading. There is a keyboard, and sets of buttons and levers. Otherwise it looks like an ordinary desk.[19] (Bush, 1945)

Bush termed this device the memex, and he explicitly modeled it on the functions of a brain.

I must remind you that this prescient vision was published in 1945, when just a few commercial television stations in the United States only broadcast

18 Naughton, John. "John Brockman: the Man Who Runs the World's Smartest Website." *The Observer,* Jan. 8 2012, www.theguardian.com/technology/2012/jan/08/john-brockman-edge-interview-john-naughton. Accessed Aug. 23, 2025.

19 Bush, Vannevar. "As We May Think." *The Atlantic Monthly,* vol. 176, no. 1, Jul. 1945, pp. 101–108. Aug. 23, 2025.

their black and white signals four hours a week, and the fax machine wouldn't be invented for another 20 years.

The paper was highly influential, if for no other reason than a summary of it was published in the September 10 issue of Life magazine that year. Copies of Life magazine were distributed around the country and to American military bases around the world. A seminal moment in computing history happened later in 1945, as a young, bored Navy radar technician came across a native bamboo hut on stilts, on the edge of the jungle on the island of Leyte, in the Philippines. Inside that hut, he found a small Red Cross reading library, which included this issue of Life magazine.[20]

The sailor's name was Doug Engelbart, and he would come to be known as one of the foremost engineering pioneers in the computing industry, developing, among other things, networked computing, hypertext, the computer mouse, and the earliest of what we know now as graphical computer user interfaces. Apple co-founder Steve Wozniak credits Engelbart *"for everything we have in the way computers work today."* Sixteen years after first reading the article, Engelbart came across Bush's essay again and felt compelled to write him a letter, telling Bush what a large influence the essay had on his thinking and work. Later, toward the end of his career, he reflected on that eureka moment in the hot, steamy hut in 1945 – it had dawned on him that information was expanding so fast that *"the complexity/urgency factor had transcended what humans can cope with… I suddenly flashed that if you could do something to improve human capability with that, then you'd really contribute something basic."*

Inspired by Bush's vision, Engelbart set out to have a career that would make the world a better place by harnessing collective human intellect to solve important problems, and he saw computers as the vehicle for doing this.[21]

Elsewhere, William Ross Ashby was the first to articulate "intelligence amplification" as an alternative to "artificial intelligence," in his 1956 book *Introduction to Cybernetics*. He muses that problem-solving is just a case of appropriate problem selection. Intelligence is just a measure of your ability to select. That power can be amplified by technology, according to Ashby. His distinction is, *"What is new is that we can now do it synthetically, consciously,*

20 Madrigal, Alexis C. "The Hut Where the Internet Began." *The Atlantic*, 16 Jun. 2015, https://www.theatlantic.com/technology/archive/2015/06/the-hut-where-the-internet-began/277551/. Accessed Aug. 23, 2025.

21 The Unfinished Revolution, Strategy and Means for Coping with Complex Problems," *Doug Engelbart's Colloquium at Stanford*, Apr. 2000, https://www.dougengelbart.org/colloquium/colloquium.html. Accessed Nov 20, 2025.

deliberately."[22] Notably, as Ashby pondered how technology could be controlled to achieve these benefits, he developed his Law of Requisite Variety, which posits that a control system must be as complex as the system it tries to control. Ashby's work on adaptation and self-regulation in complex systems was very influential in the subsequent development of early computer science and artificial intelligence.

One of the other people who read As We May Think and Ashby's work was JCR Licklider, a psychologist focusing on human factors associated with information processing. Licklider became convinced that computers and effective human–computer systems could enable Bush's ideas about extending human memory and cognition. In 1960, he published a paper titled Man-Computer Symbiosis, which has become yet another important building block of today's computer industry.

Licklider envisioned a time when machine cognition could exceed that of humans and that the key to future human success is a tight coupling of human brains and computing machines. This would manifest as "thinking centers," which were like libraries connected to each other through computer networks.[23]

> In the anticipated symbiotic partnership, men will set the goals, formulate the hypotheses, determine the criteria, and perform the evaluations. Computing machines will do the routinizable work that must be done to prepare the way for insights and decisions in technical and scientific thinking." He concedes that "in due course, electronic or chemical "machines" will outdo the human brain in most of the functions we now consider exclusively within its province. (Licklider, 2008)

Licklider mused, *"Man-computer symbiosis is probably not the ultimate paradigm for complex technological systems."*

His impact on the computing industry, including this paper and his subsequent work, was inspired, like Engelbart, by As We May Think, and today, he, like Engelbart, is credited with being an important visionary of our networked world. It is ironic that as a psychologist, Licklider carved out a career path that leans heavily on the computing and artificial intelligence side of the symbiosis, while Engelbart, the engineer, focused primarily on human factors and outcomes.

---

22 Ashby, Ross W. *An Introduction to Cybernetics*. Methuen and Company, London: 1964.

23 Licklider, Joseph C.R. "Man-Computer Symbiosis." *IRE Transactions on Human Factors in Electronics,* vol. 1, 2008, pp. 4–11.

Engelbart built on the work of Bush, Licklider, and Ashby when he explicitly dealt with intelligence augmentation in his foundational 1962 report, Augmenting Human Intellect: A Conceptual Framework. He begins the report, *"By 'augmenting human intellect' we mean increasing the capability of a man to approach a complex problem situation, to gain comprehension to suit his particular needs, and to derive solutions to problems."*[24] (This comes close to my definition of thought, by the way.) He adds, *"Increased capability in this respect is taken to mean a mixture of the following: more-rapid comprehension, better comprehension, the possibility of gaining a useful degree of comprehension in a situation that previously was too complex, speedier solutions, better solutions, and the possibility of finding solutions to problems that before seemed insoluble."*

In Augmenting Human Intellect, we see Engelbart's vision: he clearly foreshadows modern-day human interfaces by describing an architect at his desk manipulating images and data on a large screen using a computer mouse and other devices. This was when computers were interfaced via punch cards and computer tape, and data was just holes in paper.

Though their ultimate visions were similar, Licklider's vision was broader, encompassing freeing up humans from routine tasks to perform more advanced cognitive work, and focused largely on enhancing computers' ability to perform not only handling routine preformulated computations, but also addressing problems that are difficult to think through in advance. He envisioned intuitive technology that, through trial and error, could move beyond "What is the answer?" to "What is the question?" Licklider's work drew him toward advances in the potential of technology and specifically artificial intelligence.

Engelbart's vision was narrower, and his North Star was the human. He focused on directly enhancing human capabilities and potential, and improving how people work, think, and solve complex problems through advanced tools and systems. It's no surprise, for example, to see that toward the end of his career, his work centered on fostering "collective intelligence." In this framework, technologically-enabled groups of people come together to *"a form of universally distributed intelligence, constantly enhanced, coordinated in real time, and resulting in the effective mobilization of skills. I'll add the following indispensable characteristic to this definition: The basis and goal of collective intelligence is mutual recognition and enrichment of individuals."*[25]

---

24 Engelbart, Douglas C. "A Conceptual Framework." *Douglas C. Engelbart Papers,* Stanford University Libraries, web.stanford.edu/dept/SUL/library/extra4/sloan/mousesite/EngelbartPapers/B5_F18_ConceptFrameworkPt1.html. Accessed Aug. 23, 2025.

25 Lévy, Pierre. *Collective Intelligence: Mankind's Emerging World in Cyberspace.* Perseus Books, 1997.

What do we gain by looking back at people like Engelbart, other than a historical perspective? First, he foresaw three states of augmentation that we can consider:

- "those that are executed completely within the human integument, which we call explicit-human process capabilities";
- "those possessed by artifacts for executing processes without human intervention, which we call explicit-artifact process capabilities";
- "*and there are what we call the composite process capabilities, which are derived from hierarchies containing both of the other kinds.*"

Second, he popularizes "intelligence amplification." He considers Ashby's earlier comment to be perhaps a lighthearted musing that was considered and then rejected. In Engelbart's mind, the term does "*not imply any attempt to increase native human intelligence,*" but rather a better fit for human intelligence and the problems to be solved. The system – the computer and the person – will be augmented. "*We will have amplified the intelligence of the human by organizing his intellectual capabilities into higher levels of synergistic structuring.*"

Bush, Licklider, and Engelbart saw clearly into the next 50 years (though perhaps a bit optimistically), and they all foresaw us using computers to augment our native faculties by outsourcing memory and processing to devices.

3 Scouting the Fringes: Extended and Virtualized Cognition

So, what can we see if we extrapolate into the near future? How might today's emerging technology extend our thinking beyond native modes of thinking ("*explicit-human process capabilities*") into "*explicit-artifact process capabilities,*" or symbiotic "*composite process capabilities*" in ways that may not have been specifically envisioned by these three luminaries?

Let's consider some scenarios:

- Picture a pilot flying along in a modern fifth-generation jet fighter, say the new F-35 Lightning II. As she looks ahead and sees the real world through her windshield in normal, full, natural color, inside her visor, she has a digital display of all the information she needs to complete her mission – speed, heading, height, targets, warnings, etc. As her eyes move from her windshield down toward her feet, she "sees" through the plane, as feeds from various cameras around the plane now display on her visor, giving

> her the ability to see under, all around, and through the plane.[26] And this vision isn't just ordinary light and standard color wavelengths; she can toggle infrared or night vision, and upgrade or change out those cameras to see in any other necessary frequency in the electromagnetic spectrum. AI constantly scans these fields, highlighting elements that warrant increased human attention and augmenting or tagging them with necessary information, such as identifying a friendly ally or an inbound threat.

The $400,000 Rockwell Collins Gen III augmented reality capabilities I described are a quantum leap forward over the native abilities of our eyes, and how we see today, and as a result, how we perceive and think about the world.[27]

But how is this example relevant to near-future thought?

Consider radar. In World War II, the US spent $17,756 in 1943 (about $240,000 in today's money) to install the SCS-520 airborne radar on some of its aircraft.[28] Like the Rockwell Collins helmet, the SCS-520 radar was state-of-the-art, ultra-top-secret technology, and expensive. Within 10 years, however, this radar technology was used in the civilian sector to heat food in commercial settings; in 20 years, it was in use in homes across America; this technology, initially known as the RadarRange, is today's ubiquitous microwave oven.

4 The Transformative Unknown: When Tech Thinks beyond Us

Today, hundreds of millions of Americans use their smartphones to navigate using real-time GPS maps, complete with "streetview." Millions of others use their smartphone cameras and displays to overlay the names of constellations in the sky (Star Chart), or find out the names of people standing around them

26 "F-35 Gen III Helmet Mounted Display System." *Collins Aerospace*, https://www.collinsaerospace.com/what-we-do/industries/military-and-defense/displays-and-controls/airborne/helmet-mounted-displays/f-35-gen-iii-helmet-mounted-display-system. Accessed Nov. 19, 2025.

27 Davenport, Christian. "Meet the Most Fascinating Part of the F-35: The $400,000 Helmet." *The Washington Post*, 1 Apr. 2015, www.washingtonpost.com/news/checkpoint/wp/2015/04/01/meet-the-most-fascinating-part-of-the-f-35-the-400000-helmet/. Accessed Aug. 23, 2025.

28 "US Night Fighter Radars of WWII." HyperWar: U.S. Navy Reference, ibiblio.org/hyperwar/USN/ref/NightFighterRadars/index.html. Accessed Aug. 23, 2025.

(NameTag),[29] play augmented reality games like Pokémon Go,[30] translate printed words,[31] or arrange "hookups."[32] Looking just slightly forward, we see that Land Rover has demonstrated a "see-through" hood in its concept car,[33] Continental announced the availability of Augmented Reality Heads Up Displays,[34] and companies like Magic Leap challenge our imagination by blending the real and virtual world.[35] And Facebook changed its name to Meta in anticipation of a VR/AR/metaverse[36] future.[37]

What will the world and our thinking look like 10 or 20 years from now, if technologies like I described in the helmet example above become as ubiquitous as the microwave?

It's 2030. You walk into a room full of strangers and, via your intelligent glasses, can look around the room and see everyone's name under their face. The cloud-based facial recognition system interacts with your preferences and utilizes AI and machine learning (based on your past interests/behavior), parsing out and displaying other relevant data. Right next to the real physical image of the person you are looking at, you see a combination of other data, including where they work, their latest social media feeds, the value of their house, and even their stated dating status. You'll also know more about this

29 "Facial Recognition App Matches Strangers to Online Profiles." *CNET*, 7 Jan. 2014, www.cnet.com/news/facial-recognition-app-matches-strangers-to-online-profiles. Accessed Aug. 23, 2025.

30 "Be A-R-fraid: Augmented Reality Game Turns Your Home into Horror Story as Zombies, Demons Stalk You." *Daily Mail*, [date unknown], www.dailymail.co.uk/sciencetech/article-3063974/Be-AR-fraid-afraid-Augmented-reality-game-turns-home-horror-story-zombies-demons-stalk-you.html. Accessed Aug. 23, 2025.

31 "Word Lens." Wikipedia, updated 5 months ago, en.wikipedia.org/wiki/Word_Lens. Accessed Aug. 23, 2025.

32 "Singles Around Me." Google Play Store, play.google.com/store/apps/details?id=com.singlesaroundme.android%26;hl%3Den. Accessed Aug. 23, 2025.

33 "Land Rover Debuts 'See-Through' Virtual Imaging Technology in Discovery Vision Concept." *Media.LandRover.com*, 8 Apr. 2014, media.landrover.com/en-us/news/2014/04/land-rover-debuts-see-through-virtual-imaging-technology-discovery-vision-concept . Accessed Aug. 23, 2025.

34 Eyes on with Heads-up Display Car Tech PCMag, www.pcmag.com/article2/,2817,2461037,00.asp . Accessed Aug. 23, 2025.

35 "Magic Leap Explainer." *Engadget*, 25 Feb. 2015, www.engadget.com/2015/02/25/magic-leap-explainer/. Accessed Aug. 23, 2025.

36 See A. E. Saddik, F. Lamberti, S. Mann, F. G. Prattícó, R. Thawonmas and Y. Yuan, "Metaverse and Extended Universe (XV): Opportunities and Challenges for Consumer Technologies," in *IEEE Consumer Electronics Magazine*, vol. 13, no.3, pp. 25–26, May 2024.

37 I recommend the following as a primer to the dangers in this specific area: K. Michael, R. Abbas and S. Papagiannidis, "The Social Implications of XR: Promises, Perils, and Potential," in *IEEE Technology and Society Magazine*, vol. 43, no. 1, pp. 91–108, March 2024.

person than just what is in the cloud – your local telemetry, augmented by edge-based and remote AI platforms, will be able to analyze and interpret visual, auditory, and other data from this person to understand more about them than maybe even they know. Maybe they're married, but you will know their dating mood regarding the person they are talking to. When they talk, you will not only know if they are lying or not, but you might diagnose diseases or conditions they have. If they don't speak English, you can either read a real-time translation in your glasses or choose to select an audio real-time translation into your earbud. Their micro-expressions will be analyzed by AI, and you will know their inner emotions, perhaps even better than they do.

– Now consider the future of Brain-Machine Interfaces. Harvard Professor James Geary said he expects *"to see the day when robots like these have biological components (i.e. some wetware to go along with their hardware) and when human beings have internal technological components (i.e. some hardware to go along with our wetware)."* He quotes Duke neuroscientist Miguel Nicolelis, *"The body's going to be very different 100 years from now… In a century's time you could be lying on a beach on the east coast of Brazil, controlling a robotic device roving on the surface of Mars, and be subjected to both experiences simultaneously, as if you were in both places at once. The feedback from that robot billions of miles away will be perceived by the brain as if it was you up there."*[38]

These are the fringes of intelligence augmentation, yesterday, today, and the not-too-distant tomorrow. I briefly highlight how our senses can be augmented and extended in remarkable new ways to navigate various realities. This will open new and exciting fields, just as the invention of glass lenses unlocked our ability to see far (astronomy) and very near (microscopes). And this will significantly shape the way we think. These stories are the tip of the iceberg of how we will augment our senses, like electricians who can bio-hack rare earth magnets into their fingers to sense electrical current with their hands, or enhanced sensors that give us the ability to "smell" as accurately as dogs do.

And these examples don't explicitly address Baudrillard's concern that we are substituting technology for reality. Here, perhaps, we can turn to science fiction for more use cases. Many of our leading ideas in human intelligence augmentation come to us from this field. Cyborgs and highly augmented human cognition were pioneered by science fiction writers as early as the

38 Geary, James. "Brain-Machine Interface (BMI)." *Edge.org*, 21 Feb. 2025, https://www.edge.org/response-detail/11756. Accessed Aug. 23, 2025.

1920s, and more recently by authors like William Gibson,[39] Vernor Vinge,[40] and David Brin.[41] But while giants like these wrote groundbreaking books that arguably charted the trajectory of the current conversation, let us consider a couple of examples that are more in the public eye and conversation for a moment:

- *Ready Player One:* Ernest Cline's 2011 book Ready Player One[42] was made into a blockbuster film by Steven Spielberg in 2018. Set in a dystopian world of 2045, the plot centers on a virtual world, the Oasis, in which people spend many hours each day escaping the harsh reality of everyday life.[43] In their physical homes, they donned various goggles, helmets, gloves, suits, and interacted with treadmill-type equipment to physically experience and interface with their virtual avatars and surroundings in the Oasis. While the physical humans were on Earth, the Oasis featured many virtual worlds to be explored, constantly being built up like in Minecraft, as well as businesses one could start or patronize, generating real income in these virtual worlds. Things had to be bought in Oasis, using virtual currency. Virtual currency can be purchased with real-world money, earned through virtual work, or by selling virtual items you own or create in the Oasis. Many people in the real world spend their days working and recreating in the Oasis. Everything you can do in the real world, you can do in the Oasis, plus so much more.
- *Upload.* In 2020, Amazon Prime launched a TV series written by Greg Daniels.[44] Set in 2033, the show is based on the ability to upload human consciousness into digital afterlives of their choosing. People in the afterlife can see what's happening in the real world via large television screens and can call and talk to real people. Real people can visit virtually. But the ability to download the consciousness back into physical bodies is still not quite developed. Meanwhile, you can live a life filled with as much leisure as you can afford. The megacorp that runs the program maintains your account based on the amount you paid to them when you died, or as long as your loved ones continue to pay for it. The company even pays real-world employees to work as digital avatars in the afterlife, in various capacities, to attend to the needs of digital consciousnesses.

---

39 Gibson, William. *Neuromancer*. Ace Books, 1984.

40 Vinge, Vernor. "True Names." *Binary Star No. 5*, Dell, 1981.

41 Brin, David. *Kiln People*. Tor Books, 2002.

42 Cline, Ernest. *Ready Player One*. Crown Publishers, 2011.

43 *Ready Player One*. Directed by Steven Spielberg, Warner Bros. Pictures, 2018.

44 *Upload*. Created by Greg Daniels, Amazon Prime Video, 2020–2023.

These two examples are clear substitutions of the virtual simulation for reality. While it may be hard to get our head around how Baudrillard claimed we were living in a simulation of reality when he was writing in 1981, there's no question that characters here are doing so in these examples.

Baudrillard's concern wasn't about virtual reality, though, as much as our being already divorced from reality and being on a slippery slope where everything we interact with is mediated through technology. Ready Player One and Upload are plausible stops on that slope. And they're not just plausible from a philosophical argument's perspective; they are, in fact, quite plausible technically.

The gaming industry has created extensive, enduring, and realistic virtual worlds that are inhabited in real time by people from all over the world, who spend hours or even days exploring and interacting in simulated environments like Second Life, World of Warcraft, or Minecraft. The ability to and consequences of uploading consciousness are already a growing topic of philosophers, neuroscientists, and technologists. While several fundamental technical challenges face both these scenarios, I see few that won't eventually be overcome.

Before we try to tie all these fringe scenarios into a map of the landscape of emergent cognition, there's one more dimension we need to explore.

The above are extended and virtualized cognitive scenarios. They may be revolutionary and even shocking, but I'd argue they are not yet cognitively transformative.

What is cognitively transforming?

First, consider that we've looked at what technology like computing and artificial intelligence does, historically. It approximates what humans can do, but at scale. At least most of what humans do, like calculating. But they can't create or do things we can't. Except they can. For years, computer researchers have been extending the ability of technology not only to beat humans at games like chess and Go, but to innovate and create. Technologies do not just tell us when we've made a typo, now they write songs and sonnets, and create artworks optimized to various audiences and or cultural traditions.

In May of 2020, OpenAI, a San Francisco-based artificial intelligence research laboratory, launched Generative Pre-trained Transformer 3 (GPT-3). This artificial intelligence platform focuses on the deep learning of human language, and can produce written texts that are very difficult to distinguish from that written by real humans. And these systems don't need the basic training that AlphaZero required to begin its learning arc. These systems are designed to operate autonomously without the need for initial prompting or specific instructions from users and are known as autonomous agents or autonomous AI. They are programmed to perceive their environment, make decisions, and take actions independently to meet specific goals or tasks. We are rapidly transitioning from highly specialized AI/ML platforms that struggle to grasp concepts like causation and abstract notions such as human emotions, to General Purpose (GP) artificial intelligence, which can acquire new skills and even self-improve.

At the time of this writing, halfway through 2025, students everywhere ask GPT and other AIs to do their homework for them, while teachers struggle to use other AIs to combat this by spotting the computer-generated work. And in the ever-accelerating arms race of technology, students learn to combat these teacher screens by running their work through multiple AIs.[45] Et cetera. Way back in 2018, during one of my TEDx talks, I asked the audience how they would think, or what they would think about, when AIs can make PowerPoint presentations and TED Talk scripts better than we can. Well, we are already there.

Soon (in a matter of years), computing platforms will be able to do almost everything humans can do, but better (and differently). Already, OpenAI is predicting that ChatGPT 5, available in the 2026 timeframe, will be smarter than a college professor.[46] And then it will apply those skills to improve itself, in unpredictable ways. None of the scenarios described above envision technology like this. Yes, they illustrated technology that creates new worlds or provides new insights about the real world. And technology that is always there, like a servant, to help with discrete tasks, in useful and understandable ways. But in the scenarios described above, we don't have all-powerful AIs acting in new and superhuman ways, alongside humans. And that will be cognitively transforming.

The other transformative vector is how humans will think in the near future when we have outsourced cognition more fully to technology, what we referred to in the last section as the Big Transformational Opportunity. It's also the topic I was alluding to in my 2018 TED Talk audience questions. In both cases, I mention it when justifying why thought is being disrupted. Here we are looking forward to seeing if we can understand what that big transformational opportunity might look like. Which makes me ask:

---

45 E.g. "How to Detect Use of Multiple AI-Based Tools by Student?" *Academia Stack Exchange*, 15 Nov. 2023, academia.stackexchange.com/questions/203925/how-to-detect-use-of-multiple-ai-based-tools-by-student. Accessed Aug. 23, 2025.

46 Morrison, Ryan. "ChatGPT Will Have PhD-Level Intelligence in the Next 2 Years." *Tom's Guide*, 24 June 2024, www.tomsguide.com/ai/chatgpt/openai-cto-says-chatgpt-will-have-phd-level-intelligence-in-the-next-2-years. Accessed Aug. 23, 2025.

> What happens to our thoughts when technology not only remembers our children's phone numbers for us, but it fully scaffolds our reasoning, moods, and emotions, better than we can ourselves? What will we think about, and how, with the cognitive resources that are freed up?

If technology continues to expand these capabilities and moves toward general-purpose technologies that can learn and invent, what is left for humans to think about? The optimist in me, inspired by the aspirations of Engelbart, Licklider, Ashby, and Bush, suggests that we can learn to think in new, holistic, and emergent ways. We can leap forward, as we discussed a few pages ago, by considering how our thinking transformed with tool use, language, and writing. That human thought will transform in response to these new cognitive scaffoldings seems a foregone conclusion – one that is not represented in the scenarios mentioned above.

Why is it not? Because it is literally impossible for people to imagine. But the examples listed above provide us with some use cases at the fringes of human thought.

The truth is, I can't describe what that transformation will actually look like. I, too, am trapped in our existing paradigm.

My vision is the opposite of the narrative that technology is making us dumber, but I'm not techno-utopian, since I don't see very many signs that we are heading in the right direction, at least not at a material scale.

I acknowledge that in many ways, technology will replace our native faculties. But this is not a nostalgic reverie – my argument is that those native faculties are often anachronisms, evolved primarily for a period of some 250,000–300,000 years ago. With technology, some of us will lose the ability to do simple everyday things of old, like write cursive, read a map or do long division. But, frankly, most of us haven't had those things in a while. And yes, for many people, at first glance, the cumulative effect of disrupting thought may look like we are all en route to the dystopian future we saw in the movies Wall-E or Idiocracy.

But consider this book.

Technology in many forms helped pull together the material here and to discuss in videoconferences, emails, and other collaborations with people around the world. (And 99% of this was written before ChatGPT was on the scene.) The fact that I have technologies like a robot to vacuum my floors and a microwave to speed up cooking freed up some time for me, and I applied that time not (just) to watching TV, or gardening, but to reading, and thinking about the meta-meta issues like this. I was using, as Clive Thompson noted, the Internet (and other tech) to think deeply.

In a talk I gave a few years ago at Arizona State University, I mentioned that right now, some young girl in Lagos or Wollongong could sit at a table using her phone to create the next technology that will inflect the collective human potential curve upward.[47] Never before have humans had such potential at their fingertips, at this scale. My vision is for us to learn how to use technology to boost our productivity and leave us all feeling better off. That technology will give us the discretionary resources to tackle bigger problems in a different way.

How that plays out in use cases beyond what I've described is too speculative for this work, and frankly, even for me now, and within this work. I can only cast a preliminary vision, highlighting the likely opportunities for this transformation, and hope readers can use these insights as part of their own framework to outline and build positively future emergent opportunities.

Now we can pull all these pieces together to create a simplified landscape of the use cases, and it helps us to see future areas that are "off the map," the way an old globe labeled uncharted areas "here are dragons."[48] We may not predict exactly what these future use cases and characteristics will be. However, by extending the material we've covered so far just a little bit, we can deduce this future opportunity space.

As Nobel physicist Dennis Gabor said, *"Rational thinking, even assisted by any conceivable electronic computors(sic), cannot predict the future. All it can do is to map out the probability space as it appears at the present and which will be different tomorrow when one of the infinity of possible states will have materialized....The future cannot be predicted, but futures can be invented."*[49]

Is charting the future of human cognition and humanity a hard task? YES! But now consider the technologies we have to help us.

Here's a simplified illustration of the possibility map of disrupted cognition. It is what I visualize when I think about Riemer and Johnston's advice to "*The best we can achieve is to try to theorize the nature of technology-enabled indus-*

---

47 Scannell, Patrick. "The Great Irony Of Technology: Why Technology Leaves Us Feeling Worse and What to Do about It." *YouTube*, uploaded by Public Interest Technology (PIT) Colloquium, Date unknown, www.youtube.com/watch?v=gf7yGBSG1aU. Accessed Aug. 23, 2025.

48 Literally, HC SVNT DRACONES" (i.e., hic sunt dracones, *here are dragons*). Now a meme that many old maps came with descriptions like this, however one of the known specific example is the 1504 Hunt-Lenox globe: "1504 Hunt-Lenox Globe," *Wikipedia*, https://en.wikipedia.org/wiki/Hunt%E2%80%93Lenox_Globe. Accessed Aug. 23, 2025.

49 Gabor, Dennis. *Inventing the Future*. Alfred A. Knopf, 1964. [The spelling "computors" was used in the book].

try crises."[50] Please note, however, this is the glass-half-full view. I've consciously side-stepped, for now, the antithetical; the range of dystopian outcomes. I'll leave it to others to warn of the whole range of how we'll all go to hell. My mission is to start developing a blueprint for how we can avoid that. Figure 3.1, below, represents how I visualize that.

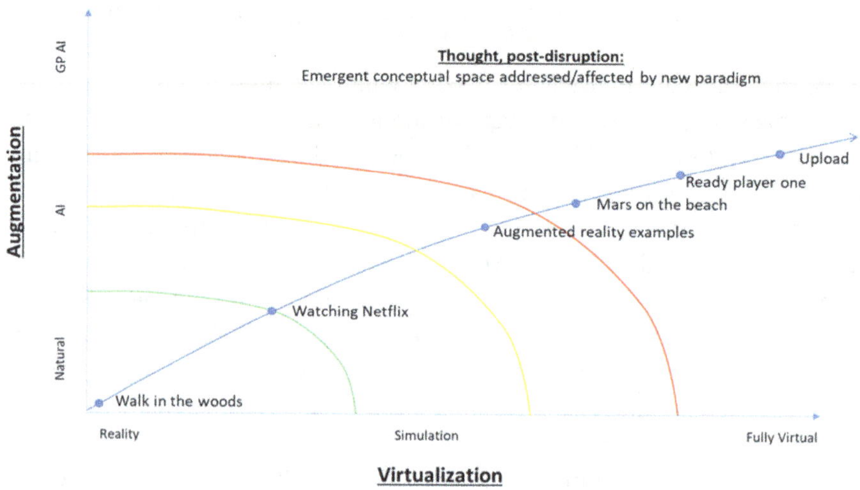

FIGURE 3.1 A conceptual rendering of the present and future landscape of thought

50 See the Disruption chapter for an exposition of Riemer and Johnston.

CHAPTER 4

The Most Important Issue Facing Humanity

I seem to see it being gradually atrophied by those two world-wide enemies of the imagination, the wireless and the cinema. To a generation for whom everything which used to nourish the imagination because it had to be won by an effort, and then slowly assimilated, is now served up cooked, seasoned, and chopped into little bits, the creative faculty (for reading should be a creative act as well as writing) is rapidly withering, together with the power of sustained attention; and the world which used to be so grand à la charté des lampes[1] is diminishing in inverse ratio to the new means of spanning it; so that the more we add to its surface the smaller it becomes.[2] (Edith Wharton (in the early 20th century))

I have a foreboding of an America in my children's or grandchildren's time – when the United States is a service and information economy; when nearly all the manufacturing industries have slipped away to other countries; when awesome technological powers are in the hands of a very few, and no one representing the public interest can even grasp the issues; when the people have lost the ability to set their own agendas or knowledgeably question those in authority; when, clutching our crystals and nervously consulting our horoscopes, our critical faculties in decline, unable to distinguish between what feels good and what's true, we slide, almost without noticing, back into superstition and darkness...

The dumbing down of America is most evident in the slow decay of substantive content in the enormously influential media, the 30 second sound bites (now down to 10 seconds or less), lowest common denominator programming, credulous presentations on pseudoscience and superstition, but especially a kind of celebration of ignorance.[3] (Carl Sagan (at the end of the 20th century))

1 The phrase *"grand à la charté des lampes"* wasn't a standard French term; it was a literary reference Wharton used to describe the diminishing sense of mystery in the modern, electrically-lit world.

2 Wharton, Edith. *The Writing of Fiction*. Charles Scribner's Sons, 1925, p. ~34.

3 Sagan, Carl. *The Demon-Haunted World: Science as a Candle in the Dark*. Random House, 1995, p. 25.

 | HTTPS://DOI.ORG/10.1163/9789004739161-007

I've been known to talk at length, in complex ways, about complex topics. Fortunately, I've never talked at length and in complicated ways about simple things.[4] However, here I will do the opposite.

1 Why the Disruption of Thought Is Humanity's Most Important Issue

My claim is that the disruption of human thought is the most important issue facing humanity. But I'm going to make the simplest case for it, because the points become self-evident when you reflect on them as you read the rest of the book.

First, let's assume my point that thought isn't working well for us. I'll make that case in Part 4 and in more detail in my book *The Great Irony of Technology*.

Our minds struggle to adapt to technology's demands, leaving most people unhappier and less healthy despite material gains. These effects vary across different groups, but few escape technology's psychological toll entirely. This creates a troubling paradox: if current technology has improved our material conditions while worsening our well-being, what happens when we deploy even more powerful technologies over the next two decades?

We face a critical moral challenge. Technology must deliver not just economic benefits, but psychological and health improvements that reach across all segments of society. The stakes could not be higher.

Second, this disruption of thought is the root cause of many of society's collective problems.

Social polarization is a real force in American society, but less visible is the long-term trend of erosion in the pillars of democracy and civilization. This includes, as Sagan foreshadowed, reduced trust in science, education, medicine, business, military, news, and the government.[5] The loss of faith in government is widespread, encompassing distrust in Congress, the Supreme Court, the presidency, the judiciary, executive branches, and even voting and democracy itself.

---

4 That comment was for the benefit of those who know me, and made it this far into this material, because, yes, I've been known to overthink and talk at length about even the most simple and superficial topics.

5 Jones, Jeffrey M. "Confidence in U.S. Institutions Down; Average Now at New Low." *Gallup*, 5 Jul. 2022, https://news.gallup.com/poll/394592/confidence-institutions-down-average-new-low.aspx. Accessed Aug. 23, 2025.

We can blame social media companies for the increasing political polarization in the United States and try to develop policy prescriptions to deal with that, but they are not root cause.

The root cause is us.

Even if we somehow have the wisdom and political will to temporarily solve or improve the negative effects that social media has on American society, the problem will just pop up elsewhere.

The ongoing disruption of thought outlined in this book is a prime driver that fosters a declining sense of well-being, an increasing sense of disorientation, and a growing sense of loss. This emotional entropy shapes not just the rage of middle-aged White (predominantly) January 6 protesters storming the US Capitol, but also an increased sense of depression in affluent suburban teens of all races, as well as everyone else. This is not a series of individual emotional crises; it is a systemic and *collective* problem.

Yes, we have a *moral* imperative to help position the individuals in our society to experience better lives. However, we also have a strategic imperative to address it for the benefit of society. Otherwise, we may face a situation where, as Yeats says, the "center cannot hold." We've taken the ideas of safety, security, America, and even civilization for granted. However, a deep review of human history and a close examination of the factors outlined in this book suggest that we may have been overconfident in our stability, and we need to reassess and address the root causes tearing us apart. So, we have a *strategic policy* imperative to address issues rooted in the disruption of thought.

Finally, we live in a complex world with complex problems. Not all of them stem from the disruption of thought. For example, as Vint told me after reading this chapter, the issues I am describing here might not matter if global warming kills our species.

If we are to survive and thrive in a world of increasingly complex problems, we will need new and more powerful ways of thinking, ones that can evolve and adapt as fast as the problems we face. The disruption of thought is not a problem to solve – it is a situation we face, from which we can seize potential. It can provide us with a significant transformational opportunity to leapfrog into more reflective, holistic, and effective forms of thought, both individually and collectively. If we do so, we should be able to not only understand (right-brain), but also control and act (left-brain) on complex local and global challenges that do NOT come from the disruption of thought, such as climate change, environmental disasters, pandemics, and cyberattacks.

The World Economic Forum puts out an annual Global Risk Landscape report. In 2020, they presented those risks in a simple chart that maps out risks by the likelihood of them happening on the vertical axis compared to the impact they will have if they happen on the horizontal axis.[6] (They've since moved to a more complicated "interconnected" web graphical format). The lower left and middle (low- to mid-level impact and likelihood) of the 2020 chart are full of events that either have root causes stemming from the disruption of thought, or this disruption significantly influences them. Examples include terrorist attacks, social instability, adverse technological effects, and state collapse. I would argue that social instability, for example, is heavily influenced by the disruption of thought, as are the risks of failures of policy and governance. The complexity of our world overcomes our ability to think, and things collapse because we think sub-optimally. Events like terrorist attacks, disruptions in food supply, and financial meltdowns are conditioned by the disruption of thought. Recognizing that disrupted thinking serves as both a root cause of certain global risks and an amplifying factor for others opens new pathways for more effective prevention and response strategies.

The chart's upper-right quadrant – representing high-likelihood, high-impact events – features climate-related disasters, climate action failures, and extreme weather. Transforming cognitive disruption into enhanced collaborative thinking and problem-solving capabilities positions us to tackle these paramount challenges.

Thus, I conclude that the disruption of thought is the most important issue facing humanity because: a) failing to address the disruption exacerbates and accelerates existing problems, often as the root cause and b) positively embracing this opportunity will strengthen our ability to resolve some of our biggest issues. The Disruption of Thought is both the most likely event to occur and the one with the biggest impact. It is the most high and right issue we face.

However, the disruption of thought is not only absent on the 2020 map, but it is also missing from 2025.[7] It's literally not on our radar. Of note, though, are social polarization, inequality, and declining health and well-being, all of which are direct results of the disruption of thought and now the most dangerous risk the world faces.

6 World Economic Forum. The Global Risks Report 2020. 15th ed., *World Economic Forum*, Jan. 15, 2020, https://www.weforum.org/docs/WEF_Global_Risk_Report_2020.pdf. Accessed Aug. 23, 2025.

7 World Economic Forum. The Global Risks Report 2025. 20th ed., *World Economic Forum*, Jan. 15, 2025, https://www.weforum.org/docs/WEF_Global_Risks_Report_2025.pdf. Accessed Aug. 23, 2025.

2 The Case for Disruption: Four Converging Dynamics

This, then, is the introduction to the case I make for the disruption of thought: four elements, each of which disrupts human thought. All occurring simultaneously and in parallel, in an evolutionary blink of an eye, leading inevitably to a stark picture of thought undergoing disruption. I cover these four and more in greater detail in Part 3, but have introduced them here to get to the crux of the book quickly, before considering the rest of the material in more detail.

By my definitions of disruption and thought, thought would be disrupted if any one of the following occurs:

- **Changes in the fundamentals of thought:** We face profound shifts across all dimensions of cognition, what we think about, how we process information, and the values driving our thought processes. Simultaneously, control mechanisms over our thinking are rapidly evolving. These converging factors precipitate a crisis wherein cognitive processes increasingly fail to serve the majority effectively.
- **Changes in the fundamentals of technology that we use to scaffold our thought:** The essential nature of technologies supporting human thought is fundamentally changing, consequently altering our cognitive architecture and information processing methodologies.
- **Changes in the nature of our relationship with reality:** Contemporary existence has become so heavily mediated that we struggle to differentiate between lived experience and objective reality. This disconnection impairs our ability to identify informational biases and errors, resulting in distorted worldviews that typically skew toward excessive negativity.
- **The big transformational opportunity:** Historical innovations – bipedalism, tool utilization, language development, and writing systems – established foundations for expanded human cognitive capabilities. As we increasingly delegate cognitive functions to sophisticated technologies, we encounter another potential evolutionary threshold. We face a critical choice: approach this transition with intentional design and strategic purpose, or allow it to unfold through reactive, uncoordinated adaptations.

These four dynamics (among others detailed in Part 3) are occurring simultaneously within a compressed timeframe, significantly disrupting established cognitive patterns. This disruption, if effectively managed, presents a substantial opportunity for cognitive advancement.

PART 2

A New(ish) Model for Human Thought

1 Purpose of This Part of the Book

> For the exogenously extended organizational complex functioning as an integrated homeostatic system unconsciously, we propose the term 'Cyborg'[1]
> (Manfred Clynes & Nathan Kline)

This part contains a new(ish) model of human thought that is extraordinarily interdisciplinary. Building on and synthesizing from across the cognitive sciences, I assemble a model that will allow us to gain critical insights into the modern niche that are not visible from any one existing specialty. The previous chapters introduced this material very lightly; here, we get down to brass tacks.

1 Clynes, Manfred E., and Nathan S. Kline. "Cyborgs and Space". *Astronautics*, Sep. 1960, p. 27.

CHAPTER 5

Stream 1: Thinking with the Brain

> The human brain is generally regarded as a complex web of adaptations built into the nervous system, even though no one knows how.[1] (Michael Gazzaniga)

> One cannot say anything about such a complicated structure as the brain without indulging in oversimplification.[2] (Paul D. MacLean)

The brain has been called the most complicated thing in the known universe.[3]

How can I describe it in a few pages, while keeping you awake and interested? More importantly, why do you need to know about the brain?

Well, this is a book about disrupting thought, so covering fundamentals here is an important foundation. But if you know this material already, and or want to take my word for it in the remaining chapters, you can skip over this chapter. However, I suggest you don't go too far forward, because the material covered in Chapters 6 and 7 is probably new to many of you. I highly recommend considering the details of the model's streams before we assess the validity of our findings when using it.

1 From Anatomy to Networks: Discovering How the Brain Works

It was the early Greeks who identified the brain as the site in the body where thinking occurs.[4] In the following 2,000 years, various studies (autopsies and dissections) examined and mapped out the anatomical features of the brain.[5]

With the invention of the microscope and other technologies, we unlocked our understanding of the differences in parts of the brain, including the distribution, density, shape, and size of cell bodies. The arrangement of neurons,

1 Gazzaniga, Michael S. *The Mind's Past*. University of California Press, 2000, p. 13.

2 MacLean, Paul D. *A Triune Concept of the Brain and Behaviour*. University of Toronto Press, 1973.

3 Watson, James D. "Discovering the brain." *National Academy of Sciences*, 1992.

4 Bay, Noel Si-Yang, and Boon-Huat Bay. "Greek Anatomist Herophilus: The Father of Anatomy." *Anatomy & Cell Biology*, vol. 43, no. 4, Dec. 2010, pp. 280–283.

5 Pevsner, Jonathan. "Leonardo Da Vinci's Contributions to Neuroscience." *Trends in Neurosciences*, vol. 25.4, 2002, pp. 217–220.

 HTTPS://DOI.ORG/10.1163/9789004739161-009

glial cells, and other cellular components in the brain's gray matter is known as cytoarchitecture, and it is crucial for understanding the brain's functional organization at the cellular level. By the end of the 19th century, scholars had identified 52 areas of the brain that performed different specialized functions, which were previously unknown to earlier scientists and anatomists.[6]

Modern tools and techniques of the last two decades – such as the development of novel preparation techniques, optical methods, quantitative approaches to analysis in 2D and 3D images, and high-performance computing – expand on that view, showing that there are at least 150 regions within the brain;[7] each is distinguished from the other by how cells are arranged and/or function.

And we no longer look just at dead tissue; we can observe it in real time in living people, as they think. These same approaches, though, also highlight that the brain does not isolate specific functions to unique localized structural/anatomical regions of the brain (e.g., nodes), but instead through connections of those regions to each other, through the edges of those functional areas, which are connected to each other through an extremely large number of micro, meso, and macro neural networks.[8]

2 The Massively Modular Brain

As we previously touched on briefly, modularity is the term that describes the organization of nodes into groups, based on their connection to one another within the brain via anatomical structural connectivity (the connectome of synaptic contacts between neighboring neurons or fiber tracks connecting neuron pools in spatially distant brain regions),[9] functional (temporal dependency of neuronal activation patterns of anatomically separated brain regions),[10] or effective connectivity (the influence one neuronal system exerts upon another, thus reflecting causal interactions between activated brain

---

6 Brodmann, Korbinian. *Brodmann's Localisation in the Cerebral Cortex*. Translated by Laurence J. Garey, Springer, 2006.

7 Amunts, Katrin, and Karl Zilles. "Architectonic Mapping of the Human Brain beyond Brodmann." *Neuron*, vol. 88, no. 6, 2015, pp. 1086–1107.

8 Sporns, Olaf. "Connectome Networks: From Cells to Systems." *Micro-, Meso-and Macro-Connectomics of the Brain*, Springer, 2016, pp. 107–127.

9 Pevsner, Jonathan. "Leonardo Da Vinci's Contributions to Neuroscience." Trends in Neurosciences, vol. 25.4, 2002, pp. 217–220.

10 Lang, Elmar Wolfgang, et al. "Brain Connectivity Analysis: A Short Survey." *Computational Intelligence and Neuroscience*, 2012.

areas).[11,12] Connections *within* a module are typically denser than those between modules.[13] Some nodes are highly connected to multiple regions of the brain and function as connector hubs.[14]

The numerous large-scale networks that span regions of the brain have been demonstrated empirically and consistently through technological advancements.[15] Low-level anatomical connectivity can be shown via labeled tracer injections or diffusion imaging data.[16] Higher-level functional networks can be demonstrated using techniques such as BOLD-fMRI or electroencephalography (EEG)/magnetoencephalography (MEG).[17]

Fodor,[18] who first developed the idea of modularity in the early 1980s (without the benefit of these modern techniques), felt it occurred in low-level peripheral processing centers (like visual processing), but not in central processing like rational thought,[19] and he continues to maintain this point in his later research. As late as 2001, he said that neuroscientists may not yet actually understand how the mind works.[20]

Largely because of that evolving imaging technology and other techniques, however, there is a growing consensus that the brain is modularized, highly interconnected, and likely massively modularized.[21] Fodor first articulated the

---

11 Lang, Elmar Wolfgang, et al. "Brain Connectivity Analysis: A Short Survey." *Computational Intelligence and Neuroscience*, 2012.

12 Park, Hae-Jeong, and Karl Friston. "Structural and Functional Brain Networks: From Connections to Cognition." *Science*, vol. 342, no. 6158, 2013, p. 1238411.

13 Newman, M. E. J. "Modularity and Community Structure in Networks." *Proceedings of the National Academy of Sciences*, vol. 103, no. 23, 2006, pp. 8577–8582.

14 Sporns, Olaf. "Connectome Networks: From Cells to Systems." *Micro-, Meso-and Macro-Connectomics of the Brain*, Springer, 2016, pp. 107–127.

15 See Biswal, Bharat, et al. "Functional Connectivity in the Motor Cortex of Resting Human Brain Using Echo-Planar MRI." *Magnetic Resonance in Medicine*, vol. 34, no. 4, 1995, pp. 537–541 and Raichle, Marcus E. "The Brain's Default Mode Network." *Annual Review of Neuroscience*, vol. 38, 2015, pp. 433–447.

16 Gamanut, Radu, et al. "The Mouse Cortical Connectome, Characterized by an Ultra-Dense Cortical Graph, Maintains Specificity by Distinct Connectivity Profiles." *Neuron*, vol. 97, no. 3, 2017, pp. 698–715.

17 Sporns, Olaf, et al. "The Human Connectome: A Structural Description of the Human Brain." *PLoS Computational Biology*, vol. 1, no. 4, 2005, p. e42.

18 It's important to note that we've been using massive modularity to describe the structure and function of the physical brain, while Fodor's unit of study is the "mind," which is informed by the physical layer.

19 Fodor, Jerry A. *The Modularity of Mind: An Essay on Faculty Psychology*. MIT Press, 1983.

20 Fodor, Jerry A. *The Mind Doesn't Work That Way: The Scope and Limits of Computational Psychology*. MIT Press, 2001.

21 Carruthers, Peter. *The Architecture of the Mind: Massive Modularity and the Flexibility of Thought*. Oxford University Press, 2006.

massive modularity idea,[22] and this was further developed by people like Pinker, Spele, and Carruthers. For example, Carruthers added that these massive modules are also encapsulated, meaning they operate most often in isolation from other modules in the mind. Why “massively?” Massive modular proponents are making the distinction and case that the number and scope of these modules are much greater than previously thought, and that this increased scale and complexity are crucial to understanding how the brain works.

Scholars within the Modular school still debate the extent of domain specificity (how much of your thinking occurs in special localized parts of the brain, as opposed to across networks that span it),[23] modularity, degeneracy, the general-purpose features of core systems, and encapsulation.

Competing models to Massive Modularity exist. One prominent alternative is the domain-general model, which contends that cognitive processes are not modular but instead rely on shared, general-purpose mechanisms that operate across multiple domains. Another approach is connectionist or neural network models, positing that cognitive processes arise from the interactions of interconnected nodes or units within distributed networks of neurons, emphasizing the importance of parallel processing and distributed representation, and the collective activity of interconnected neural populations rather than from discrete modules. Bayesian models of cognition theorize that the brain operates as a probabilistic inference engine, continuously updating beliefs and making predictions based on available evidence, integrating prior knowledge, sensory input, and uncertainty estimation to generate adaptive behaviors and decisions. Then we have Bayesian Brain,[24] Thousand Brains Theory of Intelligence,[25] Dennett's Multiple Drafts,[26] Embodied and Situated

---

22 Fodor, Jerry A. *The Modularity of Mind: An Essay on Faculty Psychology*. MIT Press, 1983.

23 The most extreme of which might claim that memory works down to the scale of a single neuron within your 86 billion neurons, which might be programmed to recognize a specific face, such as your grandmother, or Jennifer Aniston: Aur, Dorian. “Where is the ‘Jennifer Aniston Neuron’?” *Nature Precedings*, 2010.

24 Knill, David C., and Alexandre Pouget. “The Bayesian Brain: The Role of Uncertainty in Neural Coding and Computation.” *Trends in Neurosciences*, vol. 27, no. 12, 2004, pp. 712–719.

25 Hawkins, Jeff, with Sandra Blakeslee. *A Thousand Brains: A New Theory of Intelligence*. Basic Books, 2021.

26 Dennett, Daniel C. *Consciousness Explained*. Little, Brown and Company, 1991.

Cognition,[27] and fields such as Epistemic Structural Realism, which discuss the nature of reality independent of the mind.[28]

Critics still debate the extent of empirical evidence supporting massive modularity.[29] Despite the existence of various models (many of which are not mutually exclusive of Massive Modularity) and criticisms, the massive modularity approach to brain function appears to have the most current evidence on its side. So, while Fodor may have been right that we do not yet comprehend the mind, for now, there is utility in adopting the massively modular view of how the *brain* works.

But we do not just "think" within the *brain*: the brain itself is part of the central nervous system embedded within the human *body*. So, any model of the "wet" layer of human cognition would need to incorporate the modular systems between the cardiovascular, respiratory system, digestive, excretory, nervous, and sensory systems like visual and auditory apparatus, which are connected not only physically via many nerve circuits, but also through complex chemical and other pathways such as the endocrine system.

Some scholars, such as Lisa Feldman Barrett, argue that our emotions and mental states (valence) are constructed by the brain primarily based on internal assessments of various organs and other physiological parameters of the body.[30] This process, known as interospection, is distinct from your review of higher-order mental processes like thoughts and feelings, known as introspection. So, when I refer to the stream of Thinking with the Brain (TB) or Massively Modular (MM), I will do so not only to describe neural circuits of the brain, but more broadly to refer to a holistic "embodied brain."[31]

27 Clark, Andy. *Being There: Putting Brain, Body, and World Together Again.* MIT Press, 1997.

28 Worrall, John. "Structural Realism: The Best of Both Worlds?" *Dialectica*, vol. 43, no. 1–2, 1989, pp. 99–124.

29 Laland, Kevin N., and Gillian R. Brown. *Sense and Nonsense: Evolutionary Perspectives on Human Behaviour.* 2nd ed., Oxford University Press, 2011.

30 Barrett, Lisa Feldman. *How Emotions Are Made: The Secret Life of the Brain.* Houghton Mifflin Harcourt, 2017.

31 But not "embodied cognition," which as a term has come to be closely allied with externalism, and the next two streams. Three Streams Model could be considered largely compatible/consistent with and a part of the embodied cognition and extended cognition traditions; our variations are: deep emphasis on Massively Modular theory approach, modified to include the embodied brain that uses modules in and out of the brain; distinction between the social and the material; stronger emphasis on social than is traditionally applied to extended cognition; extending the modular concept to nodes at the social and material level; a deep emphasis on the exocortex components of the present and near future, including algorithms and AI; and emergent properties of the three layers when considered together.

3 "Where Modules Come From: Evolution vs. Experience"

In Pinker's 1997 book *How the Mind Works,* he explicitly uses the term "massive modularity" to refer to specialized mental mechanisms that are domain-specific, automatic, encapsulated, and universal. He frames it against the older idea of the mind as a single, general-purpose reasoning engine. Inherent in his view is the notion that brain functions are largely determined by genetically derived mechanisms, whose design evolved to solve specific problems that were present in some past ancestral environment.

This idea has become known as "evolutionary adaptedness" (or "environment of evolutionary adaptedness, often abbreviated as "EEA"). It was first proposed by Bowlby,[32] and then popularized by Tooby & Cosmides, and then Pinker,[33] Carruthers[34] and others. This evolutionary emphasis has also been referred to as the Santa Barbara school of thought.[35] Tooby and Cosmides sum it up simply that the "evolutionary function of the human brain is to process information in ways that lead to adaptive behavior."[36] However, they'd be the first to point out that evolutionary-derived cognitive adaptations occur over tens of thousands, if not hundreds of thousands of years. So, while a popular conception of the brain is that it is the apex of a long line of progressive evolution that is now complete, resulting in us – the wonderfully technologically savvy modern humans, the Santa Barbara school might counter that by saying that our brains are in fact an anachronism, adapted to past environments that no longer exist.

Massive modularity pundits not only see genetic evolution as the prime driver of our brains, some even believe it is the only driver. Pinker has argued forcefully against "blank slate" views and emphasized the primacy of evolved psychological mechanisms. In The Blank Slate, he critiques the idea that cul-

32 Bowlby, John. *Attachment and Loss.* Basic Books, 1969.

33 Pinker, Steven. *How the Mind Works.* Penguin UK, 2003.

34 Carruthers, Peter. *The Architecture of the Mind: Massive Modularity and the Flexibility of Thought.* Oxford University Press, 2006.

35 Laland, Kevin N., and Gillian R. Brown. *Sense and Nonsense: Evolutionary Perspectives on Human Behaviour.* 2nd ed., Oxford University Press, 2011.

36 Tooby, John, and Leda Cosmides. "Evolutionary Psychology and the Generation of Culture, Part I: Theoretical Considerations." *Ethology and Sociobiology*, vol. 10, no. 1–3, 1989, pp. 29–49.

ture is the primary driver of human psychology,[37] while Tooby and Cosmides believe that culture is the expression of evolved mental modules.[38]

Some critics of massive modularity object to genetically determined evolutionary influence and suggest that individual experience over a lifetime (known as ontogenesis) interacts with the brain's neuroplasticity to play a larger role than past evolutionary traits. Panksepp, for example, highlights the "*remarkable degree of neocortical plasticity within the human brain, especially during development*" and hypothesizes "*the developmental interactions among ancient special-purpose circuits and more recent general-purpose brain mechanisms can generate many of the 'modularized' human abilities that evolutionary psychology has entertained.*"[39] Other critics note the limits of the genetic argument by cataloguing the fact that much of modern human cognition is rooted in practices that have emerged in the last few millennia, centuries, and even decades, and that human cognition uniquely requires extensive and specific training. These two factors mean that cognition is not shaped primarily or exclusively through genetically derived mental models.[40]

So begins a nature/nurture[41] argument about the role of plasticity and developmental experience versus genetic determinism. Carruthers acknowledges that modules are not fully developed at birth and concedes to the likely effects of plasticity during maturation, but he does not seem to consider it important after maturity. He still stands by neural structures as being primarily shaped by genetic timescales, influenced by experience during the first two decades of one's life (during which the modules mature under gene-regulation), but not by change on the cultural timescales, or across one's adult life. Plasticity is not explained by massive modularity because modules cannot be changed, according to Carruthers.

I am not here to pick a fight with Carruthers; I highly respect his work, and strongly recommend his book, *The Architecture of the Mind*, as one of my favorite reads during this project. (Plus, he was kind enough to trade a few emails with me and let me pre-read one of his forthcoming books.) My goal

---

37 Pinker, Steven. *The Blank Slate: The Modern Denial of Human Nature*. Viking, 2002.

38 Tooby, John, and Leda Cosmides. "Evolutionary Psychology and the Generation of Culture, Part I: Theoretical Considerations." *Ethology and Sociobiology*, vol. 10, no. 1–3, 1989, pp. 29–49.

39 Panksepp, Jaak, and Jules B. Panksepp. "The Seven Sins of Evolutionary Psychology." *Evolution and Cognition*, vol. 6.2, 2000, pp. 108–131.

40 Enquist, Magnus, et al. "The Human Evolutionary Transition: From Ecological Dominance to Directional Selection." Bulletin of Mathematical Biology, vol. 72, no. 5, 2010, pp. 1006–1029.

41 Or a more nuanced "nature via nurture" argument in the case of writers like Matt Ridley.

here is to point out some prominent critiques of the tool I am recruiting, such that we can use our other tools in ways that compensate for this potential overplay of the nature over nurture hand.

Michael Arbib bridged this gap between biology and culture in his pioneering work on the origins of human language.[42] He begins his seminal and influential work, *How the Brain Got Language*, by noting that the specialization of a science can hold us back, because we each look at the evidence illuminated to us by our respective field's specialized approach. He says –

> There is an old story about a gentleman walking home one dark night when he sees a drunk searching under the light from a lamppost. He asks the drunk what he is doing, and the drunk replies, 'I'm looking for my keys. The gentleman joins in the search, but finds no trace of the keys. 'Are you sure you lost them here?' 'No, I lost them up the street, but it's dark to look for them there.[43]

The streetlights or lampposts of our fields don't allow us to see the truth outside the boundaries of our specialty. By combining four different "lampposts" Arbib was able to see that language appears to have roots in both biological and cultural evolution.[44] Language did not develop based on genetic evolution, according to Arbib.[45] Rather, the brain evolved some circuitry on top of primate grasping control centers in the brain, and that circuitry was "language-ready." The biological evolution of the grasping would have occurred consistent with the EEA, but did not result in language – it evolved to solve some older, other problem, but as such these circuits had the potential to be "hijacked," or utilized by humans for a different purpose, perhaps just 40,000 years ago. In 1982, Gould and Vrba coined the term "exaption" to describe using a genetically derived feature that previously solved one problem differ-

42 And he has been another of those many experts who've taken the time to reply to me on my quest, and has offered me several conversations by email, phone, and Zoom, which have helped shape my understanding, for which I am grateful.

43 Michael A. Arbib. *How the Brain Got Language: The Mirror System Hypothesis*. Oxford UP, 2012, p. 3.

44 Arbib is uniquely qualified to combine lampposts: his Ph.D. was in mathematics, but his teaching career included posts as a professor of computer science, as well as a professor of biological sciences, biomedical engineering, electrical engineering, neuroscience, and psychology.

45 Arbib, Michael A. *How the Brain Got Language: The Mirror System Hypothesis*. Vol. 16, Oxford University Press, 2012.

ently to address a new problem, a process that can work much faster than evolution.[46]

Discussions about how the brain thinks, especially when we invoke evolutionary arguments, usually get into popular but flawed arguments about cognitive evolution. The classic is Paul MacLean's three-part or "triune" model, originally proposed back in the 1960s. He explained it this way: *"man, it appears, has inherited essentially three brains. Frugal Nature in developing her paragon threw nothing away. The oldest of his brains is basically reptilian; the second has been inherited from lower mammals; and the third and newest brain is a late mammalian development which reaches a pinnacle in man and gives him his unique power of symbolic language."*[47] To say this model has been influential is an understatement: it is one of the most influential ideas in brain science since World War II.[48]

Much more recently, neuroscientist and technology entrepreneur (among other things, he founded Palm Computing) Jeff Hawkins simplified things by lumping the brain's pesky complicatedness into the "old brain" category, and he draws the focus instead to the "new brain" of the neocortex. *"The neocortex is the organ of intelligence,"* writes Hawkins. *"Almost all the capabilities we think of as intelligence – such as vision, language, music, math, science, and engineering – are created by the neocortex."*[49]

These and similar ideas have been roundly criticized, especially by evolutionary comparative biologists who point out that brains did not evolve from simple to complex in linear arcs. As one paper points out, *"Instead, the correct view of evolution is that animals radiated from common ancestors. Within these radiations, complex nervous systems and sophisticated cognitive abilities evolved independently many times structure and function of the brain radiated in multiple directions, with complex neural systems and sophisticated cognitive abilities evolved independently many times."*[50]

Still, models like these persist, and are commonly invoked even in current standard neuroanatomy books, which discuss how humans' special cognitive

---

46 Gould, Stephen Jay, and Elisabeth S. Vrba. "Exaptation – A Missing Term in the Science of Form." *Paleobiology*, vol. 8, no. 1, 1982, pp. 4–15.

47 MacLean, Paul D. "Man and His Animal Brains." *Modern Medicine*, vol. 32, 1964, pp. 95–106.

48 Harrington, Anne. "At the Intersection of Knowledge and Power: The Development of the 'Brain/Mind' Sciences in Modern Europe." *The History of Science*, Andrew Cunningham and Perry Williams, eds., vol. 31, 1992, pp. 76–104.

49 Hawkins, Jeff. *A Thousand Brains: A New Theory of Intelligence*. Basic Books, 2021, p. 3.

50 Cesario, Joseph, et al. "Your Brain Is Not an Onion with a Tiny Reptile Inside." *Current Directions in Psychological Science*, vol. 29, no. 3, 2020, pp. 255–260.

abilities and behavioral flexibility are derived from "newer" parts of the brain that give us the ability to control and inhibit ancient and more reflexive parts of the brain like the cerebellum, hippocampus, or amygdala.

Simplifications have their benefits, though. Hawkins proposes the neocortex as the general-purpose machinery that solves the problems it is assigned. It does not set goals and motivations; those are, according to him, generated in places like the brainstem. The interplay between the older parts of the brain – which not only generate the wants and needs, but also play host to the "database" of experience, and makes final judgements – with the neocortex's generation of ideas, possible solutions, and strategic plans, is a model, which can embrace and hold on our own thoughts and minds, without the terrifying complexity of what is actually going on.

Or as Cesario and colleagues conclude in their critique of the old brain/new brain models:

> Perhaps mistaken ideas about brain evolution persist because they fit with the human experience: We do sometimes feel overwhelmed with uncontrollable emotions and even use animalistic terms to describe these states. These ideas are also consistent with such traditional views of human nature as rationality battling emotion, the tripartite Platonic soul, Freudian psychodynamics, and religious approaches to humanity. They are also simple ideas that can be distilled to a single paragraph in an introductory textbook as a nod to biological roots of human behavior. Nevertheless, they lack any foundation in our understanding of neurobiology or evolution and should be abandoned by psychological scientists.[51] (Cesario, et al., 2020).

4 The Other Terrifyingly Complex Stuff

Libraries are filled with the vast material I've left out of this brief journey into how the brain works. Here are just a few relevant snapshots of specialized work, using different units of analysis:

- Iain McGilchrist's 600-page tome *The Master and His Emissary* is a landmark documenting the differences in structure and function of each side of the human brain.[52] This contrast, known as hemispheric lateralization, has

51 Cesario, Joseph, et al. "Your Brain Is Not an Onion with a Tiny Reptile Inside." *Current Directions in Psychological Science*, vol. 29, no. 3, 2020, pp. 255–260.

52 McGilchrist, Iain. The Master and His Emissary: The Divided brain and the Making of the Western World. Yale University Press, 2019. Note that the pop science take on this area has left a bad taste in many neuroscientists' mouths. Throughout this book, I will refer

very significant effects on how we think and behave. McGilchrist makes a convincing case that these physiological differences and interconnected functions have manifested themselves in human society, and most especially, the dominance in modern society of the left hemisphere's traits of rationality, reductionism, and materialism has come at the expense of the right side's more balanced and integrated way of understanding the world.

- Neuroscientist Mark Humphries gives us a gem in his recent book, *The Spike*. Writing with an engaging, easy-to-read yet extraordinarily well-informed style, he takes us on a journey of a single electrical spike traversing the brain's workings in just over two seconds.[53] His specialty: using computational neuroscience to investigate a single neuron's activity to encode the past, present, and future, thus guiding human behavior.
- Biological anthropologists like Chet Sherwood do comparative work on the neurological differences between human and other apes, enlightening us with discussions and data on extra-allometric neuroanatomical distinctions in human neocortical architecture and function that cannot be easily explained as correlates of brain size, and whose long list of publications includes papers like: "*Invariant synapse density and neuronal connectivity scaling in primate neocortical evolution.*"[54] His preferred unit of analysis is comparing anatomical and molecular function of the human brain to that of other great apes.
- Jim Giordano and others[55] speculate about how the brain's computation is driven by quantum interactions. He envisions a mechanism whereby quantum operations occur in tryptophan tubules, possibly using the extended coherence of nuclear spins of phosphorus. These coherent states are very small (sub-atomic/atomic to low nanoscale) and fast (sub-nanosecond)

back to McGilchrist frequently, but since modern neuroscience emphasizes inter-hemispheric integration, when I refer to the hemispheres in the context of McGilchrist, please view these as metaphors for cognitive styles rather than strict anatomical modules.

53 Humphries, Mark. *The Spike: An Epic Journey through the Brain in 2.1 Seconds*. Princeton University Press, 2021.

54 Sherwood, Chet C., et al. "Invariant Synapse Density and Neuronal Connectivity Scaling In Primate Neocortical Evolution." Cerebral Cortex, vol. 30.10, 2020, pp. 5604–5615.

55 See Penrose and Hameroff's Orchestrated Objective Reduction (Orch-OR); Giordano also references other supporting concepts as various components of this speculative framework, including Matthew Fisher's hypothesis that nuclear spins, particularly of phosphorus atoms, could maintain coherence for long periods; Edward Calabrese position that neural systems are "noisy" environments, and neurocellular domains are exquisitely sensitive to initial conditions that shape and sustain how particular physical dimensions of neural cells respond to various factors (i.e., "attractors").

events that exist in an inhospitable (warm, wet, and noisy) neural environment that lends itself to decoherence. But maybe it's the collapse of coherent quantum states into decoherence that enables translation of quantum operations to classical (electrochemical) mechanical structures and functions. While at a very small ("hormetic") scale, these infinitesimally minuscule and discrete operations can tune or condition other chains of events ("prime"), which themselves are amplified and propagate at progressively larger size and speed scales and levels across neural cells and networks, ultimately resulting in macro-level changes in thoughts/behaviors. In this way, incredibly small (sub-atomic) scale events in neural nodes and networks of the brain could be quantum superposition orchestrated as massive parallel computing at scale: quantum computing lets our brains integrate vast amounts of information almost instantaneously.[56]

- Still other scholars look not at the brain in isolation, but at its functions as an emergent property of a whole body network, particularly in symbiotic processing with our guts. These researchers include Emeran Mayer (author of The Mind-Gut Connection),[57] John Cryan & Ted Dinan (pioneers in the "psychobiotics" of the gut biome),[58] Rob Knight (microbiome expert),[59] and Sarkis Mazmanian (leading thinker at the intersection of neuroscience and bacterial microbiology).[60] Collectively, the field explores the complex communication network between the central nervous system (the brain and spinal cord) and the gastrointestinal system, which connect to each other bidirectionally through chemical neurotransmitters and hormones such as serotonin and cortisol, but also mechanically through physical networks like the vagus nerve. Working in concert, this axial pairing shapes digestion, hunger, immune response, and mood, and preconditions or primes the brain's higher-level functions.

These are just some curated facets/examples of a fast-evolving mountain-sized store of information about the human brain, the "most complicated thing in the universe."

---

56 Personal correspondence.

57 Mayer, Emeran. *The Mind-Gut Connection: How the Hidden Conversation within Our Bodies Impacts Our Mood, Our Choices, and Our Overall Health.* Harper Wave, 2016.

58 Dinan, Timothy G., et al. "Psychobiotics: A Novel Class of Psychotropic." *Biological Psychiatry*, vol. 74, no. 10, 15 Nov. 2013, pp. 720–726.

59 Knight, Rob. *Follow Your Gut: The Enormous Impact of Tiny Microbes.* TED Books, 2015.

60 Mazmanian, Sarkis K., et al. "An Immunomodulatory Molecule of Symbiotic Bacteria Directs Maturation of the Host Immune System." *Cell*, vol. 122, no. 1, Jul. 15, 2005, pp. 107–118.

But since my unit of analysis is *thought*, and not some particular part of the *brain*, I've covered just enough to give you a flavor of the wavetops of this domain, focusing most especially on the modular debate (broadly interpreted, as I'd see Giordano's speculations as sub-atomic scale nodes, and the brain–gut axis as a macro-level module), as it gives us a framework, which we can co-opt as we look at the other streams where I've found human thought.

5 My Current View of Thinking with the Brain

I accept the computational neural nature of massive modularity and am inclined to the general flavor of massive modularity proposed by Pinker,[61] Fodor,[62] Tooby and Cosmides,[63] Geary,[64] and, to a certain extent by Carruthers.[65] But I note that this school's reliance on genetic determinism may be limiting. And I find it under-emphasizes extra-cranial cognitive processes because it fails to consider modules of outsourced cognition that occur outside of an individual's body, as shared understandings with other people, or external things.

Taking a cue from Arbib's inter-disciplinary approach, let us now turn to the second stream of the model

6 Plain Language Summary of My Use of the Term Thinking with the Brain (Simplified to the Absurd)

Our brain has many parts; we can conceive of them each as nodes. These parts, although different, rarely perform specific, unique functions independently. Rather, they work by being connected via complex networks to other parts of the brain. Think of it like a very large choir where each musician is a node.

61 Pinker, Steven. *The Language Instinct: How The Mind Creates Language*. Penguin UK, 2003; and Pinker, Steven. *How the Mind Works*. Penguin UK, 2003.

62 Fodor, Jerry A. *The Modularity of Mind: An Essay on Faculty Psychology*. MIT Press, 1983.

63 Cosmides, Leda, and John Tooby. "The Psychological Foundations of Culture." *The Adapted Mind: Evolutionary Psychology and the Generation of Culture*, edited by Jerome H. Barkow, Leda Cosmides, and John Tooby, Oxford University Press, 1992, pp. 19–13.

64 Geary, David C. *The Origin of Mind: Evolution of Brain, Cognition, and General Intelligence*. American Psychological Association, 2005.

65 Note that there are significant differences of opinion and disagreement within this group. Note that Fodor's response to Pinker's book *How the Mind Works* was titled *The Mind Doesn't Work That Way*. So, I'm making a VERY GENERAL agreement here about massive modularity.

There are very few soloists, but with each song a different mix of people in the choir sings. Each unique network that connects the parts of the brain needed for different thought processes (or cluster/mini-choir) is known as a module.

Your normal daily thinking requires an extremely large number of these complex networks and is referred to as "massively modular." Your cognitive orchestra extends far beyond your skull. When you solve problems, experience joy, or read these words, you witness coordinated performance across networks spanning from prefrontal cortex to your gut microbiome. It's not just your brain thinking – it's your entire embodied being working as one unified, complex system.

So, when I use the term "Thinking with the Brain" (TB) or "massively modular," I use it to mean a massively modular "embodied" brain, incorporating not just the modules that occur within the brain, but also those modules that span across the brain to the other parts of the body.

CHAPTER 6

Stream 2: Thinking with Culture

> The great error of individualistic psychology is the supposition that man thinks. …This is an error…His mind and thoughts are the products of his social medium or the social element whence he arose and in which he lives.[1] (Ludwig Gumplowicz)

1 From Group Mind to Social Construction

Earlier, I mentioned that one of the neurobiological approach's faults is its failure to explain the significant leap forward in our thinking, which Harari and others argue occurred around 50,000 years ago.

In this chapter, we'll examine how our thinking is influenced by the people around us and those who came before us, and how this perspective helps us understand the significant disruption in thought that occurred 30,000–60,000 years ago (depending on whom you talk to). We'll also discuss the specific changes that occurred in human thinking back then, since one critique of the Savannah Hypothesis is that while many species faced the challenges of surviving in that environmental niche, only one developed humans' unique cognitive ability.[2]

I'll argue that humans have outsourced cognition into modules that span people.[3] To understand that, we will set aside the hard sciences of neuroscience and draw on several closely related fields in the "soft sciences." This encompasses diverse research areas, including socio-cultural psychology, cultural–historical psychology, linguistic relativity, sociology of knowledge, social cognition, and distributed cognition, as well as evolutionary, developmental, and ecological approaches – fields that developed to address fundamentally different questions than those pursued by neuroscientists.

1 Gumplowicz, Ludwig. *The Outlines of Sociology*. Translated by Frederick W. Moore, American Academy of Political and Social Science, 1899.

2 Enquist, Magnus, Stefano Ghirlanda, and Johan Lind. *The Human Evolutionary Transition: From Animal Intelligence to Culture*. Princeton University Press, 2023.

3 Other animals do, too; but that is beyond the scope of this work. But humans have brought it to an extraordinary new level.

HTTPS://DOI.ORG/10.1163/9789004739161-010

All these branches share common roots that go back at least to the classical period.[4] Rousseau[5] and Hegel[6] both developed the concepts that our thinking is intrinsically linked to our social environments and these ideas came to be known as "group mind." Triplett first identified an empirical performance effect when people worked together, as opposed to individually, in a classic 1898 article.[7] This phenomenon is now known as social facilitation.

The group mind model was prominent in the 20th century, influencing Émile Durkheim[8] (considered one of the founders of sociology), along with other early leading sociologists such as McDougall,[9] Ross,[10] Wundt,[11] and LeBon.[12] The idea was also reconceived as the "sociology of knowledge" – a term coined in the 1920s.[13] As the behavioral approach became mainstream, so, too, did the idea of collective social thinking, which became closely linked with behavioral approaches, for better or for worse.

McDougall was staunchly anti-behavioralist, though, and in his book The Group Mind, he laid out "principles of collective psychology," which were the basis for cohesion in groups, in his opinion.[14] McDougall had eugenic beliefs, though, and attributed the origins of shared thought to shared genes. Allport countered in the spirit of the growing influence of behaviorists, however,

---

4 Vaisey, Stephen. "Socrates, Skinner, and Aristotle: Three Ways of Thinking about Culture in Action 1." *Sociological Forum*. Vol. 23. No. 3. Blackwell Publishing Ltd, 2008.

For a non-Western classical approach of how Chinese society viewed the individual as part of a single large complex organism, see Lin, Yutang. *My Country and My People*. London, 1936; and Heinemann, William, and Donald J. Munro. *Individualism and Holism: Studies in Confucian and Taoist Values*. Donald J. Munro, editor. Center for Chinese Studies, University of Michigan, 1985.

5 Rousseau, Jean-Jacques. *A Treatise on the Social Contract*. Beckett and DeHondt, 1767.

6 Hegel, Georg Wilhelm Friedrich. *The Phenomenology of Mind*. Translated by J. B. Baillie, Allen and Unwin, 1910.

7 Triplett, Norman. "The Dynamogenic Factors in Pacemaking and Competition." *American Journal of Psychology*, vol. 9, 1898, pp. 507–533.

8 Durkheim, Émile. *The Division of Labor in Society*. Translated by Lewis A. Coser, Free Press, 1984.

9 McDougall, William. *The Group Mind: A Sketch of the Principles of Collective Psychology*. Cambridge University Press, 1920.

10 Ross, Edward Alsworth. *Social Psychology: An Outline and Source Book*. Macmillan, 1908.

11 Wundt, Wilhelm. *The Language of Gestures*. Translated by J. S. Thayer, Mouton, 1973.

12 Le Bon, Gustave. *The Crowd: A Study of the Popular Mind*. Dover Publications, 2002.

13 Scheler, Max (ed.). *Versuche zu einer Soziologie des Wissens. München und Leipzig: Duncker & Humblot, 1924*. Karl Mannheim. *Ideology and Utopia: An Introduction to the Sociology of Knowledge*. Translated by Louis Wirth and Edward Shils. Harcourt, Brace and Company, 1936.

14 McDougall, William. *The Group Mind: A Sketch of the Principles of Collective Psychology*. Cambridge University Press, 1920.

when he wrote: "*science ... studies the behavior of the individual in so far as his behavior stimulates other individuals, or is itself a reaction to this behavior.*"[15] The idea in either flavor – group mind or the sociology of knowledge – stayed linked predominantly to the behavioral movement, and as behaviorism waned in the second half of the 20th century, collective thinking models had somewhat fallen out of favor.

Is there room there between the two? For something other than cultural interactions, which are genetically pre-ordained, or just a visceral reaction to others' behaviors? Something that explains how, as Gumplowicz said, "*mind and thoughts are the products of his social medium or the social element,*" and can help support Ridley's point, that what makes us unique is "*our ability to accumulate culture and transmit information, across the seas and generations?*" A model wherein the neural and biological correlates/substrate constitute not only (or regularly) through cognitive "nodes" in each of our heads and bodies, but whose cognitive effectiveness is functionally realized through the "modules" of synchronization and connectedness of a shared understanding, maybe mediated by visual signals, language, and other means?

Where might we see evidence of this model?

Surprisingly, the next thread we pull on is Darwin. Biologist Ernst Mayr argues that Darwin's biggest contribution to science was actually his advancement of "population thinking."[16] Less famous than his statements on natural selection was Darwin's belief that experiences from one's life were also transmitted to the offspring,[17] something he called "the inherited effects of use and disuse."[18] And this is not an offhand reference; in the preface to *The Descent of Man*, he notes that his earlier and more famous *Origins* book had been misinterpreted to be strictly about natural selection, and he clarified that "great weight" must be placed on the "effects of use and disuse." This became the foundational principle in his idea of population thinking (variations within populations, not necessarily species).[19] A population of animals was not com-

---

15 Allport, Floyd Henry. "Methods in The Study of Collective Action Phenomena." *Journal of Social Psychology* 15(1) 1942, pp. 165–185.

16 Mayr, Ernst. "Typological versus Population Thinking." *Conceptual Issues in Evolutionary Biology*, edited by Elliott Sober, 3rd ed., MIT Press, 2006, pp. 325–328.

17 We'll avoid, for now, a discussion of how Darwin postulated the mechanism of experience transmission, as his gemmules transmission of use/disuse can strike one as Lamarkian. Empirically, Darwin noted the role of individual experience, and the importance of that being transmitted between generations. He got that right, but his hypothesized transmission mechanism may have been off.

18 Darwin, Charles. *The Origin of Species*. PF Collier & Son, 1909.

19 Darwin, Charles. *The Descent of Man and Selection in Relation to Sex*. American Home Library, 1874.

posed of mirror images of each other. Each behaved uniquely, and to understand a population, one needed to understand the history of the interactions of the group. This was a distinctly new approach in science. Mayr explains that Darwin's own work exemplified a deviation from the norm, from the typified thinking of his time, and by shaking up Victorian thinking, Darwin's own work is a textbook example of the variations of thought that exist within a culture.

A century later, in the late 1970s and 1980s, these ideas were picked up again and became known as Social Cognition. The field emerged after behaviorism lost favor, and scientists like Dennett became interested in "explaining" rather than "explaining away."[20] The field was a merger of social psychology and cognitive psychology,[21] and addressed how people make sense of the world, the events that befall them, other people, and themselves.[22] It examines underlying conscious and unconscious cognitive and affective processes, how they occur and evolve over time, and how they correspond to various patterns of brain activation.[23] Despite the name of the field, though, there doesn't seem to be a consensus within the field that human cognition can occur socially; rather, social environments have a strong influence on the cognition that happens in our individual brains, and vice versa. Social cognition has been applied more broadly in many fields, from social psychology to primatology, for numerous purposes.[24]

Michael Tomasello's influential 1999 work *The Cultural Origins of Human Cognition* fundamentally challenged prevailing views by arguing that human cognitive superiority stems from cultural rather than purely biological evolution.[25] Through comparative analysis with chimpanzees, Tomasello demonstrated that while primates possess basic problem-solving abilities, they lack

20 Taylor, Shelley E. "Social Cognition and Health." *Personality and Social Psychology Bulletin*, vol. 8, no. 3, 1982, pp. 549–562.

21 Brown, Curtis. "Behaviorism: Skinner and Dennett." *Philosophy of Mind*, Trinity University, 2001.

22 Fiske, Susan T., and Shelley E. Taylor. *Social Cognition: From Brains to Culture*. McGraw-Hill, 2008.

23 Taylor, Shelley E. "Social Cognition and Health." *APA Handbook of Personality And Social Psychology, Volume 1: Attitudes and Social Cognition*. American Psychological Association, 2015, pp. 339–361.

24 See for example Mitchell, Jason P., et al. "The Link between Social Cognition and Self-Referential Thought in the Medial Prefrontal Cortex." *Journal of Cognitive Neuroscience*, vol. 17, no. 8, 2005, pp. 1306–1315, and Santos, Laurie R., et al. "Rhesus Monkeys (Macaca mulatta) Know What Others Can and Cannot Hear." *Animal Behaviour*, vol. 71, no. 5, 2006, pp. 1175–1181.

25 Tomasello, Michael. *The Cultural Origins of Human Cognition*. Harvard University Press, 1999, p. 8.

"shared intentionality" – the uniquely human capacity to share mental states and coordinate goals with others. This cognitive foundation enables humans to engage in sophisticated teaching, learning, and cultural transmission that accumulates across generations. Tomasello's framework positions distinctly human abilities – language, theory of mind, and complex reasoning – as emerging from this collaborative intentionality rather than from innate genetic programming alone, fundamentally reframing human cognition as a culturally scaffolded phenomenon.

Richerson and Boyd's 2005 *Not by Genes Alone* advanced the socially constructed cognition debate by defining culture as knowledge, beliefs, values, and skills *"acquired, stored, and transmitted by a population"* through social learning mechanisms.[26] Crucially, they argued that culture operates independently of both genetic inheritance and environmental determinism – populations sharing similar genetics or environments can exhibit dramatically different cultures. Their Cultural Evolution Theory (CET) positions culture as an autonomous force that shapes human development through teaching, imitation, and social transmission, establishing a theoretical framework that treats cultural evolution as a distinct process alongside biological evolution.

It is this cultural mechanism that allowed a progressive bootstrapping of human capability. William James, widely accepted as the founder of psychology, recognized this well over a century ago, when he said: *"A remarkable parallel, which to my mind has never been noticed, obtains between the facts of social evolution and the mental growth of the race, on the one hand, and of zoological evolution, as expounded by Mr. Darwin, on the other."*[27]

But is this thought, or something that influences thought? While these social cognition and CET scholars emphasize the importance of culture as an influence on the thoughts of the individual members, few of them make the explicit argument that culture may in fact be the nodes and part of the modules of the thinking we do within our brains, and connect social thinking directly with concepts like massive modularity.[28]

Lisa Feldman Barrett comes close, in *How Emotions Are Made*, when she points out that emotions, ideas, and concepts – the building blocks of our

---

26 Richerson, Peter J., and Robert Boyd. *Not by Genes Alone: How Culture Transformed Human Evolution*. University of Chicago Press, 2008.

27 James, William. "Great Men and Their Environment." *The Will to Believe and Other Essays in Popular Philosophy*, Harvard University Press, 1979, pp. 216–254.

28 The social cognition field, in its various forms, closely examines the interactions between social environments and corresponding brain activity, but the authors have not yet seen a social cognition claim cognition occurs outside the human brain, through shared activity.

thoughts – are all socially constructed.[29] This was a key concept to my research quest because it validated ideas I was sensing and developing. This isn't some avant-garde controversial writer here: Barrett is among the top 1% of most cited scholars in the fields of psychology and neuroscience. She says these building blocks of our thought only become real when two people agree, through a process called collective intentionality.[30] You don't even need to be aware that you are agreeing. She says: *"Make something up, give it a name, and you've created a concept. Teach your concept to others, and as long as they agree, you've created something real….Then we transmit these concepts to others, wiring each other's brains for the social world. This is the core of social reality."*[31]

Let me give you a couple of real-world examples.

Take blue. The color didn't exist as a word in many ancient languages, around the world, in part because there was no easy way to create the color – there were few blue minerals, and those that are blue are chemically unstable or too hard to use. 4,000–4,500 years ago, the Egyptians invented a technology for mass producing blue by grinding sand, copper, and natron, and heating the mixture in a furnace. As this product and technique, along with later techniques developed in China and elsewhere, spread to places like Rome, the local languages developed and used the word "blue".[32]

Harold Bloom claims a similar cultural phenomenon happened as Shakespeare developed and refined his writing technique. Before Shakespeare, Bloom argues that humans were characterized in literature and thus in life. The actors' stories unfolded, but their basic nature was relatively stable and unchanging. Shakespeare did not merely characterize; he created dynamic characters that develop and change throughout the story. Often it happens after a person has self-reflection and redefined themselves. His technique and success not only shaped subsequent literature but also literally changed how people considered themselves and others, according to Bloom. So much

29 Barrett, Lisa Feldman. *How Emotions Are Made: The Secret Life of the Brain*. Houghton Mifflin Harcourt, 2017.

30 There are very interesting synergies between Barrett's analysis and the Thousands Brains model; both posit that for each concept, process, idea or feeling in our brains, there are a very large number (certainly thousands) of competing parallel processes, each of which is making predictive (Barrett's word) assessments of the element, based on partial information (Hawkins emphasis).

31 Barrett, Lisa Feldman. *How Emotions Are Made: The Secret Life of the Brain*. Houghton Mifflin Harcourt, 2017, p. 136.

32 Schenkel, Wolfgang. "Color Terms in Ancient Egyptian and Coptic." *Anthropology of Color: Interdisciplinary Multilevel Modeling*, edited by Robert E. MacLaury et al., John Benjamins, 2007, pp. 211–228.

so that his magnum opus on the topic is titled: *Shakespeare; The Invention of the Human.*[33]

Others help build the story of social and cultural constructs. While early scholars argued that knowledge is empirical and gained directly through the experience of the senses,[34] more recent research shows that much of what we know comes from other people. Gelman, like Barrett, and in the tradition of Vygotsky,[35] points out that concepts – the building blocks of ideas – are social constructs that cannot be isolated from their context. And while direct sensory experience can augment knowledge, social/culturally transmitted mechanisms *"may be the only plausible source of information for concepts that are not readily accessible by direct observation, such as how the brain works."*[36]

Steven Johnson, in looking at the diffusion of innovation within a society, says, "*A good idea is a network.*"[37] While he adds that he is not talking about a "global brain" or a "hive mind," his caveat applies to the original innovation, since collaborations are rarely capable of creativity (in his mind). "*This is not the wisdom of the crowd, but the wisdom of someone in the crowd. It's not that the network itself is smart; it's that the individuals get smarter because they're connected to the network.*"[38] But ideas travel on rails laid by institutions, infrastructures, and norms – socio-technical systems that make cognition actionable.

Across the last half-century, scholars in a field closely related to CET, known as socio-technical systems (STS)[39] have shown that cognition doesn't

33 Bloom, Harold. Shakespeare: *The Invention of the Human.* Riverhead Books, 1998.

34 Piaget, Jean, and Eleanor Duckworth. "Genetic Epistemology." *American Behavioral Scientist,* vol. 13.3, 1970, pp. 459–480.

35 Vygotsky Lev S. *Mind in Society: The Development of Higher Psychological Processes.* Harvard University Press, 1978.

36 Gelman, Susan A. "Learning from Others: Children's Construction of Concepts." *Annual Review of Psychology,* vol. 60, 2009, pp. 115–140.

37 And, as David Brin said, so too is a bad one. Our modern world is composed of networks of competing narratives that vary in veracity, intent, and various degrees of effectiveness in propagation.

38 Johnson, Steven. *Where Good Ideas Come from: The Natural History of Innovation.* Penguin, 2011, p. 58.

39 Cultural Evolution Theory (CET) explains how cultural traits, norms, and technologies spread, persist, and cumulatively change via population-level social-learning biases and selection (typically with formal models and experiments at micro/meso scales), whereas Geels's strand of STS – socio-technical transitions under the Multi-Level Perspective – analyzes how innovations reconfigure stabilized regimes through actor coalitions, institutions, infrastructures, markets, and policy under niche–regime–landscape dynamics (typically via historical/qualitative systems analysis over decades); yet they meet on a coevolutionary view in which diffusion and adoption hinge on the coupling of CET's

merely live in individual heads; it is scaffolded by socio-technical systems – interlocking webs of artifacts, infrastructures, institutions, skills, and meanings. Large-technical-systems histories[40] and Social Construction of Technology (SCOT)[41] reframed technologies as socially shaped and socially shaping; Constructive Technology Assessment[42] added intentional steering. Building on these strands, Geels' Multi-Level Perspective[43] explains how innovations incuate in niches, challenge stabilized regimes of practices and rules, and are pressured by landscape forces (e.g., climate, geopolitics). Geels' microfoundations work[44] then puts agency back in – showing how entrepreneurs, users, firms, and policymakers mobilize narratives, resources, and institutions to reconfigure regimes. Read this way, culture is not just an influence on thought; it is the operating system of socio-technical life: the shared models, norms, and routines that make individual cognition effective by coupling minds to tools, organizations, and networks.

---

micro transmission mechanisms with Geels-STS's meso/macro architectures and governance that make some variants stick.

40 Bijker, Wiebe E., Thomas P. Hughes, and Trevor Pinch, editors. *The Social Construction of Technological Systems: New Directions in the Sociology and History of Technology*. MIT Press, 1987.

41 Pinch, Trevor J., and Wiebe E. Bijker. "The Social Construction of Facts and Artefacts: Or How the Sociology of Science and the Sociology of Technology Might Benefit Each Other." *Social Studies of Science*, vol. 14, no. 3, 1984, pp. 399–441.

42 Rip, Arie, and René Kemp. "Technological Change." *Human Choice and Climate Change*. Vol. 2: Resources and Technology, edited by Steve Rayner and Elizabeth L. Malone, Battelle Press, 1998, pp. 327–399.

43 Geels, Frank W. "The Multi-Level Perspective on Sustainability Transitions: Responses to Seven Criticisms." *Environmental Innovation and Societal Transitions*, vol. 1, no. 1, 2011, pp. 24–40.

44 Geels, Frank W. "Technological Transitions as Evolutionary Reconfiguration Processes: A Multi-Level Perspective and a Case-Study." *Research Policy*, vol. 31, no. 8–9, 2002, pp. 1257–1274.

2 How Cultural Cognition Works

This is called a network effect,[45] and in the tech industry, we call this Metcalfe's Law:[46] the power or value of a network is proportional to the square of the number of connected users of the system (n^2).[47]

Socially and socio-technically scaffolded cognition does not always come through formal mechanisms such as education, or even conscious conversations or reading.[48] Socially mediated *cues*, perceived unconsciously or consciously, are important to how we understand and react to our environments.[49] Like when we pick up micro-expressions on the faces of people around us and unconsciously sense how they are feeling. These innate human cognitive functions are difficult to replicate.

Artificial Intelligence researchers, for example, struggle to build systems that can effectively scan their surroundings and interpret them in such a way as to take appropriate action. This is known as the "frame problem" origi-

45 Some readers have asked me to date when the Disruption of Thought occurs/occurred. When asked I usually paraphrase Hemingway and Tim O'Reilly: "Gradually, and then suddenly." The entire 20th century brought unprecedented structural changes to the modern human environmental niche, and this uptick started to materialize in the late 19th century. But when pressed for specifics, I rather arbitrarily point to August 9, 2002 – the date Sprint launched the first nationwide 3G mobile network in the US. I was one of four critical path program leads on this launch, and 3G set the stage for many key modern social influences in the US, including social media. More significant than Sprint's actions alone, however, is that by this time wired broadband had penetrated most American homes, Verizon had begun its partial 3G rollout, and AT&T would follow within months. The era of pervasive connectivity had begun. Without grasping Johnson's insight above, and Metcalfe's Law, one cannot fully appreciate the phase change that has occurred in human systems over the past two decades due to accelerating connectivity.

46 Hendler, James, and Jennifer Golbeck. "Metcalfe's Law, Web 2.0, and the Semantic Web." *Journal of Web Semantics*, vol. 6, no. 1, 2008, pp. 14–20.

47 This idea, too, can be traced back to Darwin, almost verbatim: "*...if some one man in a tribe, more sagacious than the others, invented a new snare or weapon...the plainest self-interest, without the assistance of much reasoning power, would prompt the other members to imitate him; and all would thus profit...If the new invention were an important one, the tribe would increase in number, spread, and supplant other tribes...In a tribe thus rendered more numerous there would always be a rather greater chance of the birth of other superior and inventive members.*" Darwin, Charles. *The Descent of Man and Selection in Relation to Sex*. 2nd ed., John Murray, 1874.

48 Olson, David R. *The World on Paper: The Conceptual and Cognitive Implications of Writing and Reading*. Cambridge University Press, 1996.

49 Dijksterhuis, Ap, and John A. Bargh. "The Perception-Behavior Expressway: Automatic Effects of Social Perception on Social Behavior." *Advances in Experimental Social Psychology*, vol. 33, Academic Press, 2001, pp. 1–40.

nally raised in AI by McCarthy & Hayes,[50] later sharpened coincidentally by Fodor, and applied to the philosophy of the mind.[51] Cognitive resources, especially attention and conscious focus, are scarce resources that need to be applied efficiently. Deciding when you have enough information to make a decision or take action is difficult. Consider programming a robot to make tea, for instance. Fodor speculates – "*How … does the machine's program determine which beliefs the robot ought to re-evaluate given that it has embarked upon some or other course of action?*" He notes cognitive scarcity as "*Hamlet's problem: when to stop thinking*". The frame problem, according to Fodor, is "*Hamlet's problem viewed from an engineer's perspective.*"[52]

How is this social?

Our environments are social constructs, and often the fastest and easiest way to understand how to act is to look at how others are acting, and to copy them, through a process broadly described as "norm psychology."[53] A key tenet of norm psychology (and echoed in the social cognition literature) is a deeply seated need to belong.[54] And to do this, one needs, at a minimum, to consider what and who is within a context (situation awareness). Endsley defines situation awareness as "*The perception of the elements in the environment within a volume of space and time, the comprehension of their meaning, the projection of their status into the near future, and the prediction of how various actions will affect the fulfillment of one's goals.*"[55]

But Howard and Cambria say this is insufficient. To make the most accurate decisions, a person also needs to understand the *intention* of the people in context. They define intention awareness as the "*process of integrating actors'*

---

50 McCarthy, John, and Patrick J. Hayes. "Some Philosophical Problems from the Standpoint of Artificial Intelligence." *Stanford University*, 1969, www-formal.stanford.edu/jmc/mcchay69.pdf. Accessed Sep. 1, 2025.

51 Fodor, Jerry A. *The Modularity of Mind.* MIT Press, 1983.

52 Fodor, Jerry A. "Modules, Frames, Fridgeons, Sleeping Dogs, and the Music of the Spheres." *The Robot's Dilemma*, Ablex, 1987, pp. 139–149.

53 Chudek, Maciej, and Joseph Henrich. "Culture–Gene Coevolution, Norm-Psychology and the Emergence of Human Prosociality." *Trends in Cognitive Sciences*, vol. 15, no. 5, 2011, pp. 218–226.

54 Fiske, Alan Page. "Relational Models Theory: A Contemporary Overview." *Relational Models Theory: A Contemporary Overview*, edited by Nick Haslam, Lawrence Erlbaum Associates Publishers, 2004, pp. 3–25.

55 Endsley, Mica R. "Design and Evaluation for Situation Awareness Enhancement." *Proceedings of the Human Factors Society 32nd Annual Meeting*, vol. 32, no. 2, Human Factors Society, 1988, p. 97.

intentions into a unified view of the surrounding environment."[56] This includes *"consistent tracking and extrapolation of objects in the user's environment, but also exploits circumstantial semantics and sentics"* which *"in fact, are key for sensemaking, in which they extend the meaningfulness that the system attributes to every single object/actor of the environment by considering it as an entity carrying specific conceptual and affective information, rather than simply a point in a space or time window."*

Not that all humans are constantly on high alert, as anyone who has tried to navigate a crowded grocery store full of people mindlessly wandering around can attest. But maybe it does not need to be that complicated. Consider, for a moment, the complex collective behavior of flocks of starlings, known as murmurations. This activity, which becomes more pronounced as the birds enter mating season in late autumn through late winter, can leave people who witness it mesmerized and awed by its beauty and complexity. To leading ornithologists of the early 20th century, the activity was so profound that they thought it must be due to some sort of telepathy.

"They must think collectively, all at the same time... a flash out of so many brains," wrote ornithologist Edmund Selous, in his book *Thought-Transference (or What?) in Birds*.[57] A century later, and we find that technologies like high-speed photography and robust mathematical models have helped us understand these natural phenomena, and while we haven't proved or disproved avian telepathy, we can show these self-organizing behaviors can occur by each bird tracking as few as seven of their flying buddies and applying three simple rules/parameters to those neighbors. Basically: watch a few people in your group and copy or react to their movements.[58]

So, we are constantly scanning our environment (or at least some subset of it), consciously and unconsciously, for social cues to which we apply significant analysis and forecasting to determine what makes the most sense in a particular circumstance. If everyone in a store started abandoning their carts and running toward the exits, you'd likely do the same, since it would probably

---

56 Howard, Newton, and Erik Cambria. "Intention Awareness: Improving upon Situation Awareness in Human-Centric Environments." *Human-Centric Computing and Information Sciences*, vol. 3, no. 1, 2013, pp. 1–17.

57 Selous, Edmund. *Thought-Transference (or What?) in Birds*. Constable, 1931. Cited in Rosenthal, Sara Brin, et al. "Revealing the Hidden Networks of Interaction in Mobile Animal Groups Allows Prediction of Complex Behavioral Contagion." *Proceedings of the National Academy of Sciences*, vol. 112, no. 15, 2015, pp. 4690–4695.

58 Rosenthal, Sara Brin, et al. "Revealing the Hidden Networks of Interaction in Mobile Animal Groups Allows Prediction of Complex Behavioral Contagion." *Proceedings of the National Academy of Sciences*, vol. 112.15, 2015, pp. 4690–4695.

be a better use of your cognitive resources than trying to figure out exactly what was happening. Given the limitations of our cognitive resources, this leads to various biases and errors in how we construct and model our social environments.[59] However accurate, though, this social awareness and process of interacting with others is a key process by which we evaluate our needs, make sense of our environment, the people around us, and ourselves,[60] use available resources to achieve goals, and to control processes used to manage resources. (As a reminder, that last bit was my definition of thought.) And as such, our thoughts are constantly mediated and enacted via social nodes, *and I conclude that these social nodes and modules are a critical, integral, material component of our thinking.*

But now I must get more precise; I've been using the terms *social* and *culture* interchangeably. "*Social*" as defined by the Oxford Pocket Dictionary of Current English, means "*relating to society or its organization.*"[61] However, empirically, it usually connotes the more general "of or relating to other people." Tylor's classic definition of culture gives us more rigor: "*that complex whole which includes knowledge, belief, art, law, morals, custom, and any other capabilities and habits acquired by man as a member of society.*"[62] I'll now try to use the term *culture* to refer to the influences and cognitive processes based on our interactions with other people. After all, we are not just scanning around a room we find ourselves in, looking at and thinking about the people we see; we are harnessing the work and influence of many generations of previous people to think, through these constructed lenses of knowledge, belief, art, laws, customs, etc., as we scan the room.

If the transmission processes described earlier account for how thoughts, ideas, beliefs, and other core cultural cognitive processes span *distance*, how do they span *time*?

59 Gilbert, Daniel T., et al. "On Cognitive Busyness: When Person Perceivers Meet Persons Perceived." *Journal of Personality and Social Psychology*, vol. 54, 1988, pp. 733–740. And Kahneman, Daniel, et al. "Experimental Tests of the Endowment Effect and the Coase Theorem." *Journal of Political Economy*, vol. 98, 1990, pp. 1325–1348.

60 E.g., Mitchell, Jason P., et al. "Encoding-Specific Effects of Social Cognition on the Neural Correlates of Subsequent Memory." *Journal of Neuroscience*, vol. 24, no. 21, 2004, pp. 4912–4917, and Mitchell, Jason P., et al. "The Link between Social Cognition and Self-Referential Thought in the Medial Prefrontal Cortex." *Journal of Cognitive Neuroscience*, vol. 17, no. 8, 2005, pp. 1306–1315.

61 "Social." *Encyclopedia.com*, www.encyclopedia.com/humanities/dictionaries-thesauruses-pictures-and-press-releases/social-0. Accessed Aug. 3, 2024.

62 Tylor, Edward Burnett. *Primitive Culture: Researches into the Development of Mythology, Philosophy, Religion, Art, and Custom*. Vol. 1, John Murray, 1871.

Daniel Wegner hypothesized something he calls transactive memory as a mechanism through which groups collectively encode, store, and retrieve knowledge within individuals, as well as the processes that update this memory.[63] This external memory requires a group to not only encode and store the root information, but also create and store the meta-information about *who* within the group knows *what*, so it can be retrieved. The result is that the group can retain more information, for longer, than any one individual.

Writing a few years after Wegner, Rubin examined narrative traditions as collective memory networks and thinks the future of technology is informed by the evolution of how human minds and collective memory have evolved.[64]

(This discovery was particularly striking – throughout my research, I had been connecting disparate ideas that seemed to fit together, but occasionally encountered frameworks like this that crystallized clusters of connections I had intuited but not yet seen formally articulated by other scholars.)

But ideas are not passed perfectly across either space or time. We gain information from others by learning and communicating, and many of our forms of learning are unique to humans, including imitative learning, instructed learning, and collaborative learning.[65]

Gradually, this vein of research became known as Cultural Evolution Theory (CET), driven by works from Dawkins,[66] Sperber,[67] Henrich,[68] and the writing pair of Richerson and Boyd. In a process consistent with natural selection, CET says cultural transmission and inheritance have better, shorter, and easier ideas winning out over others in a marketplace of ideas.[69] However, it's not just the best ideas that survive, faithfully replicated from person to person. One

---

63 Wegner, Daniel M. "Transactive Memory: A Contemporary Analysis of the Group Mind." *Theories of Group Behavior*, edited by Brian Mullen and George R. Goethals, Springer, 1987, pp. 185–208.

64 Rubin, David C. *Memory in Oral Traditions: The Cognitive Psychology of Epic, Ballads, and Counting-Out Rhymes*. Oxford University Press, 1995.

65 Tomasello, Michael, et al. "Cultural Learning." *Behavioral and Brain Sciences*, vol. 16, 1993, pp. 495–552, and Whiten, Andrew, et al. "Conformity to Cultural Norms of Tool Use in Chimpanzees." *Nature*, vol. 437, 2005, pp. 737–740.

66 Dawkins, Richard. *The Selfish Gene*. Oxford University Press, 2016.

67 Sperber, Dan. "Anthropology and Psychology: Towards an Epidemiology of Representations." *Man*, vol. 20, no. 1, 1985, pp. 73–89.

68 Henrich, Joseph, and Richard McElreath. "The Evolution of Cultural Evolution." *Evolutionary Anthropology*, vol. 12, no. 3, 2003, pp. 123–135.

69 See Lumsden, Charles J., and Edward O. Wilson. *Genes, Mind, and Culture*. Harvard University Press, 1981, and Cavalli-Sforza, Luigi Luca, and Marcus W. Feldman. *Cultural Transmission and Evolution: A Quantitative Approach*. Princeton University Press, 1981, and Boyd, Robert, and Peter J. Richerson. *Culture and the Evolutionary Process*. University of Chicago Press, 1985.

only need look at the news about the US elections to see that misbeliefs and inaccurate information can diffuse through a population,[70] in part because these *"errors"* can be adaptive.[71] Richerson and Boyd believe that most of our culturally transmitted information is neither discrete nor faithfully transmitted.[72] Information gained from others is often incomplete and fragmented[73] or even misperceived. Think of the telephone game we played as kids.

Malcolm Gladwell, in his recent book *Talking to Strangers,* points out that we all exhibit a range of visual signals that often "mismatch" with our internal state. It's hard for us to pick up on this, especially with strangers, but we still tend to believe the cues we perceive.[74] And this picture above assumes that the information we receive has not been purposefully changed, which regularly occurs in real life.[75] As such, our cultural cognitive and transmission mechanisms could be described as often inaccurate, lossy, noisy, and rife with competing information.

Interestingly, this mirrors the human brain. Pinker says that while computers are fast and reliable, human brains are slow and noisy. Barrett points out that much of our "thinking" in the brain is just millions of competing probabilistic predictions, many of which will ultimately be wrong.

CET doesn't have the exclusive on cultural cognition. Edwin Hutchins, a cognitive psychologist and anthropologist, built upon earlier work by Wundt, Vygotsky, and others to argue that the mind was in the world, rather than the world being in the mind.[76] He coined the term "distributed cognition" to refer to his model. He used the case of naval navigation to show that human work systems do not exist wholly in individuals' heads, but instead, knowledge and cognition are distributed across objects, individuals, things, and tools in

---

70 Sperber, Dan. "Culturally Transmitted Misbeliefs." *Behavioral and Brain Sciences*, vol. 32, 2009, pp. 534–535.

71 McKay, Ryan T., and Daniel C. Dennett. "The Evolution of Misbelief." *Behavioral and Brain Sciences*, vol. 32, 2009, pp. 493–561.

72 Richerson, Peter J., and Robert Boyd. *Not by Genes Alone: How Culture Transformed Human Evolution.* University of Chicago Press, 2005.

73 See Gelman, Susan A. "Learning from Others: Children's Construction of Concepts." *Annual Review of Psychology*, vol. 60, 2009, pp. 115–140, and Keil, Frank C. "Words, Moms, and Things: Language as a Road Map to Reality." *Monographs of the Society for Research in Child Development*, vol. 63, no. 1, 1998, pp. 152–158.

74 Gladwell, Malcolm. *Talking to Strangers – What We Should Know about the People We Don't Know. Little* Brown and Company, 2019.

75 Ayers, Michael S., and Lynne M. Reder. "A Theoretical Review of the Misinformation Effect: Predictions from an Activation-Based Memory Model." *Psychonomic Bulletin & Review*, vol. 5, no. 1, 1998, pp. 1–21.

76 Hutchins, Edwin. *Cognition in the Wild.* 1995. MIT press, 1995.

the environment. Of course, individuals think: they remember, perceive, have expertise, entertain hypotheses, and are biased in the evaluation of the above. But Hutchins points out that the group-system, as a whole, has an emergent property, which constitutes a cognitive or mental state that no individual member of the group manifests by themselves.

You can see how Hutchins got my attention here, validating some of my thoughts, and helping to frame my thinking in this chapter.

One way to think about how ideas and concepts develop and diffuse within a population is memetics. Dawkins popularized this approach when he coined the term "meme" to refer to ideas that can faithfully replicate, and that hop from one person's brain to another, much like genes might replicate themselves.[77] While this concept has become so mainstream that our children use the term daily, there are several scholarly criticisms of this approach, mainly because memes don't have a mechanism for reproducing, and the cultural mechanisms that result in ideas moving from person to person are so diverse that the utility of the meme model is limited.[78]

We do not have the time or the space here to review the literature on how, specifically, ideas and concepts diffuse and accumulate within a culture, just as we avoided the specifics of how neural networks actually function in the brain. I'll accept Richerson and Boyd's general statement that it happens through teaching, imitation, and other forms of social transmission. It is fair to say the CET scholarship is varied, and many of its tenets are still evolving. Current trends shaping the debate tend to these themes: a) culture accumulates, b) ideas and concepts are widely varying within groups, c) some variations are usually more consistent across populations than others, maybe through forces such as "guided variation" or "attractor" mechanisms,[79] d) culture usually evolves over time, just as a physical species might,[80] and e) multiple cultures can live alongside each other, just as different variations of a species might co-exist.

77 Dawkins does not acknowledge the earlier work done by Semon (Semon, Richard. "Engraphic Action of Stimuli on the Individual." *The Mneme*, translated by Louis Simon, George Allen & Unwin, 1921.), who developed the concept of a "mnemic trace" or "engram" to describe a hypothesized memory trace, or neural tissue change thought to occur upon stimulation of the brain.

78 Lewens, Tim. *Darwin*. Routledge, 2007. Ch. 7.

79 Scott-Phillips, Thom, and Dan Sperber. "The Mutual Relevance of Teaching and Cultural Attraction." *Behavioral and Brain Sciences*, vol. 38, 2015, pp. 44–45; and Claidière, Nicolas, and Dan Sperber. "The Natural Selection of Fidelity in Social Learning." *Communicative & Integrative Biology*, vol. 3, no. 4, 2010, pp. 1–2.

80 Claidière, Nicolas, et al. "How Darwinian Is Cultural Evolution?" *Philosophical Transactions of the Royal Society B*, vol. 369, 2014, article 20130368.

Speaking of species, what happened to our species on that dusty savannah to set the stage for this cultural transmission and accretion? After all, the research shows that other species, like chimpanzees, have culture, and that this culture influences how they use tools in complicated ways. For example, fishing for termites requires specific different kinds of sticks, which are harvested well before their use, and the sticks need to be specially prepared to work effectively. Chimps in different parts of the world prepare their sticks differently.[81] The most effective chimps use a combination of specially prepared sticks in specific complex sequences.[82] Behavioral sequences are learned by watching others, but also mainly through associative learning, in a series of trial and error, where successful steps are eventually strung together into more complex behaviors. This is a simple but powerful process that, combined with certain genetic predispositions, accounts for most animal behaviors. Learning from others may be faster, but learning by doing is better.

3 The Human Evolutionary Transition

But there is something fundamentally different about humans' cognitive processes, according to a theory put forth by Magnus Enquist and colleagues in a provocative book (it has become a cornerstone of my thinking on this topic), *The Human Evolutionary Transition*, which builds on the cultural story told by Tomasello.

This book unpacks the phenomena that humans' astounding diversity of behaviors requires extensive and often collaborative sequences. Enquist et al. point out that to do these via simple trial and error via associative learning would take literally trillions of attempts. Instead, humans store and process both sequential and episodic information. They can then reflect on this information and recombine information that has been gathered previously, possibly in many different situations. These cognitive traits are more generalizable and flexible. They allow us to build mental worlds and representations governed by rules, enabling us to construct ideas, complex collaborative languages, and fields such as math and music. But this approach is not always beneficial; it comes at great evolutionary risk.

---

81 Sanz, Crickette, et al. "Design Complexity in Termite-Fishing Tools of Chimpanzees (Pan troglodytes)." *Biology Letters*, vol. 5, no. 3, 2009, pp. 293–296.

82 Musgrave, Stephanie, et al. "The Ontogeny of Termite Gathering among Chimpanzees in the Goualougo Triangle, Republic of Congo." *American Journal of Physical Anthropology*, vol. 174, no. 2, 2021, pp. 187–200.

To understand that, and why it was unique for humans to gain this unique toolbox (which Enquist et al. refer to as "thinking"), Enquist considers the requisite changes that brought it about. Principally, thinking requires a long time to develop. Changes in human life history, whereby offspring take a long time to develop and mature, were key. This rich environment lets children learn from others, including adults. But it was biologically expensive, and they remained vulnerable to many risks during this extended rearing.

Other animals like whales and elephants, too, have extended life histories. Elephants use that time to learn complex vocal communications, and sophisticated information processing and motor control of their trunk, among other things.[83]

But along the way, humans uniquely evolved a genetic predisposition to create faithful representations and processing of episodic memory (recollecting happenings from the past that occurred in particular spatial and temporal contexts),[84] sequential information ("sequential memory"), semantic memory (recalling facts about the general world), and flexibility of mental processing.

This, combined with enhanced motor flexibility, particularly with our hands, bipedal efficiency, and throwing and with changes in motivations for learning and social interactions, creates the milieu we know of as human culture. These specific changes occurred uniquely within our species, setting in motion the flywheel of human capabilities. Like a savings account that doubles every day, Enquist and his colleagues note that most of humanity's advancements have accumulated in the last few millennia, centuries, and now even decades.

4 Language

A discussion of social thinking and the brain inevitably (and often quickly) leads to linguistics.

Language is not only itself a social construct, but it is also a key mechanism for transmitting information between individuals, and even a key influence on how we think.[85] Andy Clark sums up the power of language: *"The Linguistic scaffoldings that surround us, and that we ourselves create, are both cognition-*

83 Stoeger, Angela S., and Paul Manger. "Vocal Learning in Elephants: Neural Bases and Adaptive Context." *Current Opinion in Neurobiology*, vol. 28, 2014, pp. 101–107.

84 Squire, Larry R., and Stuart M. Zola. "Episodic Memory, Semantic Memory, and Amnesia." *Hippocampus*, vol. 8, no. 3, 1998, pp. 205–211.

85 Kahane, Henry, and Renée Kahane. "Humanistic Linguistics." *The Journal of Aesthetic Education*, vol. 17, no. 4, 1983, pp. 65–89.

enhancing in their own right, and help provide the tools we use to discover and build the myriad other props and scaffolding whose cumulative effect is to press minds like ours from the biological flux."[86]

One central debate is whether language *influences* how we think (the weak version of the Sapir-Whorf hypothesis)[87] or if language *determines* how we think (the strong form). However, this is quite a complex and multi-faceted debate, which, while interesting, does not shape the utility of our model. So, we will leave it to others to describe and progress the debates within the interplay between language and cognition, as well as fascinating but distracting aspects of language such as the distinction between the communications styles and mechanisms between low-context (in which communication occurs predominantly through explicit statements in text and speech) cultures, and those with high-context (when pertinent communication includes other cues such as body language and the use of silence).[88]

5 Are People Things?

Why not lump people (Thinking with Culture) with things (Thinking with Technology) in our embrace of extended cognition? Why three streams, instead of just two (thinking in our head and thinking outside our body, or as Hutchins says, in the world)?

Because:

a) People are a special case of things, with significant differences from objects.
b) We literally think about the two different classes differently.

Fiske and Taylor, who literally wrote the book on Social Cognition, outline these ways that people differ from objects, which encapsulate points made earlier by Endsley and Howard & Cambria:

- People have intentional influences on their environment.
- People perceive back (social cognition is mutual cognition).
- Social cognition tells you about you, since you can learn about yourself from how others judge you.
- The act of being observed changes the object, as they may adjust to your observation.

---

86 Clark, Andy. *Supersizing the Mind: Embodiment, Action, and Cognitive Extension.* Oxford University Press, 2008, p. 60.

87 Chandler, Daniel. *The Sapir-Whorf Hypothesis.* 1994.

88 Hall, Edward T. *Beyond Culture.* Doubleday, 1976.

- What you can't see about a person matters a lot; what you can't see about objects may be less important.
- People change over time, more than objects.[89]
- It's harder to determine your accuracy in judging people; it's easier to test your accuracy with objects.
- People are incredibly complex, so observers have to make many reductive choices to model a person; this is less true of objects (related to Fodor's frame problem).
- Human complexity requires social explanation; it is more important to know why a person is frail than why a cup is.

Humans place different importance on features of our social environment, and the maps they build out of the things they pay attention to are called salience.[90] We use salience maps of the characteristics above, shaped heavily by prior experience, immediate context, and various attentional tasks (your goal, the vividness of things in your visual field, or instructions you have) and, like the starlings, pay attention to a small percentage of the information available to us[91] and use that minute information to decide how to act.[92]

One way we could build that map is bottom up – by direct sensory observation of the people we are interacting with. But we are cognitively constrained, so much of our sense-making of other people uses top-down approaches, leveraging our prior experience to lump people into categories (e.g., stereotypes) that are conceptually driven, so we often attend to or use more efficient (and lossy) top-down approaches.[93] Both types of information are further shaped by other cognitive reductions, such as what we infer from what we observe, and the attributions we apply to the observations.[94]

Studies using technologies such as functional Magnetic Resonance Imaging (fMRI) have repeatedly shown that we use our brains differently to think about people than other aspects of our environment.

---

89 Note – here Fiske and Taylor are referring to physical objects; in this book I also accept that Materially Modular Cognition includes digital items like SW, data, algorithms, etc. that can change much faster; this will be introduced in the following stream, and further developed later in the book.

90 Fiske, Susan T., and Shelley E. Taylor. "Social Cognition: From Brains to Culture." 2020, pp. 1–672.

91 Kahneman, Daniel. *Attention and Effort*. Prentice-Hall, 1973.

92 McArthur, Leslie Z. "What Grabs You? The Role of Attention in Impression Formation and Causal Attribution." *Social Cognition: The Ontario Symposium*, vol. 1, 1981.

93 Macrae, C. Neil, and Galen V. Bodenhausen. "Social Cognition: Categorical Person Perception." *British Journal of Psychology*, vol. 92, no. 1, 2001, pp. 239–255.

94 Harris, Lasana T., et al. "Attributions on the Brain: Neuro-Imaging Dispositional Inferences, Beyond Theory of Mind." *NeuroImage*, vol. 28, no. 4, 2005, pp. 763–769.

In 2000, for example, a study had participants watching moving geometric shapes. When asked to attribute human qualities to the shapes, imaging showed that the participants activated and recruited different parts of the brain than they had used previously, including the medial prefrontal cortex (mPFC), superior temporal sulcus (STS), and fusiform gyrus (FFA). Fiske and Taylor note that the mPFC and STS repeatedly show up in studies of social cognition, particularly the mPFC.[95] The FFA seems to play an active role in specialized recognition, and is engaged both when looking at faces, but also when viewing objects in an expert's field (e.g., birds to birdwatchers). Fiske and Taylor conclude, "*The main point…is the dissociation between the social and nonsocial neural patterns.*"[96]

6 Pulling It All Together

What picture emerges when we weave together these threads to explore cultural cognition?

Consider for a moment your laptop or computer pad. It consists of many subcomponents and subsystems made from silica, copper, precious metals, plastics, and other various ingredients. This hardware includes a central processing unit (CPU), long-term memory, short-term memory, power supply and management, graphics processing chips, various input and output devices like keyboard, mouse, camera, Bluetooth devices, speakers, display, and a motherboard to tie it all together. But the equipment by itself does nothing; it requires an operating system (OS) to manage the computer's resources, keep everything safe, and instruct the various semiconductors how to function and perform as an integral unit. That operating system then runs various software applications that perform different tasks for you, including providing access to other remote software platforms you can connect to via the Internet to accomplish all the things you want your laptop to do for you.

When it starts up, the CPU resets and begins executing a little code built into it (this special class of code is known as firmware), called Basic Input/Output System (BIOS), which then performs startup checks on the sub-components, like memory, disk drives, and other critical hardware. Then it goes looking for a bootable device, like a hard drive, from which it loads software known

95 Amodio, David M., and Chris D. Frith. "Meeting of Minds: The Medial Frontal Cortex and Social Cognition." *Nature Reviews Neuroscience*, vol. 7, no. 4, 2006, pp. 268–277.

96 Here Fiske and Taylor cite Mitchell, Jason P., Todd F. Heatherton, and C. Neil Macrae. "Distinct Neural Systems Subserve Person And Object Knowledge." *Proceedings of the National Academy of Sciences*, vol. 99.23. 2002, pp. 15238–15243.

as a bootloader into memory. The bootloader is neither firmware nor software, like the operating system. It's a small pre-operating system program whose job is to load the operating system kernel into memory and start its execution.

Once started, the OS sets up the ability to talk to the hardware through a process known as abstraction, and brings the device drivers and system processes online, before moving on to initiating the system services and background processes. Only after all of this is online and working, does the operating system present a user login prompt or graphical login screen. When the user is successfully authenticated, the OS loads their preferences, permissions, and functions. After this, the local operating system is up and running, ready to perform various computer tasks.

Once logged into the computer, the user can begin to load applications, go online, and log in to various technologies, such as social media, the web, or artificial intelligence platforms, which then invoke other data, including their own activities, to function. This whole process is known as bootstrapping, whereby a small set of hardware and software start up and then bring online a much larger and more complicated set of architecture, assets, resources, and functions.

What we call here the OS is the regime: the tacit rules and standards that make daily life "boot" without errors. Innovation begins in niches – incubators where odd ideas can live long enough to learn – until landscape pressures make an upgrade unavoidable.

This gives us a robust but somewhat flawed analogy of "how we think," individually and collectively.

The basic point made in this chapter is that you don't just function via the hardware of your brain, and the traits you are born with: you rely on socially constructed operating systems (e.g., "civilization," "norms," "community guidelines"), and applications (e.g., "ideas," "language," "education."

The more complex and important point, though, is not the picture we get at any point, but the understanding we gain considering how our human laptop-as-(flawed)-example-of-human-thought booted up, collectively and individually.

The hardware present in a newborn *Homo sapiens* 300,000 years ago in what is now present-day Ethiopia is virtually identical to that of one born today. That includes the structure and function of the brain, as well as the "onboard" genetically programmed hardware and operating system.

However, as the human brain booted up, we collectively began to develop and then bootstrap new operating systems and applications over time. While many variations in the operating system and applications were essentially "bugs" that may not have helped those individuals and tribes, some were effective, and allowed early humans to extract more resources from their environments, live longer, and have children with a higher probability of surviving until they themselves could have children. Those socially constructed features might have been more likely to persist than others.

One of the flaws in this analogy is that we weren't just bringing new innovations to bear, like upgrading the computer's speakers or microphones; instead, we were both developing and learning how to integrate these assets and tools into our lives.

We were literally developing the airplane as it flew. We had much of the same hardware (at birth), as Enquist indicated, in terms of our ability "to make faithful representations and processing of episodic memory and flexible mental processing," but we were developing and refining the operating system through trial and error and writing and using new software. And the use of these socially constructed tools and resources changed our brains over time, as we age and experience life culturally.

So significant differences would be found between an adult brain from 300,000 years ago and the one you are using to read this paragraph. Using language, words, and other socially constructed elements has changed your brain. The "booting up" of your brain during your lifetime has changed your hardware.

As Enquist and his team point out, the world is a complex place, and obtaining the resources to live is hard. Exploring how to live in the world through trial and error is a time-consuming and risky process. Humans have adapted a unique ability to think (remember sequential and episodic information, reflect on and recombine that information, and apply it to new situations), but thinking in the modern world is both the result of and inexorably intertwined and dependent on culture.

This mode of thought is a significant element of the human cognitive domain – it describes not only an important part of how humans think, but also provides a framework for how, especially with the emergence of robust language perhaps ~50,000 years ago, knowledge begins to collect and transfer across space and time.

Mithen argues that language catalyzed the cultural "explosion."[97] This framework, as opposed to the genetic evolution of human cognitive hardware, provides an explanation for how knowledge bootstrapped and accumulated in human populations so rapidly as to lead to Harari's Cognitive Revolution. And

97 Mithen, Steven J. *The Prehistory of The Mind: A Search For The Origins Of Art, Religion and Science*. Thames and Hudson, 1996.

increasing connectivity between humans and other cultural elements is a key driver of both cultural acceleration and the disruption of thought.

To distinguish this specific conception from several similarly named but distinctively different schools of thought, I will call this stream Thinking with Culture (TC). I will treat culture and technology as two analytic streams, but read them as coupled dynamics in which culture configures use, adoption, and legitimacy. We'll talk more about technologies in the chapter. Keeping the strands separate lets us isolate the distinct forces and trends going on within each, and name mechanisms without losing sight of their entanglement. Here, I emphasize those elements of culture that Tylor defined as existing in the social (i.e., non-material) sphere, via shared understandings between people.

However, thinking culturally does not fully explain why information has accumulated so rapidly, especially in the modern era. After all, writing and other material nodes and modes of thought are major contributors to how we span both distance and time. Material thought has the special feature that lets us *scale* our reach over time and space, much faster, wider, and more effectively than we can often do in our interactions with other people. And the codification of that knowledge, beliefs, arts, laws, customs, etc., is often stored in the material world. In fact, Merlin Donald believes that the creation of the ability to physically store mental concepts in the things around us caused us to enter a new stage of human cognition, one that accounts for much of the modern world.

Let's explore that dimension of thought together in the next chapter.

7 My Take on Thinking with Culture

While we intuitively assume thinking occurs exclusively within our brains, in this chapter I make the case that human cognition actually extends through social networks, incorporating culturally transmitted elements like language, beliefs, and customs passed down across generations. These cultural components, along with surrounding individuals, function as cognitive nodes – analogous to how brain regions collaborate to generate thoughts – but operating at a collective level.

Consider the brain as hardware requiring culturally derived "software" to function effectively. This metaphor illuminates how human thought has undergone collective "booting up" over millennia, gaining significant momentum around 50,000 years ago with robust language emergence. This cognitive startup process continues accelerating as we rapidly modify our cultural software. The metaphor applies both to species-level cognitive evolution and individual development within a lifetime.

Though not yet mainstream in popular discourse or scientific consensus, this extended cognition framework builds upon and synthesizes substantial scholarly foundations. Using this perspective will offer provocative insights into human cognitive architecture that transcend traditional boundaries between individual and collective intelligence, which we'll explore later in the book.

CHAPTER 7

Stream 3: Thinking with Technology

> The Internet is the first thing that humanity has built that humanity doesn't understand, the largest experiment in anarchy that we have ever had.[1] (Eric Schmidt)

1 The Material Evolution of Human Thought

The third stream of the model embraces and develops Donald's claim that cognition exists materially.

After all, Enquist, in his helpful treatise on the importance of culture in humans, notes that alongside mental flexibility and other inherited traits that humans possess is our remarkable motor flexibility. We learned (culturally) how to use that motor flexibility to help us manipulate physical things to more efficiently extract resources from our environment.

Donald is a neuroanthropologist[2] who wrote two influential books on the evolution of human thought – *Origins of the Modern Mind*[3] and *A Mind So Rare*.[4] In Origins, he describes the evolution of the human mind in three stages:

- First – a transition from animals' episodic experience (simply living within the moment) to one shaped by mimetics – our early ancestors learning to imitate and represent ideas to others.
- Second – maybe centered around 30,000 years ago and enabled gradually through the emergence of increasingly robust language, was an oral-mythic culture, which may have organized information into narrative structures, likely with representations of ideas, and replete with meta-representations, including supernatural forces, myths, and legends.

---

1 Schmidt, Eric. Address at Internet World Trade Show. New York, 18 Nov. 1999. Qtd. in *Oxford Dictionary of Quotations*, 8th ed., Oxford University Press, 2014.

2 Another example of the diverse fields that fall under the umbrella of "cognitive sciences."

3 Donald, Merlin. *Origins of the Modern Mind: Three Stages in the Evolution of Culture and Cognition*. Harvard University Press, 1991.

4 Merlin Donald. *A Mind So Rare: The Evolution of Human Consciousness*. W. W. Norton & Company, 2001.

HTTPS://DOI.ORG/10.1163/9789004739161-011

– Third – the transition to a culture of external symbol storage and theory.[5]

This last transition led to civilization and to us.

Like language, the invention of graphical representation of information unfolded gradually over tens of thousands of years (still significantly faster than the evolutionary timescales)[6] but enabled the new function of external memory *storage* of ideas. Donald calls this a "*hardware change…albeit a nonbiological hardware change.*"[7] Unlike biological hardware, which relies on built-in memory processes, these new external systems require new cognitive processes to enable people to retrieve information.

This transition required a corresponding cognitive change in people's heads, as they shifted from having to remember *something* to having to remember *where* that something resided, and *how*[8] to decode it.

This transition to external symbol and memory storage, though, also unlocked another, much more important change in how humans thought: it led to the rise of theoretical thinking. Crucially, that shift depended not just on external records but on a sociotechnical regime – schools, scribal practices, standards, archives, and verification routines – that made records retrievable, comparable, and testable, channeling external memory into sustained theory-building.

Theoretical thinking starts with the reduced need to remember things, which results in the freeing up of cognitive resources to begin reflecting on those ideas, changing and refining them.

Donald also refers to Bruner, who categorized modern human thought in two modes: narrative stories and histories, common to us all as young children, prevalent in the arts; and paradigmatic thought, which seeks truth, is only available to us after years of systemic training and is common in science

5 Fairservice had done similar, earlier work in this vein: Fairservice Jr., Walter A., *The Threshold of Civilization.* Scribner, 1975.

6 I'm using this term in the popularly recognized sense; see for example the work of Peter and Rosemary Grant on Darwin's finches, and the rapid pace of evolution in that inspirational species (Grant, Peter R., and B. Rosemary Grant. *How and Why Species Multiply: The Radiation of Darwin's Finches.* Princeton University Press, 2007), or Cochran and Harpending's thoughts on the more recent evolutions in humans (Cochran, Gregory, and Henry Harpending. *The 10,000 year Explosion: How Civilization Accelerated Human Evolution.* Basic Books, 2009).

7 Donald, Merlin. *Origins of the Modern Mind: Three Stages in the Evolution of Culture and Cognition.* Harvard University Press, 1991, pp. 307–308.

8 See Rowland's descriptions of how an ancient Peruvian kvinu official uses his biology to decode the knots that constitute his store of records in Rowlands, Mark. *The Body in Mind: Understanding Cognitive Processes.* Cambridge University Press, pp. 134–136.

and law.[9] Donald's theoretical thinking functionally equals Bruner's paradigmatic.

Today's paradigmatic thoughts are stored in writing and other advanced symbol systems like math. Donald coined the term "exogram"[10] to capture this encoding, a term meant to contrast with an earlier term, "engram"[11] – the physical (biophysical or biochemical) substrate of memory in the brain.[12] Donald's exograms were the physical counterpart: "*external memory records of ideas, made symbolic inventions that may have undergone a process of repeated examination, testing, and improvement.*"[13] The difference was that exograms were "*likely to be permanent, are unconstrained and reformattable, can be of any medium, have virtually unlimited capacity and size, and can be subjected to unlimited iterative refinement.*"[14] These are the advantages material cognition has over thinking with just the brain.

---

9 See Bruner, Jerome. "*Two Modes of Thought.*" *Actual Minds, Possible Worlds,* Harvard University Press, 1986, pp. 11–43.

10 The idea of thoughts stored externally was first proposed by Gregory (Gregory, Richard L. *The Intelligent Eye.* Weidenfeld and Nicolson, 1970.), and Bednarik did similar work in 1987, where he developed similar ideas, but used the term psychogram. The idea that this is a fungible substitute for our brains, dates as far back to Plato – "if men learn this [writing], it will implant forgetfulness in their souls: they will cease to exercise memory because they will rely on that which is written, calling things to remembrance no longer from within themselves but from external marks" (Phaedrus, 274e–275a); and the idea of a "surrogate cortex" was developed further by Goody (Goody, Jack,. *The Domestication of the Savage Mind.* Cambridge University Press, 1977.), and subsequently by Carruthers (Carruthers, Mary. *The Book of Memory.* Cambridge University Press, 1990; and Carruthers, Mary. *The Craft of Thought.* Cambridge University Press, 1998).

11 Semon, Richard. "Engraphic Action of Stimuli on the Individual." *The Mneme,* translated by Louis Simon, George Allen & Unwin, 1921.

12 The search specifically for localized engram substrate has proven fruitless. Lashley spent a career searching unsuccessfully for a single biological locus of memory, concluding, in a pre-MM way that memories are widely distributed throughout the cerebral cortex. (Lashley, Karl S. "The Behavioristic Interpretation of Ionsciousness." *Psychological Review,* vol. 30, Part I, 237–272; Part II, 1923, 329–353; and Lashley, Karl S. 1950. "In Search of the Engram." *Society of Experimental Biology,* Symposium, vol. 4, 1950, pp. 454–482.) Later researchers came to similar conclusions. See Thompson, Richard F. *Foundations of Physiological Psychology.* Harper & Row, New York, 1967.

13 Donald, Merlin. *Origins of the Modern Mind: Three Stages in the Evolution of Culture and Cognition.* Harvard University Press, 1991, p. 308.

14 Donald, Merlin. *Origins of the Modern Mind: Three Stages in the Evolution of Culture and Cognition.* Harvard University Press, 1991, p. 309. See also Donald, Merlin. "Précis of Origins of the Modern Mind: Three Stages in the Evolution of Culture and Cognition." *Behavioral and Brain Sciences,* vol. 16, no. 4, 1993, pp. 737–748; Donald, Merlin. *A Mind So Rare: The Evolution of Human Consciousness.* W. W. Norton, 2001; and Bednarik, Robert

Viewed socio-technically, writing and numeracy didn't spread simply because individuals off-loaded memory; they became part of a socio-technical regime – scripts, devices, schools, scribal professions, archives, standards, and authorities – that made external memory reliable, searchable, and legitimate. Innovations trialed in niches (e.g., counting boards, abaci, positional notation) gradually reconfigured the regime, enabling the leap from storage to theory-building you describe.

This third transition overlaid a theoretic, paradigmatic layer on top of narrative practices, freeing cognitive resources for analysis, formalization, and theory-building while narrative capacities continued to do essential work. This mode brings with it analysis, logic, and science, and then it spawns, according to Donald – *"formal arguments, systemic taxonomies, induction, deduction, verification, differentiation, quantification, idealization and formal methods...The highest product of analytic thought, and its governing construct is theory, an integrative device that is much more than the symbolic invention; it is a system of thought and argument that predicts and explains."*[15]

With this, at a time perhaps just several thousand years ago,[16] we have the emergence of a new higher-level architecture of human cognition, which entailed new representational changes in the brain. Donald emphasizes technological innovations as cultural practices of external symbolic storage rather than genetic changes.

In *A Mind So Rare*, Donald suggests, like others,[17] that the cognitive demands of living in progressively larger and more complex social groups over hundreds of thousands or millions of years drove genetically determined evolution in cognitive improvements in attention, memory, metacognition, and retrieval capacity. These socially derived capabilities then became exapted by technological innovations to be harnessed for new purposes,[18] just as motor grasping circuitry in the brain became part of the physical substrate for lan-

G. "Doing with Less: Hominin Brain Atrophy." *HOMO: Journal of Comparative Human Biology*, vol. 65, no. 6, 2014, pp. 433–449.

15 Donald, Merlin. *Origins of the Modern Mind: Three Stages in the Evolution of Culture and Cognition*. Harvard University Press, 1991, p. 273.

16 The archeological debate about exactly when this developed is outside the scope and purpose of this paper. For a more detailed discussion, see Coolidge, Fred, Wynn, Thomas and Overmann, Karenleigh. "The Evolution of Working Memory." *Working Memory: The Connected Intelligence*, edited by Tracy Packiam Alloway and Ross G. Alloway, Psychology Press, 2013.

17 Dunbar, Robin I. M. "The Social Brain Hypothesis." *Evolutionary Anthropology*, vol. 6, 1998, pp. 178–190.

18 Note the parallels with language: we can observe that genetic evolution created a brain that was "theory-ready."

guage. According to Donald, human brain capacity cannot be understood or realized without its social and symbolic context. It also cannot be sufficiently described solely by thinking with the brain or thinking with culture.

Donald is building off earlier traditions here.

The 18th-century philosopher Giambattista Vico predated Donald by a couple of centuries, but gave the transition between Donald's second and third modes of thought a name: *sapienza poetica* – "poetic wisdom." Vico argued that preliterate humans thousands of years ago didn't think in abstract categories but through imaginative universals: thunder wasn't atmospheric pressure but Jove's anger; seasons weren't orbital mechanics but divine narrative. This wasn't ignorance, it was a coherent epistemology optimized for oral transmission and communal cohesion.

> Rational metaphysics teaches that man becomes all things by understanding them...imaginative metaphysics shows that man becomes all things by not understanding them...; and perhaps the latter proposition is truer than the former, for when man understands he extends his mind and takes in the things, but when he does not understand he makes the things out of himself and becomes them by transforming himself into them.[19] (Vico, 1744)

What Vico understood – and what matters for us – is that the shift to literacy wasn't simply adding a capability. It fundamentally restructured cognition itself. Poetic wisdom gave way to theoretical thought, metaphor to logic, and embodied narrative to abstract analysis. We gained immense analytical power but lost something too: the holistic, sensory-rich understanding that McGilchrist associates with right-hemisphere cognition.

This historical precedent matters because we're living through another such transition. If literacy displaced poetic wisdom, what are AI and other tech displacing now? Understanding past cognitive disruptions helps us navigate current ones with conscious agency.

Vico's insights, though developed in the 18th century, anticipated later thinking about extended cognition. Steiner argues that Peirce was developing an approach similar to Donald's in 1958. He notes that Peirce, in criticizing Descartes' view of the mechanical mind based in the pineal gland, "*was one*

19 Vico, Giambattista. *The New Science of Giambattista Vico*. Translated by Thomas Goddard Bergin and Max Harold Fisch, 3rd ed., Cornell University Press, 1984, para. 405.

of the first philosophers to develop the idea that human intelligence is artificial because it is constantly extended across the use of artefacts."[20]

Other contemporary writers, like Mithen, agree with Donald and see the use of external storage of ideas (what Mithen calls epistemic artifacts) as the primary cause for the cultural explosion and acceleration of rich and varied cultures that started possibly around 50,000 years ago, all without significant change in our genomes.

You can imagine how excited I was to come across this vein of scholarship – with the last chapter, we began to address the problem I had with how neuroscience by itself doesn't explain that "stone age mind" that writers usually associate with something around 50,000 years ago. Neuroscience guides our understanding of the modern mind, while the last chapter explained the cognitive revolution that started Harari's journey.

But with Mithen and Donald, I see another leap forward, several thousand years ago, as things like the Incas' knotted string *khipus* or the etchings on clay tablets of Mesopotamia. This one is brought about by technological innovation – something I understand. This begins to connect the dots and fill in the frustrating gaps in the answers to my research questions.

A key feature of this cultural explosion was, according to Mithen, humans' use of "cognitive fluidity" – the ability to process and reconcile different information in new and creative ways. The modern human mind arrived, according to Mithen, when cognitive fluidity combined with tools and richly stored information as epistemic artifacts, which spread and were widely transmitted to create the building blocks that led to modern civilization.[21]

This was good stuff.

In a 2013 book, *Working Memory*, Coolidge, Wynn, and Overmann explore this topic further, suggesting that Enhanced Working Memory (EWM) "*was one of the key evolutionary acquisitions in human cognition, and indeed may even be the smoking gun of modernity.*"[22]

It helped bring about, according to Coolidge et al., the central executive functions of the brain core to most complex modern thought, such as "*attention, active-inhibition, decision-making, planning, sequencing, temporal tag-*

20 Steiner, Pierre. "CS Peirce and Artificial Intelligence: Historical Heritage and (New) Theoretical Stakes." *Philosophy and Theory of Artificial Intelligence*, Springer, 2013, pp. 265–276.

21 Mithen, Steven J. *The Prehistory of The Mind: A Search For The Origins Of Art, Religion And Science*. Thames and Hudson, 1996.

22 Coolidge, Frederick L., Thomas Wynn, and Karenleigh A. Overmann. "The Evolution of Working Memory." *Working Memory: The Connected Intelligence*, edited by Tracy Packiam Alloway and Ross G. Alloway, Psychology Press, 2013.

ging, and the updating, maintenance, and integration of information....and that this development was the final piece – a final quantum leap in cognitive evolution, if you will – in the evolution of human executive reasoning ability, language, and culture."[23]

While Mithen notes the cultural explosion was not powered by or associated with genetic evolution, and Donald characterizes this transition as not being a biological hardware change, others may beg to differ. Bednarik points out that the celebrated brain *expansion* that occurred during human evolution overlooks another important trend – the *decrease* in human brain size that occurred in the Holocene period of the last 10,000 years; a trend that continues to the present. (Keep in mind that size does not correlate with the number of neurons.) Citing Henneberg,[24] Bednarik argues that the 17% *reduction* of average human brain size over the last 10,000 years was 37 times faster than the famous and dramatic expansion of brain size during the Pleistocene. And the rate of decreasing brain size is exponentially increasing, including a 15% reduction since the Bronze Age (3,000–5,000 years ago), and a 9.7% reduction in the few centuries since the Middle Ages.[25]

This extends an earlier trend in smaller brain sizes that started back 40,000–50,000 years ago (maybe not so coincidentally around the advent of language). This trend is a human universal and is not explained by changes in body sizes or weights, or factors like genetic drift, according to Bednarik. If the growth of human brains was due to the increasing social complexity of the last few million years, Bednarik makes a solid case that the increasing cultural and technological complexity of the last several thousand years should have *increased* the demand on the brain, and it should have continued to get bigger.

23 Coolidge, Frederick L., Thomas Wynn, and Karenleigh A. Overmann. "The Evolution of Working Memory." *Working Memory: The Connected Intelligence*, edited by Tracy Packiam Alloway and Ross G. Alloway, Psychology Press, 2012, pp. 37–60.

24 See Henneberg, M. "Brain Size/Body Weight Variability in Homo Sapiens: Consequences for Interpreting Hominid Evolution." *HOMO Journal of Comparative Human Biology*, vol. 39, 1990, pp. 121–130; and Henneberg, M. "Decrease of Human Skull Size in the Holocene." *Human Biology*, vol. 60, 1988, pp. 395–405; and Henneberg, M. "The Rate of Human Morphological Microevolution and Taxonomic Diversity of Hominids." *Studies in Historical Anthropology*, vol. 4, 2006, pp. 49–59; and Henneberg, M., and A. Saniotis. "Evolutionary Origins of Human Brain and Spirituality." *Anthropologischer Anzeiger*, vol. 67, 2009, pp. 427–438.

25 Bednarik, Robert G. "Doing with Less: Hominin Brain Atrophy." *HOMO: Journal of Comparative Human Biology*, vol. 65, no. 6, 2014, pp. 433–449.

Large brains are a negative: the brain is an expensive tissue (it consumes almost 20% of the body's energy);[26] the size of the head is a frequent source of birth problems in humans, whose pelvic girdles are adapted more narrowly to optimize efficient bipedalism;[27] and the head is physically vulnerable – a large and heavy object at the top of a tall bipedal creature can be risky and unstable, especially for the first year or two of life.[28] If we can reduce the size of the brain, and do more with less, there would be evolutionary advantages.

Bednarik considers that partial explanations for the trend in smaller brains could be found with increasingly better nutrition, and denser and/or more compact neuronal structures.[29] But a more graceful and parsimonious explanation to him is that the increasing use of external symbols *reduced the need for the brain.* On net, we domesticated ourselves,[30] developing the hallmarks of our culture – increasingly complex external cognition, which became a "*substitute for the need of more cranial space and brain power.*"[31]

Implied in Bednarik's case is that there is a genetic evolutionary adaptation to our self-constructed niche.[32] Anthropologists Cochran and Harpending make the explicit case that our technological innovations of the last few thousand years triggered evolutionary adaptations that continue to the present. In this, they specifically take aim at the accepted position, as articulated by thinkers like Stephen J. Gould, who they quote, "*There's been no biological*

---

26 Brady, Scott, et al., eds. *Basic Neurochemistry: Molecular, Cellular and Medical Aspects.* Elsevier, 2005.

27 Wittman, Anna Blackburn, and L. Lewis Wall. "The Evolutionary Origins of Obstructed Labor: Bipedalism, Encephalization, and the Human Obstetric Dilemma." *Obstetrical & Gynecological Survey*, vol. 62, no. 11, 2007, pp. 739–748.

28 Zimmerman, Robert A., and Larissa T. Bilaniuk. "Pediatric Head Trauma." *Neuroimaging Clinics of North America*, vol. 4, no. 2, 1994, pp. 349–366B.

29 Semendeferi, Katerina, et al. "Prefrontal Cortex in Humans and Apes: A Comparative Study of Area 10." *American Journal of Physical Anthropology*, vol. 114, 2001, pp. 224–241.

30 Bednarik, Robert G. "The Domestication of Humans." *Anthropologie*, vol. 46, 2008, pp. 1–17; and Bednarik, Robert G. "An Aetiology of Hominin Behaviour." *Homo*, vol. 63, no.5, 2012, pp. 319–335.

31 Bednarik, Robert G. "Doing with Less: Hominin Brain Atrophy." *HOMO: Journal of Comparative Human Biology*, vol. 65, no. 6, 2014, pp. 433–449. See also Taylor, Timothy. *The Artificial Ape: How Technology Changed the Course of Human Evolution.* St. Martin's Press, 2010; and Wrangham, Richard. *Catching Fire: How Cooking Made Us Human.* Basic Books, 2009.

32 The alternative explanation is that brain size develops not as a result of genetic programming, but due to ontogenetic influences that occurs during the secondary altriciality: the brain grows and adapts to its use as the person grows; use it less and it will grow less. This hypothesis could be invalidated through any number of approaches, including a review of the brain size of those with severe mental impairment, or comparing those with little or no stimulation to those with average or excessive stimulation as they developed.

change in humans in 40,000 or 50,000 years. Everything we call culture and civilization we've built with the same body and brain."[33]

Cochran and Harpending make the opposite argument; they propose that agriculture began a chain of events that caused an enormous increase in the rate of human evolution. These adaptations, including physical and biological features such as skin color, disease resistance, reduced strength, and lactose intolerance, were not universal, largely because the advent of agriculture (including animal domestication) was not universal.[34] Cochran and Harpending agree that as we domesticated our crops and animals, we also domesticated ourselves, with evolutionary adaptations with specific consequences for how we think. Humans have become more docile, less aggressive, and have seen increased abilities for enhanced long-term planning – features that formed the scaffolding for growing collaborative trade, governments, and urbanization.

Scholars like Coolidge and Wynn tend to agree with Cochran and Harpending that even after we were anatomically modern, we likely experienced genetic mutations that evolved and enhanced our working memory and executive functions, setting the stage or at the very least influencing the human cognitive revolution.[35]

Bednarik's claim is controversial: it is not commonly agreed that human brains have been shrinking, while others might highlight the cultural evolution of thc last 10,000, downplaying the effects of human evolution on such a relatively short time frame.[36]

To understand how quickly genetic evolution can happen, consider Peter and Rosemary Grant's work on the Galapagos finches inspired Darwin's gro-

---

33 Gould Stephen J. "The Spice of Life." *Leader to Leader*, vol. 15, 2000, pp. 14–19.

34 Critics offer alternate explanations that do not require "evolution" per se. Lactose intolerance for example is a single nucleotide polymorphism (SNP), which is a difference in a single DNA building block. These are a naturally occurring and very common type of genetic variation among people. Rather than evolving some complex adaptation, it rewards those who have a single letter transcription. Likewise, disease resistance can vary with changing gut biota, and other factors can be explained simply by using the body and muscles differently.

35 Wynn, Thomas, and Frederick L. Coolidge. "Implications of A Strict Standard For Recognizing Modern Cognition in Prehistory." *Cognitive Archaeology And Human Evolution*. 2009, pp. 117–127.

36 See, for example: Diamond, Jared. *Guns, Germs, and Steel: The Fates of Human Societies*. W. W. Norton, 1997, in which he says, "*The assumption that more recent expansions are ALL driven by cultural factors is based on the notion that modern humans everywhere have essentially the same abilities.*" (Emphasis added by me; the implication is that NO genetic effects account for change in humans during the Holocene).

undbreaking idea of evolution. As seminal as *Origin of Species* was, it didn't actually describe the origin of species. The Grants set out on what they thought would be a two-year study of the finches, and their next 40+ year study has provided key insights into how evolution works. They found that natural selection is not a linear, stable, and incremental process that occurs over long stretches of time. Instead, it is a variable and constantly changing process, which can actually be very fast, occurring in less than two years. The direction of the adaptation can go in either direction. In some years, a population of finches might favor those with larger beaks. But as the environment changes, perhaps with less rain in some years than others, causing the seeds to be smaller, the selection pressures change, and birds with smaller beaks prevail.

These last few paragraphs are not intended to make the claim that humans are experiencing rapid and ongoing evolutionary changes to their brains. There are enough other provocative and likely controversial claims I intentionally make in this book. Here I am following Riemer and Johnston's advice to "scout the fringes" of the current paradigm of thought. You can't use the Grants' work as proof of Bednarik's claim. Among other factors, finches have much shorter lifespans than humans and thus would evolve faster. But what I'm looking for here is convergence and perspective on how the evolution of technology affects how we think. And Bednarik's claims appear worth considering.

Others look at different facets of the equation. Returning to Tomasello, he notes that the shared intentionality and cultural roots of cognition build up in part via external symbols, which leads to our ability to bootstrap knowledge over time, which he calls the "ratchet effect."[37] Tomasello notes, like Cochran and Harpending, that this is made possible through uniquely human traits, such as the capacities for sharing attention with other persons, understanding others intentions, and imitation (not just copying what others do, but also copying what you perceive others intend to do); features which emerge early in human ontogeny, and are nurtured through our unusually long maturation and development. Tomasello does not link the emergence of these unique traits to recent evolution (and in fact much of his case is based on a long line of observations of human evolution and primate culture), but neither is his point completely inconsistent with the constantly evolving human as described by Cochran & Harpending.

37 Tomasello, Michael. *The Cultural Origins of Human Cognition*. Harvard University Press, 2009; and Tennie, Claudio, et al. "Ratcheting Up the Ratchet: On the Evolution of Cumulative Culture." *Philosophical Transactions of the Royal Society B: Biological Sciences*, vol. 364, no. 1528, 2009, pp. 2405–2415.

Sterelny makes the point that the capacity to learn is the main enabler of evolutionary change,[38] while Whiten and others combine the points to suggest that social learning is the key foundation for culture.[39]

2 The Extended Mind Thesis

Philosophers Andy Clark and David Chalmers conceived of this cultural ratcheting-up process as an explicitly cognitive approach in 1998, partly in response to ongoing philosophical debates about externalization and situated cognition.[40] They started with an example: a person could look at a diagram of Tetris-like shapes on a screen and rotate them in their mind to see if they fit together. Now, imagine that they could use a mouse or even a brain-computer interface to rotate them, presumably faster. Did this constitute cognition?

Yes, according to Clark and Chalmers, because, as they say, *"a part of the world functions as a process which, were it done in the head, we would have no hesitation in recognizing as part of the cognitive process, then that part of the world is (so we claim) part of the cognitive process."*[41]

This has come to be known as the Parity Principle.[42] This is a form of active externalization, where the brain and physical material are coupled:

> the human organism is linked with an external entity in a two-way interaction, creating a coupled system that can be seen as a cognitive system in its own right. All the components in the system play an active causal role, and they jointly govern behaviour in the same sort of way that cognition usually does. If we remove the external component the system's behavioural competence will drop, just as it would if we removed part of its brain. Our thesis is that this sort of coupled process counts equally well as a cognitive process, whether or not it is wholly in the head. (Menary, 2006)

38 Sterelny, Kim. "From Hominins to Humans: How Sapiens Became Behaviourally Modern." *Philosophical Transactions of the Royal Society B: Biological Sciences*, vol. 366, no. 1566, 2011, pp. 809–822.

39 Whiten, Andrew, et al. "Culture Evolves." *Philosophical Transactions of the Royal Society B: Biological Sciences*, vol. 366.1567, 2011, pp. 938–948.

40 Robbins, Philip, and Murat Aydede, editors. *The Cambridge Handbook of Situated Cognition*. Cambridge University Press, 2009.

41 Clark, Andy, and David J. Chalmers. "The Extended Mind." *Analysis*, vol. 58, no. 1, 1998, p. 8.

42 Menary, Richard. "Attacking the Bounds of Cognition." *Philosophical Psychology*, vol. 19, no. 3, 2006, pp. 329–344.

Now let's think back to the beginning of the book, where I said that technology has changed our lives so much that we no longer need to remember our children's phone numbers. I asked one of my co-authors, the primate neuroscientist, if he knew his wife's phone number – yes, he did. Like me, he has two teen daughters – does he know their phone number? No! Does this constitute cognition – that we have stored long-term memory in our phones, and not in our heads? Yes, according to both of us.

Clark and Chalmers, as philosophers, are making the exact same case: cognitive processes remain cognitive regardless of their physical location. The critical test is whether the process would qualify as cognition if it occurred in our heads – external storage doesn't disqualify it from being genuine thinking.

This coupling of using external things as part of our thought process entails Epistemic actions (as distinct from Mithen and others use of "epistemic artifacts"), which *"alter the world so as to aid and augment cognitive processes such as recognition and search."*[43,44]

You perform an epistemic action when you consciously act to create a shortcut icon on your phone's home screen (or perhaps configure your voice assistant) to dial the phone numbers of your children, whose numbers you

---

43 Clark, Andy, and David J. Chalmers. "The Extended Mind." *Analysis*, vol. 58, no. 1, 1998, p. 8.

44 Returning to our computer analogy, when you log on and start browsing the web or using cloud-based AI, your thinking no longer depends solely on your local machine. Instead, you're drawing on vast networks of servers, databases, and processing power scattered across the globe. This represents an extraordinary scaling of human cognition.

Writing offered a similar expansion – and its impact endures today in these very words, whether you're reading them in a centuries-old paper format, on a digital screen, or hearing them as synthesized speech. But the scope of what you can now access as extensions of your mind defies comprehension. We can instantly tap into humanity's accumulated knowledge, real-time global information, and artificial intelligence systems that can process and synthesize information at superhuman speeds. The expansion has become so vast and rapid that we've lost our ability to grasp its magnitude – like trying to perceive the universe's expansion while standing on Earth.

This framework has powerful implications for the questions I've been wrestling with throughout this book. In Part 3 we'll explore what this multi-dimensional model reveals about human cognition.

With this third stream of thinking – our extended mind reaching into technology – the pieces finally began clicking into place. The insights came first in flashes, brief moments of clarity, then gradually solidified into more stable and useful understanding. Years of research were finally paying dividends.

I had to work through all the perspectives we've covered – neurobiology, culture, and now technology – plus countless others that didn't make it into these pages. Without seeing all these individual elements, the larger pattern would have remained invisible. Sometimes you need to step back far enough to see how the dots connect.

haven't memorized because you know that it is in your phone and saved to your cloud contact list. Clark distinguishes these from pragmatic actions, which bring one closer to a goal (like actually dialing your child's phone number or clicking on that shortcut to dial them), but epistemic actions might bring one closer to the goal but are intended primarily to extract or uncover information.

Clark and Chalmers feel this cognitive extension occurs in both cognitive *processes*, such as remembering, as well as in cognitive *states*, such as memories. Clark and Chalmers feel it is not just information or thought processes that have been encoded externally, but part of the *mind*, since the external information contains beliefs, and even components of *self*, as one's written thoughts, mementos, notebooks, and other objects are a central part of your identity as a cognitive agent. Again, we will unpack and explore the implications of this for modern life in the next sections of this book, but here we are establishing the philosophical and academic foundations that will be useful to us later.

Clark and Chalmers are building on earlier work in areas such as how pencil and paper aid in long multiplication,[45] physical re-arrangements of letter tiles enhance cognitive performance in Scrabble,[46] and the use of a slide rule to aid in navigation.[47]

Clark, writing in 2003, explicitly acknowledged the importance of the changes brought about by this cognitive evolution, in terms of meta-representational skills developed to take advantage of external storage. In this, he echoes Wegner's transactive memory retrieval skills, and Donald's comments about learning to remember and model where information is stored externally, that the brain develops (biologically via cultural mechanisms enabled by technological innovations), to prefer "*meta-knowledge* over *baseline knowledge*."[48] Meta-knowledge is "*knowledge about how to acquire and exploit information, rather than basic knowledge about the world. It is not knowing so much as knowing how to find out*."[49] It's likely that to acquire meta-knowledge skills, one would need to be developing a suite of epistemic expertise. Empirical

45 McClelland, James L., et al. "The Appeal of Parallel Distributed Processing." *Parallel Distributed Processing, Volume 2*, edited by James L. McClelland and David E. Rumelhart, MIT Press, 1986; and Clark, Andy. *Microcognition*. MIT Press, 1989.

46 Kirsh, David. "The Intelligent Use of Space." *Artificial Intelligence*, vol. 73, 1995, pp. 31–68.

47 Hutchins, Edwin. *Cognition in the Wild*. MIT Press, 1995.

48 Clark, Andy. *Natural-Born Cyborgs: Minds, Technologies, and the Future of Human Intelligence*. Oxford University Press, 2003, p. 67.

49 Clark, Andy. *Natural-Born Cyborgs: Minds, Technologies, and the Future of Human Intelligence*. Oxford University Press, 2003, p. 67.

research supports this philosophical hypothesis – we use less of our brain to store information, but more of our brain to unpack where the information can be found or how to retrieve it.[50]

Think of it as the difference between knowing an idea and knowing how to obtain that idea.

While Clark doesn't explicitly support meta-knowledge as "top-down," and baseline as "bottom-up,"[51] it seems fair to point out that the parallels here reflect on how Hutchins describes how we navigate a complex physical world of information,[52] and how we use different approaches to try to make sense of social situations and information.[53]

Clark points out that in the real world, we often blur the distinction between the two types of knowledge: if someone asks if you know the time, you say yes, then look at your watch or phone.

You don't actually know what time it is, but you know where to find it, and you think that's the same thing.[54] Except now we are not just looking at our watch, we are leveraging a global network of material things when we ask our watch to map out the best route to a new destination (leveraging real-time networks of traffic speeds and remote satellites tracking precise locations), or prompt-engineer some AI to answer a complex question.

The result is a "large jump" in human cognitive evolution that is described by Clark as he says, *"the distinctive way human brains repeatedly create and exploit various species of cognitive technology so as to expand and re-shape the space of human reason. We – more than any other creature on the planet – deploy nonbiological elements (instruments, media, notations) to complement our basic biological modes of processing, creating extended cognitive systems whose computational and problem solving profiles are quite different from those of the naked brain."*[55]

---

50 Sparrow, Betsy, et al. "Google Effects on Memory: Cognitive Consequences of Having Information at Our Fingertips." *Science*, vol. 333, no. 6043, Jun. 14, 2011, pp. 776–778.

51 Sarter, Martin, et al. "The Cognitive Neuroscience of Sustained Attention: Where Top-Down Meets Bottom-Up." *Brain Research Reviews*, vol. 35, no. 2, 2001, pp. 146–160.

52 I speculate that spatial and temporal memory circuitry that evolved over millions of years to support omnivorous seasonal foraging may undergird the circuitry we exapted to perform this meta-knowledge.

53 Sarter, Martin, et al. "The Cognitive Neuroscience of Sustained Attention: Where Top-Down Meets Bottom-Up." *Brain Research Reviews*, vol. 35, no. 2, 2001, pp. 146–160.

54 The same could be said of the phone numbers in your phone, as in the earlier example of one of the authors not knowing his children's phone numbers. He thinks he knows them, but really, he just knows where to find them.

55 Clark, Andy. "Natural Born Cyborgs?" Edge.org, 28 Dec. 2000, https://www.edge.org/conversation/andy_clark-natural-born-cyborgs. Accessed 19 Mar. 2026.

He gives mathematics as an example. Our unique and distinctive modern math processing capabilities "*depends on a complex web of biological, cultural and technological contributions.*"

Others have waded into this debate. For example, Robert Rupert distinguishes between the Hypothesis of Extended Cognition – in which "*cognitive processes literally extend into the environment surrounding the organism*" and the Hypothesis of Embedded Cognition – where cognitive processes "*depend very heavily, in hitherto unexplained ways, on organismatically external props and devices.*"[56,57] The former extends thought into the environment, while the latter represents a further degree of the idea, with thought being highly contingent on the environment.

Scholars like Menary, Hurley, Sutton,[58] Ross, and Ladyman have written papers that develop clarifications, minor extensions, or defenses of Clark and Chalmers' Extended Mind Thesis (EMT) and are countered by critics like John Preston and, maybe most notably, Adams and Aizawa.

3 Responses and Refinements

Dennett broadly endorses extended cognition, but his emphasis falls less on real-time parity-based coupling (à la Clark & Chalmers) and more on developmental installation of "thinking tools": culturally designed symbols, practices, and institutions that, over learning and enculturation, reconfigure our internal competences. On this view, external inscriptions and artifacts matter because they engineer the learner's environment – they scaffold habits of attention, representation, and reasoning that become part of what the agent can do *without* constant tight coupling to the artifact. So Dennett agrees that cognition reaches outside the skin, but he frames the extension diachronically (how culture builds minds over time) rather than primarily synchronically (how an external element currently completes a process).[59]

---

56 Rupert, Robert D. "Challenges to the Hypothesis of Extended Cognition." *The Journal of Philosophy*, vol. 101, no. 8, 2004, pp. 389–428.

57 This spectrum resembles debates that occur at the intersection of linguistics and cognition: examining if language can shape or influence the way its speakers perceive and think about the world. This has become known as the Sapir-Whorf hypothesis, and it comes in a weak version (language influences thought but does not entirely determine it) and a strong version (language entirely determines thought).

58 Sutton, John. "Remembering." *Cambridge Handbook of Situated Cognition*, edited by Philip Robbins and Murat Aydede, Cambridge University Press, 2009, pp. 217–235.

59 Dennett, Daniel C. *Kinds of Minds*. Basic Books, 1996; and Dennett, Daniel C. "Learning and Labeling." *Mind and Language*, vol. 8, no. 4, 1993, pp. 540–547.

Sterelny, too, agrees on the importance of external devices, but believes there is more heuristic value in considering our use of tools as a "*special cases of niche construction,*[60] *cases in which human competences depend intimately on the environment being scaffolded to support adaptive decision making.*"[61] Sterelny also says that the increased cognitive demands of remembering where data is can be significant, and suggests decoupling to reduce this load.

Rowlands extends Donald's work to claim that most of human memory resides in our environment, and not our minds, and that our cognitive abilities rely on our ability to connect with a wide range of collective memory networks.[62] Rowlands, like Clark, uses notebooks as a classic example of this, but goes on to other use cases, such as examining how martial arts and dance reflect embodied skills and kinesthetic awareness. He also points to cultural practices, such as religious ceremonies or sports events that embody experiences and collective participation, and at navigation as an extended cognitive process, integrated with movements, orienting to landmarks and other external environmental factors.

In Rowlands' view, "*external operations take the form of action: manipulation, exploitation and transformation of environmental structures – ones that carry information relevant to the accomplishing of a given task.*" And "*At least*

60 This term, "niche construction," or as I've previously referred to as in terms like our "self-constructed niche" is commonly used in biology to apply across species, and is not specific to humans. Humans of course do it at a prodigious scale in today's world; see for example, Tim Taylor's The Artificial Ape, in which he explains how human evolution was shaped not just by biology but by our reliance on technology, especially tools, which became an extension of our bodies and fundamentally redefined what it means to be human. However, when viewed through a deep and long view, we see we haven't had as profound an effect on our world compared to other species. For example Howard Bloom, in his provocative *The Case of the Sexual Cosmos*, outlines how the first life on Earth, simple single cell organisms, recruited non-living resources eventually at such scale that they changed the climate from a "poisonous rock" into the habitable world we live in today. Similarly, he points to the lowly earthworm, which chows through dirt, and literally, in his words, "shit earth," which the rest of us depend for plants which provide us with the oxygen in the air we depend on.

61 Sterelny, Kim. "Minds: Extended or Scaffolded?" *Phenomenology and the Cognitive Sciences*, vol. 9.4, 2010, pp. 465–481. He refers also to Laland, Kevin N., et al. "Niche Construction, Biological Evolution and Cultural Change." *The Behavioral and Brain Sciences*, vol. 23, 2000, pp. 131–175; and Odling-Smee, F. John, et al. "Niche construction: the neglected process in evolution. 37." *Princeton University Press Olson M (1992) Intuition and the logistic equation. Trends Ecol Evol*, vol. 7, 2003, p. 314).

62 Rowlands, Mark. *The Body in Mind: Understanding Cognitive Processes*. Cambridge University Press, 1999, pp. 119–147.

some of the internal processes are ones concerned with supplying the subject with the ability to appropriately use relevant structures in its environment."

In developing a view of extended cognition, Clark and Chalmers consider social extensions, too, but they seem to consider it rare, perhaps in an "*unusually interdependent couple,*" where there exists a "*high degree of trust, reliance, and accessibility.*"[63] Compare this position with that of scholars like Lund et al., who might claim that all thought is social,[64] and we can see the need for an integrated model that examines thought in all of its planes.

Edwin Hutchins, writing in 1995, before Clark and Chalmers, embraced the idea that technology spans all three domains – individual, social,[65] and material/technological, when he set out to study the workplace. Looking at the complex processes used in naval navigation, for example, Hutchins observed the development, use, and transformation of representational models across individuals, groups, and their respective media, technologies, and artifacts.

There is no individual expert in these contexts: no one person knows all, controls the intelligent process, or has a blueprint or plan for every part. There is knowledge, but it is built into the entire process, and it is an emergent property of the whole, which needs to be the fundamental unit of analysis, according to Hutchins, if one wants to understand how thinking occurs. In these settings, cognition occurs via "*a collection of individuals and artifacts and their relations to each other in a particular work practice.*"[66]

He builds on and extends the ideas of cultural transmission (e.g., as articulated by Richerson and Boyd, Wegner, and alluded to by Ridley) to discuss how organizations learn and how that process is an "*engine of cultural change.*"

While Hutchins set out to describe "cognition in the wild" – outside laboratory settings that study isolated thought processes – his scope remains quite narrow. He describes his work as "cognitive ethnography,"[67] which focuses specifically on task-related cognitive work within discrete occupational settings rather than examining human cognition comprehensively in natural environments. His studies analyze the thinking required to perform specific

---

63 Clark, Andy, and David Chalmers. 1998. *"The Extended Mind."* Analysis 58 vol. 1, pp. 7–19.

64 Lund, Kristine. "Modelling the Co-elaboration of Knowledge: Connecting Cognitive, Linguistic, Social and Interactional Systems." *Language Is a Complex Adaptive System: Explorations and Evidence*, edited by Kristine Lund et al., Language Science Press, 2022, pp. 35–48.

65 Hutchins, Edwin. *Culture and Inference: A Trobriand Case Study.* Harvard University Press, 1980.

66 Hutchins, Edwin. *Cognition in the Wild.* MIT Press, 1995, p. xvii.

67 Another cognitive sciences niche sub-specialty.

jobs: ship navigation, airline piloting, air traffic control,[68] software development.[69]

This approach deliberately excludes the broader cognitive landscape. The wandering thoughts of a sailor during watch, a pilot's reflections on family between radio calls, or how these professionals think when off duty – all fall outside Hutchins' framework. His "wild" is actually quite domesticated: it's cognition within structured work environments, bounded by professional tasks and organizational goals.

Others, writing after Clark and Chalmers, extend the debate in provocative ways. Malafouris, for example, agrees and insists on an integrated view of the material, social, and cognitive dimensions, but, in an almost Nutcracker way, he breathes cognition and almost life into material things more so than previous authors. He leaves behind EMT and builds on top of it his Material Engagement Theory (MET).[70] Malafouris's complaint with Donald is that he didn't go far enough.

Material things have "*a dynamic cognitive biography,*" not the passive role of external storage that Donald described, according to Malafouris. Within MET, we "*think through things, in action.*"[71] In this view, the highly plastic human mind is an unfinished, constantly evolving process, subject to "*drastic, deep reorganization by incorporating new technological innovations and prostheses and thus, potentially, in a permanent state of ongoing creative evolution.*"

This occurs in a two-way dynamic between us and our rapidly changing self-constructed niche, in which "*conventional meaning of adaptation gives way to a more extensive and bidirectional view of plasticity – we may call it metaplasticity.*"[72] Though the brain is an important part of the process, Mala-

68 Hutchins, Edwin. *"How a Cockpit Remembers Its Speeds."* Cognitive Science, vol. 19, no. 3, 1995, pp. 265–288.

69 Flor, Nick V., and Edwin Hutchins. "Analyzing Distributed Cognition in Software Teams: A Case Study of Team Programming during Perfective Software Maintenance.*" Proceedings of the Fourth Annual Workshop on Empirical Studies of Programmers,* edited by Jurgen Koenemann-Belliveau, Thomas G. Moher, and Scott P. Robertson, 1991, pp. 36–59.

70 Malafouris, Lambros. "The Cognitive Basis of Material Engagement: Where Brain, Body and Culture Conflate." *Rethinking Materiality: The Engagement of Mind with the Material World,* edited by Elizabeth DeMarrais, Colin Gosden, and Chris Renfrew, McDonald Institute for Archaeological Research, 2004, pp. 53–62.

71 Malafouris, Lambros. "The Cognitive Basis of Material Engagement: Where Brain, Body and Culture Conflate." *Rethinking Materiality,* edited by Elizabeth DeMarrais, Colin Gosden, and Chris Renfrew, McDonald Institute for Archaeological Research, 2004, pp. 53–62. The quoted passages are drawn from Malafouris, Lambros. *How Things Shape the Mind: A Theory of Material Engagement.* MIT Press, 2013.

72 Malafouris, Lambros. "Thinking as 'Thinging': Psychology with Things." *Current Directions in Psychological Science,* vol. 29, no. 1, 2020, pp. 3–8.

fouris develops the strongest argument that things around us have a life, history, their own agency, and ecology of thought. To understand how a thing affects human thought, we need not only to examine it at the level of the individual's interaction with the thing, but at the level of that thing's role in broader *"cultural practices and material engagement."*

More recently, the extended mind approach has been applied specifically to our modern technologies, such as Paul R. Smart's "Web-extended mind hypothesis," wherein *"the technological and informational elements of the Web can (at least sometimes) serve as part of the mechanistic substrate that realizes human mental states and processes."*[73]

The extended mind (as set forth variously by Donald, Hutchins, Mithen, and Dennett, popularized by the 1998 Clark and Chalmers paper, and then further developed by later EMT scholars)[74] has faced criticism for several reasons:

- external cognitive processes work fundamentally differently than internal cognitive processes (you can't compare apples and oranges)[75]
- the model could extend the mind too far into the world[76]
- the core role of the brain will be lost in this view.[77]

But it has generally stood the test of time and been modestly cited,[78] predominantly by those debating the philosophical parts of the phenomenology of the

73 Smart, Paul R. "The Web-Extended Mind." *Metaphilosophy*, vol. 43, no. 4, 2012, pp. 426–445; and Smart, Paul R. "Human-Extended Machine Cognition." *Cognitive Systems Research*, vol. 49, 2018, pp. 9–23; and Halpin, Harry. "Does the Web Extend the Mind?" *Proceedings of the 5th Annual ACM Web Science Conference (WebSci '13)*, ACM, 2013, pp. 139–147.

74 Such as Robert Wilson (Wilson, Robert A. *Boundaries of the Mind: The Individual in the Fragile Sciences.* Cambridge University Press, 2004), Richard Menary (Menary, Richard. *Cognitive Integration: Mind and Cognition Unbounded.* Palgrave Macmillan, 2007), and John Sutton (Sutton, John. "Exograms and Interdisciplinarity: History, the Extended Mind, and the Civilizing Process." *The Extended Mind*, edited by Richard Menary, MIT Press, 2010, pp. 189–225).

75 Adams, Frederick, and Kenneth Aizawa. "The Coupling-Constitution Fallacy." *The Bounds of Cognition*, Wiley-Blackwell, 2008, pp. 88–105.

76 Sprevak, Mark. "Extended Cognition and Functionalism." *Journal of Philosophy*, vol. 106, no. 9, 2009, pp. 503–527.

77 See Adams, Frederick, and Kenneth Aizawa. *The Bounds of Cognition.* 2nd ed., Blackwell, 2010., and Chalmers, David. "Foreword to Andy Clark's *Supersizing the Mind.*" *Supersizing the Mind: Embodiment, Action, and Cognitive Extension*, Oxford University Press, 2008.

78 For a good survey of the range of developments, criticisms and defenses of the extended cognition, see Menary, Richard, editor. *The Extended Mind.* MIT Press, 2010.

mind[79] and by some psychologists.[80] Interestingly, extended cognition has been of growing relevance in the technology fields of both human-computer-interfaces (HCI)[81] and artificial intelligence (AI).[82]

4 Extending the Extended Mind

How far can we take the extended mind? Clark, himself, admits, "*Just how far we should then press this notion of cognitive extension, however, remains unclear.*"[83] My notebook may be an extension of my thoughts, mind, and self, but does the phone book count?

This common worry – often called the "cognitive bloat" objection – is that the extended-mind thesis risks letting too much of the world count as part of a person's mind (the objection is most associated with Adams and Aizawa). Rowlands doesn't originate that label; he responds to it by arguing that what matters is ownership/integration: a process or resource counts as mine when it is functionally integrated into my cognitive economy (reliably accessible, relied upon, and endorsed), not merely nearby.

He adds a functional criterion – an external element must make available information that was otherwise unavailable and be integrated in a way that changes my cognitive competence. To illustrate, he uses a digestion analogy: the fact that some digestive operations can be located outside the body for a time doesn't make them any less mine if they still perform my digestive function; likewise, external operations can be cognitive if they perform my cognitive function.

---

79 Thompson, Evan. *Mind in Life: Biology, Phenomenology, and the Sciences of Mind.* Harvard University Press, 2010; and Gallagher, Shaun, and Dan Zahavi. *The Phenomenological Mind.* 2nd ed., Routledge, 2020.

80 Shapiro, Lawrence. *Embodied Cognition.* 2nd ed., Routledge, 2019.

81 Agre, Philip E., and David Chapman. "What Are Plans For?" *AI Memo 1050a,* Artificial Intelligence Laboratory, MIT, 1989; Clancey, William J. "Becoming a Rover." *Simulation and Its Discontents,* edited by Sherry Turkle, Cambridge University Press, 2009, pp. 107–127; and Suchman, Lucy A. *Plans and Situated Actions: The Problem of Human-Machine Communication.* Cambridge UP, 1987, among others.

82 Smart's theory of Human Extended Machine Cognition (HEMC) in Smart, Paul R. "Human-Extended Machine Cognition." *Cognitive Systems Research,* vol. 49, 2018, pp. 9–23; see also Lindblom, Jessica, and Tom Ziemke. "Social Situatedness of Natural and Artificial Intelligence: Vygotsky and Beyond." *Adaptive Behavior,* vol. 11, no. 2, 2003, pp. 79–96.

83 Clark, Andy, and David Chalmers. 1998. "The Extended Mind." *Analysis* 58 vol 1; 7–19, p. 11.

On Rowlands' view, the key is integration into the subject's psychology (practically and, ultimately, phenomenologically), which draws a principled line between genuine extensions and indiscriminate "bloat." He draws the line at ownership: "*The entries in the telephone directory, and the pages on the internet, are not, at least as they figure in the cognitive bloat objection, owned by anyone. And anything that is to count as cognitive, I shall argue, must be owned by someone or some thing.*"[84] But then he goes on, helpfully, to describe some other dynamics about what counts and doesn't. The functioning of a material cognitive node is to "*make available information that was previously unavailable.*" A thing may never fulfill this role, or may perform multiple roles, or do this only in certain contexts, according to Rowlands. He develops a comparison with another biological example – digestion. If one could externalize your food's digestion, he notes "*The processes do not stop being mine just because they are, for a time, located outside my body.*"

The key here is the function: "*a digestive process is mine if it breaks down food I have ingested and releases energy into my body.*"[85] Ownership here means integration into one's life, particularly one's psychology. Rowlands wrestles with the questions of ownership that affect subconscious processes, before tentatively concluding that "*integration is ultimately integration into the conscious life of a subject.*"

Clark seems to have a much broader view, and to explain it, he uses the biomechanical efficiency of a bluefin tuna. According to studies performed by MIT bioengineers, the tuna is too weak to swim as fast as it can by a magnitude of about 7. But the reason it can accelerate and swim as fast as it does is that "*these fish actively create and exploit additional sources of propulsion and control in their watery environments. For example, the tuna use naturally occurring eddies and vortices to gain speed …The real swimming machine, I suggest, is thus the fish in its proper context: the fish plus the surrounding structures and vortices that it actively creates and then maximally exploits.*"[86] Here we see not so much "ownership," as much as use and or control.

Writing in 2010, Clark muses again on how far you can extend your mind. He notes the 1998 articulation only applied to cognitive processes and beliefs,

84 Rowlands, Mark. *The Body in Mind: Understanding Cognitive Processes*. Cambridge University Press, 1999, pp. 119–147.

85 The authors note this is a not too hypothetical example: Richard Wrangham makes the case that the technical innovation of fire and cooking performs the function of predigestion of food, prior to ingestion: Wrangham, Richard. Catching *Fire: How Cooking Made Us Human*. Basic Books, 2009.

86 Clark, Andy. *Being There: Putting Brain, Body, and World Together Again*. MIT Press, 1996, p. 219.

and then asks, *"What about extended desires, extended reasoning, extended perception, extended imagination, and extended emotions? I think that there is something to be said for each of these. Perhaps the camera on my iPhone can serve as an extended perceptual mechanism. And perhaps one might have something akin to an extended mood, if not an extended emotion, when one's environment is always nudging one toward happiness or sadness."*[87]

So, for now, we leave these issues here. How far a mind and thought can be extended remains unclear, and up for debate, particularly when that thing we use to extend our thinking is in the public domain, and is, as Sterelny suggests, open and contested by others' interactions.[88]

Be assured, though, I will pick this theme back up later in the book.

5 Exocortex

What we casually call "the cloud" is a socio-technical system: data centers, protocols, platforms, capital, standards, and governance (from app stores to regulators) that couple with our devices and practices; these architectures shape what feels cognitively "at hand" as much as any neural limit – exactly the kind of regime/transition dynamics Geels analyzes.

The question of what we own, control, use, or exploit, and what is a material "thing" becomes even murkier when we look at modern technology. The simple act of finding a nearby restaurant with your phone requires engagements with satellites to fix your location, cloud-based algorithms performing massive calculations to interpret the question you put to it, machine learning processes, and the retrieval of various bits of electronic data from literally all over the country or world, to assemble to you exograms rendered on your phone, which you decode to be a well-liked lunch spot around the corner.

Sure, the phone may count as a cognitive device here, because we can touch it, but do the algorithms, software systems, AI, ML, and data count? Are they physical, and/or part of material cognition?

In 2000, software developer Ben Houston coined the term "exocortex" to describe how we utilize technology to enhance our thinking, and I believe this

87 Chalmers, David J. Foreword. *Supersizing the Mind: Embodiment, Action, and Cognitive Extension*, by Andy Clark, Oxford University Press, 2008, pp. ix–xvi.

88 Sterelny, Kim. "Minds: Extended or Scaffolded?" *Phenomenology and the Cognitive Sciences*, vol. 9.4, 2010, pp. 465–481. He refers also to Laland, Kevin N., et al. "Niche Construction, Biological Evolution and Cultural Change." *The Behavioral and Brain Sciences*, vol. 23, 2000, pp. 131–175

phrase and construct are apt.[89] (Though the metaphor of a cortex is not enough, since as we've shown in early chapters that thinking, even within our own body, is FAR more holistic than just the conscious thoughts being centered in our cortex.)

While much of our technology is increasingly virtual and seems to be immaterial, it physically exists in the world, just perhaps not proximate to you. You might not touch that server rack in the Google datacenter or hold the fiber that transmits part of the process between your phone and the cloud, but these exist and are real.[89] And it is an increasingly important mode of modern human thinking that manifests physical changes in, among other places, our own embodied brains.

In a series of experiments published in Science in 2011, Sparrow, Liu, and Wegner conclude: *"when people expect to have future access to information, they have lower rates of recall of the information itself and enhanced recall instead for where to access it. The Internet has become a primary form of external or transactive memory, where information is stored collectively outside ourselves."*[91]

These are the eddies and vortexes we use to propel ourselves in our information environments, and because we use them to influence our decisions, even though they are not owned by us, and exist in an open and potentially contested state, we use them to shape our representations of the world, and as such constitutes modes of our thought.

6 My View of Thinking with Technology (TT)

I accept Clark and Chalmers' view of the extended mind, but I place more emphasis on rooting cognition in both the wet, embodied brain than they do (likely because they felt that branch was more fully articulated elsewhere), and certainly more emphasis on cultural cognition than Clark and Chalmers.

I support Rupert's Hypothesis of Extended Cognition over the more extreme Hypothesis of Embedded Cognition. Hutchins gets it right, on balance, but only regarding a specific setting, task-orientation, and sub-culture

89 Houston, Ben. *Exocortex Definition*. May 2000. www.exocortex.org/ben/exocortex. Accessed Aug. 24, 2025.

89 Trust me – I live in Northern Virginia, where we have over 30 million square feet of hulking data centers.

91 Sparrow, Betsy, Jenny Liu, and Daniel M. Wegner. "Google Effects on Memory: Cognitive Consequences of Having Information at Our Fingertips." *Science*, vol. 333, no. 6040, 2011, pp. 776–778.

constrained by task, geography, organization, and time frame. He too narrowly frames the context for thinking as it applies to a single problem or task, when, as we'll outline later, our thinking "in the real wild" engages many tasks/values concurrently. The prioritization of those tasks (and related thinking) changes greatly over time and in different environments.

Malafouris develops interesting ideas of material agency, but we need not ascribe a life of their own to the material things around us; we gain heuristic value from considering them as components of our cognition. His metaplasticity, on the other hand, hits the nail on the head in terms of describing the highly dynamic, ever-evolving, and even chaotic interplay between our thinking and our modern environments.

I have a generous view of how far one can extend the mind, and while I agree with Rowlands' point about ownership and his slightly broader integration, I think something might count as a material node of your thought even if it is just integrated into your unconsciousness. And Clark is at the threshold of the door to the future of thought, when he muses that external cognition is not just about extended cognitive processes and beliefs, but might also include, with new technologies, extended wants, reasoning, perception, imagination, and emotions. I'll illustrate how that happens later in this book.

While the scholars above generally accept *material* externalization of cognition, I explicitly embrace the encoding of externalization of material cognition as cognition to abstract (i.e., not obviously physically manifested) cloud-based digital technologies, such as AI, algorithms, software, and data. These are ultimately physically stored and/or processed on physical substrate outside of human bodies.[92] Moreover, this area will become an increasingly important area of focus in the future ways humans will think soon, which we'll discuss later. I feel Houston's "exocortex" is a useful shorthand analogy to use to refer to the material nodes of thought represented by the digital technologies that we use, not just to scaffold our thoughts, but as integral parts of our modern cognitive processing, but the analogy isn't perfect.

7 Plain Language Summary of Thinking with Technology

We literally think *with and through* the things around us – and what counts as "around us" has been expanding so rapidly that it's reshaping how we think in profound ways. These "things" don't always feel physical – like a website or a

92 And sometimes embedded within – e.g., cochlear implants.

text on our phones – but they're all part of the extended world we recruit to think for us.

When humans learned to write, we didn't have to memorize everything anymore – but we did have to learn how to read. With new technologies, we have to learn how to navigate digital spaces. Take a self-driving car, for example. It performs mental work your brain used to do: paying attention, navigating, reacting. In doing so, it becomes part of your thinking process. But when technology thinks *for* us, it often demands new skills in return. We don't have to constantly focus on driving – but we *do* need to understand how to operate and maintain the self-driving car.

This ability to offload cognitive tasks onto the world around us – this way of thinking *through* technology – has been a cornerstone of human progress. By freeing our minds from remembering every detail, we unlocked the capacity for higher-order thinking: science, law, philosophy, and more.

In essence, thinking with technology isn't just a modern convenience, it's a revolutionary force that has driven, and continues to drive, the transformation of human thought.

CHAPTER 8

Moderation by Experience

> Your experiences today will influence the molecular composition of your body for the next two to three months, or, perhaps, for the rest of your life. Plan your day accordingly.[1] (Steven Cole)

Your brain is different now because you've read this sentence.

Literally.

Which brings us to the final part of the model that needs to be addressed. It is not one of our three Streams (Thinking with Brains; Thinking with Culture; or Thinking with Technology).

Because it undergirds and affects each stream, we can consider *experience* to be the terrain shaping our streams.

It has been well established that experience changes the brain, and we can see that in the context of language,[2] reading,[3] music,[4] and behavior.

Through experience, we see constant nanoscale changes to the structure of adjoining neurons, in response to new things, like the introductory sentence just above, or to repeated firings when you get exposed to some stimulus multiple times.[5]

And these nanoscale changes can lead to macro-scale structural adaptations. This is illustrated in a classic study of London cab drivers. To become a cab driver in London, one must memorize an extensive map of 25,000 miles of road and thousands of points of interest in and around the city. The process

---

1 Cole, Steven W. "The Social Life of Genes." *Pacific Standard*, in *Stories – The Social Life of Genes – Showcase*, Apr. 6, 2025, https://psmag.com/social-justice/the-social-life-of-genes-64616/. Accessed Nov. 27, 2025.

2 Sakai, Kuniyoshi L. "Language Acquisition and Brain Development." *Science*, vol. 310, no. 5749, 2005, pp. 815–819.

3 Dehaene, Stanislas. *Reading in the Brain: The New Science of How We Read.* Penguin Group USA, 2010.

4 See Schlaug, Gottfried, et al. "Effects of Music Training on the Child's Brain and Cognitive Development." *Annals of the New York Academy of Sciences*, vol. 1060, 2005, pp. 219–230; and Besson, Mireille, et al. "Brain Waves Associated with Musical Incongruities Differ for Musicians and Nonmusicians." *Neuroscience Letters*, vol. 168, 1994, pp. 101–1; and Pantev, Christo, et al. "Increased Auditory Cortical Representation in Musicians." *Nature*, vol. 392, 1998, pp. 811–814.

5 Bartol, Thomas M., Jr., et al. "Nanoconnectomic Upper Bound on the Variability of Synaptic Plasticity." *eLife*, vol. 4, 2015, article e10778.

HTTPS://DOI.ORG/10.1163/9789004739161-012

takes many years, and during it, the drivers develop what is called The Knowledge. Each year, many people try and fail to pass the series of tests necessary to become a licensed cabbie.

Katherine Woollett and Eleanor Maguire performed MRI scans on 79 aspiring drivers before and after their training, comparing the 39 who passed both to those who didn't, as well as to a baseline of non-cab drivers examined during the same periods. Maybe it should be no surprise that those who passed did better on memory tests than those who failed (the cause and effect arrow may have led those who are better memorizers to get selected), but the researchers also showed that those who passed had grown the size of their hippocampus during their study. The hippocampus is a key structure in the brain associated with memory.

Specifically, they found significant growth in the posterior hippocampus. They were unable to determine, using the imaging she employed, whether the observed gross anatomical growth resulted from the formation of new neurons, new synapses, or an increase in non-neuronal glial cells (which support and protect neurons). They felt, however, that the results lend evidence to the posterior hippocampus' being clearly associated with spatial navigation and memorization,[6] either as a storage node for that information or as a processing hub.[7]

Malcolm Gladwell developed and popularized the 10,000-hour rule in his book Outliers, which suggests that the key to achieving world-class expertise in any skill is largely a function of practicing correctly for 10,000 hours.[8] Gladwell relies on the research of Anders Ericsson and his colleagues[9] but Ericsson takes exception to Gladwell's rule. His emphasis was deliberate and goal-

6 I'd emphasize again that other researchers debate about localized centers of thought processes and would contend that circuits throughout the brain would be involved in long-term memory to function correctly, but that there may be more localized functions within the hippocampus that play a relatively more critical role, perhaps as a central hub for those brain-wide networks. As with any single item in this book, there are rabbit holes of nuance and disagreement that I'm trying to gloss over, but not so much that you don't forget the caveats. To plumb the depths of this particular area and debate, for example, just bring up Jennifer Aniston to a neuroscientist.

7 See Woollett, Katherine, and Eleanor A. Maguire. "Acquiring 'The Knowledge' of London's Layout Drives Structural Brain Changes." *Current Biology*, vol. 21, no. 24, 2011, pp. 2109–2114; and Maguire's earlier work: Maguire, Eleanor A., Richard S. J. Frackowiak, and Christopher D. Frith. "Recalling Routes around London: Activation of the Right Hippocampus in Taxi Drivers." *Journal of Neuroscience*, vol. 17, no. 18, 1997, pp. 7103–7110.

8 Gladwell, Malcolm. Outliers: *The Story of Success*. Little, Brown, 2008.

9 Ericsson, K. Anders, and Neil Charness. "Expert Performance: Its Structure and Acquisition." *American Psychologist*, vol. 49, no. 8, 1994, pp. 725–747; and Ericsson, K. Anders. "Training History, Deliberate Practice and Elite Sports Performance: An Analysis in Response to

seeking practice, and not some specific amount of time, pointing out significant advantages accrue well before 10,000 hours. He acknowledges that the "*process of writing an engaging popular article requires considerable simplification.*"[10] (He is throwing Gladwell a bone here, basically saying Gladwell has to "dumb it down" to reach a broader audience, so Gladwell may have to skip over some important details.)

The point, though, is that with training, you do better. And that's not just the growth of new pathways in your brain; it's using different parts of your brain, something illustrated in the conscious competency model.

The term conscious competency was used by Howell in 1982, and is often attributed to Noel Burch (Gordon Training International, 1970s).[11] The model simply states that as you learn a skill, you progress through four states: unconscious incompetence (you are messing up, but you don't know it), conscious incompetence (you are messing up, and you know it, and are probably frustrated), conscious competency (as long as you focus on it, you can do it right), and finally, if you work at it hard enough, you are unconsciously competent (you do it right without thinking about it).[12]

This process will be familiar to anyone who has tried to learn to play golf or any other skilled sport. You can apply this to any skill you've tried to learn, such as music, cooking, art, public speaking, etc. This final state-of maximum performance with no conscious effort-has parallels with what Csikszentmihalyi calls a state of flow.[13] But while many of us may rarely experience the expert level of his ideal state of flow, most of us can relate to driving down the road while having a conversation or eating. We are so unconsciously competent with our driving (at least we think so) that we can free up our brain to address another task that requires conscious resources. The conscious competency model has since been applied in many contexts, and there is a solid base of research into how this model informs how people learn new skills and grow through experience.

Tucker and Collins Review – What Makes Champions?" *British Journal of Sports Medicine*, vol. 47, no. 9, 2013, pp. 533–535.

10 Ericsson, K. Anders. "Guest Post: The Danger of Delegating Education to Journalists." *Radical Scholarship*, 3 Nov. 2014. Archived by Wayback Machine, https://web.archive.org/web/20190320062202/https://radicalscholarship.wordpress.com/2014/11/03/guest-post-the-danger-of-delegating-education-to-journalists-k-anders-ericsson/. Accessed Aug. 24, 2025.

11 Burch, Noel. *Conscious Competence Learning Model.* Gordon Training International, 1970s.

12 Howell, William Smiley. *The Empathic Communicator.* Wadsworth Publishing Company, 1982. (It is not clear whether he invented it, or is just mentioning it).

13 Csikszentmihalyi, Mihaly. *Flow: The Psychology of Optimal Experience.* 1990. Harper & Row, 1990.

Cannon discussed this in terms of the complexity paradox: any simulation or training must not be too simple, or the feedback will not be accurate. However, the more realistic the simulations, the harder it is for the subject to see the cause and effect.[14] When faced with complex environments, subjects use simplification tactics like Strategic Chunking, Sequential Elaboration, Organizational Specialization and Coordination, and Intermediate Measures of Performance. As Otto Neurath said, "*To remember simplified pictures is better than to forget accurate figures.*"[15]

Consider Ericsson's comments about Gladwell's *"considerable simplification"* again, in this light.

You can see how these ideas that aren't just theoretical or philosophical – they are actually informing what I'm writing to you, right now. I realize that the material I'm writing has necessary information, I think you will need to know as we move through the important insights and conclusions from the model. I can't simplify it too much, or I won't give you an accurate picture of this important but complex topic. But if I overwhelm you with facts, researchers, and quotes, I'll lose you.[16] My reviewers and friends remind me to keep it simple. But another tactic I can use is to "chunk" the material into smaller bites.

I can do this by breaking a big, long chapter into several smaller ones, or by occasionally taking a break from sending you a stream of facts and speaking to you directly, perhaps about how the material affects me or how I apply the information in real life, like now.:)

Cannon and colleagues followed up on information overload in 2009, and noted that organizations and people struggled under information overload, and adopting simplification tactics to overcome the complexity of their information environments.[17] This then leads to the simplicity paradox, because the

---

14 Cannon, Hugh M. "Dealing with the Complexity Paradox in Business Simulation Games." *Developments in Business Simulations and Experiential Exercises*, vol. 22, 1995, pp. 96–102; and Cannon, Hugh M., and Alvin C. Burns. Framework for Assessing the Competencies Reflected In Simulation Performance, *Developments in Business Simulation and Experiential Learning, vol.* 26, 1999.

15 Neurath, Otto. "Museums of the Future (1933)." *Empiricism and Sociology*, edited by Marie Neurath and Robert Cohen, Reidel Publishing Company, 1973, pp. 218–223.

16 To be honest, I may have lost most readers by this point, because these are quite complex ideas to hold in one's head and to try to synthesize. But this was the only way I could help you understand the material, rather than asking you to believe me. But for those of you who have made it this far, I hope it has been a useful journey. As Bruce Calvert says – *"Believing is easier than thinking. Hence so many more believers than thinkers."* quotery.com/authors/bruce-calvert. Accessed Nov. 27, 2025.

17 Cannon, Hugh M., Daniel P. Friesen, Steven J. Lawrence, and Andrew H. Feinstein. "The Simplicity Paradox: Another Look at Complexity in Design of Simulations and Experien-

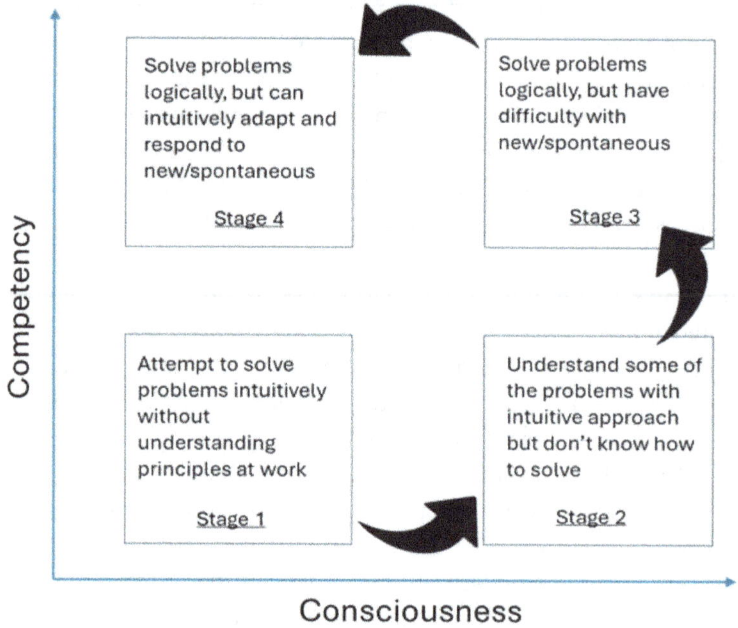

FIGURE 8.1 Howell's conscious competency model

process by which one simplifies is in itself highly complex, balancing information overload and uncertainty: "*The resolution of the paradox is to attack the uncertainty dimension by developing higher-level thinking skills.*"[18]

Consider the parallels between this statement and the meta-representational skills required to retrieve information stored socially or encoded in our surroundings. Cannon and his colleagues examine the different cognitive processes needed under various information complexity environments as a progression of stages, from content knowledge (facts and figures) to procedural knowledge (how to perform tasks or extract knowledge), meta-cognitive knowledge (self-awareness), and finally toward epistemic knowledge (understanding the nature of a particular domain).

The conscious competency model then shows us the cognitive processes associated with each stage of learning:

---

tial Exercises." *Developments in Business Simulations and Experiential Exercises*, vol. 36, 2009, pp. 243–250.

18 Cannon, Hugh M., Andrew H. Feinstein, and Daniel P. Friesen. "Managing Complexity: Applying the Conscious-Competence Model to Experiential Learning." *Developments in Business Simulation and Experiential Learning: Proceedings of the Annual ABSEL Conference*, vol. 37, 2010.

These stages map cleanly onto deliberate practice: targeted tasks, immediate feedback, and progressive challenge that migrate performance from conscious control to automaticity.

1 Experience and the Three Streams

How does this model help us understand and use the Three Streams model?

First, consider that conscious thought is a scarce resource. While the human mind is massively parallel, our conscious processing is notorious for not being a multitasker. If we can shift processing to unconscious modules, we free the brain's conscious resources to attend to other needs. Humans multitask through one of two mechanisms. One way is to switch back and forth between multiple tasks, sometimes very quickly, which gives the impression that you are doing two things at once. However, this approach erodes your performance at either one or both tasks.[19] The second way is to go into autopilot and engage in one of those tasks unconsciously, using learned and automated behaviors, such as when you drive a car and talk with your friend at the same time.[20]

The other reason to perform tasks unconsciously is that it is faster. Your eye will blink reflexively in response to a perceived stimulus in its environment in about 32.5–44 milliseconds. This process engages very few parts of the brain, involving minimal processing or thinking, which helps keep our eyes safe by blinking before damage can occur.[21] But conscious voluntary blinking takes up to 600–800 milliseconds, because the extra time is needed to engage and traffic more circuits of the brain (latency), and increased processing time as you evaluate the stimulus and control the action.[22]

Third, conscious competency reveals the specific cognitive processes at play during skill acquisition, offering a pathway to potentially outsource these pro-

19 Wickens, Christopher. D., and Daniel Gopher. "Control Theory Measures of Tracking as Indices of Attention Allocation Strategies." *Human Factors*, vol. 19, no. 4, Aug. 1977, pp. 349–365.

20 Allport, D. Alan, et al. "On the Division of Attention: A Disproof of the Single Channel Hypothesis." *Quarterly Journal of Experimental Psychology*, vol. 24, no. 2, 1972, pp. 225–235.

21 Shahani, Bhagwan. "The Human Blink Reflex." *Journal of Neurology, Neurosurgery & Psychiatry*, vol. 33, no. 6, Dec. 1970, pp. 792–800.

22 Nakanishi, Masaki, Yasue Mitsukura, Yijun Wang, Yu-Te Wang, and Tzyy-Ping Jung. "Online Voluntary Eye Blink Detection Using Electrooculogram." *Proceedings of the 2012 International Symposium on Nonlinear Theory and Its Applications* (*NOLTA 2012*), Palma, Mallorca, Spain, October 22–26, 2012. IEICE Proceeding Series 1, 2012.

cesses and accelerate skill acquisition. This enables us to navigate increasingly complex environments more quickly and effectively.

Another word for experience is "nurture." We gain experience through using things and by interacting with people. The result of these experiences shapes our brains throughout our lives.

Matt Ridley wrote a fascinating book called *Nature via Nurture* that examined the role of experience in our actions, biology, and beliefs. Ridley's core claim is that the nature–nurture split is a false dichotomy: genes are responsive systems whose expression is tuned by experience, so learning, social cues, and culture shape development by switching genes on and off across the lifespan. In his words, "*Nurture depends on genes, and genes need nurture... [genes] absorb formative experiences, react to social cues and even run memory. They are consequences as well as causes of the will.*"[23]

In practical terms, this means experience (nurture) isn't cosmetic – it's causally potent in shaping modern human behavior, because it operates through the very machinery of gene regulation and neural plasticity that builds minds.

Experience switches on and off certain genetic expressions. Exposure to poor/low nutrition while you are in the womb means you are more likely to have diabetes later in life or give birth to low-weight babies. He concludes, with strong emphasis: "*Genes are at the mercy of our behavior, not the other way around....Memory is 'in the genes' in the sense that it uses genes, not in the sense that you inherit memories. Nurture is affected by genes just as much as nature is.*"[24]

Quartz and Sejnowski lay out a solid case for the developmental influence on the brain, strongly suggesting that the problems, resources, and opportunities we face throughout our lifecycle significantly shape the brain.[25]

O'Leary goes so far as to say that the brain should be looked at not as some static nonstationary model that Pinker would have us consider, but instead as a "protocortex,"[26] which responds over the course of our lifetimes to, according to Clark, "*The whole sensory, linguistic and technological environment in*

23 Ridley, Matt. *Nature via Nurture: Genes, Experience, and What Makes Us Human*. HarperCollins, 2003, p. 6.

24 Ridley, Matt. *Nature via Nurture: Genes, Experience, and What Makes Us Human*. HarperCollins, 2003, pp. 6–9.

25 Quartz, Steven R., and Terrence J. Sejnowski. "The Neural Basis of Cognitive Development: A Constructivist Manifesto." *Behavioral and Brain Sciences*, vol. 20, no. 4, 1997, pp. 537–556.

26 O'Leary Dennis D. "Do Cortical Areas Emerge from a Protocortex?" *Trends in Neurosciences*, vol. 12.10, 1989, pp. 400–406.

which the human brain grows and develops."[27] In other words, it's nature via nurture: genes lay down biases and boundaries; experience – sensory, social, linguistic, technological – does the finishing work, tuning what those areas become good at.

Dennett says that exposure to cultural constructs such as language acts like installing a new virtual machine in the neural wetware by affecting "*myriad microsettings in the plasticity of the brain.*"[28] Griffiths and Stotz, along with several prehistoric anthropologists, make the case that our uniquely long human childhood creates a window of opportunity during which cultural scaffolding can "*change the dynamics of the cognitive system in a way that opens up new cognitive possibilities.*"[29]

So, culturally constructed and nurtured experience dramatically shapes the brain; repeated use of materials does, too, and both processes take advantage of the brain's plasticity (ability to change) to shape thought beyond the functions derived from the actions of genes alone. But as Malafouris spells out, our environments and things respond and adapt to our use of them, in a chaotic, unfinished, ever-changing, and bidirectional dynamic he called metaplasticity.

2 Plain Language Summary of Moderation by Experience

Our brains are continually changing in response to our daily experiences. They are shaped by our culturally derived constructs, like language, and by our environment, diet, behaviors, and habits. This neuroplasticity allows us to progress through stages of skill development, from initial incompetence to eventual mastery, where complex tasks become automatic and unconscious.

Our individual and collective thoughts, as manifested in culture and technology, are constantly changing in a kind of metaplasticity, as our brains, culture, and technology continuously adapt to each other. Much of this is also done unconsciously.

Therefore, while our brains at birth may resemble those of an infant born 300,000 years ago, before robust language and symbol use, our brains as adults are significantly different from those of our ancestors.

We are constantly being shaped by the things we do, what we attend to, and what we think about.

---

27 Clark, Andy. *Natural-Born Cyborgs: Minds, Technologies, and the Future of Human Intelligence*. Oxford University Press, 2003.

28 Dennett, Daniel C. *Consciousness Explained*. Little, Brown and Co., 1991, p. 210.

29 Griffiths, Paul E., and Karola Stotz. "How the Mind Grows: A Developmental Perspective on the Biology of Cognition." *Synthese*, vol. 122, no. 1–2, 2000, pp. 29–5.

CHAPTER 9

Three Streams Model Module Examples: Now and Soon

> Our wills and fates do so contrary run
> That our devices still are overthrown,
> Our thoughts are ours, their ends none of our own.[1] (The Player King, Hamlet (Shakespeare))

American computer scientist and Pulitzer Prize-winning author Douglas Hofstadter describes the brain as a medium supporting complex patterns that imperfectly mirror the world.[2]

In the last few chapters, I've shown how the brain is but one substrate from which we think. It also exists encoded in culture and technology.

Now that we've established the model components, how do they work together? Let's illustrate that with some hypothetical use cases – two in the present, and one heading for us soon.

1 Modules Today

1.1 Social Media Notifications

(1) You consciously curate your social media feed and phone settings to give you certain notifications, so you might pursue a nested hierarchy of consciously developed and unconsciously influenced values.
(2) Unconsciously (in part because companies control algorithms) that work surfaces various notifications to you (your phone makes various noises/vibrations to signal you to act).
(3) You unconsciously reach for it.
(4) What you read affects your thinking.

1 Shakespeare, William. *Hamlet.* Act 3, Scene 2. Edited by Barbara A. Mowat and Paul Werstine, Folger Shakespeare Library, 1992.
2 Hofstadter, Douglas R. *Gödel, Escher, Bach: An Eternal Golden Braid.* 20th anniversary ed., Basic Books, 1999.

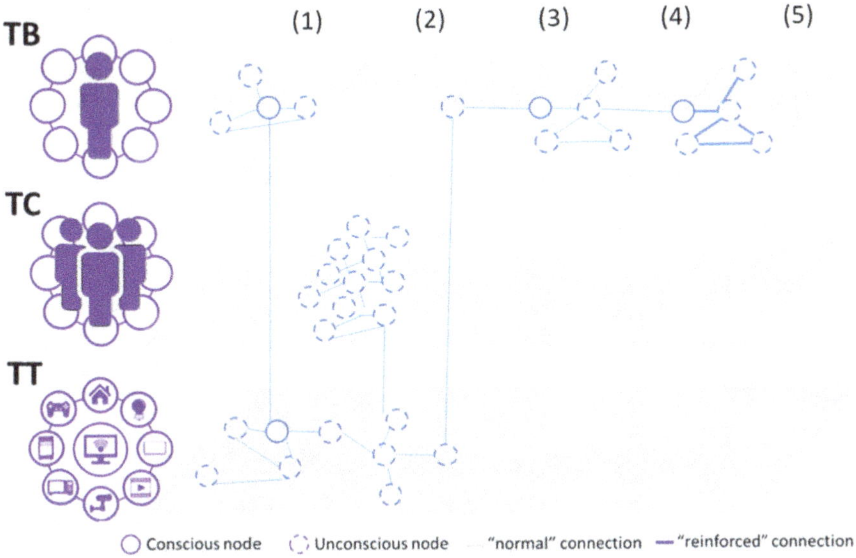

FIGURE 9.1 Conceptual rendering of Three Streams Model in simple present-day social media use case

(5) Habituated over time, your experience with social media changes the circuits in your brain (again unconsciously). The algorithms and your phone and other tech track your actions and adjust themselves accordingly.

A greatly simplified graphical representation of that brief interaction of nodes, representing the engagement of various Three Stream Model modules, might look like this:

1.2 Netflix

1. You sit down to watch Netflix, maybe out of habit, with a goal of "zoning out" (consciously, likely as an escape from the conscious and unconscious thoughts you are processing at the end of the day, and/or unconsciously to trigger reflective daydreaming, or what the neuroscientists might say is prospective activation of Default Mode Network in the brain[3]) before bed.
2. "Recommended for you" content is surfaced to you based on an algorithm (unconscious) which looks at things you've watched in the past,

3 Uslu, Handan. *Understanding the Relationship between Media Use and Maladaptive Daydreaming*. Diss. Georgetown University, 2015.

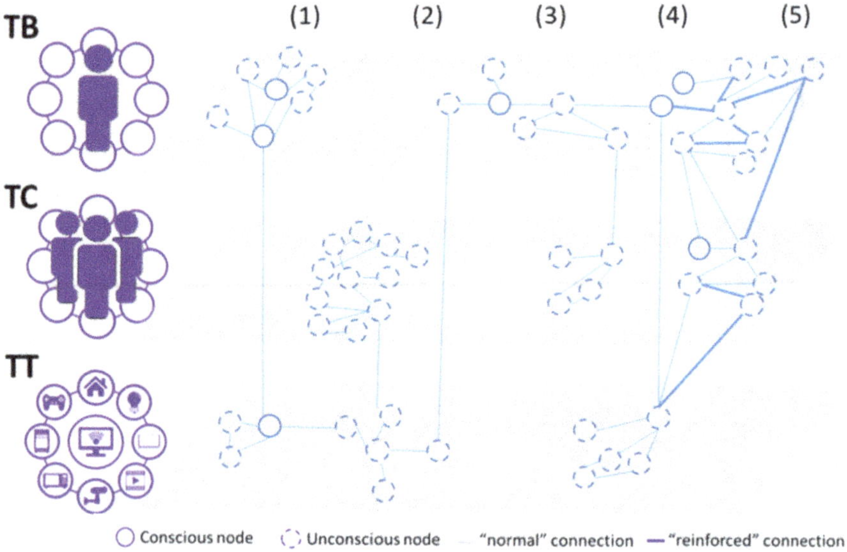

FIGURE 9.2 Conceptual rendering of Three Streams Model in simple present-day streaming use case

how much you've watched (all or some), correlations with other content that viewers like you have seen (unconscious to you), as well as various promotional consideration from certain content providers, including Netflix themselves.

3. You scroll through the feed "mindlessly," pausing on various action-oriented or sexually appealing graphics, stopping when you see a show you've heard people in the office are watching.
4. You choose that show, unconsciously, because you want to take part in the office discussions tomorrow morning (again, unconsciously–consciously you rationalize your choice to yourself as "the show looks interesting").
5. You sit back and watch your selection, mindlessly.[4]

Simplified representation of this scenario using the model might look like this:

---

4 In a conscious effort to streamline the book, I've tried hard to avoid corners of cognitive complexity; one of those areas is the "Default Mode Network" (DFM), which tends to be activated during certain activities like watching television. There are some remarkable relationships between higher order executive thinking and DFM, but this area is still subject to debate. I would encourage people to dive into the topic if you are interested in what the brain is doing when you "zone out." See for example: Raichle, Marcus E. "The Brain's Default Mode Network." *Annual Review of Neuroscience*, vol. 38, 2015, pp. 433–447.

2 Modules: Near-Future

2.1 Near-Future Parenting Morning

1. Sophia's smart home began to wake her up at 4:35, even though her alarm was set for 6. Lights gradually brightened over the next 30 minutes, shades were drawn open, and just after her last REM cycle (as gauged by her biometric watch and smart bed), her phone beeped to bring her fully awake at 5:05. She blinks and puts on her prescription intelligent glasses. She reads a note from her personal AI that she is awake 55 minutes early, because her daughter Emily is running a fever, and slept poorly last night (according to her daughter's watch).
2. Sitting on the toilet, Sophia reads prioritized work emails on her glasses, including one from her boss emphasizing the importance of her 9-AM meeting with the new client. Based on Sophia's rising stress rate (increased heart rate and decreased Heart Rate Variability on her watch, & increased cortisol levels detected in the toilet), Sophia's agent curates her social media feed toward more positive and uplifting stories, which she reads as she brushes her teeth.
3. As she walks into her daughter's room to wake her, Sophia's i-glasses display a diagnostic checklist, updated with the latest CDC guidelines from last year's still lingering pandemic, and the real-time active cases of various illnesses in the school's data trust and Emily's friend group's social media feeds. Checking her daughter over, she quickly suspects strep. Or rather, that's what she reads, as the glasses provide an augmented reality view of her daughter's condition and likely causes.
4. "What should I do, now?" she says aloud. The agent responds, *"We have an appointment at the doctor at 6, I ordered you a self-driving share ride, which will be here in 18 minutes. In the meantime, I'm having the house prepare just the right breakfast for you both, I've changed your 6 AM spin class to a 7:30 am yoga class because of your stress levels – trust me, you need it girl- and I've checked with the new client's calendar to confirm a reschedule of the 9 AM meeting to 9:30, to give you some more time to prepare."*
5. *"I've also notified the school she will be absent today and arranged to have your normal sitter come over for the day. On your rides to the doctor, gym, and work, I've prepared briefings for you on the client, his firm, and his new GMO technology"*.
6. Sophia sighs and says, "*What would I do without you?*"

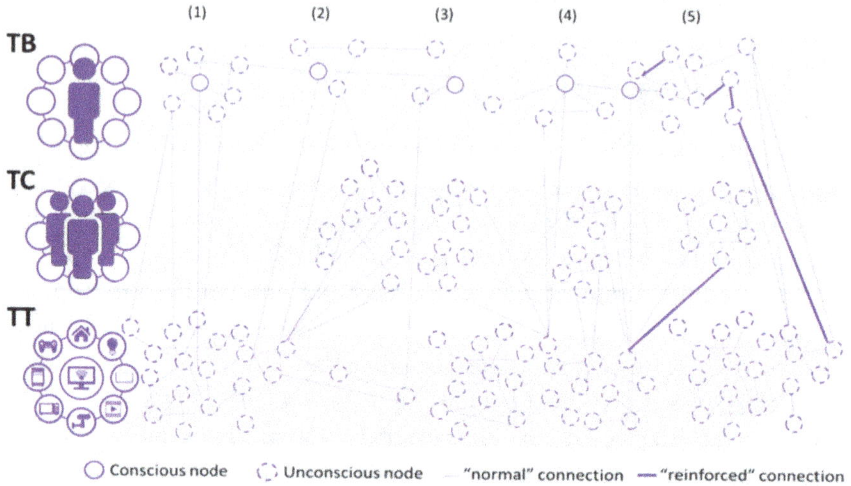

FIGURE 9.3 Conceptual rendering of Three Streams Model in likely near-future "parenting morning" use case

3 Plain Language Summary of How the Three Streams Work Together

The previous chapters explained what the Three Streams are. This chapter shows how they actually operate in everyday life.

Consider something as simple as checking your phone when it buzzes. Your brain (Stream 1) unconsciously reaches for it, shaped by habits formed through repeated experience. The notification comes from algorithms (Stream 3) designed by engineers who understand social dynamics (Stream 2), but also informed by constant feedback of others participating in these frameworks. We are all quietly cooperating to rank content, and then in turn be shaped by others influence. What you read triggers thoughts influenced by cultural contexts you've absorbed (Stream 2), while the technology tracks your response and adjusts (Stream 3). Meanwhile, this entire interaction subtly rewires your brain's circuitry (Stream 1).

The same pattern repeats everywhere: choosing what to watch on Netflix, managing a sick child's morning, navigating work demands. In each case, biological cognition, cultural frameworks and feedback, and technological systems operate simultaneously – often unconsciously – creating a cognitive process no single stream could achieve alone.

What's coming: These interactions are accelerating. The near-future scenarios aren't science fiction; they're extensions of systems already being built. Smart homes, AI assistants, biometric monitoring, and algorithmic curation will weave the three streams even tighter, automating more of our thinking while demanding new forms of metacognitive awareness about what we've outsourced and what we've kept.

CHAPTER 10

On the Neural Correlates of Thought

> To change ourselves effectively, we first had to change our perceptions.[1] (Stephen R. Covey)

Why use this very extended model of thought over the traditional models that look exclusively within the skull? Couldn't I just acknowledge that the brain is the sole domain for thinking, and that it exists within a context or milieu of culture and technology, which in turn affects the brain?[2]

For example, computer architect and neuroscientist Jeff Hawkins studies the brain to understand human intelligence, and he roots our intelligence not only within the brain, but specifically within the neocortex:[3]

> The neocortex is the organ of intelligence...Almost all the capabilities we think of as intelligence – such as vision, language, music, math, science, and engineering – are created by the neocortex. When we think about something, it is mostly the neocortex doing the thinking.... If we want to understand intelligence, then we have to understand what the neocortex does and how it does it. An animal doesn't need a neocortex to live a complex life. A crocodile's brain is roughly equivalent to our brain, but without a proper neocortex. A crocodile has sophisticated behaviors, cares for its young, and knows how to navigate its environment...but nothing close to human intelligence. (Hawkins, 2021)

Hawkins describes the neocortex as a wrinkly sheet, which has a thickness of a nice cocktail napkin, 2–3 mm thick, wrapping around the outside of the brain, that is especially enlarged in humans.[4]

Isn't what I'm trying to describe – the disruption of thought – just an "updating of the napkin?" Alternatively, is the disruption I'm describing just the inability of the brain and neocortex to adapt to rapidly changing contexts in which it finds itself?

---

1 Covey, Stephen R. *The 7 Habits of Highly Effective People.* Free Press, 1989, p. 18.

2 I'm grateful to my friend and colleague Carl Pabo for asking these insightful and important questions.

3 Hawkins, Jeff. *A Thousand Brains: A New Theory of Intelligence.* Basic Books, 2021.

4 In a neo-triune brain way, Hawkins refers to the human neocortex as the "new brain," while everything else in the brain he calls the "old brain" as it more closely resembles those structures and functions in other mammals, reptiles, and so on.

HTTPS://DOI.ORG/10.1163/9789004739161-014

No. And...yes.

I prefer the extended model because, as I tried to understand cognition in the wild, in the real modern world, I followed thought to where I found it. And it exists within and without the skull.

To explain this simply, I'll use a metaphor. Like all metaphors, it has limitations, but I've found it helps illustrate the distinctions between my model and traditional cranially-bound approaches.

Consider a modern car with its many advanced features. A late-model traditional (non-autonomous) vehicle contains hundreds of sensors monitoring the engine, doors, lanes, blind spots, backup proximity, and even tire pressure. It includes backup cameras, audio systems, microphones for calls and voice commands, satellite connectivity for radio and emergency services, and often cellular connections. As vehicles progress through the six SAE levels of autonomous driving (from Level 0 with no assistance to Level 5 with complete autonomy), they incorporate additional sensors like LIDAR and more cameras, while harnessing low-latency connectivity, edge computing, and both local and cloud-based AI.[5]

They do not think, or at least most people would not consider their car's computing to be thinking. This is just one of the many flaws of the metaphor.

But consider the vehicle's ability to sense an oncoming object in the driver's blind spot. Where in the vehicle's electrical circuits is the oncoming object? Sure, we can track the circuitry inside the car required to sense the object, but the object that triggered the reaction is outside the car.

The picture gets a lot fuzzier, however, as you look at other functions. GPS, for example, suggests a better route on the dashboard screen; where in the car's circuitry did that computing correction happen? The vehicle had

---

5 It might be useful to consider the development of connected platforms like this from an engineering and business perspective. Every time you add a sensor to a device, be it a car, thermostat, or stuffed bear, you add cost, but also design considerations. Sensors take power, and impact battery life, which might give off more heat. The data sensed by the new sensor must be processed, either locally or in the cloud. Adding sensors means adding either more computing resources, or network communications to process the data elsewhere. The increased processing and communications further drive the power requirements, and the increased power across all of these components generates heat, which needs to be dissipated. As these elements and their batteries, heat sinks, and other by products proliferate, they take up more space and the designers needs to miniaturize the hardware components to make everything fit. And all of this increases costs. In a car, which is larger and more expensive than, say, a stuffed teddy bear, these factors can be absorbed. This is why we see these features in vehicles today, but as cost/performance curves in processing, communications, and miniaturization continue to improve, these dynamics will begin appearing in low cost everyday items. We'll return to these factors in the conclusion.

received real-time updates of congestion on the initial route, and its map platform used various algorithms and other cloud-based technology to decide that the new route was better. Perhaps this was based in part on a few settings or preferences established locally by the driver, which were then entered into the car's computer, either verbally or manually. But where is the computing (our proxy for thinking in this metaphor) when that information is sent to the cloud, and used with other remote data from other sources to result in the new route displayed on the car's screen?

Sure – we know that the cars on the roads around us, including on our planned and alternative routes, almost always have cell phones inside them. These phones have GPS-assisted location, and when the cars slow down, so too do the phones inside. Cell phone companies track phone location and movement data and this data, along with smartphone OS/app telemetry and connected-vehicle data gets sold to data brokers (unbeknownst to most of their customers). Among the people buying that data are the companies that use it to plot traffic congestion (among other purposes). In our case, services like Google Maps and Waze buy data, process it together with lots of historical and real-time data, including the driver's locally entered preferences regarding tolls, highway use, and other factors, to recommend the new route. This is then sent over cellular and other networks back to the car in our story.

We can see one of the outputs of that computing on the screen in the car, but the processes occur mostly outside of the car's shell. We can find electrical and semiconductor basis for some of the processes within the car and note some changes to those elements because of the entire end-to-end chain, but if we shut off everything outside the car, it wouldn't work nearly as well, if at all. To understand the computing processes and dynamics holistically, we'd have to track all the pieces, inputs, data centers, backhaul networks, computing processes, and rules that govern the function. When I ask, "where does the navigational calculation take place?", the answer requires one to step back and see this holistic system.

Sounds abstract and theoretical, doesn't it? How does the vehicle-semiconductor-correlate model help us pragmatically understand the disruption of thought?

To square that circle, consider the performance of German performance artist Simon Weckert, who in February 2020 loaded 99 working cellphones into a small children's hand-pulled red wagon, like the Radio Flyer some of us had as kids. On a quiet Saturday morning, he walked slowly through the deserted streets of Berlin.[6] Accordingly, Google Maps and other traffic moni-

6 Kan, Michael. "Artist Manipulates Google Maps Traffic View with Cart Full of Smartphones."

toring services informed nearby drivers of the ongoing traffic gridlock and routed them around the city center.

Could we have understood this by stopping one of the re-routed cars, say VW Golf, and pulling apart its circuitry, examining any changes to the internal sensors, software, data, or onboard processing? That might give us a hint as to some of the conclusions of the entire computing dynamic and ecosystem, but it certainly wouldn't help us understand that the root cause for the rerouting was to be found in a child's wagon many blocks away.

It's not just route-finding navigation, of course. Transport agencies and automakers in Finland have piloted connected-vehicle hazard warnings (e.g., black ice) notifying cars before they hit the dangerous spots. When a vehicle's anti-slip sensors detect and adapt to black ice (usually so quickly and subtly that the driver isn't even aware it happened), that action is recorded and sent to a cloud-based service provider, which then adds that data as an event that gets pushed out as a notification to other drivers travelling the same road.[7]

Similar events happen millions of times a day when commuters are notified about accidents, cars stopped on the road, or speed traps. Other drivers (or hopefully passengers) can geotag police cars or road dangers, which is then published as relevant information to other people traveling the same routes.

Where does this hazard-detection computing happen? In the car? No, and thinking that it does is a grossly misleading simplification.

Staying with our Berlin-based VW Golf sedan for a minute, let's assume that the new route was much further than the original one, and sensors in the fuel tank detect low levels, and consequently, the low fuel signal appears on the dashboard. Cars don't operate independently from their environment – they require fuel and parts as they wear out. Given that we've located this setting in Berlin, the driver likely needs to find some diesel. If they fill up with gas instead, lots of alerts and alarms will go off; it stalls or damages the engine/fuel system, stopping the engine. If you continue to try to run the computing without the engine running, the battery will drain, and the computing will stop.

Yes, this tortured metaphor is trying to use this VW Golf proxy to illustrate that the human brain doesn't operate independently from the body's other organs, like the lungs, stomach, and small intestines, etc.; these extracranial functions are dependent on still other inputs from outside the body. If we are

PC Mag, Feb. 2, 2020, https://www.pcmag.com/news/artist-manipulates-google-maps-traffic-view-with-cart-full-of-smartphones. Accessed Aug. 24, 2025.

7 Mihai, Andrei. "New Finnish invention allows cars to easily detect black ice in real time." *ZME Science*, 27 Nov. 27, 2017, https://www.zmescience.com/science/news-science/black-ice-detection-car-27112017/. Accessed Aug. 23, 2025.

to closely examine the brain or the neocortex of a young student experiencing sudden academic declines, we might identify micro areas of neuroinflammation or cell death consistent with a root cause of low blood sugar. But unless we stepped back and looked at the whole body as a system, along with the nutritional inputs coming in from outside the body, we wouldn't be able to understand that their brain is deteriorating because it exists within, say, a body experiencing malnutrition because of food insecurity brought about by low employment affecting their parents' economic status.

Fortunately, after topping up the tank with the correct fuel, the unexpected detour the car undertook is actually the correct response to the data the system received. Yes, it's being manipulated by an outside actor (Weckert), but at least all the components are functioning properly, even if one of the inputs is flawed or being hijacked. But what happens when one of the external pieces doesn't work correctly?

Let's start with old or outdated data. If the driver hasn't updated the GPS data in this metaphorical car, they might end up on the news after blindly driving into a harbor, like two people did in Hawaii in 2023.[8] The driver might become so reliant on the data coming through the screen on their dash that they fail to pay attention to the data that's coming to them visibly via the windshield.

Outdated data can cause serious problems beyond driver error or amusing memes. In September 2022 in Hickory, North Carolina, a driver followed Google Maps onto a bridge that had collapsed years earlier and died in the fall.[9] Months prior, a neighbor had used the "Suggest an Edit" feature to inform Google that their data was out of date, but still, the tech platform provided outdated information.

Where in this accident victim's car circuitry was the flaw that led to his death? (I can hear some readers pointing to the driver, who exists within the car, as not paying attention, and one can say that's another reason metaphors are problematic; but even then, the driver, together with the technology inside his car, failed because they relied on normally dependable and effective computing information coming from outside, supporting the case I'm trying to make).

---

8 Bharade, Aditi. "Tourists in Hawaii Followed the GPS and Drove Their Car into Water." *Business Insider*, 3 May 3, 2023, https://www.businessinsider.com/tourists-hawaii-gps-drove-car-into-water-2023-5. Accessed Aug. 24, 2025.

9 Masnick, Mike. "Can Google Be Held Liable for Man Who Died Following Google Maps over a Collapsed Bridge?" *Techdirt*, Sept. 25, 2023, https://www.techdirt.com/2023/09/25/can-google-be-held-liable-for-man-who-died-following-google-maps-over-a-collapsed-bridge/. Accessed Aug. 24, 2025.

What about the outcomes when certain actors intentionally jam, hack, or spoof elements of the computing system, for non-artistic reasons? This happened on both sides[10] of the Ukrainian war and by the Israeli Defense Force's actions in Gaza.[11] The result is that an Israeli driving in the Haifa suburbs may find that their GPS thinks they are in Beirut, and suddenly changes language to Arabic, and their 10-minute drive to the store now appears to be a six-hour expedition.[12]

Problems need not come from technological glitches or humans: Mother Nature has a way of intruding on the networked technologies, which we increasingly depend on as we drive,[13] and these outages are likely to get worse with other external trends like climate change. As wildfires burn down cell towers or floods inundate the cell tower base stations and their backup generators, networks can go down. Gen Z drivers, used to navigating via their cars' built-in navigation systems, may find themselves at a distinct disadvantage as they try to evacuate over unfamiliar roads without their trusted augmentation. But the anomalies aren't local to the car; they span hundreds of square miles of the physical terrain the car finds itself in. The only thing that can be found in the circuits of the car is "signal not found."

If the causal chain that produces the driver's behavior is distributed, then so is the explanation – and so are the levers of intervention.

Sticking with the metaphor: how safe is a Tesla self-driving car navigating the suburban environments of California, Texas, or Virginia, or cruising the rural highways of Wyoming, Georgia, or Oklahoma? Safe and some claim safer

10 See Russia's efforts https://www.businessinsider.com/gnss-hacking-spoofing-jamming-russians-screwing-with-gps-2019-4, and Ukraine's – https://www.wired.com/story/gps-jamming-interference-russia-ukraine/ Accessed Aug. 24, 2025.

11 "Your GPS Thinks You Are in Cairo, but This Is Not a Cyberattack." *Ynetnews*, Dec. 3, 2023, https://www.ynetnews.com/article/sygirj5bp. Accessed Aug. 24, 2025.

12 Iddon, Paul. "What Is GPS Jamming and Why Is Israel Using It?" *The New Arab*, Apr. 22, 2024, https://www.newarab.com/analysis/what-gps-jamming-and-why-israel-using-it. Accessed Aug. 24, 2025.

13 Weise, Elizabeth, and Santucci, Jeanine."AT&T Cell Service Outage Disasters." *USA Today*, Feb. 22, 2024, https://www.usatoday.com/story/news/nation/2024/02/22/att-cell-service-outage-disasters/72697379007/. Accessed Aug. 24, 2025.

than humans.[14] Others may point out that these environments are already safer than other driving situations, though.[15]

But what if you took that Tesla Model S and dropped it into an off-road track in Moab, Utah, or the double-track roads in the high forested mountains of Colorado? It's likely that the onboard intelligence and circuitry aren't programmed for this new context and would struggle to perform. It would probably need the driver to take over to safely and effectively get from point A to point B. Even then, it is almost certain to experience damage along the way, due to the disruption in its normally designed operations. Its wiring, software, and reference data, in isolation, aren't built for that route. And the external supporting computing processes weren't designed and optimized for this new context, so they are not only of little help, but they might be counter-productive, since their programming was built to some older use cases.

Some of you point out that given time, Tesla will evolve the ability to handle more environments like this, and that software updates, some given on the fly, continually help the car to adapt to changing conditions. But this line of thinking supports the metaphor I'm developing. Here, Tesla, the manufacturer, pushes an authorized software update over the air via cellular connectivity to improve the functioning of the vehicle in new and emerging use cases. This upgrade comes from outside the shell of the car, though.

While it may be a bit of a leap for some to think about the human brain running software, or experiencing "software updates," that idea has held sway in everything from popular discourse[16] to academic settings.[17] While some of this reorganization of the structure, function, chemistry, and other physical parts of the brain results from single, intentional, and benevolent activities like cognitive behavioral therapy, in the "wild," the modern technology-infused brain is constantly experiencing unauthorized attempts to reprogram it. These can be social (cues received from friends, family, and other people),

---

14 Fox, Eva. "Tesla Autopilot Is Over 9 Times Safer than Human Drivers in Cars of Other Manufacturers." *Tesmanian*, 2023, Jan. 10, 2023, https://www.tesmanian.com/blogs/tesmanian-blog/tesla-autopilot-is-x9-safer-tesla-vehicles-are-x2-safer-than-cars-of-other-brands?srsltid=AfmBOopq91L7HcPnPCk_FuAeGagEffihykIlfhzyJt9TFvE7nWhCexpu. Accessed Nov. 19, 2025.

15 Lee, Timothy L "Are Self-Driving Cars Already Safer Than Human Drivers?" *Hacker News*, Sep. 9, 2023, https://news.ycombinator.com/item?id=37349533.

16 Carr, Nicholas. *The Shallows: What the Internet Is Doing to Our Brains*. W. W. Norton & Company, 2020.

17 Moller, Aage R., editor. *Reprogramming the Brain*. Vol. 157, Elsevier, 2006.

or technological, like social media, video games, and common Internet usage,[18] or the ubiquitous advertising platforms that undergird much of our tech.

We'll talk more in Part 3 about the consequences of this, especially in Chapter 17's review of how much agency we have within our thoughts. For now my point is to show that these updates, whether car or cognitive, originate outside the vehicle. While we can observe the implementation of how programming manifests within local structures, we can't understand the deeper root cause, intention, and context without stepping back and looking outside the onboard circuitry.

Remember that the auto metaphor is a gross simplification. The use case is primarily focused on transportation from point A to B, with secondary applications in entertainment and information. The computing needed to navigate that, whether it is occurring in or outside the car, is MUCH simpler than the thought processes we use to navigate modern life. But hopefully, this metaphor provides an accessible way of understanding why we can't look just inside the skull at the neurons and synapses of the brain to really understand human thought in the modern context.

The future Tesla software updates that enable off-road functioning provide us with a natural transition to a related question that comes up when looking into the ongoing disruption of thought: if thought is being disrupted, how are humans responding? Or, what processes does the brain use to overcome its inability to adapt to new and rapidly changing contexts in which it finds itself?

Is the hardware changing, or is our software being updated? Are we *evolving* to respond to the modern self-constructed niche, learning new adaptive behaviors, or developing and adapting new social, cultural, and technological tools to better respond?

That's the topic we'll touch on briefly in the next chapter.

---

18 Korte, Martin. "The Impact of the Digital Revolution on Human Brain and Behavior: Where Do We Stand?" *Dialogues in Clinical Neuroscience*, vol. 22, no. 2, 2020, pp. 101–111.

CHAPTER 11

Disruption vs Evolution vs Adaptation, vs ?

> Every success story is a tale of constant adaptation, revision and change.[1] (Richard Branson)

Most of us are doing ok these days. We are managing despite the dramatic pace of change described in this book.

(Let's leave aside for now the findings of my other book, The Great Irony of Technology, which finds that while most Americans are doing better objectively than their great grandparents of 100–150 years ago, largely due to technology, yet subjectively they are struggling, again primarily due to accumulating tech. Despite their gains, they feel like the world is "going to hell." Americans are increasingly fat, broke, busy, stressed, and unhappy. But for this exercise, let's all pretend, and we'll accept the "delusions of adequacy" most of us adopt to get ourselves through the day).

So, how are we doing it? Are our brains evolving to respond to emerging technology? Or are we learning to adapt to new tech? Or something else? Or all the above?

Despite the mechanism(s), does this affect or manifest somewhere in our brains, and update the "napkin" of the neocortex, or other parts of the brain? Or do the changes in how we perceive and navigate this fast-changing world manifest in other places or via other processes?

I've learned that simple questions like these tap into a classic evo-devo debate, spawning years of research, and nuanced (long-winded) multi-part answers that can fill books or become doctoral dissertations. Here, I will just spend a few pages recording my current assessment of these questions.

1 First, the Bottom Line

1. Are we, as Hawkins might say, updating the napkin? Are our brains physically changing in response to the technology in our lives?

---

1 Branson, Richard. *"Every success story is a tale of constant adaptation, revision and change. A company that stands still will soon be forgotten. Trying to provoke positive change is a principle we've embedded across the Virgin family for more than four decades." Virgin.com,* Jul. 22,2015. Accessed Aug. 24, 2025.

HTTPS://DOI.ORG/10.1163/9789004739161-015

- Yes – as I alluded to in the last chapter. In both good and bad ways, we're still in the early days – causes and effects are shifting fast, while research struggles to keep up, making it hard to gauge the full impact. Innovation happens, people adopt it; it may begin shaping their brains. Think of our brains as living cities, constantly being reshaped by the digital architecture we introduce. But here's our scientific challenge: we're trying to study a moving target. Just as researchers develop the perfect experiment to understand how a particular technology affects our minds, that technology morphs into something new, with different effects. Also, our subjects live in the real world, where many other variables are changing. It's difficult to recreate a broader, more inclusive version of Woollett & Maguire's classic four-year study on how the volume of the hippocampus (dependent variable) changes according to the time cab drivers spend memorizing London's streets (independent variable). In the real modern world, we face a far more complex and dynamic web of cause-and-effect variables.

2. Is it because we are genetically evolving to better co-exist with technology?
   - Unlikely as a major driver at present. Much of the tech driving the disruption of thought is rooted in the growth of Internet connections, and these innovations are most poignant in the last twenty years – within one human generation – too short for strong, population-wide selection effects. That said, natural evolutionary pressures may operate at the margins; I'm not prepared to claim they are absent – only that they are not (yet) a material factor in how we adapt (again, skipping right over those pesky rabbit holes of epigenetics and techniques such as CRISPR-Cas9).
3. Are the operative changes manifested in our brain the result of plastic reorganization and other changes to the brain that occur through repeated use or exposure to certain stimuli, or because of experiences we have as we grow, develop, and age?
   - Yes – very likely. But there are many factors at play; each observable change is small, evidence is mixed in places, and technologies evolve quickly – making clean causal attributions hard. The cumulative outcome may be like subtly "scrambling" the brain in many small, nuanced ways. And each of us in different ways and combinations.
4. Are we learning to live with technology?
   - Yes – learning occurs (not all of it good), but adoption and proliferation often outpace collective learning.

5. Are we harnessing tech to augment how we adapt to our self-created niche(s)?
   – Yes – though, again, acceleration may outpace augmentation, and the front widens constantly.

In summary: we are mostly "muddling through." Brains and thoughts are changing, but in diffuse, constantly shifting ways, via multiple processes, yielding a mix of net-positive and net-negative transformations.

Let me elaborate on these answers briefly.

2 Technologically Induced Brain Changes

Despite the rapid pace of technological change, the prolonged usage required to impact brain physiology, and the time-consuming nature of rigorous research, studies examining specific technological interactions show that the brain can undergo changes when interacting with particular technologies. These are canaries in the coal mine; however, no study yet captures the combined long- and short-term effects of all modern technology on the brain – only specific use cases.

The changes show up across cognitive domains. Frequent media multitasking correlates with differences in gray-matter density and connectivity in regions governing cognitive control (like the anterior cingulate cortex).[2] Persistent notifications and task-switching challenge sustained attention and executive control, engaging the frontoparietal control network[3] – though some video games improve selective attention and visual processing. Reliance on smartphones as external memory alters encoding and retrieval strategies; even a phone's mere presence imposes cognitive costs on working memory tasks,[4] though studies show mixed results as people adapt toward remembering *where* information lives rather than the information itself.[5]

---

2 E.g. Loh, Kep Kee, and Ryota Kanai. "Higher Media Multi-Tasking Activity Is Associated with Smaller Gray-Matter Density in the Anterior Cingulate Cortex." *PLOS ONE*, vol. 9, no. 9, 2014, e106698.

3 Moisala, Mona. *Brain Activations Related to Attention and Working Memory and Their Association with Technology-Mediated Activities*. Unigrafia, 2017.

4 Liebherr, Magnus, et al. "Smartphones and Attention, Curse or Blessing? – A Review on the Effects of Smartphone Usage on Attention, Inhibition, and Working Memory." *Computers in Human Behavior Reports*, vol. 1, 2020, 100005.

5 Liebherr, Magnus, et al. "Smartphones and Attention, Curse or Blessing? – A Review on the Effects of Smartphone Usage on Attention, Inhibition, and Working Memory." *Computers in Human Behavior Reports*, vol. 1, 2020, 100005.

Beyond cognition, technology reshapes physiology and learning. Evening blue-light exposure suppresses melatonin, disrupting sleep timing and quality with downstream effects on recovery and plasticity.[6] Higher social-media use correlates with shifts in attentional capacity, memory processes, social cognition, self-concept, and self-esteem – effects that vary widely across individuals and platforms.[7] Yet well-designed digital learning tools and games can promote positive neuroplasticity, enhancing problem-solving and cognitive flexibility when matched to learner goals and feedback.[8]

3 Human Evolutionary Responses to Technology

Evolution is tricky business. Not everyone agrees it has occurred, and others struggle to accept that it may have occurred since the emergence of humans, perhaps 300,000 years ago, much less in the last few decades.

Part of the problem is that genetic variation often occurs in subtle, hard-to-detect ways that many people might not consider significant, and these changes manifest differently in various pockets of humanity. This book doesn't argue that technology significantly drives modern human evolution, but it is still important to understand the cause-and-effect relationship between technology and (natural) evolution.

What follows is a quick sketch of some examples of this interplay:

3.1 Past

Technologies shaped human evolution even before *Homo sapiens* – fire's improved caloric yield from cooking supported increased energy budgets that literally fueled larger brains.[9] More recent examples include Tibetans developing genetic advantages for high-altitude oxygen uptake over just 30,000–

6 See Pundir, Meenakshi, et al. "Emerging Biotechnologies for Evaluating Disruption of Stress, Sleep, and Circadian Rhythm Mechanism Using Aptamer-Based Detection of Salivary Biomarkers." *Biotechnology Advances*, vol. 59, 2022, 107961; and Snyder, Cynthia K., and Anne-Marie Chang. "Mobile Technology, Sleep, and Circadian Disruption." *Sleep and Health*, Academic Press, 2019, pp. 159–170.

7 Firth, Joseph, et al. "The 'Online Brain': How the Internet May Be Changing Our Cognition." *World Psychiatry*, vol. 18, no. 2, 2019, pp. 119–129.

8 Kühn, Simone, et al. "Playing Super Mario Induces Structural Brain Plasticity: Gray Matter Changes Resulting from Training with a Commercial Video Game." *Molecular Psychiatry*, vol. 19, no. 2, 2014, pp. 265–271.

9 Wrangham, Richard. *Catching Fire: How Cooking Made Us Human*. Basic Books, 2009.

40,000 years (with some populations adapting in mere centuries).[10] Agriculture drove genetic changes in lactose tolerance, starch digestion,[11] disease resistance,[12] and alcohol metabolism[13] as these traits dispersed through farming communities. Cooking and food processing shrank human jaws and oral cavities, leading to crowded teeth, sleep apnea, and decreased respiratory effectiveness.[14] Brain size itself decreased about 150–200 cc (roughly 10%) from the Upper Paleolithic to the present[15] – attributed variously to smaller body size, dietary changes,[16] or the externalization of knowledge through complex societies and language that reduced the need for larger brains.[17]

3.2 Present

It's even more controversial to claim genetic evolution occurred in the last century, or even in the last 500 years. It's more supportable to attribute observable changes in human populations during these timeframes to cultural, environmental, medical, and nutritional factors rather than genetic evolution.

But non-human biological studies provide several well-known cases showing significant abrupt genetic evolution in the face of sudden environmental disruptions. Italian wall lizards were introduced to a new island in 1971, and within thirty years (twenty generations), adapted to the new niche with larger

---

10 Moore, Lorna G. "Human Genetic Adaptation to High Altitudes: Current Status and Future Prospects." *Quaternary International*, vol. 461, 2017, pp. 4–13.

11 Leonardi, Michela, et al. "The Evolution of Lactase Persistence in Europe: A Synthesis of Archaeological and Genetic Evidence." *International Dairy Journal*, vol. 22, no. 2, 2012, pp. 88–97.

12 Pearce-Duvet, Jessica MC. "The Origin of Human Pathogens: Evaluating the Role of Agriculture and Domestic Animals in the Evolution of Human Disease." *Biological Reviews*, vol. 81, no. 3, 2006, pp. 369–382.

13 Li, Ting-Kai, et al. "Genetic and Environmental Influences on Alcohol Metabolism in Humans." *Alcoholism: Clinical and Experimental Research*, vol. 25, no. 1, 2001, pp. 136–144.

14 Walter, Jens, and Ruth Ley. "The Human Gut Microbiome: Ecology and Recent Evolutionary Changes." *Annual Review of Microbiology*, vol. 65, no. 1, 2011, pp. 411–429.

15 Henneberg, Maciej. "Decrease of Human Skull Size in the Holocene." *Human Biology*, vol. 60, no. 3, 1988, pp. 395–405.

16 Leach, Helen M. "Human Domestication Reconsidered." *Current Anthropology*, vol. 44, no. 3, 2003, pp. 349–368.

17 DeSilva, Jeremy M., et al. "When and Why Did Human Brains Decrease in Size? A New Change-Point Analysis and Insights from Brain Evolution in Ants." *Frontiers in Ecology and Evolution*, vol. 9, 2021, 742639.

heads, stronger bites, and a new digestive structure.[18] Another striking example is England's peppered moth, which rapidly evolved from 2% to 95% dark-colored forms in response to Industrial Revolution pollution.[19] When air quality improved, this effect was reversed. Even Darwin's vaunted finches show us the ability to evolve quickly: during a 1977 drought, the average beak depth increased by about 5% in just one generation.[20]

With humans facing dramatic shifts in selection pressures, it is not just logical but imperative to consider the genetic impact, potentially visible in much less than millennia.

One of the first factors to consider is epigenetics – environmental factors that influence how human genes are expressed within an individual's lifetime, without changing the underlying DNA sequence. Studies have shown gene expression is affected by many aspects of modern life, including changing diet and nutrition, exposure to chemicals and microplastics, varying physical environments, and altered stress or physical activity.[21] And those effects can get passed on to our children, in a neo-Lamarckian way.[22]

These genetic changes show up in unexpected places: brain size reduction may have stabilized recently (possibly from improved nutrition);[23] Cape Verde islanders evolved malaria resistance in just twenty generations;[24] women are evolving – later age at first birth and delayed menopause;[25] Estonian studies found DNA differences explained educational and occupational achievement

18 Herrel, Anthony, et al. "Rapid Large-Scale Evolutionary Divergence in Morphology and Performance Associated with Exploitation of a Different Dietary Resource." *Proceedings of the National Academy of Sciences*, vol. 105, no. 12, 2008, pp. 4792–4795.

19 Cook, Laurence M., et al. "Selective Bird Predation on the Peppered Moth: The Last Experiment of Michael Majerus." *Biology Letters*, vol. 8, no. 4, 2012, pp. 609–612.

20 Grant, Peter R., and B. Rosemary Grant. "Unpredictable Evolution in a 30-Year Study of Darwin's Finches." *Science*, vol. 296, no. 5568, 2002, pp. 707–711.

21 Kanherkar, Riya R., et al. "Epigenetics across the Human Lifespan." *Frontiers in Cell and Developmental Biology*, vol. 2, 2014, 49.

22 Trerotola, Marco, et al. "Epigenetic Inheritance and the Missing Heritability." *Human Genomics*, vol. 9, no. 1, 2015, 17.

23 Villmoare, Brian, and Mark Grabowski. "Did the Transition to Complex Societies in the Holocene Drive a Reduction in Brain Size? A Reassessment of the De Silva et al. (2021) Hypothesis." *Frontiers in Ecology and Evolution*, vol. 10, 2022, 963568.

24 Hamid, Iman, et al. "Rapid Adaptation to Malaria Facilitated by Admixture in the Human Population of Cabo Verde." *eLife*, vol. 10, 2021, e631.

25 Chan, Shirley, et al. "Is Menopause Still Evolving? Evidence from a Longitudinal Study of Multiethnic Populations and Its Relevance to Women's Health." *BMC Women's Health*, vol. 20, 2020, pp. 1–15.

following the Soviet Union's collapse;[26] the Framingham Heart Study documented genetic adaptation for lower blood pressure and cholesterol in response to Western diets;[27] and smoking addiction's genetic component is declining in later generations as selection pressure removes alleles associated with withdrawal difficulty.[28]

3.3 Future

Selection pressures are those things that affect our long-term reproductive success, such as biological factors (life span, diet, and reproductive cycles), but also factors influencing our social success and ability to attract mates, as well as factors that influence our offspring's health and well-being (and future long-term reproductive success).

When you consider how fast all these foundational elements are changing in our culture today (outlined in Chapter 15), it's hard not to see powerful forces being unleashed on our DNA, now. These changes range from lab-based reproduction (already well-established) to disrupting factors yielding social and economic success. Attractive mates are no longer strong, capable people skilled at making things with their hands, who build products and families and communicate well in small, compact communities, but "influencers" who can successfully recruit large followers they've never met. All the while, everything else that shapes our lifecycle roars on virtually unchecked, including education, diet, physical environment, our increasing ability to eliminate those things that killed our grandparents' generation, and our continued discovery of new ways to kill ourselves.

And that's before we begin applying new technologies like CRISPR to our gene pool.[29]

---

26 Rimfeld, Kaili, et al. "Genetic Influence on Social Outcomes during and after the Soviet Era in Estonia." *Nature Human Behaviour*, vol. 2, no. 4, 2018, pp. 269–275.

27 Levy, Daniel, et al. "Evidence for a Gene Influencing Blood Pressure on Chromosome 17: Genome Scan Linkage Results for Longitudinal Blood Pressure Phenotypes in Subjects from the Framingham Heart Study." *Hypertension*, vol. 36, no. 4, 2000, pp. 477–483.

28 Wrensch, Margaret R., et al. "CYP1A1 Variants and Smoking-Related Lung Cancer in San Francisco Bay Area Latinos and African Americans." *International Journal of Cancer*, vol. 113, no. 1, 2005, pp. 141–147.

29 The sweeping effects of CRISPR and other related fast-emerging gene-editing technologies on human cognition are too speculative at this point, and thus outside the scope of this work.

4 Ontogenetic or Experience-Based Changes

As we discussed briefly in Chapter 8, a baby born today has a brain much like babies from 300,000 years ago, but the modern adult brain – and the body supporting it – differs greatly due to lifelong experiences and vastly changed environments.

Language and literacy fundamentally reshape the brain.[30] People who speak different languages show different brain physiology, with multilinguals displaying enhanced executive functions and cognitive control.[31] Language ability correlates with changes in gray matter and lateral asymmetry, especially when learned early in life.[32] Reading increases white matter connectivity,[33] enhances neural processing speed,[34] expands hippocampus use and volume,[35] and creates specialized regions in the left temporal lobe – the "visual word form area" that responds specifically to written words.[36]

Digital technology shows opposing effects. Heavy social media use correlates with reduced gray matter in attention-controlling areas,[37] increased amygdala activity,[38] reduced capacity for detailed long-term memories due to shallow processing,[39] enhanced multitasking at the cost of depth,[40] and altered neural pathways for social information processing. The Flynn Effect –

30 Damasio, Antonio R., and Hanna Damasio. "Brain and Language." *Scientific American*, vol. 267, no. 3, 1992, pp. 88–109.

31 Halsband, Ulrike. "Bilingual and Multilingual Language Processing." *Journal of Physiology-Paris*, vol. 99, no. 4–6, 2006, pp. 355–369.

32 Zhang, Haobo, et al. "Grey Matter Correlates of Three Language Tests in Non-Demented Older Adults." *PLOS ONE*, vol. 8, no. 11, 2013, e80215.

33 Yeatman, Jason D., et al. "Development of White Matter and Reading Skills." *Proceedings of the National Academy of Sciences*, vol. 109.44, 2012, pp. e3045–e3053.

34 Schlaggar, Bradley L., and Bruce D. McCandliss. "Development Of Neural Systems For Reading." *Annual Review Neuroscience*, vol. 30.1, 2007, pp. 475–503.

35 Marcelle, Enitan T., et al. "The Role Of The Hippocampus in Working Memory and Word Reading: Novel Neural Correlates of Reading Among Youth Living in The Context of Economic Disadvantage."

36 Dehaene, Stanislas, and Laurent Cohen. "The Unique Role of the Visual Word Form Area in Reading." *Trends in Cognitive Sciences*, vol. 15, no. 6, 2011, pp. 254–262.

37 He, Qinghua, et al. "Brain Anatomy Alterations Associated with Social Networking Site (SNS) Addiction." *Scientific Reports*, vol. 7, no. 1, 2017, 45064.

38 He, Qinghua, et al. "Excess Social Media Use in Normal Populations Is Associated with Amygdala-Striatal but Not with Prefrontal Morphology." *Psychiatry Research: Neuroimaging*, vol. 269, 2017, pp. 31–35.

39 Firth, Joseph, et al. "The 'Online Brain': How the Internet May Be Changing Our Cognition." *World Psychiatry*, vol. 18, no. 2, 2019, pp. 119–12.

40 Gazzaley, Adam, and Larry D. Rosen. *The Distracted Mind: Ancient Brains in a High-Tech World*. MIT Press, 2016.

IQ increases of 3 points per decade last century[41] – has reversed in Western countries,[42] attributed to environmental toxins, nutrition changes, reduced long-form reading, and increased digital media consumption alongside its positive effects from improved nutrition, education, and medicine.[43]

Physical changes extend beyond the brain. Westerners' bodies have shifted through behavioral and environmental factors: higher body fat, earlier puberty, epidemics of lifestyle diseases (diabetes, hypertension, heart disease), declining fertility, lower testosterone, spine and neck posture problems, increased myopia, and reduced impulse control – with technology use implicated in many of these. Life expectancy nearly doubled in the US (from 39.4 years in 1880 to 79.25 in 2024), granting us an "extra life" through vaccines, antibiotics, sanitation, and engineering advances.[44] Yet this trend leveled off in 2010 and reversed for two years starting in 2020, driven by lifestyle disease epidemics and "deaths of despair" (suicide, drug- and alcohol-related deaths) hitting hardest among white non-Hispanics without college degrees in rural areas – populations most disrupted by technological automation, offshoring, and the subjective psychological toll of rapid cultural change.[45]

5 Learning in the Face of Technological Change

So yes, we're changing. Our brains rewire through use (11.2), our genes respond to selection pressures (11.3), and our developmental environments shape cognition in ways our ancestors never experienced (11.4). But these are slow, diffuse, largely unconscious processes – brain plasticity takes years, genetic adaptation takes generations, and ontogenetic changes depend on stable developmental environments we no longer provide. Meanwhile, technology compounds monthly. The question becomes: can conscious learning close the gap?

41 Flynn, James R. "The Mean IQ of Americans: Massive Gains 1932 to 1978." *Psychological Bulletin*, vol. 95, no. 1, 1984, p. 29.

42 Dutton, Edward, et al. "The Negative Flynn Effect: A Systematic Literature Review." *Intelligence*, vol. 59, 2016, pp. 163–169.

43 Bratsberg, Bernt, and Ole Rogeberg. "Flynn Effect and Its Reversal Are Both Environmentally Caused." *Proceedings of the National Academy of Sciences*, vol. 115, no. 26, 2018, pp. 6674–6678.

44 Johnson, Steven. *Extra Life: A Short History of Living Longer.* Riverhead, 2021.

45 Case, Anne, and Angus Deaton. *Deaths of Despair and the Future of Capitalism.* Princeton University Press, 2020.

We are learning, to some extent, how to adapt to technology. A quick chat with generative AI on healthy tech habits yields advice on growth mindset, lifelong learning, curiosity, information literacy, and mindfulness.

I even have a few suggestions in this book, of ways I've learned to better adapt to emerging technologies; I share those in Chapter 21.

We have a long history of learning about new advances and how to deal with them. The printing press made books and ideas more widely available, leading to an uptick in people learning to read and possibly thinking more deeply. The steam engine unlocked new opportunities, enabling innovators to transform tasks and industries. Consequently, people replaced by machines had to be retrained into new roles. Our workplaces were transformed by the personal computer, prompting us to retool tasks and personal toolkits to adapt.

Heck, most of us have learned to ride bicycles, which in their day were amazing transportation innovations.

However, empirically, I don't see us collectively learning well about how to live with emerging technologies, at least not yet. We've had mobile phones for two decades, and I see more, not less, people driving around while looking at their phones. Parents observe their children's screen time with regret and confusion, while all of us wonder about how to integrate AI into our lives and what it will be like in the next few years.

Yes, there are exceptions to this general statement. For example, personally, I've taken the time to learn how to eliminate robo-calls and robo-texts from my phone, disable most notifications on the devices in my life, subscribe to services that help protect my privacy and digital and network security, and I've grasped that social media, television news, clickbait headlines, and most ad-based platforms usually undermine my physical and mental well-being. I also invest time in understanding features and risks associated with most of the tech in my life. That means I read through and consider closely the terms and conditions that most people ignore and/or just click through. Yes, it is a chore to read the EULAs of basic things like Ring Doorbell or Nest Thermostat, but trust me, it's eye-opening and relevant. Considering the short- and long-term consequences of services like 23-and-me, TikTok, or Alexa/Siri/Bixby takes work. But this is my ongoing educational path.

And yes, many of us have learned to adapt to new technologies by limiting or eliminating them. But that's like saying we've learned how to co-exist with our new bike by parking it in the garage.

The strongest tool in learning how to live with emerging technology may come from recruiting technology itself, which is what we will probe next.

6 Using Tech to Help Us Cope with Technological Change

Spoiler alert: In Chapter 21, one of my prime suggestions for adapting to emerging technologies is through the intentional adoption of other technologies to help us. At its current acceleration rate, the only thing that can evolve as fast as technology is technology.

In daily life, I see more people using existing technologies to adapt to technology than I see people displaying new learned behaviors that are not technologically-enabled.

I find these aids fall into one of two buckets: those developed by companies to proactively address criticisms regarding harm associated with their products and innovations being developed by firms to address other opportunities to improve our lives.

It is important to recognize that technology has given us more time. Working from home eliminates time in traffic, as does online shopping, food delivery services, mobile banking, remote doctor visits, and even dating apps and smart doorbells. These technologies have collectively transformed how Americans manage their time, offering convenience, efficiency, and new ways to handle everyday tasks. Thus, technologies help us cope, providing us with more time. And as Steve Jobs said in his 2005 Stanford address, "the most precious resource we all have is time."[46]

But what we do with that extra time is on us. If your microwave or Roomba oven saves you 10 minutes in your day, how do you use that time? Now scale that up to the "extra life" we've been given through expanded life expectancy.

6.1 Defensive Technologies: Damage Control

Technology companies increasingly mitigate their products' harms – not from altruism, but because criticism threatens adoption. Screen time monitors (Apple's iOS tracking, Android's Digital Wellbeing) let users set app limits and schedule breaks. Modern cars fight distracted driving with eye-tracking cameras and steering-wheel sensors that alert wandering attention. ISPs bundle parental controls – content filters, activity reports, scheduling tools – into home routers, automatically updating as threats evolve.

These are reactive measures, concessions to backlash. They help, but they're designed by the same companies profiting from the behaviors they're

---

46 Jobs, Steve. *"You've Got to Find What You Love,' Jobs Says."* Stanford University Commencement Address, Jun. 12, 2005. *Stanford News*, https://news.stanford.edu/stories/2005/06/steve-jobs-2005-graduates-stay-hungry-stay-foolish. Accessed Nov. 19, 2025.

now helping you limit. It's worth asking: how effective is self-regulation when the regulator's revenue depends on your continued use?

Beyond corporate offerings, a parallel ecosystem has emerged: privacy-focused search engines (DuckDuckGo), encrypted messaging (Signal), VPNs (NordVPN), and distraction blockers (Freedom). These tools represent user-driven demand for protection – a market signal that digital self-defense has become a necessity, not a luxury.

6.2 Enhancement Technologies: Amplifying Capability

Other technologies aim not to protect against harm but to unlock human potential. The pandemic accelerated adoption of tools that now feel essential: remote healthcare (telehealth platforms), asynchronous communication (Ma rco Polo's video messages), and mental wellness apps spanning mindfulness (Calm, Headspace), focus gamification (Forest), and therapeutic support (Sanvello's CBT tools).

Education has been revolutionized. YouTube tutorials teach obscure skills; Khan Academy supplements formal schooling; MOOCs (Coursera, edX) reimagine higher education. AI powers adaptive learning (DreamBox for math, Duolingo for languages), while assistive technologies serve neurodivergent learners through text-to-speech, real-time captioning, and personalized pacing. Industry certifications from Google, Microsoft, and AWS often outweigh university degrees for technical roles.

Productivity ecosystems have matured: work management platforms (Notion, Asana), AI transcription (Otter.ai), automation tools (Power Automate), and collaborative workspaces (Slack, Teams, Zoom). Wearables track sleep, activity, and biometrics – quantifying what was once invisible, creating feedback loops for improvement.

Even our capacity for cooperation has been amplified. Sharing platforms (Airbnb, TaskRabbit) extend reciprocal altruism to strangers. Crowdfunding (GoFundMe), citizen science, and microfinance platforms demonstrate technology's potential to foster genuine collective action.

The pattern: Technology accelerates faster than any single brain can adapt, so we're recruiting other technology to keep pace. It's bootstrapping – using one tool to handle another tool's consequences – in an endless escalation. The question isn't whether these tools help (they do), but whether we're gaining ground or just running faster on the same treadmill. **This is muddling through at the individual level.**

7 Industry-Specific Productivity: The Escalation Paradox

These adaptations look impressive on paper – and they are. Agricultural innovation averted Malthusian collapse. Medical advances granted us an "extra life." Manufacturing efficiency lifted billions out of poverty. But each solution introduces new complexity that demands yet more adaptation. Precision agriculture requires farmers to master GPS, sensors, and data analytics – skills that didn't exist a generation ago. Smart factories demand programmers where machinists once stood. Medical advances create patients who survive childhood diseases only to face lifestyle epidemics we've engineered through abundance.

The pattern: We solve one problem by creating systems that generate new problems requiring new solutions. It's not that these technologies fail – it's that they **succeed at making us dependent on even more technology.**

The personal adaptations above – defensive tools and enhancement technologies – operate at individual scale. But the deeper adaptive dynamics play out in the industrial systems that underpin civilization itself. Four sectors reveal the pattern:

- **Precision agricultural:** Agricultural technology has evolved through three major waves that followed technology breakthroughs: First, the Green Revolution (1960s-1980s) dramatically increased yields through high-performing crop by incorporating new technologies such as high-yielding varieties of cereals (particularly dwarf wheat and rice, developed through a mix of genetic breeding and genetic modification), widespread use of chemical fertilizers (to produce their high yields, the new seeds require far more fertilizer than traditional varieties), pesticides, controlled irrigation, and increased mechanization.[47] Second, genetic and precision technologies (1990s-2010s) enabled targeted crop improvements and resource-efficient farming through GPS, sensors, and biotechnology.[48] Most recently, digital agriculture and automation (2010s-present) have brought AI, robotics, and data analytics to farming, while also emphasizing sustainability through techniques like vertical farming and precision irrigation. Together, these

47 Evenson, Robert E., and Douglas Gollin. "Assessing the Impact of the Green Revolution, 1960 to 2000." *Science*, vol. 300, no. 5620, 2003, pp. 758–762.

48 James, Clive. "Global Status of Commercialized Biotech/GM Crops: 2011." *ISAAA Briefs*, no. 43, International Service for the Acquisition of Agri-biotech Applications, 2011. https://www.isaaa.org/resources/publications/briefs/43/. Accessed Nov. 19, 2025.

advances[49] have helped us avoid the Malthusian trap,[50] and transformed agriculture from a largely manual, experience-based practice to a high-tech, data-driven industry capable of feeding billions while potentially reducing environmental impact.

- **Manufacturing:** We are in the Fourth Industrial Revolution (following the innovations of steam power, electricity, electronics, and computers).[51] This time, digital technologies like artificial intelligence, IoT, robotic automation, cyber-physical systems (CPS), cloud computing, and cognitive computing are dramatically transforming traditional assembly lines into on-demand precision smart factories. 3D printing, digital twins, and other innovations have revolutionized prototyping, innovation, time to market, and small-batch manufacturing. Simultaneously, data exchange processes create highly coupled, responsive, just-in-time supply chains. These effects have reshaped the entire global manufacturing landscape, from the ground up, creating new jobs in programming and systems integration while making production more sustainable and customizable than ever before.
- **Medicine:** Recent medical technology came in three technologically-driven waves: First, advanced diagnostics like AI-powered imaging and genetic testing enabled unprecedented insight into the human body, catching diseases earlier and with greater precision.[52] Second, treatment innovations, including robotic surgery, personalized medicine, and groundbreaking therapies like mRNA vaccines[53] and CAR T-cell treatments,[54] have dramatically improved patient outcomes. Finally, as noted above in the Healthcare and Wellness section, digital health technologies have democratized care through telemedicine, wearables, and remote monitoring, while connecting vast networks of medical data to drive future breakthroughs.[55] Together,

49 Benke, Kurt, and Bruce Tomkins. "Future Food-Production Systems: Vertical Farming and Controlled-Environment Agriculture." *Sustainability: Science, Practice and Policy*, vol. 13, no. 1, 2017, pp. 13–26.

50 Kögel, Tomas, and Alexia Prskawetz. "Agricultural Productivity Growth and Escape from the Malthusian Trap." *Journal of Economic Growth*, vol. 6, 2001, pp. 337–357.

51 Schwab, Klaus. *The Fourth Industrial Revolution*. Crown Business, 2017.

52 Topol, Eric J. *Deep Medicine: How Artificial Intelligence Can Make Healthcare Human Again*. Basic Books, 2019.

53 Pardi, Norbert, et al. "mRNA Vaccines – A New Era in Vaccinology." *Nature Reviews Drug Discovery*, vol. 17, no. 4, 2018, pp. 261–279.

54 June, Carl H., et al. "CAR T Cell Immunotherapy for Human Cancer." *Science*, vol. 359, no. 6382, 2018, pp. 1361–1365.

55 Keesara, Sirina, Andrea Jonas, and Kevin Schulman. "Covid-19 and Health Care's Digital Revolution." *New England Journal of Medicine*, vol. 382, no. 23, 2020, p. e82.

these advances are pushing medicine beyond traditional boundaries, making healthcare more precise, accessible, and effective than ever before.

- **Environmental:** Among the technological consequences we must figure out how to adapt to, is anthropogenic climate change. Here, too, we can look at technology as a source (but not a cure-all) for coping. Environmental technology has evolved over the last few decades from simple pollution controls to sophisticated climate solutions, with breakthroughs in renewable energy making solar and wind power cheaper than fossil fuels in many regions.[56] Advanced innovations like direct air capture,[57] electric vehicles,[58] and smart grids[59] are creating a pathway to decarbonization, while AI and satellite tracking provide unprecedented insight into environmental changes.[60] The precision agriculture innovations mentioned above have significant environmental benefits. Smart city systems that optimize resource use offer hope that technology can help humanity limit and respond to environmental damage, thereby enhancing resilience.[61]

These industry transformations illustrate both our adaptive capacity and its limits. We've escaped the Malthusian trap through agricultural innovation – but now face climate disruption from industrial farming. We've extended lifespans through medical technology – but created epidemic lifestyle diseases from sedentary abundance. We've automated manufacturing – but hollowed out communities built around craft labor.

Each wave of technological adaptation solves yesterday's problems while constructing tomorrow's challenges. The question isn't whether we're adapting – clearly we are, across biology, culture, and technology itself. The ques-

---

56 International Renewable Energy Agency (IRENA). *Renewable Power Generation Costs in 2024*. IRENA, 2024. https://www.irena.org/Publications/2025/Jun/Renewable-Power-Generation-Costs-in-2024. Accessed Nov. 19, 2025.

57 Li, Shuangjun et al. "Critical Review on Mobile Direct Air Capture: Concept Expansion, Characteristic Description, and Performance Evaluation." *Matter*, vol. 7.3, 2024, pp. 889–933.

58 Zaino, Rami, Vian Ahmed, Ahmed M. Alhammadi, and Mohamad Alghoush. "Electric Vehicle Adoption: A Comprehensive Systematic Review of Technological, Environmental, Organizational and Policy Impacts." *World Electric Vehicle Journal*, vol. 15, no. 8, 2024, art. 375.

59 Rajaperumal, T. A., and C. Christopher Columbus. "Transforming the electrical grid: the role of AI in advancing smart, sustainable, and secure energy systems." *Energy Informatics*, vol. 8.1, 2025, 51.

60 Stern, Nicholas. "Green and Intelligent: The Role of AI in the Climate Transition." *npj Climate Action*, 2025.

61 Bibri, Simon E., et al. "Environmentally Sustainable Smart Cities and Their Converging Models: Mapping the Evolution of AI, IoT, and Big Data Solutions." *Energy Informatics*, no. 6, 9, 2023.

tion is whether we're gaining ground or simply replacing one set of dependencies with another, more complex set.

This is muddling through at civilizational scale. We adapt, but reactively, diffusely, without coordinated intention. Biology scrambles to keep up. Culture transmits adaptive techniques too slowly. Technology bootstraps itself in an escalating arms race. Each mechanism works – but none fast enough, none comprehensively enough.

In Chapters 21 and 22, I'll explore what CONSCIOUS participation in this process might look like. But first, we need to understand why our current adaptive mechanisms – impressive as they are – remain structurally inadequate for the pace of change we've unleashed.

PART 3

Insights from the Model

1 Purpose of This Part of the Book

A model's purpose is to distill complexity into a simplified representation that can be grasped and applied to generate insights or value that would be difficult or impossible to achieve when dealing with the whole system directly.

This part introduces key insights gained by examining thought through the Three Streams Model outlined in Part 2.

 HTTPS://DOI.ORG/10.1163/9789004739161-016

CHAPTER 12

So What?

> Ideas that require people to reorganize their picture of the world provoke hostility.[1] (James Gleick)

Daniel Kahneman won the 2002 Nobel Prize in Economic Sciences for integrating psychology into economics – especially how we judge and decide under uncertainty. His later, widely taught distinction between "System 1" and "System 2" made those insights easy to work with.

In a 2011 talk, Kahneman stressed that "System 1" and "System 2" are explanatory devices, not literal brain systems. He uses them because they're helpful. *"They don't exist. I don't believe there is such a thing as System 1 and System 2. Don't look for them in the brain, because they are not two systems in the brain of which one does one and the other does the other."*[2]

Why would a Nobel Laureate use models that aren't accurate representations of the real thing, and don't even exist?

For one, it is simpler.

As theorists from Borges to Baudrillard remind us, we often work with simulations because the world is unwieldy: a map is portable in ways a mountain is not. But every model is a lossy reduction of the thing it represents.

It's also what Cannon and colleagues describe: simplification can be a useful tactic for coping with information overload, complexity, and uncertainty.

I don't claim that the Three Streams Model is *actually* how people think. There it is.

However, using that model to investigate how people think is *helpful*. By that, I mean how people think in the whole of the wild, not in the lab. In Chapter 12, I explained how I followed thought to where I found it, and that looking outside the skull was a better representation, personally, of how we process information than models that started and stopped within the brain. In this way, I found it more helpful than the traditional biological models. Although not as simple, it provides a better ability to assess the true root cause, from which we can develop more comprehensive solutions.

---

1 Gleick, James. *Chaos: Making a New Science*. Viking Press, 1987.

2 Kahneman, Daniel. "Thinking, Fast and Slow | Daniel Kahneman | Talks at Google." *YouTube*, uploaded by Talks at Google, https://www.youtube.com/watch?v=CjVQJdIrDJ0. Accessed Aug. 25, 2025.

I also mean this model is better for understanding how we actually think – not how we optimally should think, but how we messily, unconsciously, and often maladaptively do think in the wild. For example, philosopher Michael Wheeler (embodied/embedded cognition) frames the target for cognitive science as explaining how embodied, situated systems generate intelligent behavior that is adaptive, context-appropriate, and flexible across environments.[3]

But how does one define intelligent? And if we are only looking at things that are adaptive and appropriate, what is all the other stuff going on in our thinking? We can approach Wheeler objectively by recognizing that he is trying to optimize thinking, but we can't optimize a system until we understand how it works.

Let's review the insights gained using the model and see how it helps us understand the whole of our thinking, and perhaps to help prepare us for Wheeler's intention.

But before cataloging what's disrupting us, let me be clear about the trajectory of this work: understanding disruption isn't defeat – it's the prerequisite for conscious construction of new thought-forms. The insights that follow might initially feel like an inventory of forces working against us. They are. Yet by mapping exactly how brain, culture, and technology operate at different speeds, pursuing mismatched values, controlled by distributed actors – we gain the leverage to intervene. Awareness precedes agency. This diagnostic section ultimately equips us to become architects of our own cognitive evolution, which we'll explore in Part 4.

1 Insights

Three Streams is a complex model that is far less parsimonious than others we have reviewed so far. Perhaps what I call an eclectic synthesis is an "all-of-the-above, kitchen-sink" approach. But what do we gain in the tradeoff by adding additional planes of cognition? What is the heuristic utility of Three Streams? What insights does this model provide that are not offered by competing models examining brain-based thinking?

The strength of the model lies in its ability to reveal changes in our thinking that we are unaware of when observing cognition from a purely biological perspective.

---

3 Wheeler, Michael. *Reconstructing the Cognitive World: The Next Step*. MIT Press, 2005.

Taking this approach lets us isolate six major insights, most of which are not completely visible when using the traditional tunnel-vision perspective of any one traditional scientific discipline.

Those insights are that modern thought is:

1) highly unconscious, at every level
2) highly complex; within each plane and in their emergent whole
3) a pursuit of many values, that are radically changing and becoming anachronistic
4) subject to varying rates of change, in each plane
5) subject to a complex web of actors and institutions who control our thinking
6) is, in its current state, sub-optimal and perhaps even maladaptive. It isn't working for us.

Let's go through and unpack those insights, one by one. **You've built the model. Now see what it reveals.**

CHAPTER 13

Thought Is Highly Unconscious, at Every Level

> The mind operates most efficiently by relegating a good deal of high-level, sophisticated thinking to the unconscious, just as a modern jumbo jetliner is able to fly on automatic pilot with little or no input from the human, 'conscious' pilot.[1] (Timothy Wilson)

1 Thinking with the Brain…

…is massively unconscious. By its very nature, we cannot perceive it, so it's hard to understand how much of our thinking is unconscious. Miller said of consciousness, "*It is the result of thinking, not the process of thinking, that appears spontaneously in consciousness.*"[2]

In a now classic experiment, researchers put four identical sets of pantyhose on a table and asked shoppers to pick their favorite. The majority picked the one on the right. When asked why, they indicated things like "*the color*" or "*the texture*".[3] None answered correctly: that they were right-handed and had an unconscious right-sided bias.[4] These researchers conclude that how we think is not accessible to our conscious introspection – many effects and processes occur outside of consciousness.

How much of our wet-brain thinking is unconscious? Scientists don't like to speculate, but one, Emanuel Donchin, director of the Cognitive Psychophysiology Laboratory, University of Illinois, goes on the record: "*An enormous portion of cognitive activity is non-conscious, figuratively speaking, it could be 99 percent; we probably will never know precisely how much is outside awareness.*"[5]

---

1 Wilson, Timothy D. *Strangers to Ourselves: Discovering the Adaptive Unconscious*. Harvard University Press, 2002, p. 6.

2 Miller, George A. *Psychology: The Science of Mental Life*. Harper & Row, 1962.

3 Nisbett, Richard E., and Timothy D. Wilson. "Telling More than We Can Know: Verbal Reports on Mental Processes." *Psychological Review*, vol. 84, no. 3, 1977, p. 231.

4 Joseph, Rhawn. *The Right Brain and the Unconscious: Discovering the Stranger Within*. Springer, 2013.

5 Donchin, Emanuel, as quoted in Dhority, Lynn. *The ACT Approach: The Use of Suggestion for Integrative Learning*. PLS Verlag, 1991. Primary source unverified.

HTTPS://DOI.ORG/10.1163/9789004739161-018

If we can't model how *much* of thought is conscious, can we look at the key cognitive processes, and see *which type* of processes is unconscious?

Beliefs represent a key class of unconscious thoughts to scholars like Fodor[6] and Dennett,[7] and this is important because they characterize thought as being constructed upon these predominantly unconscious systems.

Pinker notes, *"The mind is organized into modules or mental organs, each with a specialized design that makes it an expert in one arena of interaction with the world… Much of what they do goes on beneath the level of consciousness, and their operations are driven by evolutionary adaptations to ancestral problems."*[8]

Carruthers goes further in making the claim that consciousness is an illusion, that we don't have a special class of cognitive processes that can be defined as conscious. He points to the two schools of consciousness – the Global Workspace, as articulated Baars[9] and later by Dehaene,[10] which argues that to be conscious, things must be in working memory and available to other conscious cognitive processes; he compares that to an alternative view known as Higher Order Thought (HOT), associated with Graziano,[11] Rosenthal,[12] and others,[13] that conscious mental states are simply those you know. Neither approach would consider most decisions and judgments as conscious, since *"they are not accessible in working memory, nor are we directly aware of them. We merely have what I call 'the illusion of immediacy' – the false impression that we know our thoughts directly."*[14]

Carruthers acknowledges that some subset of reasoning processes can occur using our inner voice, accessing items in our working memory, such as Kahneman's System 2 thinking, and seem to be conscious. But even in

6 Fodor, Jerry A. *The Language of Thought*. Harvester Press, 197.

7 Dennett, Daniel C. "Beyond Belief." In Woodfield, Andrew, *Thought And Object: Essays On Intentionality*. Clarendon Press, 1982.

8 Pinker, Steven. *How the Mind Works*. W. W. Norton & Company, 1997, pp. 21–27.

9 Baars, Bernard J. *A Cognitive Theory of Consciousness*. Cambridge University Press, 1988.

10 Dehaene, Stanislas. *Consciousness and the Brain: Deciphering How the Brain Codes Our Thoughts*. Viking, 2014.

11 Graziano, Michael S. A. *Consciousness and the Social Brain*. Oxford University Press, 2013.

12 Rosenthal, David M. *Consciousness and Mind*. Oxford University Press, 2005.

13 See, for example: Lau, Hakwan, and Richard E. Passingham. "Relative Blindsight in Normal Observers and the Neural Correlate of Visual Consciousness." *Proceedings of the National Academy of Sciences*, vol. 103, no. 49, 2006, pp. 18763–18768.

14 As quoted in Ayan, Steve; "There is No Such Thing as Conscious Thought", *Scientific American,* Dec.20, 2018; https://www.scientificamerican.com/article/there-is-no-such-thing-as-conscious-thought/ Accessed Nov. 17, 2025.

those specific cases, most of the pre-work filtering and content generation is done by unconscious parts of the brain, according to Carruthers.[15]

This was shown in experiments based on Benjamin Libet's work, which started in the 1980s. Using EEGs, Libet and colleagues measured brain activity before decisions.[16] They noted that significant activity occurred well before conscious decisions. The high level of unconscious activity, as much as half a second before conscious decision-making, led Libet and later researchers to believe the brain was pre-processing unconscious signals while building up a "readiness potential" for decision-making.[17]

As Dennett and Fodor have unconscious beliefs, and Pinker and Carruthers would say we make most of our decisions due to unconscious processes, writers like Damasio[18] and Feldman[19] highlight that much of our affective cognition – emotions – occurs unconsciously. These are not necessarily contrary or the opposite of conscious thought, but rather "*They are best seen as older forms of reason, assembled by biological evolution and not by conscious deliberation.*"[20] Our emotions are a set of stone-age cognitive processes, not readily available to us, consciously.

As Donchin indicated earlier, we may never know objectively how much of our brain is unconscious. But let's perform a little thought experiment to get an additional handle on the question.

---

15 Carruthers, Peter. *The Centered Mind: What the Science of Working Memory Shows Us about the Nature of Human Thought.* Oxford University Press, 2015.

16 It should be noted that since Libet's now classic experience, other scholars have acknowledged his results, but pushed back on his interpretations. While Sapolsky, for example, might argue there is no free will (your behaviors are determined by "*biology over which you had no control, interacting with the environment over which you had no control, made you you."(Sapolsky, Robert M. Determined: A Science of Life without Free Will.* Penguin Press, 2023), others like Kevin Mitchell responds, "*yes, there is some machinery that we use to make decisions; but it's machinery we use to make decisions. We're making the decisions."* (Cho, Adrian. "Yes, We Have Free Will. No, We Absolutely Do N*ot." Nautilus,* Feb. 7, 2024, nautil.us/yes-we-have-free-will-no-we-absolutely-do-not-4319; Accessed Aug. 25, 2025).

17 See Libet, Benjamin. "Unconscious Cerebral Initiative and the Role of Conscious Will in Voluntary Action." *Behavioral and Brain Sciences*, vol. 8, no. 4, 1985, pp. 529–556 and Libet, Benjamin, et al. "Time of Conscious Intention to Act in Relation to Onset of Cerebral Activities (Readiness-Potential): The Unconscious Initiation of a Freely Voluntary Act." *Brain*, vol. 106, 1983, pp. 623–642.

18 Damasio, Antonio R. *Looking for Spinoza: Joy, Sorrow, and the Feeling Brain.* Houghton Mifflin Harcourt, 2003.

19 Barrett, Lisa Feldman. *How Emotions Are Made: The Secret Life of the Brain.* Houghton Mifflin Harcourt, 2017.

20 Damasio, Antonio. "Neural Basis of Emotions." *Scholarpedia*, vol. 6, no. 3, 2011, p. 1804.

2 Thought Experiment

Let's try to quantify the level of unconscious activity in the brain. To do that, we'll consider the brain as if it were a digital computer.[21]

Around you, right now, in the real external world, there is a nearly infinite amount of information, some of which your senses can extract. Most of you can see light in the human visible light spectrum, for example, but not infrared, which snakes can see, or ultraviolet, which rats can perceive.[22] And while a dog may smell if there is a person in another room (and their gender),[23] detect explosives in a bag on the floor,[24] and even if you have cancer,[25] you might only notice someone has been cooking fish recently.

Internally, you can get a rich bounty of information about your own biological state to process and reflect on, in addition to the external information.

However, your body can't take in all that much external and internal information. Even if your sense organs can perceive certain stimuli, your sensory apparatus has limitations on what it can transmit and receive. Your brain is made of "*limited input, vision-dominated 100-megabit asynchronous control systems. ...On the order of* 10^8 *signals reach the brain*"[26] per second, says neuroscientist Dennis McBride.

How much is that? 58 megabytes (MB) per second. To put that in comparison, the standard HD movie typically streams at about 5 MBs per second (Mbps).

---

21 Let's be clear, this thought experiment is rife with problems. Both neuroscientists and computer scientists will be standing on their chairs shouting that these things are incompatible, and the sources for the numbers widely vary, as do the interpretations. Nonetheless, it is a thought experiment that attempts to make clear the scale of unconsciousness, not an authoritative and definitive formula and claim.

22 Osorio, Daniel, and Misha Vorobyev. "Photoreceptor Spectral Sensitivities in Terrestrial Animals: Adaptations for Luminance and Colour Vision." *Proceedings of the Royal Society B: Biological Sciences*, vol. 272, no. 1574, 2005, pp. 1745–1752.

23 Czarnik, Anthony W. "A Sense for Landmines." *Nature* vol. 394.6692, 1998, pp. 417–418.

24 Gazit, Irit, and Joseph Terkel. "Domination of Olfaction over Vision in Explosives Detection by Dogs." *Applied Animal Behaviour Science*, vol. 82, no. 1, 2003, pp. 65–73.

25 Moser, Emily, and Michael McCulloch. "Canine Scent Detection of Human Cancers: A Review of Methods and Accuracy." *Journal of Veterinary Behavior*, vol. 5, no. 3, 2010, pp. 145–152.

26 McBride, Dennis K. "The Quantification of Human Information Processing." *Quantifying Human Information Processing*, edited by Dennis K. McBride and Dylan Schmorrow, Rowman & Littlefield, 2005, p. 33.

The typical brain can store about 1 petabyte of information in memory.[27] How much is that? It has been estimated that Facebook handles four petabytes of data per day.[28] But what you are actively retrieving and working with in your short-term memory is, of course, a very small fraction of that petabyte.

The brain is, according to McBride, a powerful processor, one that is "highly self-referential." While the brain can *take in* 10^8 signals, it actually *processes* something like 10^{14}–10^{16} signals per second; far more processing than what it takes in. That's because your brain spends a lot of time thinking about the signals that have come in, playing them out against experience and memories, the other competing signals available, and forecasting those signals into various future scenarios. It's mulling things over. In McBride's terms, it's referring to itself. To the tune of 100 trillion 10 quadrillion signals per second. At 4.7 bits per signal[29] that's 470 trillion to 47 quadrillion bits, which is 58.75 terabytes to 58.75 petabytes (over a million gigabytes). Per second.[30]

If you are not a computer scientist, the sheer scale of this probably escapes you, so it's useful to do another comparison, one of comparing the average human brain to the world's best supercomputer.[31] Computer processing is measured in FLOPs – FLoating Operations Per Sec, which is a comparison to an industry baseline, and we can't use McBride or Bartol et al. for this.

Fortunately, Eric Drexler has modeled this for us. He cites Sandberg and Bostrom's estimate[32] that the average human brain processes a range of $0.9 - 33.7 \times 10^{16}$ FLOPs.[33] With numbers this big, they come up with new names, and so we'll now use Peta-FLOPs (PFLOPs), which equate to one thousand million million flops (10^{15}). Drexler's numbers convert to 9–337 PFLOPs processing performance for the human brain, and then compare it

---

27 Bartol, Thomas M., Jr., et al. "Nanoconnectomic Upper Bound on the Variability of Synaptic Plasticity." *eLife*, vol. 4, 2015, e10778.

28 EdgeDelta Blog. "In 2023, Facebook Produced around 4 PB on Average per Day." *EdgeDelta*, Mar. 11, 2024. https://edgedelta.com/company/blog/how-much-data-is-created-per-day. Accessed Nov. 18, 2025.

29 Bartol Jr, Thomas M., et al. "Upper Bound on The Variability of Synaptic Plasticity." *Elife* 4, 2015, e10778.

30 (10^{14}–10^{16} spikes/s and ~4.7 bits/spike): Bits/s: $4.7\times10^{14} \rightarrow 4.7\times10^{16}$. Bytes/s (÷8): $5.875\times10^{13} \rightarrow 5.875\times10^{15}$. Decimal units: ≈58.75 TB/s → ≈5.875 PB/s.

31 Again, I have many colleagues who will object to this simplification, arguing that brains and computers operate so fundamentally differently that any comparison is misleading. While they have a point, play along with me, as I try to make an understandable point, rather than making an incomprehensible one that accurately explains the differences in detail, distracting us from the main point.

32 Sandberg, Anders, and Nick Bostrom. *Whole Brain Emulation: A Roadmap*. Technical Report #2008-3, Future of Humanity Institute, University of Oxford, 2008.

33 Drexler, K. Eric. *Reframing Superintelligence*. 2019.

to the then most powerful supercomputer in the world. Fugaku, the world's top supercomputer as of November 2020, operates at a max of 442 PFLOPs,[34] which is better, but not significantly more than our max of 337 PFLOPs.

3 Putting the Thought Experiment in Perspective

Now let's connect these staggering numbers to your everyday experience.

Right now, consider everything flooding your senses – every sound, sight, smell, and sensation in and around you. How much of this rich sensory world captures your conscious attention? You're processing the equivalent of about 10 HD movies simultaneously, yet you consciously notice perhaps 1% of this information. More likely, it's closer to 1/100th of 1%.

Think about your memories. Your brain stores roughly a quarter of Facebook's entire daily global data flow. When you consciously recall a childhood moment or search for someone's name, you're accessing just a tiny fraction of this vast library. How much of that immense storage feels available to your conscious mind?

Consider your thinking itself: On your brain's worst days, it operates at roughly 1/40th the capacity of the world's most powerful supercomputer in 2020. On exceptional days, maybe 2/3rd. Yet how much of even this reduced capacity reaches your conscious awareness? Or using McBride's numbers, every minute your brain is processing almost as much data as Facebook does every day.

Despite the methodological flaws in this thought experiment, it illustrates a remarkable conclusion: while an expert estimates unconscious processes handle "only" 99% of our mental activity, the reality appears far more extreme.

Our most treasured mental experiences – our beliefs, decisions, emotions, and sense of self – emerge almost entirely from unconscious mechanisms.

And by our albeit fuzzy math and problematic comparisons, rather than 99%, unconscious processes likely account for far more than 99.99999% of everything happening in your mind.

4 Thinking with Culture

It should be easier to see that nearly all our Thinking with Culture is unconscious.

---

34 "TOP500." *Wikipedia*, Wikimedia Foundation, en.wikipedia.org/wiki/TOP500. Accessed Apr. 13, 2023.

Psychologists Susan Fiske and Shelley Taylor describe people as "activated actors," meaning that much of our social cognition is triggered automatically by cues in the environment, often without deliberate thought. For example, your emotional state may shift instantly if you see a friend across the room answer a call and burst into tears – an unconscious reaction shaped by social meaning. Yet if you saw a stranger or a marginalized individual in distress, you might react differently, revealing how unconscious associations and cultural frames guide perception and response.[35]

Cultural psychology shows that these differences extend across societies. Richard Nisbett found that East Asian students tend to approach problems more holistically, while Western students more often emphasize analytic and causal reasoning.[36] Similarly, Stanislas Dehaene demonstrates how linguistic structures – such as the regular base-ten number system in Chinese – affect how people represent and manipulate numbers, producing subtle advantages in memory and calculation, even though the underlying biological limits are comparable across cultures (Dehaene). Language itself shapes thought: speakers of different tongues perceive time, space, and number in ways that reflect linguistic categories. Neuroscience research supports this picture: functional imaging shows that cultural background modulates brain activity during tasks of perception and decision-making, providing biological evidence for the shaping power of culture (Han et al.).

Unconscious imitation is another striking example. Tanya Chartrand and John Bargh documented the "chameleon effect": in conversation, people unconsciously mimic others' gestures, postures, and expressions, and this mimicry increases social rapport.[37] Cecilia Heyes reviewed decades of evidence on "automatic imitation" and found it to be a robust phenomenon across cultures and experimental designs.[38] Neuroscientists have linked these effects to mirror neuron systems, which activate both when performing an action and when observing others perform it, suggesting deep evolutionary roots.[39]

---

35 Fiske, Susan T., and Shelley E. Taylor. *Social Cognition: From Brains to Culture*. Sage, 2013.

36 Nisbett, Richard E. *The Geography of Thought: How Asians and Westerners Think Differently ... and Why*. Free Press, 2003.

37 Chartrand, Tanya L., and John A. Bargh. "The Chameleon Effect: The Perception–Behavior Link and Social Interaction." Journal of Personality and Social Psychology, vol. 76, no. 6, 1999, pp. 893–910.

38 Heyes, Cecilia. "Automatic Imitation." *Psychological Bulletin*, vol. 137, no. 3, 2011, pp. 463–483.

39 Iacoboni, Marco. *Mirroring People: The Science of Empathy and How We Connect with Others*. Picador, 2009.

Why might such unconscious mimicry be adaptive? Take fish, for example, which imitate others as part of an adaptive behavior called schooling. To stay relatively safe from predators, fish stay together. This very complex coordination of thousands of fish is produced by a simple rule: copy your neighbor.[40] Similarly, when you mirror someone else, you establish a nonverbal and predominantly unconscious rapport with them.[41] How is this adaptive?

Peter Turchin makes the case in his book *Ultrasociety* that human achievement over the last 10,000 years rests on one unique human characteristic: our ability to coordinate and collaborate with others, especially strangers.[42] If Turchin is right, it would seem these cultural attributes are exaptions built up on unconscious social modules.

Fazio and Olson review the literature on how making decisions is subject to socially cued priming, either through socially constructed stereotypes or by perceived social attitudes of those in our environment, and conclude that what we think about is deeply affected by social priming.[43] Greenwald et al. similarly review cognitive constructs (stereotype and self-concept) and affective constructs (attitude and self-esteem) and come to similar conclusions.[44]

Research in implicit social cognition reinforces this conclusion. Social neuroscientist Deborah Stanley notes that stereotypes, biases, and associations often guide behavior automatically, shaping decisions without conscious oversight.[45]

Together, these findings highlight how much of our culturally grounded thinking is absorbed once, then deployed automatically, rarely revisited or reconsidered. Generations of cultural shaping, combined with evolved reflexes, mean that most of our culturally influenced cognition occurs unconsciously.

---

40 Grimm, Volker, et al. "Pattern-Oriented Modeling of Agent-Based Complex Systems: Lessons from Ecology." *Science*, vol. 310, no. 5750, 2005, pp. 987–991.

41 Manusov, Valerie. "Mimicry or Synchrony: The Effects of Intentionality Attributions for Nonverbal Mirroring Behavior." *Communication Quarterly*, vol. 40, no. 1, 1992, pp. 69–83.

42 Turchin, Peter. *Ultrasociety: How 10,000 Years of War Made Humans the Greatest Cooperators on Earth*. Beresta Books, 2016.

43 Fazio, Russell H., and Michael A. Olson. "Implicit Measures in Social Cognition Research: Their Meaning and Use." *Annual Review of Psychology*, vol. 54, no. 1, 2003, pp. 297–327.

44 Greenwald, Anthony G., et al. "A Unified Theory of Implicit Attitudes, Stereotypes, Self-Esteem, and Self-Concept." *Psychological Review*, vol. 109, no. 1, 2002, p. 3.

45 Stanley, Deborah A. "Getting Implicit about the Unconscious: Do We Really Know What Social Neuroscience Has to Offer Implicit Social Cognition Research?" *Journal of Social Issues*, vol. 72, no. 1, 2016, pp. 107–118.

5 Thinking with Technology

Does anything need to be said?

I've shown that you think differently, consciously or unconsciously, when you outsource thinking to things. You might say you know the time when you don't, but you know it should be available to you on your watch. These meta-representations about where information is, or what resources could be harnessed to achieve certain computing needs, occur in the head, but as Clark and Chalmers indicate, if something does some process the brain could do, then it is a substitute for the brain, and therefore isn't occurring in the brain; Q.E.D.[46] – it is happening unconsciously.

When you use a calculator, you might use it to do something you could do, but if you aren't consciously doing it, it's unconscious. Your manipulation of the calculator might be conscious, but the calculations it performs are opaque and unconscious to you.

The same logic applies to search engines, navigation systems, recommender algorithms, and AI assistants. These systems do enormous amounts of processing outside of awareness, surfacing only when their results are presented.

Similarly, you might need to use conscious thought, even a lot of conscious thought, to know where information can be uncovered and to extract and decode that information usefully, but the information stored or processed outside of your body is still done so unconsciously.

This simple logical conclusion rapidly becomes even more self-evident as we shift the lens from a calculator to the algorithm-fueled Technology that permeates most of our daily life today. Those processes occur outside our consciousness, only coming into play when humans connect with external processes, which is a very small percentage of the technologies that we use to think on our behalf.

Nothing in this section should be taken to mean that these processes aren't part of your thinking, because they happen outside of your head. We've addressed this point earlier. Here, I just mean to point out that they are almost exclusively unconscious.

We'll come back to the importance of this in Chapter 20, and in the Conclusion, but at least this helps us better understand *how* we think.

---

46 Q.E.D. or QED is an initialism of the Latin phrase "quod erat demonstrandum," literally meaning "what was to be shown." Source – Wikipedia; colloquially often used to say "thus it is shown."

CHAPTER 14

Thought Is a Highly Complex, Distributed System

> The brain is the last and grandest biological frontier, the most complex thing we have yet discovered in our universe. It contains hundreds of billions of cells interlinked through trillions of connections. The brain boggles the mind.[1] (James D. Watson)

> Read books. Try to understand the reason why things happen, why they are as they are. If you see only the surface phenomena, then the world becomes extremely confusing, ever more unsettling. But if the reasons are understood there's a kind of simplicity that emerges.[2] (David McCullough)

1 Before You Begin

This chapter sketches out the complexity not only of the brain but also of the other two streams of thought and the surprising emergent effects when we consider them together (1+1+1=5). These domains are rarely examined side by side, which makes the exercise demanding but worthwhile. In a world of tweets and scholarship essays capped at 140 characters, it is tempting to simplify to the point of distortion.

But as Cannon's Simplicity Paradox reminds us, oversimplification can actually increase confusion by stripping away the context needed to judge accuracy and relevance. I aim instead for moderate depth: compressed, but not dumbed down.

This chapter maps the actual terrain of human thought – not as abstraction, but as measurable reality. The brain. Culture. Technology. And critically, how they amplify each other.

Fair warning: this condenses several thousand pages of research into a couple dozen pages of dense material. But without this foundation, the disrup-

1 Watson, James D. Foreword. *Discovering the Brain*, by Sandra Ackerman, National Academies Press, 1992, p. iii.

2 McCullough, David. *The American Spirit: Who We Are and What We Stand For*. Simon and Schuster, 2017, p. 58.

HTTPS://DOI.ORG/10.1163/9789004739161-019

tions described in Part 4 will seem like science fiction rather than documented fact. If you invest 30 minutes here, you'll save yourself years of reading across neuroscience, cultural anthropology, and computer science – and you'll understand exactly why modern thought is so fragile.

Think of this as your base camp before the summit. The view from here matters.

2 Introduction to the Complexity of Thought

Both brain science and our broader models of thought confirm one point: thought is not just complex, but perhaps incomprehensibly so. Each advance in neuroscience reveals additional layers of intricacy. The more we learn, the more we realize how much lies beyond easy grasp.

Why should we care?

As Vint noted in the foreword, technological acceleration and dependency have created *"a sophisticated but potentially fragile ecosystem of networks and platforms."* When we unpack thought itself, we see similar dynamics: astonishing benefits, hidden vulnerabilities, and an increasing opacity that can feel like a spell woven around us. In Part 4, I describe this trend as "The Return of Magic." In that section, I'll also show how increasingly complex systems are also inherently more prone to dramatic disruptions and even failure. So much so that when they happen not only should we not be surprised but we ought to consider them normal. We can't discuss those dynamics later without laying the foundation for that discussion here.

The Three Streams model suggests that thought across brain, culture, and technology is not three times as complex as biology alone, but potentially orders of magnitude more so. Due to the rapid evolution of each level (discussed in the next chapter), various levels of control (discussed in chapter 17), the dynamic feedback loops between planes (chapter 16), and the combined emergent properties and capabilities of the whole three-layered system, these streams combine into a system far beyond the already daunting intricacies of the brain alone.

With that in mind, let's review the complexity currently associated with each of the 3 modes of thought, as they exist in isolation, and then some of the emergent considerations.

3 Complexity within Thinking with the Brain

How do we describe "the most complicated object in the known universe?"[3]

3.1 Scale and Architecture

Start with scale. I've hinted at some of the facts earlier, especially with McBride's quantitative summary in the last chapter. The brain has been identified, if not as the sole place where cognition takes place in the body, as at least the primary hub where most "wet" cognition takes place. It weighs about three pounds on average (about 2% of a typical human mass) but requires 20% of our daily calories to power it. That is why it is called an "expensive tissue" metabolically. Humans do not have the largest brains on earth – larger animals like whales (20 lbs.) and elephants (10 lbs.) have that distinction. But we do have the largest brain that one would expect from any animal, given our size, diet, and other lifestyle factors.[4]

Contemporary neuroscience reveals the brain as a system of extraordinary complexity. With approximately 86 billion neurons forming an estimated 100 trillion synaptic connections, the human brain represents one of the most complex biological systems known.[5] However, raw quantification fails to capture the practical reality of this complexity.

The gross architecture of the brain is relevant, of course. It has three main parts (cerebrum, cerebellum, and brainstem), two hemispheres which have significantly differentiated functions, and four main lobes (frontal, temporal, parietal, and occipital). Each of these is further subdivided into various layers (most notably the cortex) and deeper structures.[6] Not only does this architecture change in volume and structure throughout a person's life,[7] there are differences in how the brain is organized between men and women.[8] When we go from gross examination to microscopic, we observe the various unique

3 Michio Kaku, *The Future of the Mind: The Scientific Quest to Understand, Enhance, and Empower the Mind.* Doubleday, 2014.

4 Jerison, Harry J. *Evolution of the Brain and Intelligence.* Academic Press, 1973.

5 Herculano-Houzel, Suzana. "The Human Brain in Numbers: A Linearly Scaled-up Primate Brain." *Frontiers in Human Neuroscience*, vol. 3, 2009, pp. 31–41.

6 Nolte, John. *The Human Brain.* Mosby/Elsevier, 1993.

7 Toga, Arthur W., et al. "Mapping Brain Maturation." *Focus*, vol. 29, no. 3, 2006, pp. 148–390; and Creasey, Helen, and Stanley I. Rapoport. "The Aging Human Brain." *Annals of Neurology*, vol. 17, no. 1, 1985, pp. 2–10.

8 See – Kiesow, Hannah, et al. "10,000 Social Brains: Sex Differentiation in Human Brain Anatomy." *Science Advances*, vol. 6, no. 12, 2020. Almost any point in this chapter can be picked apart, or couched in deeper nuance. In this case, as an example, the specific conclu-

cytoarchitectures (areas where the cellular composition is relatively distinct from the other areas) that we discussed earlier.

3.2 Networks, Cells, and Connections

Brain networks operate across multiple scales simultaneously. At the microscale, individual neurons integrate signals from thousands of synaptic inputs before generating output signals. At the mesoscale, neural assemblies coordinate activity across brain regions. At the macroscale, large-scale networks like the default mode network and executive control network coordinate complex cognitive functions.[9]

This multi-scale organization enables remarkable functional flexibility. The same neural substrate can support perception, memory, reasoning, and motor control through dynamic reconfiguration of network activity patterns. Such flexibility proves crucial for understanding how brains integrate with external cognitive resources.

Interestingly, the vaunted cerebral cortex, often associated with higher intelligence and conscious thought, while accounting for approximately 40% of adult human brain mass, only contains 19% of the neurons. Meanwhile, the cerebellum ("little brain"), the structure located at the lower back of the brain, while holding approximately only 10% of the brain's volume, contains over 50% of the brain's total neurons (that's because the cerebellum contains vast numbers of granule cells, the smallest neurons in the brain, explaining this high neuronal density).[10]

The brain is often described as having two types of matter – white and gray. Roughly speaking, white matter (heavily myelinated long-range tracts) supports distributed communication; gray matter (dense local circuitry) handles much of the local computation.

White matter is on the order of 50–60% of brain volume; gray matter is the rest. White matter gets its name because many of those neurons have axons sheathed in white fatty proteins called myelin. Neurons here usually have long axons, which function to transmit signals over distance. This is typically found around parts of the cortex, optic nerves, lower parts of the brain stem, and around the spinal cord – areas that are primarily involved with the communi-

---

sion might be spun thusly: "There are population-level averages that differ in some measures, amid large overlap; interpretation remains debated."

9 Sporns, Olaf. *Networks of the Brain.* MIT Press, 2011.

10 Azevedo, Frederico A. C., et al. "Equal Numbers of Neuronal and Nonneuronal Cells Make the Human Brain an Isometrically Scaled-up Primate Brain." *Journal of Comparative Neurology*, vol. 513, no. 5, 2009, pp. 532–541.

cations networks within and around the brain. Gray matter makes up the other 40% of the brain, is found in the surface (cortex) of the cerebellum, deep within the substructures of the cerebellum, as well as various other parts of the brain. The parts of the brain typically associated with processing are mostly gray matter, whereas the communications areas are mainly composed of white matter.

Glial cells, which we touched on briefly in Chapter 10 discussing the changes to a cabbie's brain during extensive memorization, provide crucial support, insulation, and protective functions that are vital for the overall well-being and functionality of the brain. Like neurons, they are highly differentiated. Most common are astrocytes (star-shaped cells that provide structural support for neurons and help regulate the chemical environment – ions, neurotransmitters, and nutrients – around them), oligodendrocytes (responsible for producing myelin, the insulating layer most commonly found in white matter, allowing for faster transmission of nerve impulses along axons), microglia (immune and clean up cells that also help moderate inflammation), and ependymal cells (these line parts of the brain and produce and circulate the cerebrospinal fluid, which cushions and encases the brain).

Glial cells may have functions beyond mere support and contributions to learning and memory. They have been identified to being involved in various neurological disorders, hinting at other yet unknown functions.[11] While our understanding of the sophisticated glial cell functions has progressed significantly, ongoing research is likely to uncover new aspects of their roles in the nervous system.[12]

Neurons vary greatly. They are normally grouped into three major categories that have different functions: sensory neurons, motor neurons, and interneurons. Additionally, neurons come in many shapes (unipolar, pseudounipolar, bipolar, and multipolar) and operate in different ways. The two most dominant families of neurons are excitatory (a common one is the pyramidal neuron, which is one of the most common types of neurons in areas like the cerebral cortex, which performs much of higher order cognitive processing) and inhibitory. These different families are further differentiated by the neurotransmitters they use to communicate – excitatory release glutamate, while inhibitory use *g*-amino-butyric acid (GABA).

11 Salter, Michael W., and Beth Stevens. "Microglia Emerge as Central Players in Brain Disease." *Nature Medicine*, vol. 23, no. 9, 2017, pp. 1018–1027.

12 Fields, R. Douglas, et al. "Glial Biology in Learning and Cognition." *Neuroscientist*, vol. 20, no. 5, 2014, pp. 426–431.

Neurons themselves are quite complex, in addition to being quite diverse. They are made up of various substructures (e.g., dendrites and axons), which themselves vary dramatically from neuron to neuron, and they vary in shape and size. They operate through a sophisticated operation of building up a discharge, often known as a synaptic spike or action potential. They maintain connectivity with thousands to tens of thousands of synapses to other adjacent neurons.[13]

Many people refer to these spikes that signal across these boundaries as binary; they either happen or it doesn't, but it is possible that neurons can modulate and filter incoming signals and might themselves each have an intrinsic plasticity that is harnessed to create learning.[14] Synapses, which come in two different types, electrical and chemical, come in dozens of subtypes. And neurons don't just blindly react to a single neighboring impulse before spiking and sending the signal down the road; they may have to receive hundreds or even thousands of spike signals from corresponding neurons, before they have the "confidence" to spike on their own and pass on the message.

Mark Humphries estimates that a person will have 34 billion spikes in a lifetime, constituting their lived experience.[15] He also introduces us to the ideas of "dark neurons," which compose a lot of our brain but don't spike at all (these are sometimes seen as dead or dying neurons that aren't counted in the brain's totals, but there could be other reasons these are not being activated during recording periods; the truth is we still don't know, and there is even debate about what to call them) and spontaneous spikes (which happen without any cause or stimuli).

The organization of neurons into networks via connected series of synaptic pathways has been referred to as a "connectome," Sporns et al. and Hagmann describe it as a comprehensive structural description of the network of ele-

13 Huttenlocher, Peter R., and Arun S. Dabholkar. "Regional Differences in Synaptogenesis in Human Cerebral Cortex." *Journal of Comparative Neurology*, vol. 387, no. 2, 1997, pp. 167–178.

14 Ohtsuki, Gen. "Modification of Synaptic-Input Clustering by Intrinsic Excitability Plasticity on Cerebellar Purkinje Cell Dendrites." *Journal of Neuroscience*, vol. 40, no. 2, 2020, pp. 267–282.

15 Humphries, Mark. *The Spike: An Epic Journey through the Brain in 2.1 Seconds*. Princeton University Press, 2021.

ments and connections forming the human brain.[16,17] If there are 86 billion neurons in the brain, and an average of 1,000 synapses to each other, the result is a profoundly complex arrangement of 100 trillion synapses of myriads types,[18] arranged into a nearly infinite number of networks, all changing constantly as time, environment, resources, and other factors vary.

How much of this "activity" across these synapses constitutes "thought" or cognition? Well – all of it. It is the processing that we do in our brains. Is some of it unconscious? Yes, most of it; as we discussed in the last chapter. Is some of it spurious noise? Yes, that is how the brain works.

A very compelling paper published in 2020 used some sophisticated resting state fMRI scanning to conclude that an individual might have 6,000 "thoughts" in a day.[19] Is all of our brain activity required to generate those thoughts? If not, can/should we eliminate or disregard the rest? This is analogous to opening the hood of your car, and trying to decide which wire you want to cut; since you don't understand what it does, perhaps it isn't needed…? If it happens in the brain, it is either thought, or a crucial process for supporting thought; until we know better, it may be best to treat all of what happens in brain as a component of biological thought.

As I discussed earlier, the brain is heavily embodied in the body, and leverages physical nerve networks as well as chemical pathways to sense internal and external stimuli. Some of these internal sensations may in fact be, as Feldman pointed out, the basis for our valence and emotions, and these may be an older cognitive mode. It is through these emotional, valence, and sensory systems that the brain interacts with the body and through that the external world. Even here we see unrecognized forces at play.

---

16 Sporns, Olaf, Giulio Tononi, and Rolf Kötter. "The Human Connectome: A Structural Description of the Human Brain." *PLoS Computational Biology*, vol. 1, no. 4, 2005, e42. See also Hagmann, Patric. *From Diffusion MRI to Brain Connectomics*. PhD thesis, Université de Lausanne, 2005.

17 Hagmann, Patric. *From Diffusion MRI to Brain Connectomics*. 2005. École Polytechnique Fédérale de Lausanne, PhD dissertation, 2005.

18 Azevedo, Frederico A. C., et al. "Equal Numbers of Neuronal and Nonneuronal Cells Make the Human Brain an Isometrically Scaled-Up Primate Brain." *Journal of Comparative Neurology*, vol. 513, no. 5, 2009, pp. 532–541.

19 See Tseng, Jennifer, and Jordan Poppenk. "Brain Meta-State Transitions Demarcate Thoughts across Task Contexts Exposing the Mental Noise of Trait Neuroticism." *Nature Communications*, vol. 11, 2020, p. 3480. What they are actually doing is an fMRI estimate based on state-transition markers. This idea that equates to a "thought" is highly controversial. Again an example of the nuance and detail I'm glossing over a bit in this chapter, just to give you a sense of scale of the whole, without going too much into the weeds.

3.3 Sensing the World: Beyond Five Senses

For example, we don't have just five senses. With apologies to Aristotle, who goes to great length explaining why there cannot be more than five senses,[20] we have more like 20. In addition to hearing, sight, taste, touch, and smell, you have several other senses we depend on but don't often talk about. No, we are not just taking the obvious and breaking them down into the underlying components, although there is value to taking sight, for example, and breaking it down into how we sense color (cones) and brightness (rods). Also, there are different evolutionary (and culinary) reasons to discriminate taste into sweet, salty, sour, bitter, and umami. No, we are talking about different senses, with different sensors connected to the brain via different pathways, allowing us to react to a number of variables. Let's consider them in two simple groups: external-facing and internal-facing senses.

We have far more than five senses. External modalities include distinct systems for touch, pressure, temperature, and pain (nociception) – each with specialized receptors and neural pathways. Internal senses (interoception) include hunger, thirst, proprioception, time perception, and even vestiges of magnetoception.[21] These interoceptive signals, processed in brain regions like the anterior insula, form the substrate of emotion itself.[22] The distinction between these sensory modalities becomes relevant in conditions where one system may be compromised while others remain intact, such as in certain neuropathies where pain sensation is lost while touch remains functional, or vice versa.

Turning inward, however, we open up a whole new field called interoception, and it is thought that these senses, some of which are rooted in specific areas of the brain (anterior insula, anterior cingulate cortices, posterior cingulate cortex) are the basis for emotions.[23] This includes hunger, thirst, equilibrium (a.k.a. vestibular), itch (yes, this is a distinct sense), proprioception (where your respective body parts are), thermoception (how hot/cold you feel), tension (e.g. how tight your muscles feel), stretch (how much distension

---

20 Aristotle. *On the Soul*. Translated by J. A. Smith, *The Internet Classics Archive*, classics.mit.edu/Aristotle/soul.3.iii.html. Accessed Aug. 25, 2025.

21 Brandt, Thomas, et al. "Human Senses and Sensors from Aristotle to the Present". *Frontiers in Neurology*, vol. 15, article 1404720, 3 July 2024.

22 Chen, Wei-Guang, et al. "The Emerging Science of Interoception: Sensing, Integrating, Interpreting, and Regulating Signals within the Self." *Trends in Neurosciences*, vol. 44.1, 2021, pp. 3–16.

23 Critchley, Hugo D., et al. "Neural Systems Supporting Interoceptive Awareness." *Nature Neuroscience*, vol. 7, no. 2, 2004, pp. 189–195.

is occurring in your lungs, bladder, etc.), chemoreceptors (an adjunct to taste, this nervous pathway is wired directly to your medulla, and can trigger a vomit reflex), your ability to sense time,[24] and finally, possibly even vestiges of magnetoception (ability to orient oneself with respect to a magnetic field, such as the earth's).[25]

If we are to look forward later in the book into ways in which we might augment or adapt cognition, we should have an awareness of how it works at this level of granularity.

Chapter 8 covered how the human brain reacts to experience and thus its ability to change over time. Any understanding of how we understand a person may think with their embodied brain is really just a snapshot of time; it is constantly changing.

As is our sense of self, which evolves over time, and through experience.

3.4 The Society of the Mind

A typical popular view of the brain is that it is some type of single monolithic actor, consulting the charts, and steering the ship. In the literature the term for this is homunculus – literally the metaphor of a little person in your head, controlling you much as the little Arquillian alien controlled the humanoid shell in the original Men in Black movie.[26]

Kaplan disagrees with this view, stating:[27] "*in truth the individual is a congress of selves, each pursuing values to which the other selves may be indifferent or hostile – if, indeed, they are even 'aware of the pursuit.'*"

Artificial intelligence pioneer Marvin Minsky had a similar idea. He developed a model of the human mind as being composed of a very large number of "agents," each of which was mindless. The interactions between the agents constituted what he called the "society of the mind."[28]

This idea that the individual is composed of multiple "selfs" or subpersonalities has gained traction in clinical settings. Internal Family Systems (IFS) psychotherapy, for instance, sees the self:

> as a dynamic, evolving, adapting, and open system of subpersonalities or 'parts' that operate in ever-changing relationships with one another and the outside

24 Craig, Arthur D. "How Do You Feel? Interoception: The Sense of the Physiological Condition of the Body." *Nature Reviews Neuroscience*, vol. 3, no. 8, 2002, pp. 655–666.
25 Baker, Robin. *Human Navigation and Magnetoreception*. eBook Partnership, 2017.
26 Which everyone knows is better than the sequels.
27 Friedrich, Carl J., editor. *Rational Decision*. Atherton Press, 1964.
28 Minsky, Marvin. *Society of Mind*. Simon and Schuster, 1988.

> world. The self in IFS is not a singular or bounded thing but, rather, a process that emerges from a complex of internal relationships among constantly changing 'parts.' Agency, accordingly, is not localized within a fixed, singular core but is distributed and mutable, shifting among different subpersonalities depending on the setting, circumstances, and interactions at hand.[29] (Lester, 2017)

If Kaplan, Minsky, and IFS are right, the "*individual*" (which is derived from Latin "individuus," meaning not divisible), is actually and ironically subdivided into many different parts, most of which operate unconsciously, and many of which may be working at odds with each other.

Interestingly, the software industry has a term – "persona" that is useful to us here. In Latin, "personas" is the word used to describe masks worn by actors to portray different characters. Personas was adopted by software developer Alan Cooper in the 1980s, as a concept to help guide software design.[30] Essentially Cooper's personas were fictional characters that might use the software and were defined by their goals.

So, I will adopt the term "person" in lieu of "individual," and "personas" to represent the myriad of sub-identities we have within us, divided, and described based on the goals they are trying to achieve.

This is a level of complexity that far exceeds the Arquillian analogy. The Pixar movie Inside Out gave us a better analogy, though still greatly simplified, when they had the 11-year-old Riley ruled by five characters in her head, representing the emotions of Joy, Sadness, Fear, Disgust, and Anger. The arguments, fights, alliances, and other dynamics in that plot of five characters are much better suited to describe our cognitive complexity than the Arquillian model. We must remember that it is not just five emotions fighting it out, but a large congress of personas, or "society of Riley."

This short tour barely scratches the surface – I've simplified a system that operates simultaneously across molecular, cellular, network, and system levels, each with emergent properties invisible to the layers below. The brain isn't just complicated; it's layered complexity compounding into higher-order complexity. And this is only Stream 1."

---

29 Lester, Rebecca J. "Self-Governance, Psychotherapy, and the Subject of Managed Care: Internal Family Systems Therapy and the Multiple Self in a US Eating-Disorders Treatment Center." *American Ethnologist*, vol. 44, no. 1, 2017, pp. 23–35.

30 See Cooper, Alan. *The Inmates Are Running the Asylum*. Sams, 1999; and Cooper, Alan, et al. *About Face 3: The Essentials of Interaction Design*. Wiley, 2007.

4 Complexity within Culture

Culture is complex, yes?

But have you ever stopped to try unraveling that ball of yarn? If so, you'll find that, as a global construct, it approaches or even exceeds the complexity of the brain – the "most complicated thing in the universe."

4.1 The Culturalome: Our Social Network

When you think about your culture, you might think initially of your town, state, country, or maybe even your football club as having some type of culture to which you singularly and exclusively associate. But the truth is, each of your many different sub-identities – your personas that make up the "society of you" – associates with a different culture, and they have many to choose from. That's because there are ~8.2 billion living people on the planet,[31] plus a legacy of effects and influence from the ~117 billion who have lived on Earth at some point.[32] People are currently organized into 195 countries, within 4–9 civilizations (widely debated) and an infinite number of cultures.

Consider for a moment, of the person like the neuron: while the average neuron might have thousands of connections to other neurons, each person today may *know* (in the real world) around 300[33] – 600[34] other people. We've also *seen* a lot of people in real-life that we may not know, but to whom we feel a cultural connection.

Consider the schoolgirl looking through the windows of the bus at the girls from other schools who she doesn't know. She quickly decodes things she associates with and doesn't; unconsciously flagging new innovations she will adopt when she gets back home, like a particularly interesting new way to distress her jeans. How many people are you exposed to this way? Most of us

---

31 United Nations Department of Economic and Social Affairs, Population Division. "World Population Prospects 2024." United Nations, 2024, population.un.org/wpp/. Accessed Feb. 17, 2025.

32 Population Reference Bureau. "How Many People Have Ever Lived on Earth?" Population Reference Bureau, 2022, www.prb.org/articles/how-many-people-have-ever-lived-on-earth/. Accessed Feb. 17, 2025.

33 McCarty, Christopher, Peter D. Killworth, H. Russell Bernard, Eugene C. Johnsen, and Gene A. Shelley. "Comparing Two Methods for Estimating Network Size." Human Organization, vol. 60, no. 1, 2001, pp. 28–39.

34 Estimates range from 450–750, with significant gender differences, as well as cultural and age effects – See McCormick, Tyler H., et al. "How Many People Do You Know?: Efficiently Estimating Personal Network Size." *Journal of the American Statistical Association*, vol. 105, no. 489, 2010, pp. 59–70.

live in urban areas now, so this could be tens if not hundreds of thousands over our lifetime.

And that's living nonfictional people we know or have been exposed to directly, in real life. That doesn't include the people you've had a "mediated" exposure to; people who are living or dead, fictional, and nonfictional. You have the equivalent of a virtual synapse between you and every author you've read, every speech you've heard a recording of, every character you've seen on television, and every person who pops up in your social networks.

The numbers are staggering: between screens, social media, books, movies, and advertising, the average American is exposed to over 60 million distinct people and characters across their lifetime – approaching the complexity of the brain's 86 billion neurons.[35] Like neural networks, these cultural connections operate at micro (dyadic relationships), meso (communities), and macro (civilizational) scales. And like the brain, they're highly plastic, constantly reorganizing.

That's a "culturalome" approaching the complexity of the brain's neurons. And, just like in the brain, people in this network tend to have multiple net-

35 We can begin to quantify this: in 2024–2025 the average American reads 12 books a year, watches 3–4+ hours a day of some form of television, and spends about 5–7 hours a day with their smartphone and screens, which goes up to over 12 hours if you count multitasking (e.g. scrolling your phone while you watch TV). During their life, the average American will have seen 3,600–5,000 movies and over 30,000 TV shows. Your online time, between your phone and your computer usually includes two and a half hours of social media, where the average person has 8 social media accounts, and on Facebook, the most popular platform in 2025, the average user has 338 friends, each of whom post and forward information about dozens if not hundreds of people they know, but you don't. Along the way, you are being exposed to characters in advertising, as well – at the rate of a whopping 5,000–10,000 ads per day. And emerging platforms which feature short form content are now significant – American children watch 113 minutes a day on TikTok. Totaling it all up, and an average American adult might have been exposed to over 60 million people in advertising, one hundred and fifty thousand people in TV shows, perhaps around 75,000 while watching online videos, 25,000 in movies, 72,000 in books, and two- to three thousand in social media. Combine that with the real-world of unmediated exposure to say 100,000 people we may have seen IRL and felt a cultural response to (consciously or unconsciously), and the 300–600 people we actually know. I think this is VERY conservative, but the numbers are all over the map, ranging between small cluster with deep ties, and large exposure to "weak ties," with variation across platforms (e.g. does Tinder count?) and demographics. See for example The Manifest. "One-Third of Americans Like, Share, or Post on Social Media More Than 10 Times Per Day." *PR Newswire*, Jan. 17, 2019, www.prnewswire.com/news-releases/one-third-of-americans-like-share-or-post-on-social-media-more-than-10-times-per-day-300780034.html. Accessed Sep. 5, 2025. And "Primates on Facebook." *The Economist*, Feb. 26, 2009, economist.com/science-and-technology/2009/02/26/primates-on-facebook. Accessed Sep. 5, 2025.

works organized at the micro (a very large number of dyadic relationships, specific just to you and one other person), meso (many "community" and family networks), and macro (religion/national/civilizational affiliations). And they are highly plastic, changing over time.

4.2 Cultural Evolution and Digital Acceleration

But our connections with the past aren't just to people we know or read about in books or have seen on a screen. Artifacts like tool use, language, writing all come down to us from previous generations. So, too, do the socially constructed and shared abstracts like symbols, myths (including words, and the concepts of money, nations, corporations) and knowledge. If we make these analogous to the glial cells, then these socially constructed nodes seem to vastly outnumber the people or characters we know, and therefore are an order or two of magnitude greater than the ratio of glial to neurons in even the most glial-rich parts of the brain.

To really drive that home, look around you right now. I'm surrounded in my office by thousands of books and dozens of technology devices. On my desk are papers, a coffee cup, notebooks, and a container holding pens and pencils. Now think about how a single, simple pen is itself composed of maybe 100 supporting technologies.[36] That is just a mundane thing, one of the thousands of objects within sight of me here, now. Many of the others are FAR more complex. Our culture is scaffolded, layer upon layer, by a nearly infinite amount of socially and culturally constructed nodes – so much so that we become oblivious to their supporting roles, and we don't notice the water in which we swim.

The complexity of our social cognition is not just due to this static quantitative view of the cultural nodes of people and socially constructed artifacts; within each of the cultures we associate with, from dyad to civilization, there is a distribution of access to resources; and this is widely studied by scholars.

Maguire, for example, would look at cultures through the lenses of heterogeneity (distribution of populations between social groups) and inequality (differential access to material and social resources) within a society.[37] Another approach uses stratification as a proxy for cultural complexity through

---

36 We'll talk more about this, and the attribution in Chapter 16, when we talk about Kevin Kelly's pen.

37 McGuire, Randall H. "Breaking Down Cultural Complexity: Inequality and Heterogeneity." *Advances in Archaeological Method and Theory*, Academic Press, 1983, pp. 91–142.

either class (people's relationship to the means of production)[38] or more multidimensional models, which look at stratifications as a function of how each person interacts with the social forces of inequality, including power, education, age, sex, ethnicity, and class.[39]

Just as brains change over time, so, too, do cultures and socially constructed nodes. A person could move geographically from one place to another, and as such be a cultural transplant that has bidirectional effects on both the person as well as the culture they transplanted out of or into. Enslaved people brought to America represent a classic example of this. They also show how these cultural diasporas have multi-generational affects, as they transformed from slaves in one generation through multiple generations, The Great Migration, and subsequent generations into modern era African Americans.[40] And that term itself evolves back to "Blacks."[41] This doesn't have to be a collective diaspora, either, as we permanently move our homes around the country or world, individually, or temporarily travel for work or pleasure, we are like neurons moving around the brain, constantly re-organizing ourselves, and exposing us to new cultural networks.

Note, too, how the socially constructed nodes, like language, evolve. For example, the ritual of meeting someone for the first time is often facilitated using "phatic expressions" which serve a function to link two people in communication in a way that is distinct from their actual informational content.[42]

It is a pattern of speech that enhances the efficiency of your communication with the other person. You are bonding, through shared ritual. "Nice to meet you" might be a phatic expression intended to convey certain functions to a new colleague at work. You don't know yet whether it will be nice to know them; you are just saying an automatic greeting that the other may recognize and appreciate.

A classic phatic expression is when someone thanks you, and you respond, "You are welcome." This expression has noticeably changed among younger generations in recent years to "No problem." Because its content is different

---

38 People's relationship to the means of production.

39 See Lenski, Gerhard Emmanuel. *Power and Privilege: A Theory of Social Stratification.* UNC Press Books, 1984, and Jeffries, Vincent, and H. Edward Ransford. *Social Stratification: A Multiple Hierarchy Approach.* Allyn and Bacon, 1980, and Berreman, Gerald Duane, and Kathleen M. Zaretsky. *Social Inequality: Comparative and Developmental Approaches.* Academic Press, 1981.

40 Thanks to Paul Budde for suggesting this.

41 Associated Press. "Explaining AP Style on Black and White." *AP News*, Jul. 20, 2020, apnews.com/article/archive-race-and-ethnicity-9105661462.

42 Coulmas, Florian, editor. *Conversational Routine: Explorations in Standardized Communication Situation and Prepatterned Speech.* Mouton, 1979.

than what an older generation might use (though still serving the same phatic utility), an older person might translate it from a phatic function into a content-specific function and take issue with it.[43]

The Internet has been a major force for social evolution, driving the adoption of new language devices, terms, and even new concepts of work, education, and money. Gretchen McCulloch, an Internet linguist, devoted a whole book to how the Internet has not only created new forms of communications in the forms of emojis and textspeak (e.g., LOL), and new words (e.g., "meatspace" -the physical world, opposite of online, or as I occasionally use here – IRL), but also new forms of grammar and use of existing words, such as "because," or phatic expressions, like "No problem."[44]

As language evolves, so does the idea of money or currency. Yes, money became digital, and you are now as likely to Venmo money as you are to write a check, but the idea of capital has shifted, as social media influence, in the form of followers and likes, has begun to compete with currency as a form of accumulated wealth. And during the pandemic, many of us learned new modes of acquiring both, in the form of rapidly changing work and education.

The final level of social complexity for us to consider is that the Internet now connects us in new and powerful ways at an unprecedented scale. These fresh links/social synapses unlock social transformation at a pace and reach that is unlike anything humans have seen before.

In 1990, only a very small fraction of people, perhaps less than half a percent were using the Internet globally.[45] A decade or so later, approximately half of America was online,[46] and today, almost all of America is online,[47] and 91%

43 "Why 'No Problem' Can Seem Rude: Phatic Expressions." *Arizona Show*, azshow.info/watch/eGnHoKAXhCw/why-no-problem-can-seem-rude-phatic-expressions.html. Accessed Aug. 25, 2025.

44 McCulloch, Gretchen. *Because Internet: Understanding the New Rules of Language*. Riverhead Books, 2020.

45 World Bank. "Internet Users for the United States (ITNETUSERP2USA)." *FRED, Federal Reserve Bank of St. Louis*, updated Mar. 26, 2025, https://fred.stlouisfed.org/series/ITNETUSERP2USA. Accessed Sep. 5, 2025.

46 See for example: Perrin, Andrew, and Sara Atske. "31% of American Adults Are Online Almost Constantly." Pew Research Center, 26 Mar. 2021, www.pewresearch.org/short-reads/2021/03/26/about-three-in-ten-u-s-adults-say-they-are-almost-constantly-online/. Accessed Sep. 5, 2025.

47 World Bank. "Internet Users for the United States (ITNETUSERP2USA)." *FRED, Federal Reserve Bank of St. Louis*, updated 26 Mar. 2025, https://fred.stlouisfed.org/series/ITNETUSERP2USA. Accessed Sep. 5 2025.

have smartphones.[48] Globally, over 5.6 billion people have access to the Internet,[49] and at least 4.6 billion people carry a smartphone of some kind.[50] That means people are now just a few finger movements away from connecting potentially with anyone else in the world, basically for free. Ideas can hatch in one person's head in Bangladesh and move across the world in seconds or minutes. Recall Metcalfe's Law: the power or value of a network is proportional to the square of the number of connected users of the system (n^2).

All the while, we have seen the idea of "our network" restructured and reconceived not as the people who know personally, but rather the people we have ready access to via the Internet, in new ways. These new links can include anonymous followers who can be catalyzed in an instance, all over the world; to new "colleagues" on LinkedIn; a vast pool of potential mating partners on platforms like Tinder; a Doordash or Uber driver; new customers following you on Mercari, Poshmark and Ebay; and new anonymous resources ready to do your work, in places like Amazon Turk or other gig economy apps such as Lugg or Favor.

The global digital network represents humanity's first collective brain – one that flickered to life in 1990 as scattered neurons began forming their initial connections. Over three decades, this neural web has grown exponentially, linking billions of individual minds across continents through fiber optic synapses and wireless dendrites. Each new user, platform, and device adds another neuron to this sprawling cortex, creating pathways that reshape themselves with every click, share, and search.

Today, this digital consciousness processes the thoughts, dreams, and daily experiences of nearly eight billion people simultaneously. It learns constantly, adapting its connections as trending topics surge through its networks like electrical impulses across brain tissue. Yet despite its vast processing power, this global mind remains fundamentally adolescent – capable of remarkable creativity and connection, but also prone to confusion, contradiction, and the occasional spectacular mistake.

Our digital collective consciousness continues experimenting with its capabilities, sometimes inspiring unprecedented collaboration, other times amplifying our deepest divisions. Like any 35-year-old, it is still asking itself what it

48 Pew Research Center. "Mobile Fact Sheet." *Pew Research Center: Internet & Technology*, Nov. 13, 2024, https://www.pewresearch.org/internet/fact-sheet/mobile/. Accessed Sep. 5 2025.

49 DataReportal. "Digital around the World – DataReportal – Global Digital Insights." *DataReportal*, 2025, datareportal.com/global-digital-overview. Accessed Sep. 5, 2025.

50 Backlinko. "Smartphone Usage Statistics for 2025." *Backlinko*, Jan. 31, 2025, backlinko.com/smartphone-usage-statistics. Accessed Sep. 5, 2025.

wants to be when it grows up. The question remains: what kind of intelligence will emerge?

While this last page deals explicitly with technologies like cellular networks, smartphones, and social platforms, it was only to note their enabling role in our emerging cultural dynamics.

Let's now examine the complexity of our final layer – thinking within Technology.

5 Complexity within Technology

5.1 The Scale of Our Exocortex

It is daunting to try to conceive the sheer quantitative scale of what is available to us, today, outside of our skin, to aid in our cognitive tasks.

Our existing "exocortex"[51] is a highly modular and increasingly networked environment of more than 15 billion connected and programmable devices around the world in 2023, representing millions of kinds of hardware and software. That number is expected to double by 2030.[52]

One of the world's most common connected devices is an iPhone, the newest model of which rivals in some ways the computing capacity of some of the top 1990s supercomputers.[53] They provide instant access to most of the

51 Remember this is shorthand slang; Ben Houston's term is a metaphor for the fact that technology has become the cortex outside of our skin. However, I've endeavored to show that ALL parts of the brain are critical to thought, and that even then, we have to consider the embodied brain and the rest of the body's participation in thought processes. But exocortex is a cool simplifying catchphrase for all of that.

52 Cisco. "Cisco Annual Internet Report (2018–2023) White Paper." *Cisco*, 2020, www.cisco.com/c/en/us/solutions/collateral/executive-perspectives/annual-internet-report/white-paper-c11–741490.html. Accessed on Sep.5, 2025.

53 On on-device AI tasks, the iPhone 16 Pro Max's A18 Pro includes a 16-core Neural Engine reported at roughly 35 trillion operations per second, which puts its raw operations-per-second figure in the same numeric ballpark as ASCI Red's landmark ~1 teraflop (≈1 trillion floating-point ops/sec) LINPACK result in 1997–98. The comparison is apples-to-oranges – phones excel at low-precision AI inference while ASCI Red was built for double-precision scientific math. But hey, the iPhone fits in your back pocket, costs <$1,500, and takes very little energy, while ASCI RED filled a large room, cost $55 million to build, and drew enough electricity to power ~1,400–1,800 typical homes. See "Apple Debuts iPhone 16 Pro and iPhone 16 Pro Max." Apple Newsroom, Sep. 6, 2024, https://www.apple.com/newsroom/2024/09/apple-debuts-iphone-16-pro-and-iphone-16-pro-max/, and D'Angelo, Anthony. "Apple Announces Its New A18 and A18 Pro iPhone Chips." TechCrunch, Sep. 9,2024, https://techcrunch.com/2024/09/09/apple-announ

world's accumulated knowledge (much of it in real-time, and free, hosted by over a billion websites[54] – perhaps 200 million are active[55] – stored on 100 million servers around the world).[56] In addition to browsing the web, the typical phone has an average of 90 apps installed,[57] and perhaps 30 are used at least once per month.[58] Many of those apps are free, but even more complexly, they are ad-supported. We'll return to that again in a minute. The phone also has a rich suite of hardware that includes not only robust microphones and multiple cameras, but also highly sophisticated proximity, accelerometer, ambient light, moisture, gyroscope, compass, barometer and touch (e.g., fingerprint) sensors and other technical features like haptic feedback drivers.

Behind each of these powerful perception sensors is a portfolio of increasingly powerful processing tools, many of which actually further outsource processing to the cloud, over increasingly fast 4G, 5G, and Wi-Fi connection, to enable the camera to turn into a powerful facial recognition platform, for example, or to notify you of those in your proximity that are in your network, such as those looking to hookup near you.

5.2 Intimate Monitoring: When Technology Gets Personal

If you are one of the tens of millions of people with an Apple Watch, things get more intimate, as it has the ability to perform sophisticated analysis of your heartrate throughout the day, and notify you with specific and highly accurate feedback about what is going on unconsciously in your own body, for reasons such athletic performance, health alerts if you get an irregular heart rate, and even notification of your estimated stress levels, through analysis of a stress proxy called Heart Rate Variability (HRV).

---

ces-its-new-a18-iphone-chip/, and "ASCI Red." TOP500, 1997, https://top500.org/resources/top-systems/asci-red-sandia-national-laboratory/, and "Reflecting on the 25th Anniversary of ASCI Red and Continuing Themes for Our Heterogenous Future." *HPCwire*, Apr. 26, 2022, https://www.hpcwire.com/2022/04/26/25th-anniversary-of-asci-red-and-continuing-themes-for-our-heterogenous-future/.

54 Kumar, Naveen. "Big Data Statistics 2025 (Growth & Market Data)." *DemandSage*, Jun. 24, 2025, demandsage.com/big-data-statistics/. Accessed Nov. 19, 2025.

55 "How Many Websites Are There?" *Siteefy*, siteefy.com/how-many-websites-are-there/. Accessed 2Aug. 25, 2025.

56 Sire, Tom. "How Many Servers Are There in the World?" *Tech Pulsion*, techpulsion.com/how-many-servers-are-there-in-the-world/. Accessed Aug. 25, 2025.

57 Olsson, Sam. "App Downloads Statistics." *Kurve*, kurve.co.uk/blog/app-downloads-statistics. Accessed Aug. 25, 2025.

58 BuildFire. "Mobile App Download Statistics & Usage Statistics (2025)." *BuildFire*, Dec. 31, 2024, https://buildfire.com/app-statistics/. Accessed Sep. 5,2025.

5.3 The Ubiquity Problem: Technology Everywhere

While the most common or visible technology in our lives may be the smartphone, technology has infused itself into the world around us and our way of life.

The average American had 21 connected devices in 2023.[59] Looking at those other connected devices in our homes, we see televisions that can listen to us, or watch our movements and interface those movements into environments like video games. We have new kinds of devices like Amazon Echo, with platforms like Alexa or Microsoft's Cortana, built specifically around interacting with our voice. These are constantly listening to help us with any questions we have, or to place an order for anything our minds might desire, from products on marketplaces like Amazon, to food from services like DoorDash, to music, podcasts, and lessons from other online content sources. If you have the right kind of car today, you can just get in, tell it where you want to go, and it will drive you there, and park itself, all the while listening to and watching you, interacting with your needs, attention, and intentions.

Our devices don't just listen, passively, and try to discern the words and intentions. Like we see on the iPhone Face ID, powerful cloud-based AI and ML can deeply analyze your written, spoken, visual signals and gestures, and do things like accurately determine everything from emotions and lying[60] to diagnosis of key diseases.[61]

The decreasing costs of this technology has now pushed it beyond the traditional platforms into classically un-technology platforms such as dolls, ther-

---

59 Deloitte. "Connectivity & Mobile Trends 2023: Connected Consumer Study." *Deloitte Insights*, Sep. 5, 2023, https://www.deloitte.com/us/en/insights/industry/telecommunications/connectivity-mobile-trends-survey/2023/connectivity-mobile-trends-survey-full-report.html. Accessed Sep. 17, 2025.

60 Shulevitz, Judith. "Alexa, Should We Trust You." *The Atlantic*, Nov. 2018, https://www.theatlantic.com/magazine/archive/2018/11/alexa-how-will-you-change-us/570844/. Accessed Dec. 2, 2025.

61 Chan, Justin, et al. "Contactless Cardiac Arrest Detection Using Smart Devices." *NPJ Digital Medicine*, vol. 2, no. 1, 2019, pp. 1–8; and Arora, Siddharth, et al. "Detecting and Monitoring the Symptoms of Parkinson's Disease Using Smartphones: A Pilot Study." *Parkinsonism & Related Disorders*, vol. 21, no. 6, 2015, pp. 650–655; and Maor, Elad, et al. "Voice Signal Characteristics Are Independently Associated with Coronary Artery Disease." *Mayo Clinic Proceedings*, vol. 93, no. 7, Elsevier, 2018.

mostats, door locks, lights and switches, appliances large and small, doorbells, pet feeders, mirrors, and soon into our toilets.[62]

5.4 The Hidden Complexity: A Case Study in Smart Thermostats

To understand how this ubiquity creates unexpected complexity, consider what happens when we upgrade seemingly simple devices.

What happens to the actual complexity when you make simple changes like upgrading a doorbell or thermostat from a traditional "dumb" appliance to a new-fangled "smart" one?

Take your humble Google Nest Thermostat. Sure, it's a great improvement over the old standard analog temperature controls in our houses. We can feel more comfortable, spend less money, and use less energy.[63] Another clear case of objective gains that not only benefits us, but also plays a small part in bettering the world and addressing climate change.

These improvements are due in part to the amazing amount of data these devices can observe about you and your household – they can monitor and record the current temperature, humidity, motion, and ambient light wherever they are installed. They also transmit that data back to the cloud, where Google aggregates this information with all of the other information it gets from your use of Google services and products. They use this information to better target ads to you, but they also share it with third parties to deliver an enhanced suite of services that make these smart thermostats significantly better than the old ones. This seems pretty benign – after all, who cares what your temperature preferences are, or what time you get up and move around the house?

Two University of London legal scholars who examined Nest's "terms of use," "end user license agreements," and privacy policy, and learned that for a consumer to really understand where that data goes and how it will be used, they would have to eventually read through nearly a thousand pages of contracts from all of the byzantine network of parties who have access to that data.[64] Even if you did all that detective and legwork and were fully informed

---

62 Armitage, Hanae. "Smart Toilet Monitors for Signs of Disease." *Stanford Medicine News Center*, Apr. 1, 2020, med.stanford.edu/news/all-news/2020/04/smart-toilet-monitors-for-signs-of-disease.html. Accessed Aug. 25, 2025.

63 "Energy Savings from the Nest Learning Thermostat: Energy Bill Analysis Results." *Nest Labs*, Feb. 2015, https://storage.googleapis.com/nest-public-downloads/press/documents/energy-savings-white-paper.pdf. Accessed Sep. 5, 2025.

64 Noto La Diega, Guido, and Ian Walden. "Contracting for the 'Internet of Things': Looking into the Nest." *European Journal of Law and Technology*, vol. 7.2, 2016.

before you installed the device, many of those parties reserve the right to unilaterally update those policies, without giving you any notice, and you'd have to continually check back in on all those policies to really know how your data is being used.

Remember that comment a couple of pages ago, noting that we have access to all that amazing data and apps, often for free, because they are ad -supported? It makes one rethink that complexity.

This device is a useful reference point in part because it is so simple – it doesn't (yet) have a microphone or a camera. But your Nest doorbell has both, and increasingly, so do dozens of other devices in your home, in the form of your phones, laptops, and tablets, to your Alexa devices and smart TVs, and now even in your car and toys. These devices have various sensors, including many more that record your life. Can you imagine how many companies and people may have access to all of that? Or how many contracts you would have to read to be aware of all the ways all this data about your life is being used?

5.5 Security Vulnerabilities and Systemic Risks

These privacy concerns represent just one dimension of a broader security challenge, of how your data gets used.

Or misused. No less an authority than the FBI has stated, *"Beyond the risk that your TV manufacturer and app developers may be listening and watching you, that television can also be a gateway for hackers to come into your home. A bad cyber actor may not be able to access your locked-down computer directly, but it is possible that your unsecured TV can give him or her an easy way in the backdoor through your router."*[65]

You know that Kasa smart plug you bought on Amazon so that you could shut off the Christmas tree just by talking with Alexa? Maybe Amazon flagged the 4-pack for just $23.99 as the "Overall Pick"[66] for you. You remember how easy it was to install? Just connect to your home Wi-Fi network, enter the network password, and connect it to Alexa, and voila. At a recent national security convention, a prominent cybersecurity expert told me that Chinese companies primarily manufacture these smart plugs. Installed in millions of homes across the country, they literally have all the Wi-Fi access, passwords,

---

65 "FBI Smart TV Security." *TechCrunch*, Dec. 1, 2019, techcrunch.com/2019/12/01/fbi-smart-tv-security/. Accessed Aug. 25, 2025.

66 "Kasa Smart Plugs." Amazon. https://www.amazon.com/kasa-smart-plugs/s?k=kasa+smart+plugs. Accessed Dec. 2, 2025.

connectivity, and power to control whatever connected devices you have in your home. Talk about an expanded attack surface.[67]

We'll touch on this aspect again in the next chapter.

5.6 Accelerating Change: The Pace Problem

While the current technological landscape is already daunting, it is the accelerating pace of change and adoption that represents the true complexity challenge.

Connected devices engage with a number of platforms, like cloud-based social networking. Today, many professionals use low-cost AI agents to coordinate their schedule with others,[68] while others use Amazon's recently launched quantum computing on demand to engineer newer, more powerful technologies of the future.[69]

That point brings us to our final level of complexity. As daunting as the picture of technology in our lives today is, the real complexity lies in the accelerating pace of change and adoption.

We've already noted that the number of connected devices will double by the end of the decade. In 2025, the world will have about 180 Zettabytes (ZB) of data; that's almost double what we had in 2022.[70] At least one source says that it will balloon to over 600 ZB by 2030.[71]

How fast will other dimensions evolve? While we'll talk a little bit more about this in Chapter 16, it's useful to look back a few years to sense how well we can forecast this stuff.

In 2017, the World Economic Forum released a report: Deep Shift – 21 Ways Software Will Transform Global Society, which has since been widely cited. That survey of leading technology executives found that most thought by 2025 that 10% of people will wear clothes connected to the Internet; driverless cars

---

67 Bonaventura, Davide, Sergio Esposito, and Giampaolo Bella. "The IoT Breaches Your Household Again." *arXiv preprint arXiv:2407.12159*, 2024.

68 "xAI." *xAI*, x.ai/. Accessed Aug.25, 2025.

69 "Amazon Braket – Get Started with Quantum Computing." *AWS*, aws.amazon.com/blogs/aws/amazon-braket-get-started-with-quantum-computing/. Accessed 25 Aug. 2025.

70 See, for example: Valkhof, Bart, et al. "Data Volume Is Soaring. Here's How the ICT Sector Can Sustainably Handle the Surge." *World Economic Forum*, May 22, 2024. Accessed Sep. 5, 2025.

71 UBS Chief Investment Office. *How the Data Universe Could Grow More Than 10 Times from 2020 to 2030*. Mar. 21, 2024, https://www.ubs.com/us/en/wealth-management/insights/market-news/article/_jcr_content.0116445243.file_PS9jb250ZW50L2RhbS9pbXBvcnRlZC9jaW9yZXNlYXJjaC9wZGYvMTUvOTYvMzIvOS8xNTk2MzI5L2VuL3VzLzE1OTYzMjkucGRm/1596329.pdf. Accessed Sep. 5, 2025.

will account for 10% of all vehicles on the road, we will be transplanting 3D=printed organs and creating 3D-printed production cars; and an Artificial Intelligence technology will sit on a corporate board of directors.[72]

5.7 Phase Change: On Post-Late-Stage-Grasshoppers and Caterpillars

But this acceleration isn't simply about doing familiar things faster – we're experiencing fundamental qualitative shifts.

Yes, while those projects were a little premature, the discussion of an AI board member illustrates that technology change isn't just "more of the same." We are not just processing more, faster, and connecting to more things.

We are experiencing a phase change.

New technologies will do things that never existed before. Yes, your new $1,500 Lululemon workout mirror may look, at first glance, like your old school Jane Fonda VHS workout video behind a slightly reflective glass panel. But with cameras, the addition of a "virtual you," connected to stored and live workouts in which you share experiences with people from around the world, the mirror blurs the line between fitness, shopping, entertainment, and even social experiences.

Similarly, while the first Google Glass wasn't a hit, and 2024's Vision Pro has already been discontinued, we will doubtlessly be seeing an explosion in augmented reality and virtual reality glasses, certainly by the end of this decade. These layer whole new strata of technology entanglement to our current cognitive experience. With functional augmented reality, you'll soon be able to walk into a room, and by combining facial recognition with other platforms, you could look around a room full of strangers and see everyone's name, social media profile, job title and company, and value of their home, all overlaid over their face as you see it in the "meatspace"' of the real world through the glasses. This is the F-35 helmet we discussed in Chapter 4, coming to life at the consumer level. In just a few years.

And those very same glasses and goggles allow us to explore an infinite and incalculable number of virtual worlds that are axiomatically unprecedented in human history.

---

72 "Tech Life Predictions for 2030." *World Economic Forum*, 2017, weforum.org/agenda/2017/10/tech-life-predictions-for-2030/. Accessed Aug. 25, 2025.

5.8 Humanity's Phase Change

Let's end this section by discussing my "phase-change" claim. In the next chapter, we will acknowledge that nearly every generation may have felt it was experiencing an unprecedented level of change. How do I explain or substantiate my bold and potentially hyperbolic phase-change claim?

When I've tried to explain it before, knowing what I know now about deep human evolution and our current technological ecosystem niche, I tried to use mechanical system analogies: it doesn't take much energy, for example, to convert plain water at the freezing threshold to move it from liquid to solid, or vice versa. But that little nudge changes the fundamental characteristics of the H_2O. However, it may be inappropriate to apply such framing to human biological systems.

I turned next to the butterfly metaphor. At first, I rejected it as too common – every caterpillar that lives long enough eventually becomes a butterfly; it wasn't context-dependent. And it can be an overused example of how things transform. But when you really examine the butterfly's metamorphosis, it's anything but ordinary. The caterpillar literally dissolves into "imaginal soup" – a nutrient-rich cellular goo with no continuity of brain, memory, or identity. While the caterpillar and the butterfly have the same DNA, the butterfly is essentially a different organism built from the dissolved remains of the first. It's not transformation; it's complete reconstruction of identity itself.

But still, the metaphor was limited; I needed something contextual.

The answer was locusts.

Grasshoppers and locusts are essentially the same insects, but they are dramatically different in key characteristics such as individual attributes, collective behavior, and ecological impact. The transformation is so complete that until the 1920s, scientists believed solitary and gregarious locusts were entirely different species.[73]

The transition is situationally dependent.

Grasshoppers are shy, antisocial insects that live independently, avoid others, jump away when disturbed, and display muted green or brown colors that help them blend into vegetation. They eat about half their weight per day,

---

73 Mullen, Sean. "When Grasshoppers Go Biblical: Serotonin Causes Locusts to Swarm." *Scientific American*, Feb. 20, 2024, www.scientificamerican.com/article/when-grasshoppers-go-bibl/. Accessed Sep. 23, 2025.

See "Grasshopper Behavior." *BioExpedition*, Jun. 28, 2024, www.bioexpedition.com/grasshopper-behavior/. Accessed Sep. 23, 2025.

mostly grass, leaves, and flowers, and will typically travel a few yards to a few hundred meters daily. Grasshoppers have an elaborate mating song produced by rubbing their hind legs on their forewings.[74]

Locusts, however, are gregarious and eat about their total body weight each day, mostly consuming agricultural crops like cereals. They become highly mobile, actively seek company, and develop distinctly different colors and large muscles that equip them for prolonged flights, shifting from solitary creatures into coordinated swarms that can stretch across vast distances.[75] They travel over 100 kilometers per day in coordinated swarms and have been known to fly from West Africa to the Caribbean (3,100 miles) in just 10 days, and have different mating songs, rituals, and life cycles than grasshoppers.

What triggers this Jekyll and Hyde transformation?

Population density. When the threshold level is hit – induced by several contacts per minute over a four-hour period – the behavioral shift, which manifests as a threefold increase in serotonin levels in their nervous systems, can begin in as little as two hours.[76]

Am I saying that the Internet, 5G, Wi-Fi, and social media are going to increase our "contacts per minute" beyond a threshold for a prolonged period and cause us all to change color, develop large muscles for extended flight, double our daily diet, and alter our mating patterns? Of course not; this is a metaphor.

But it's part of an apt metaphor for the human transformation occurring today.

We're BOTH late-stage grasshoppers at the locust threshold, but instead of changing colors and growing different muscles, we are also late-stage caterpillars, dissolving into our own form of imaginal soup. The question is whether we emerge, and if we do, will it be as something that flies in formation or devours everything in its path.

---

74 See "Grasshopper Behavior." *BioExpedition*; and Piercy, Katie. "How do Grasshoppers Reproduce?" *Meadowia*, April 2, 2021, meadowia.com/how-do-grasshoppers-reproduce/. Accessed on Dec. 2, 2025.

75 "Locusts, Facts and Photos." *National Geographic,* 3 May 2021, www.nationalgeographic.com/animals/invertebrates/facts/locusts. Accessed Dec. 2, 2025.

76 See "A Brain Chemical Changes Locusts from Harmless Grasshoppers to Swarming Pests." *University of Cambridge*, www.cam.ac.uk/research/news/a-brain-chemical-changes-locusts-from-harmless-grasshoppers-to-swarming-pests; Accessed Dec. 1, 2025; and "Locust." *Wikipedia*, Aug. 2, 2025, en.wikipedia.org/wiki/Locust. Accessed Dec. 1, 2025.

Technology is now directly interfacing with our cognitive processes rather than just augmenting them. Previous revolutions changed *what we could do*. This one is changing *how we think*.

How do I substantiate that? Largely through this book, which, primarily over the next three chapters, demonstrates changes over the last 100–150 years that are as dramatic as the grasshopper-to-locust transition, and then cultural and technological changes over the last 10 years that are as abrupt as the behavioral change in the triggered grasshopper.

5.9 Summary of Complexity of Thinking with Technology

Thinking with Technology demonstrates complexity that rivals – and maybe even surpasses – that of the human brain. Yet this represents only one Stream. The true magnitude of complexity emerges when we examine how they function together as a unified whole.

6 Emergent Complexity

The Three Streams Model is not a static platform: our thinking is dynamic, plastic, and adaptive at every level. That change is explored in Chapter 16, but here we examine the emergent complexity arising from interactions between the streams.

For every change in each layer, there is an effect on and likely adaptation by the other layers, which then creates cascading effects back on the original layer. Picture a person getting pushed into an overcrowded train in Japan – the closest passengers step back and then those behind them shuffle, creating ripple effects throughout the car.[77] As technological proliferation expands exponentially, these interactions between layers generate emergent cognitive and cultural effects.

As cost/proliferation curves expand exponentially in nearly every area of human existence,[78] the follow-on interactions between the layers create a deluge of emergent technology and cognitive effects.

---

77 Japanese rail lines actually employ people to do this – their formal job title is "People Pusher."

78 This is a broad sweeping claim; we'll explore this statement more in Chapter 16.

As a simple illustration of that, consider the cultural innovation of micro-banking and microfinance in Kenya. Mobile money saw strong adoption in Kenya, and it did more than digitize transfers:[79] Rigorous analyses link it to higher consumption and poverty reduction, illustrating how technical rails can catalyze cultural and economic adaptation. This led to more people having mobile phones. This supported increased microfinance adoption, which further hastened economic consumption and poverty reduction.

Culturally, innovation and entrepreneurship flourished, based largely on these self-reinforcing technologies, and this led to broader wealth creation, new educational opportunities, and facilitated broader networks of ideas and money. However, these trends also brought with them new epidemics of lifestyle diseases,[80] showing how gains intersect with other trends in non-linear ways, and co-travel with new risks.[81]

Even if we stopped technological innovation completely right now, we would have years of continuing adaptations to the changes rendered so far.

Technological innovation is characterized by a long tail of consequences; some positive and some negative. Put another way is the earlier quote from Melvin Kranzberg's first law of technology: "*Technology is neither good nor bad; nor is it neutral.*"[82] After each technological launch, there is a quickening initial adoption and adaptation, but that's also followed by a long cascade of additional effects, including lock-in, over time. And each of those effects themselves has positive feedback loops and bootstrapping effects, in zero-sum and non-zero-sum open complex systems.[83]

79 See Suri, Tavneet, and William Jack. "The Long-Run Poverty and Gender Impacts of Mobile Money." Science, vol. 354, no. 6317, 2016, pp. 1288–1292; and Jack, William, and Tavneet Suri. "Risk Sharing and Transactions Costs: Evidence from Kenya's Mobile Money Revolution." *American Economic Review*, vol. 104, no. 1, 2014, pp. 183–223; and Suri, Tavneet. "Mobile Money." *Annual Review of Economics*, vol. 9, 2017, pp. 497–520, www.annualreviews.org/doi/10.1146/annurev-economics-063016-103638. Accessed Aug. 19, 2025.

80 "Lifestyle Diseases Pose New Burden on Africa." *UN Africa Renewal*, Dec. 2016, un.org/africarenewal/magazine/december-2016-march-2017/lifestyle-diseases-pose-new-burden-africa. Accessed Aug. 25, 2025.

81 Mensah, Julia, et al. Combating Noncommunicable Diseases in Kenya: An Investment Case. World *Bank*, https://documents1.worldbank.org/curated/en/428881586197529642/pdf/Combating-Noncommunicable-Diseases-in-Kenya-An-Investment-Case.pdf, 2020. Accessed Dec. 2, 2025.

82 Kranzberg, Melvin. "Technology and History: 'Kranzberg's Laws.'" *Technology and Culture*, vol. 27, no. 3, 1986, pp. 544–560, www.jstor.org/stable/3105385. Accessed Aug. 19, 2025.

83 Holland, John H. *Emergence: From Chaos to Order*. Perseus Publishing, 1998.

Accelerating change means a cascading typhoon of accelerating responses and adaptations to those changes, which then, in turn, affect technology development in a not-so-virtuous and continuously accelerating whip snap of change, whose long-term consequences on human systems are, at best, poorly understood. This is an amplified version of the dynamic that Hagel, Brown, and Davison observed starting in 2009: systems never have a chance to return to stability, because changes are introduced faster than adaptations.[84]

Like Lorenz's butterfly effect,[85] where minuscule changes produce large-scale results over time, each technological innovation may trigger years of cognitive and cultural transformation.[86]

It is quite possible that the level of complexity here equates to something more profound – perhaps with each new technological innovation, the effect on human thought is something like *"a single butterfly wing flap causing years of weather change, and decades of changes in human culture."*

As Andy Clark has argued, our mature mental routines are not merely self-engineered – they are massively, overwhelmingly, almost unimaginably so.[87,88] We are continuously introducing changes to this already complex system.

7 Plain Language Chapter Summary

Thought seems deceptively simple – it just flows, and it works. But it's more like a newborn baby: when you step back and consider its complexity, you marvel at the miracle that it could even exist. Without understanding this complexity, you won't grasp thought's needs and vulnerabilities, how to nurture it, or what threatens its well-being.

84 See the Disruption chapter in the appendix for a more fulsome discussion of this. And Hagel, John, et al. "The Big Shift: Measuring the Forces of Change." *Deloitte Center for the Edge*, 2009, www2.deloitte.com/us/en/pages/center-for-the-edge/articles/big-shift-measuring-forces-of-change.html. Accessed 1Aug. 19, 2025.

85 To be clear, Lorenz was discussing mathematical models; here we are making a metaphorical leap into conceptual not literal mathematical models.

86 Lorenz: "Predictability," *AAAS 139th Meeting*, 1972, https://www.scribd.com/document/754797224/Butterfly-effect-Wikipedia. Accessed Dec. 2, 2025.

87 As Andy Clark has argued, our mature mental routines are not merely self-engineered – they are massively, overwhelmingly, almost unimaginably so.

88 Readers might want to refer to the Optimistic Perspectives on Human Progress/Candles in the Dark section of the Suggested Reading appendix, for the subjective experience of "eureka" moments that occasionally happen within the long research process.

CHAPTER 15

Desiderata: Thought Is a Highly Nested Pursuit of Many Changing Values

> Ultimately, man should not ask what the meaning of his life is, but rather must recognize that it is he who is asked.[1] (Viktor Frankl)

> Life consists in what a man is thinking of all day.[2] (Ralph Waldo Emerson)

This section may be the most important insight of this book.

Why bury it so deep? Because understanding these points requires the foundation we've built through everything covered so far.

Values are things we think are important or needed, and valued outcomes are things or experiences we pursue to help meet those needs.[3] These were central to the definition of thought we set out in the introduction to this work (emphasis added to this chapter's relevance):

Thought is the processes and tools a person uses (consciously or unconsciously) to:

- evaluate their needs and goals,
- make sense of their environment (including other people) and themselves,
- consider how to use available resources to achieve **valued outcomes** (goals), and control processes used to manage resources to achieve their valued outcomes

At the beginning of this Insights section, I made the point that values – that which we hold dear and pursue – are becoming rapidly anachronistic, subject to radical changes and manipulations, and have even now become maladaptive. Now, I will dive into this claim and examine the evidence, layer by layer.

1 Frankl, Viktor E. *Man's Search for Meaning*. Translated by Ilse Lasch, Beacon Press, 2006, p. 9.

2 Emerson, Ralph Waldo. *Emerson in His Journals*. Edited by Joel Porte, Harvard University Press, 1982, p. 377.

3 It's critical to remember that most of this is unconscious, and that there is no singular determinant of what is wanted or needed. We have a congress of selves, each competing for its own, often competing, needs and wants. We are also not going to delve too deeply into the differences here between "needs" and "wants," which while important in most contexts, here, they each represent a different degree of value being pursued by competing selfs.

 HTTPS://DOI.ORG/10.1163/9789004739161-020

1 Disrupted Values in the Brain

As we've discussed, Pinker and the Santa Barbara School correctly recognize that our evolutionary past shapes contemporary behavior, but they miss important elements. First, genetics alone doesn't determine human nature – culture, individual experience, and material environment exert significant influence on our behaviors, even shaping genetic expression through epigenetic effects and mechanisms like telomeric fragility. Second, their focus on the Environment of Evolutionary Adaptedness (EEA) ignores critical evolutionary periods both before and after this timeframe.

Humans didn't emerge 200,000 years ago from nowhere. Our cognitive evolution extends across tens of millions of years of primate development, offering specific insights into our current condition. We'll return to this deeper history shortly.

Many scholars, including leading figures, reference our "stone age past" while overlooking recent evolutionary developments within the last 50,000 years. Technologist Richard Boyd jokes that humans haven't received an upgrade since the Pleistocene,[4] yet Cochran and Harpending show genetic evolution has continued – and accelerated – into the present.

These genetic misconceptions reveal how our self-created modern environment makes many inherited traits anachronistic and often maladaptive.

Understanding our primate heritage gives us a framework for seeing modern life in a new light.

When biologists discover new species, they systematically catalog key characteristics that drive behavior and shape cognition.[5] Social structure proves fundamental: group size, foraging patterns, leadership hierarchies, and mating dispersal strategies all influence how a species processes information and makes decisions, and generally exist within their ecosystem.

Individual traits carry equal weight in this analysis. Researchers examine primary sensory systems, body size, sexual dimorphism (difference in body size between genders), locomotion patterns, communication repertoires, and dietary preferences. These physical and behavioral characteristics directly correlate with cognitive capabilities and social complexity.

---

4 Boyd, Richard. "Richard Boyd, Co-Founder & CEO of Tanjo Inc – Interview Series." *Interview by Antoine Tardif, Unite.AI*, 16 Apr. 2020, updated Dec. 9, 2022, https://www.unite.ai/richard-boyd-the-co-founder-ceo-of-tanjo-inc-interview-series/. Accessed Sep. 6, 2025.

5 See, for example – Mayr, Ernst. *Species, and Evolution: an Abridgment of Animal Species and Evolution*. Vol. 19. Harvard University Press, 1970.

Daily activity budgets reveal how species allocate time between foraging, rest, and play – each reflecting different cognitive demands. Dietary analysis goes deeper still, examining not just what the animals eat, but how they obtain food: the breadth of their diet, the balance between fruits, proteins, and vegetation, the sophistication of their extractive foraging techniques, and their ability to adapt these strategies seasonally. Most significantly, researchers document innovations in feeding behavior, as these often signal the cognitive flexibility that underlies cultural transmission and technological development.

Cultural transmission is a critical factor. Biologists identify learned behaviors that vary between groups despite similar environments – these cultural practices distinguish one troop from another and demonstrate the species' capacity for social learning. Reproductive patterns reveal additional complexity: mating frequency, gender-specific strategies, and birth rates all influence group dynamics and resource allocation.

Family structures encompass the entire lifecycle, from birth through senescence (death from old age). Researchers examine offspring-rearing practices, developmental milestones, and aging processes to understand how knowledge transfers across generations. Parental investment varies dramatically – some species rely on single-parent care while others involve both parents, extended family, and the entire group in raising young. And some provide none. The mechanisms of learning prove equally diverse: knowledge may pass directly from parent to offspring, spread horizontally through peer groups, or develop through individual trial and error.

Finally, mortality pressures shape cognitive evolution. Predation threats, environmental hazards, and resource competition drive the development of defensive strategies, risk assessment capabilities, and cooperative behaviors. These survival pressures often correlate with the cognitive complexity required for social coordination and technological innovation

Now, let's examine those same characteristics in the species *Homo sapiens*.

In my book on the co-evolution of cognition and technology in the human species,[6] I isolate over 60 major changes that shaped human cognitive evolution. For the last 50,000 years of human history, many of these have remained relatively stable, with certain inflections occurring with agriculture and the advent of civilization.

However, in the last few centuries, and particularly the last 100 years, these biological characteristics that have shaped our cognitive goals have been rad-

6 Forthcoming.

ically overhauled. This includes key characteristics affecting our survival, protection from the elements, food acquisition and processing, mating, safety against predators and others, alliance building, and other aspects of small group living, progeny rearing, intergenerational knowledge transfer, sleep, home range, daily range, method of moving, and daily activity budget. Here, we'll review just five representative examples of those changes, as they are explored more fully in my and others' work.

1.1 Food: From Scarcity to Abundance

Food: For most recorded history and all pre-recorded, the foods you ate, whether you hunted, foraged, traded, or grew, generally came from within 100 miles of you, and varied based on seasonal availability. Yes, there were major changes in that food, such as after agriculture and animal husbandry, but the proximity did not change dramatically. At least not when compared to the changes that have occurred within the last 100 years.[7] Now, one can walk into a modern American high-end grocery store like Wegmans and choose from a very diverse choice of over 50,000 items from all over the world, including many flown in the day before from South America and other continents.[8] As recently as only 30 years ago, the standard grocery store might carry only 7,000 items.[9]

To put this in perspective, let's jump back to my work on primate cognitive evolution. Consider the vegetarian lemur. A Grey Mouse Lemur might weigh 100 g and have a brain that weighs 2 g. A similarly sized tarsier species, say a Philippine Tarsier, would have a brain that is about twice as big as the lemur's.[10] The Lemur must have spatial and temporal awareness of its envi-

---

7 One reviewer pointed out the changes that happened to the European diet (and other parts of the world) through the technological innovations of shipping and trade, bringing foods from the 'new world', spice islands and other places. But I'd argue these were primarily luxury goods, and these new imported foods represented a small % of the ordinary person's diet. By contrast, with commodity trades moving around the world on bulk carrier ships, and the export of other packaged good via container ships and air, the majority of the everyday American's diet likely comes from more than 100 miles away.

8 Acosta, Gina. "How Wegmans Keeps Winning." *Progressive Grocer*, 18 Sept. 2020. Accessed 6 Sept. 2025.

9 Ruhlman, Michael. Grocery: The Buying and Selling of Food in America. Abrams, 2017.

10 See Stephan, Heinz, et al. "New and Revised Data on Volumes of Brain Structures in Insectivores And Primates." *Folia Primatologica*, vol. 35(1), 1981, pp. 1–29.

ronment to know which food is available when, and how to harvest it.[11] Tarsiers are the only living primate that is entirely carnivorous, catching invertebrates by jumping at them and opportunistically preying on various arboreal animals including beetles, arachnids, and small vertebrates.[12]

Some primatologists conclude that since the brain is an energetically expensive organ, a higher quality diet – such as the nutrient-dense prey the tarsier feeds on – enables that larger brain.[13]

Consider also the greatly increased cerebellum of the gorilla, which we discussed briefly at the beginning of the book.[14] They spend 10–12 hours a day extracting foods from difficult plants,[15] which require sophisticated handling, and the ecological niche adaptation is the neural evolution of the advanced cerebellum.

Finally, consider the evolutionary split between the common chimpanzee and their bonobo cousins. They evolved into separate species about a million years ago when chimp populations became separated by the Congo River. Those on the south side of the river evolved into bonobos.[16] The two species have different physical bodies, but more importantly, completely different cultures, with the chimps being warlike and male-dominated, and the bonobos being peaceful and female-dominated.[17] Given they share the same climate and environment, consume the same foods, and evolve from the same gene pool, what accounts for the difference? One theory is that chimps have gorillas on their side, which eat the foods the chimps would fall back on in

---

11 Richard, Alison F., and Robert E. Dewar. "Lemur Ecology." *Annual Review of Ecology and Systematics*, 1991, pp. 145–175.

12 See Jablonski, Nina G., and Robin H. Crompton. "Feeding Behavior, Mastication, and Tooth Wear in the Western Tarsier (Tarsius bancanus)." *International Journal of Primatology*, vol. 15.1, 1994, pp. 29–59, and "Tarsier." *Wikipedia*, Jul. 7, 2025, en.wikipedia.org/wiki/Tarsier. Accessed July 28, 2025.

13 DeCasien, Alex R., Scott A. Williams, and James P. Higham. "Primate Brain Size Is Predicted by Diet But Not Sociality." *Nature Ecology & Evolution, vol.* 1.5, 2017, pp. 0112.

14 Barks, Sarah K., et al. "Brain Organization of Gorillas Reflects Species Differences in Ecology." *American Journal Of Physical Anthropology, vol.* 156.2, 2015, pp. 252–262.

15 Doran, Diane M., et al. "Western Lowland Gorilla Diet and Resource Availability: New Evidence, Cross-Site Comparisons, and Reflections on Indirect Sampling Methods." - *American Journal of Primatology: Official Journal of the American Society of Primatologists, vol.* 58.3, 2002, pp. 91–116.

16 Takemoto, Hiroyuki, Yoshi Kawamoto, and Takeshi Furuichi. "How Did Bonobos Come to Range South of the Congo River? Reconsideration of the Divergence of Pan paniscus from Other Pan Populations." *Evolutionary Anthropology: Issues, News, and Reviews*, vol. 24.5, 2015, pp. 170–184.

17 Hohmann, Gottfried, and Barbara Fruth. "Culture in Bonobos? Between-Species and Within-Species Variation in Behavior." *Current Anthropology*, vol 44.4, 2003, pp. 563–571.

tough times, and the chimps have had to become aggressive to compete for that food in times of famine or drought. Their culture results from varying access to food.[18]

What do all these examples have in common, and how are they relevant to modern life, our research questions, and this chapter?

My answer is that they provide appropriate context of timeframes and dietary change. Diet affects cognition, and small changes in diet can have profound cognitive, behavioral, and cultural implications.

While the dietary adaptations above occurred over millions of years of specialization, my position is that we've made much more profound changes than the ones described above, in mere decades. While traditionally, we can't expect genetic evolution in that short time (remember that Cochran and Harpending feel it may be going even just since the advent of agriculture, and the Grants felt it could occur within a 2-year period in Galapagos finches), that doesn't mean that the changes of food won't affect selection pressures, behavior, or other factors that affect our lives within our lifespans.

Our own bodies ARE currently optimized to an evolutionary past – that fitment shapes the size and organization of our bodies, as well as our culture and behavior. As a result, we are hardwired for a time when resources were scarce. Now we live in a world of our own making where resources are abundant. Walking Wegmans aisles, we are biologically primed to seek salt, fats, and sweets, even though those are maladaptive to our current niche. While you may not have ever noticed the special joy of receiving an orange in your stocking at Christmas, your grandparents may remember that excitement – you couldn't walk to an orange tree in the winter from places like Ohio or New York. It was a very special treat. Now, to you, it is simply a year-round commodity that gets a little more expensive after certain weather events.

Our inherited biases, shaped by biological evolution, cause us to value and prioritize the wrong aspects of our diet. One clear result is that we have growing epidemics of obesity throughout the world. And this is just a simple, visible example of the anachronism.

18 White, Frances J., and Richard W. Wrangham. "Feeding Competition and Patch Size in the Chimpanzee Species *Pan paniscus* and *Pan troglodytes*." *Behaviour*, vol. 105, nos. 1–2, 1988, pp. 148–64.

1.2 Tools: The Shift from Physical to Digital

Let's be clear – tool use is rare in the wild. That's why we make such a big deal when we see the otter that uses stones to crack open shellfish,[19] or a crow uses a hooked twig to get insects out of a dead tree.[20] Chimps are the most commonly cited example of animal tool use, especially as their tools require advanced planning, tool preparation, a kit of different tools each of which must be gathered and prepared far in advance and requires learning multiple complex skills, which takes years to master (with females learning 1–2 years faster than males),[21] and those skills vary culturally across chimpanzee groups.[22]

These nuanced details matter, as subtle improvements in termite fishing add critical minerals and calories to the diet of chimpanzees, thus enhancing their ability to thrive in their environment and flexibly adapt to periods of change or challenge.

This is why it's such a big deal that in the last 100 years, large portions of the human population have moved from using physical tools to digital tools. According to the Bureau of Labor Statistics, 100 years ago, 74% of the jobs in America were farming and industrial, and now those jobs, which rely on physical tools, make up less than a quarter of our workforce, while 2/3 of Americans perform white-collar jobs relying primarily on digital tools.[23] These shifts also saw major changes in gender participation.

We have never seen such a profound shift in tool use in a species before. Note too that within the shift from physical to virtual, we also have changing gender roles, expectations, and behaviors. The standard narrative of our evolutionary past was that men did the dangerous and arduous hunting while women tended to the children and foraged for food.[24] This narrative arc carried into the 20th century, where men were expected to go do the dangerous

19 Fujii, Jessica A., et al. "Ecological Drivers of Variation in Tool-Use Frequency across Sea Otter Populations." *Behavioral Ecology*, vol. 26, no. 2, 2015, pp. 519–526.

20 Hunt, Gavin R. "Manufacture and Use of Hook-Tools by New Caledonian Crows." *Nature*, vol. 379, no. 6562, 1996, pp. 249–251.

21 Lonsdorf, Elizabeth V. "Sex Differences in the Development of Termite-Fishing Skills in the Wild Chimpanzees, Pan troglodytes schweinfurthii, of Gombe National Park, Tanzania." *Animal Behaviour*, vol. 70, no. 3, 2005, pp. 673–683.

22 Boesch, Christophe, et al. "Chimpanzee Ethnography Reveals Unexpected Cultural Diversity." *Nature Human Behaviour*, 2020, pp. 1–7.

23 "Women in the Labor Force: a Databook." *U.S. Bureau of Labor Statistics*, Mar. 2006, bls.gov/opub/mlr/2006/03/art3full.pdf. Accessed Aug. 25, 2025.

24 This narrative has been challenged on many fronts. See Lee, Richard B., and Irven DeVore, editors. *Man the Hunter.* Aldine Publishing Company, 1968, and critiques,

and hard work in factories, while women tended to the children and kept house.[25] However, today there are no rational reasons to expect gender differences in the increasingly white-collar jobs. One place this changing representation shows up is in college: women accounted for less than 20% of college degrees in 1900,[26] and now there are more women than men earning college degrees.[27] This is a cultural role shift enabled by changing tool use, among other factors.

These shifts require different things of our brains and our bodies. Jobs that took years to learn, and whose skills were passed on from generation to generation, have been replaced with technology, leaving those workers having to tackle the cognitive shock and cognitive transition to new skills, for which they might not be as well suited. Examples of the continuation of this trend can be seen in trucking: one of the most popular blue-collar jobs in America, with almost 3.5 million people (mostly male) making their living driving trucks.[28] However, the global forecast for self-driving trucks is 4.2 million units, and the US is expected to be the largest market.[29] What of the cognitive change that results when these drivers are laid off, and need to adapt to new jobs?

While many species employ tools to access desired resources, humans are the ultimate tool users. Tool use shifts within populations – especially changes in how tool expertise develops and spreads – produce major cognitive behavioral changes. Some established practitioners become untethered when their

---

such as Wall-Scheffler, Cara, et al. "The Myth of Man the Hunter: Women's Contribution to the Hunt across Ethnographic Contexts." *PLOS One*, vol. 18, no. 6, 2023, e028710.

25 Again, I'm expressing typical normative expectations and myths. Women and even children were early members of the industrial workforce (though even here, there were stereotypical roles for them), and women's participation in manufacturing during WWII showed that gender roles were more culturally than biologically derived).

26 *National Center for Education Statistics (NCES)*. "Table 318.10. Degrees Conferred by Postsecondary Institutions, by Level of Degree and Sex of Student: Selected Years, 1869–70 through 2031–32." *Digest of Education Statistics 2023*, U.S. Dept. of Education, Oct. 2023, https://nces.ed.gov/programs/digest/d23/tables/dt23_318.10.asp. Accessed Sep. 6, 2025.

27 *NCES Condition of Education*. "Undergraduate Degree Fields." 2024, https://nces.ed.gov/programs/coe/indicator/cta. Accessed Sep. 6,2025.

28 U.S. Bureau of Labor Statistics. "May 2023 National Occupational Employment and Wage Estimates." OEWS, U.S. Dept. of Labor, Apr. 3, 2024, https://www.bls.gov/oes/2023/may/oes_nat.html. Accessed Sep. 6, 2025.

29 "Autonomous Vehicle Market Demand to Reach 4.2 Million Units by 2030." *PR Newswire*, May 6, 2020, prnewswire.com/news-releases/autonomous-vehicle-market-demand-to-reach-4-2-million-units-by-2030-grand-view-research-inc-301035671.html. Accessed Aug. 25, 2025.

hard-won skills no longer deliver results, while others gladly leave behind the outdated traditional tools to adopt new tools or pursue new values.

1.3 Movement: Expanding Range, Shrinking Activity

Animal locomotion and range is critical to their behavior. Humans are well-known for being uniquely bipedal and, as discussed above, very unique tool makers and users. Kingdon hypothesized that human bipedalism developed through biological adaptations that occurred over millions of years, where our ancestors spent their lives squatting and foraging for food on the ground.[30] During this transition to the ground, early bipedal humans wouldn't have been as fast or able to travel as far and efficiently as our 4-legged ancestors; they would have had smaller ranges and stayed close to trees.

However, after we had "fully evolved" an efficient bipedal anatomy, we became able to travel great distances efficiently, letting us migrate further than many earlier primates, and respond better to shifting climate patterns. More importantly, bipedalism freed up our hands, which evolved to become dexterous and multifunctional. We could not only manipulate better, but now we could carry resources, and shift them in both time and place. Carrying resources back to the group led to changes in culture and social standing.[31]

These profound changes happened over millions of years, yielding us our unprecedented human versatility and adaptability.

But even more profound are the major changes in how we as modern humans move, compared to just 100 years ago.

Then, the most common way for our grandparents to move around their environment was walking. And that meant most people spent most of their time within a range of things they could easily walk to. Yes, horses were used, but not nearly as commonly as walking, and access to horses was skewed to certain populations. Trains had existed for 50 years, but they were also expensive, and traveling by train was an "occasion," much like traveling by plane was in the mid-20th century. Cars, too, existed, but were still an expensive luxury, and could only go to those places over limited paths and roads that were designed originally for horses.

Consequently, people tailored their expectations and wants to what they could walk to. We saw that in our Ingalls family anecdote earlier.

---

30 Kingdon, Jonathan. *Lowly Origin: Where, When, and Why Our Ancestors First Stood Up*. Princeton University Press, 2003.

31 See Wrangham, Richard. *Fire: How Cooking Made Us Human. Basic Books,* 2009.

Today, the average American spends almost an hour a day in a car, traveling around 30 miles.[32] Most of the time, we sit, for more than 6 hours a day. Our sitting is ramping up, even before the pandemic: adolescents went up from 7.0 to 8.2 hours a day just between 2007 and 2016, while adults increased from 5.5 hours to 6.4 hours a day during the same period.[33] We don't ride horses much anymore, but we've expanded our light-rail and bus transportation options, and not only introduced flying, but we have dramatically overhauled the economics to make it accessible to most people. This means that while we are both more sedentary than ever before, we have also increased our overall "range" to far greater than 100 years ago. But walking, the activity that previously constituted the primary mode of our movement/transportation, is now so rare that we adopt technology to count our steps and encourage us to do more of it.

Beyond the changes in behavior and distance, economics have been radically altered, letting goods move around the world at lower costs than ever before. And, as "goods" are increasingly dematerialized into digital data and experiences, the distribution of valued products is even further altered radically.

Changing movement in an ancient pack or tribe would have changed what values they sought and could access. The changes of the last century, especially in the last 20 years, give modern humans an unprecedented opportunity to have and therefore want new things. I'm sure my children do not have the same smile on their faces when finding an orange in their Christmas stocking as my grandparents had.

1.4 Social Groups: From Tribes to Megacities

Much is made of the fact that early humans, like most primates, evolved and existed in groups of perhaps 150.[34] The size and complexity of these social groups are often cited as a key selection factor that led to the evolution of

32 "National Household Travel Survey Daily Travel Quick Facts." *U.S. Bureau of Transportation Statistics*, bts.gov/statistical-products/surveys/national-household-travel-survey-daily-travel-quick-facts. Accessed Aug. 25, 2025.

33 Yang, Lin, et al. "Trends in Sedentary Behavior among the US Population, 2001–2016." *JAMA*, vol. 321, no. 16, 2019, pp. 1587–1597.

34 Dunbar, Robin I. M. "The Social Brain: Psychological Underpinnings and Implications for the Structure of Organizations." *Current Directions in Psychological Science*, vol. 23, no. 2, 2014, pp. 109–114.

larger and more capable brains over millions of years.[35] The task of keeping track of this many peers rewarded people who had the cognitive capacity to do so. In these small tribes, your ancestors' brains evolved and became optimized for group living.

Urban living demands fundamental changes. Brains hardwired to track 150 individuals must develop new tools for interacting with strangers, lacking shared experience or contextual baselines. Traits like altruism and behaviors like collaboration become adaptive when you must interact with strangers. The selection for these human tendencies over the last 10,000 years is what Peter Turchin focuses on in his book *Ultrasociety*. Without cooperation between strangers, we could never accomplish civilization, much less the pinnacle human achievements like the International Space Station, for example.[36] Turchin notes that societies must transform from the small egalitarian groups of directly mediated relationships to hierarchical large-scale systems, which enforce conformance, belonging, and norms through a variety of tactics to maintain cohesion. In his opinion, the principal engine of this cultural evolution was, ironically, warfare.

Turchin's work on the period of the last 10,000 years makes it critical to look at the increased urbanization of the world in the last 100. Cities have been around for thousands of years, growing larger and more common over much of that time. There's nothing new there.

Except there is: according to the UN, in 1920, only 14% of the world lived in urban areas,[37] but by 2018, it had grown to 55%, and is expected to reach 60% by 2030. In places like Europe and America, that number is already over 80%. Within that global 60%, one out of three people in the world will live in a city of at least half a million.[38] Megacities, with populations of over 10 million, didn't exist until New York and Tokyo reached that level in the 1950s, but there are already 33 around the world, and there are expected to be 43 by 2030.

If working in groups of 150 or less for millions of years greatly shaped our cognition, and our system of human societies was shaped culturally and

35 Byrne, R. W., and A. Whiten. *Machiavellian Intelligence: Social Expertise and the Evolution of Intellect in Monkeys, Apes and Humans*. Clarendon Press, 1988.

36 Turchin, Peter. *Ultrasociety: How 10,000 Years of War Made Humans the Greatest Cooperators on Earth*. Beresta Books, 2016.

37 "Growth of the World's Urban and Rural Population, 1920–2000." *United Nations*, 1969, population.un.org/wup/Archive/Files/studies/United%20Nations%20(1969)%20-%20Growth%20of%20the%20World's%20Urban%20and%20Rural%20Population,%2-01920-2000.pdf. Accessed Aug. 25, 2025.

38 "The World's Cities in 2018: Data Booklet." *United Nations*, 2018, un.org/en/events/citiesday/assets/pdf/the_worlds_cities_in_2018_data_booklet.pdf. Accessed Aug. 25, 2025.

maybe even genetically over the last 10,000 years by a modest increase in collective effort by groups over 150, and almost all of us lived in rural communities until 100 years ago, what are the cognitive effects on the individual when we rapidly scale up urbanization so quickly, and rewire our cultural networks as discussed in the last chapter?

That remains to be seen. But our valued behaviors and outcomes come from living in these small rural groups, and we have had very little time to learn and adapt to the ecological demands that come from the increasingly urban social structures we are creating for ourselves.

1.5 Lifecycle: Reinventing Birth, Death, and Everything in Between

Lifecycle: Scholars note how slowly human children develop relative to other animals. It requires a huge parental investment and significant evolutionary trade-offs. The payoffs include that the human brain can grow outside of the womb (overcoming the obstetric dilemma in which larger brains are good, but when they get too large, both the baby and mother are at risk of dying in childbirth), and the extended rearing is necessary to gain the enhanced learning required for them to survive in their complex evolutionary niche.[39]

Life cycle analysis also examines reproduction. Before birth control, and with human females' unusual tendency to be sexually receptive throughout the year (i.e., they don't have select periods of time in which they are sexually receptive, e.g., "heat," like many other animals),[40] human females gave birth early and often.[41]

One classic theory of human development that examines the impact of changing life cycles in human history is called the "grandmother hypothesis." This was originally developed in the 1960s by Kristen Hawkes and other scholars to explain why humans are among the very few species that live past reproductive age. They make the case that menopause evolved in humans because giving birth later in life was risky for the mother and the children. And this had consequences for her older children, who could be left without a mother. Menopause was biologically adaptive because it kept the mother

---

39 Bogin, Barry. "Evolutionary Hypotheses for Human Childhood." *American Journal of Physical Anthropology*, vol. 104, no. S25, 1997, pp. 63–89.

40 See Hrdy, Sarah Blaffer. *The Woman That Never Evolved.* Rev. ed., Harvard University Press, 1999., and Dixson, Alan F. *Primate Sexuality: Comparative Studies of the Prosimians, Monkeys, Apes, and Humans.* 2nd ed., Oxford University Press, 2012.

41 Haines, Michael R. "Fertility and Mortality in the United States, 1790–2000." *A Population History of North America*, edited by Michael R. Haines and Richard H. Steckel, Cambridge University Press, 2000, pp. 139–230.

around to raise the previous children. Hawkes also theorized that a long post-reproductive life span resulted in increased retention of knowledge and skills, and these older women were useful in activities like specialized food gathering and collectively raising children.[42]

In particular, it is speculated that these older females were a reservoir of difficult food acquisition skills, which were used during periods of food scarcity. As such, they would have been key pillars of group resiliency. Hawkes and colleagues draw a direct line between the evolution of menopause and the increased lifespan of post-menopausal women and the social accretion of skills and resources, which bootstrapped human cultural survival and progress.

Debates like these usually examine biological factors from deep human evolution, dating back hundreds of thousands or millions of years. They often overlook the remarkable changes in human life cycles over the last century or two, failing to consider their cognitive implications.

But now I will.

The first point to consider in the modern era is the significant extension of human lifespans. In 1920, the average life expectancy was about 34 years globally, and now it is over 73. Life expectancy in developed countries, such as those in Europe and America, has more than doubled, and is now closer to 80.[43] Much of this is due to decreased infant and child mortality.[44]

With higher confidence in child survival, women around the world are waiting longer to have children and are having far fewer of them. And they CAN choose to do that, with steady progress in education, science, medicine, and invention and adoption of world-changing technologies like birth control.[45]

In 1790, American women gave birth, on average, to 8 children; 100 years ago, that number was 3.8 children per woman.[46] It dropped to almost two during the Great Depression, shot up again during the baby boom, and is now down again to 2.1. Global rates mirror this – in 1965, there were over five babies born on average to each mother, and that number has more than

42 Hawkes, Kristen. "The Grandmother Effect." *Nature*, vol. 428, no. 6979, 2004, pp. 128–129.

43 "Life Expectancy." *Our World in Data*, ourworldindata.org/life-expectancy. Accessed Aug. 25, 2025.

44 Hertz, Erica, James R. Hebert, and Joan Landon. "Social and Environmental Factors and Life Expectancy, Infant Mortality, and Maternal Mortality Rates: Results of a Cross-National Comparison." *Social Science & Medicine*, no. 39.1, 1994, pp. 105–114.

45 "Access to Contraception." *American College of Obstetricians and Gynecologists*, Jan. 2015, acog.org/clinical/clinical-guidance/committee-opinion/articles/2015/01/access-to-contraception. Accessed Aug. 25, 2025.

46 For 1905, 1920, and 1930, see Haines, Michael R. "American Fertility in Transition: New Estimates of Birth Rates in the United States, 1900–1910." *Demography, no.* 26.1. 1989, pp. 137–148.

halved to 2.4 today, and is expected to drop to 1.7 by 2100.[47] In 1910 America, ¾ of women had their first birth before the age of 25;[48] the first birth age for America and the OECD countries is currently at just over 30, and is an increase of 2–5 years for most countries since 1970.[49] In 1910, less than 10% of all births occurred after 30.

Having fewer babies later in life means a woman is more likely to have more resources (capital, food, educational, etc.) per child, over a longer period of the child's life, which is likely to mean more cognitive development. With all generations having greatly expanded lifespans, the resources the mother can provide for the children are now also likely to include more resources from longer-living grandparents.

If menopause improved culture and human progress, what are the ramifications of this last century's doubling of lifespans, and funneling an increased set of resources into the fostering of a smaller number of children? The range of choice in wanting children, needing children, or even what you want for the future of your children has never been greater. And while this may seem to be a small matter of domestic impact, Hawkes' hypothesis gives us a chance to see how a small change in what we want and need across our life cycle can have major long-term repercussions for our species.

Except, we haven't made a few small changes in these areas, we've made many unprecedentedly massive ones. And we don't have the luxury of hundreds of thousands of years to adapt and figure it out.

1.6 Summing up the Disruption of the Values the Brain Evolved to Optimize

Typically, science advances as a chain: building one strong link at a time, working on a solid base of data, within a specialized area of strength and confidence, building on a base of past scholarship.

We find our lampposts, sit down, and work diligently.

However, this approach is ineffective when dealing with multiple streams of imperfect data from different disciplines that employ different tools and

47 Vollset, Stein Emil, et al. "Fertility, Mortality, Migration, and Population Scenarios for 195 Countries and Territories from 2017 to 2100: a Forecasting Analysis for the Global Burden of Disease Study." *The Lancet*, vol. 396, no. 10258, 2020, pp. 1285–1306.

48 "First Births to Older Mothers, 1972–2006." *CDC National Center for Health Statistics*, Apr. 2010, cdc.gov/nchs/data/databriefs/db68.pdf. Accessed Aug. 25, 2025.

49 "Age of Mothers at Childbirth". *Organisation for Economic Co-operation and Development*, oecd.org/els/soc/SF_2_3_Age_mothers_childbirth.pdf. Accessed Aug. 25, 2025.

methodologies. It doesn't allow us to see or work outside the lit area, nor can we integrate multiple areas that are lit but poorly.

Alison Wylie coined the term "cabling and tacking" to describe her approach to archaeological digs, which feature the same dynamics. She would have to braid together scant evidence from different fields to create the discipline necessary to come to solid conclusions.[50]

Conclusively dating an artifact embedded in the ground might start with a combination of stratigraphy (the principle that older layers of soil or sediment are typically found below younger layers), seriation (arranging artifacts or features based on their stylistic or typological similarities), and cross-dating (comparing artifacts or features from different sites or contexts with known chronological sequences), before moving on to techniques like radiocarbon dating (carbon-14 dating) and dendrochronology (tree-ring dating). And that's just to date an artifact. To understand the item, archaeologists use a range of methods, including typology, analysis of context, material, use-wear, residue, and experimental attempts to recreate and use the artifacts.

They also utilize emerging technologies like ancient DNA (aDNA). Each technique has its own shortcomings, biases, and requirements. Determinations require "tacking" back and forth between the different fields. Although each strand may be incomplete, taken together, she could weave them into a compellingly strong and rigorous conclusion. This approach follows and represents the essential value of the post-positivist approach to science: uncovering that which cannot be seen without triangulation from multiple perspectives.

Similarly, above, I've woven together material from multiple disciplines that seems to conclude that many of the most important biological factors that shaped human evolution and cognition, including the few we have described above, have been radically altered in just the last few generations.

Neuroscientist Sébastien Bohler points to the reward centers of our brain, optimized to pursue five objectives (values) that evolved to optimize the survival of the species: eat, reproduce, acquire power, extend our territory, and impose ourselves in front of others. Bohler points out though that these centers are never satisfied and always demand more. As a result, he feels that

50 Wylie, Alison. "Archaeological Cables and Tacking: the Implications of Practice for Bernstein's 'Options beyond Objectivism and Relativism.'" *Philosophy of the Social Sciences*, vol. 19, no. 1, 1989, pp. 1–18.

our inherited brain is no longer our resource or ally; it is becoming our worst enemy.[51]

As a result, we likely have many values and goals embedded in our inherited unconscious thought processes, which, as Bohler points out, are no longer optimal in this self-created niche, and may in fact be actively working against our long-term success.

In his short story "Stones of Significance," David Brin imagines asking the wisest caveman leader what they would want for their descendants. The answer, he suggests, would be-[52]

> I wish for my descendants freedom from care about the big carnivores, plus all the salts, sugars, fats and alcohol they could ever desire.'
>
> Rich irony, indeed. To a cave person, those four foods were rare treats. That is why we crave them to this day.
>
> Could the sage ever imagine that her wish would someday come true, beyond her wildest dreams? A time when destiny's plenitude would bring with it threats unforeseen? When generations of her descendants would have to struggle with insatiable inherited appetites? The true penalty of success? (Brin, 2000)

2 Disrupted Values in Culture

Fortunately, humans need not wait for the long process of evolution to take place: we've created the ability to adapt within our cultures, like Turchin pointed out in *Ultrasociety*.[53] Cultural evolution lets us adapt much faster than genetics. But does it adapt fast enough?

Using Tylor's definition – *"Culture, or civilization, taken in its wide ethnographic sense, is that complex whole."*,[54] which we introduced in Chapter 6; how fast is it changing, or can it change?

---

51 Bohler, Sébastien. *Le Bug Humain: Pourquoi Notre Cerveau Nous Pousse à Détruire la Planète et Comment l'en Empêcher* (*Bohler, Sébastien. Le Bug Humain: Pourquoi Notre Cerveau Nous Pousse à Détruire la Planète et Comment l'en Empêcher. Robert Laffont, 2019*). Robert Laffont, 2019.

52 Brin, David. "Stones of Significance." *Analog Science Fiction and Fact*, vol. 120, no. 1, Jan. 2000. Reprinted in *Insistence of Vision*, Stillpoint Digital Press, 2016.

53 Turchin, Peter. *Ultrasociety: How 10,000 Years of War Made Humans the Greatest Cooperators on Earth*. Beresta Books, 2016.

54 Tylor, Edward B. *Primitive Culture: Researches into the Development of Mythology, Philosophy, Religion, Language, Art, and Custom*. John Murray, 1871, p. 1.

In the last 100 years, US (as an example) cultural changes include:[55]

- Transition from horse-drawn carriages and telegraph to self-driving cars, space tourism, and the Internet
- In 1920, only 35% of homes had electricity (nearly all located in towns or cities),[56] now essentially all homes not only have electricity, but also radio, TV, cable, and the Internet.
- Medicine goes from pre-penicillin, to curing many of the world's diseases, and using techniques like fMRI and robotic surgeries[57]
- Introduction of Civil Rights[58]
- Introduction of Equal Rights[59]
- Since 2015, same-sex marriage has been legal nationwide, and public support has climbed above two-thirds[60]
- Introduction of birth control; legalization of abortion[61]
- Transition from physical/material culture to digital[62]
- Rise of consumerism[63]

---

55 A note to the methodology of this chapter. First, take into account Wylie's Cabling and Tacking point from the last chapter. But even with that in mind, let me self-critique the material we are about to go through: a) I jump around, including apples and oranges, and fields which are often incommensurate; b) I don't review the nuance in each situation, including (with some small but relevant exceptions I strive to note in the footnotes), for example, the counter-trends, and c) I don't provide the complete rebuttal to each of the points I'm making. In that respect, a reviewer might call this subjective "cherry-picking." But this is my empirical lived experience summary of 10 years of extensive research, and I think it paints a very significant and important trend toward disruption, which in the light of the other material in this book, is critically important to acknowledge, if we are to plan for the future of society. I think the authenticity and import of that overcomes the methodological critiques that will come with this approach, and any effort to skip the nuance was done out of a nod to conciseness, given the extremely wide range of material we are already covering here. The nuance and counterpoints are picked up in the cited material, and the sources they cite, and I'll leave it to them to cover that.

56 Waxman, Olivia B. "The Electricity Lobby Was Behind the First National Christmas Tree Lighting." *Time*, Dec. 1, 2016. *Time.com*.

57 Bud, Robert. *Penicillin: Triumph and Tragedy*. Oxford University Press, 2007.

58 Civil Rights Act 1964; Voting Rights Act 1965; Title IX 1972; See *Obergefell v. Hodges*, 576 U.S. 644. Supreme Court of the United States, 2015. Legal Information Institute, Cornell Law School, www.law.cornell.edu/supremecourt/text/14-556. Accessed Feb 26, 2024.

59 Buechler, Steven M. *Women's Movements in the United States: Woman Suffrage, Equal Rights, and Beyond*. Rutgers University Press, 1990.

60 Ball, Carlos A., ed. *After Marriage Equality: The Future Of LGBT Rights*. NYU Press, 2019.

61 Post-2022, abortion access depends on state law after Dobbs overturned Roe.

62 See, for example, Bell, Daniel. *The Coming of Post-Industrial Society: a Venture in Social Forecasting*. Basic Books, 1999.

63 Cohen, Lizabeth. *A Consumers' Republic: The Politics of Mass Consumption in Postwar America*. Knopf, 2003.

- Massive changes in industries, livelihoods, and where people live and how they are educated[64]
- Significant change in family structure: from extended to nuclear, fewer children, greatly increased divorce, blended and "nontraditional" families[65]
- Massive increases in physical (roads)[66] and digital networks connecting us to each other and the world[67]
- Rise of social movements, fueled in part by Internet and digital communications, but also by rising cultural change[68]
- Tripling of population[69]
- Life expectancy increased 65%[70]
- Profound changes in recreational drugs[71]

---

64 See Carter, Susan B., et al., editors. *Historical Statistics of the United States: Millennial Edition.* Cambridge UP, 2006.
Encyclopedic data (5 vols./online), and Fischer, Claude S., and Michael Hout. *Century of Difference: How America Changed in the Last One Hundred Years.* Russell Sage Foundation, 2006.

65 See Cherlin, Andrew J. "The Deinstitutionalization of American Marriage." *Journal of Marriage and Family*, vol. 66, no. 4, 2004, pp. 848–861. (with the caveat that multigenerational living has been rebounding since 2007).

66 USA went from mostly unpaved local roads (<4% roads paved prior to 1900) to a fully national, paved, multilane system with the Interstate network (≈48k miles) and an explosion in usage (vehicle-miles traveled grew from ~0.1 billion in 1900 to trillions today). See National Academies of Sciences, Engineering, and Medicine. "History and Status of the U.S. Road System." *Assessing and Managing the Ecological Impacts of Paved Roads*, National Academies Press, 2005, pp. 49–65.

67 "*Broadband is the indispensable infrastructure of the digital age – the 21st-century equivalent of what canals, railroads, highways*" See Federal Communications Commission. Connecting America: The National Broadband Plan. FCC, 2010, https://transition.fcc.gov/national-broadband-plan/national-broadband-plan.pdf. Accessed Sep. 6, 2025.

68 See Heimans, Jeremy, and Henry Timms. *New Power: How Power Works in Our Hyperconnected World–and How to Make It Work for You.* Doubleday, 2018; and Inglehart, Ronald, and Christian Welzel. *Modernization, Cultural Change, and Democracy: The Human Development Sequence.* Cambridge University Press, 2005.

69 U.S. Census Bureau. "Table A. Resident Population – Selected Years 1790 to 2020." *Historical National Population Estimates*, www.census.gov. (Shows U.S. ≈106M in 1920 vs. ≈331M in 2020).

70 Arias, Elizabeth, et al. "United States Life Tables, 2021." *National Vital Statistics Reports*, vol. 72, no. 12, National Center for Health Statistics, Nov. 7, 2023.

71 To be clear, Americans have long used psychoactives, but the last few decades mark a break: the market shifted from mainly plant-derived drugs and diverted pills to ultra-potent synthetics (especially fentanyl), moved onto e-commerce/darknet channels, and produced unprecedented overdose mortality. See Pardo, Bryce, et al. *The Future of Fentanyl and Other Synthetic Opioids.* RAND Corporation, 2019, www.rand.org/pubs/research_reports/RR3117.html. Accessed Jul. 16, 2025. Also National Conference of State

If that's not enough, consider the major changes in US culture in just the last *10 years*:

- Rise of the gig and sharing economies[72]
- Rise in work from home, online learning[73]
- Prominence of social media (governing through tweets)[74]
- Increasing digitization of products[75]
- Increased climate change effects[76]
- Significant changes in where we get our news[77]
- Smartphones dominance in daily our lives[78]

Legislatures. "Cannabis Overview." *NCSL*, updated 2025, https://www.ncsl.org/civil-and-criminal-justice/cannabis-overview. Accessed Sep. 6, 2025.

72 See Sundararajan, Arun. *The Sharing Economy: the End of Employment and the Rise of Crowd-Based Capitalism. MIT* Press, 2016. And, Schor, Juliet B. *After the Gig: How the Sharing Economy Got Hijacked and How to Win It Back.* University of California Press, 2020. Also Katz, Lawrence F., and Alan B. Krueger. "The Rise and Nature of Alternative Work Arrangements in the United States, 1995–2015." *ILR Review*, vol. 72, no. 2, 2019, pp. 382–416.

73 See U.S. Census Bureau. "The Number of People Primarily Working From Home Tripled Between 2019 and 2021." *US Census Bureau*, 15 Sept. 2022, https://www.census.gov/newsroom/press-releases/2022/people-working-from-home.html. Accessed Sep. 6, 2025; and National Center for Education Statistics. "Undergraduate Enrollment." *Condition of Education 2023, U. S. Dept. of Education*, 2023, https://nces.ed.gov/programs/coe/indicator/cha. Accessed Sep. 6, 2025; and Pew Research Center. "How COVID-19 Changed U.S. Workplaces." *Pew Research Center*, Feb. 12, 2025, https://www.pewresearch.org/politics/2025/02/12/how-covid-19-changed-u-s-workplaces/. Accessed Apr. 23, 2025.

74 Hilbert, Martin. "Digital Technology and Social Change: the Digital Transformation of Society from a Historical Perspective." *PLOS ONE*, Jul. 15, 2020, www.ncbi.nlm.nih.gov/pmc/articles/PMC7366943/. Accessed Nov. 19, 2025.

75 Verhoef, Peter C., et al. "Digital Transformation: a Multidisciplinary Reflection and Research Agenda." *Journal of Business Research*, vol. 101, Nov. 2019, pp. 889–913.

76 Davenport, Frances V., et al. "Escalating Costs of Billion-Dollar Disasters in the US: Climate Change Necessitates Disaster Risk Reduction." *Global Environmental Change*, vol. 78, Jan. 2023, http://www.sciencedirect.com/science/article/pii/S2667278222000906. Accessed Feb.19, 2025.

77 From 1925 to 2025, Americans shifted from newspapers to radio (1920s-1940s), then television (1950s-1990s), and finally to digital platforms where many now get news from social media, entertainment sources like late-night talk shows, and online influencers rather than traditional journalism. See "News Platform Fact Sheet, 2024." *Pew Research Center*, Sep. 17, 2024; and SUNY American Government Courses. "The Evolution of the Media." *SUNY American Government*, courses.lumenlearning.com/suny-amgovernment/chapter/the-evolution-of-the-media/. Accessed Oct 5, 2025.

78 See Turkle, Sherry. *Alone Together: Why We Expect More from Technology and Less from Each Other. 3rd* ed., Basic Books, 2017. And Twenge, Jean M. *iGen: Why Today's Super-Connected Kids Are Growing Up Less Rebellious, More Tolerant, Less Happy – and*

- Decline in the number of people who place a high value on having children[79]
- Increasing wealth gap[80]
- First Black president, first major party female presidential candidate, first female (and non-White) vice president
- Rise of the "cord-cutter"; changing communication and television landscape and consumption modes (e.g. bingeing)[81]
- Sharp decline in mental health, especially among youth; decline of happiness[82]
- Rapid increase in online shopping[83]

Completely Unprepared for Adulthood – and What That Means for the Rest of Us. Atria Books, 2017.

79 The research demonstrates that the share of US adults younger than 50, without children, who say they are unlikely to ever have kids rose 10 percentage points between 2018 and 2023 (from 37% to 47%) Most significantly, among those under 50 who are unlikely to have children, 57% say a major reason is they just don't want to, representing the highest cited factor. See Minkin, Rachel, and Kiley Hurst. "The Experiences of U.S. Adults Who Don't Have Children." *Pew Research Center*, Jul. 25, 2024.

80 The top 1% of US income earners now own $49.2 trillion, up from ~$30 trillion a decade earlier. By contrast, the total held by the bottom 50% is now only $4 trillion, and over the decade, they have experienced a steady decline in their share of the nation's total net worth. See *Visual Capitalist,* 8 Feb. 2025, http://www.visualcapitalist.com/visualized-the-1s-share-of-u-s-wealth-over-time-1989–2024/. Accessed Sep. 2, 2025; "A Visual Breakdown of Who Owns America's Wealth." *Visual Capitalist*, 4 Mar. 2025, www.visualcapitalist.com/a-visual-breakdown-of-who-owns-americas-wealth/. Accessed Sep. 2, 2025.

81 *"The share of Americans who say they watch television via cable or satellite has plunged from 76% in 2015 to 56% this year, representing a dramatic 20 percentage point drop in traditional TV subscriptions over just six years."* See Rainie, Lee. "Cable and Satellite TV Use Has Dropped Dramatically in the U.S. since 2015." *Pew Research Center*, Mar. 17, 2021, www.pewresearch.org/short-reads/2021/03/17/cable-and-satellite-tv-use-has-dropped-dramatically-in-the-u-s-since-2015/. Accessed Sep. 2, 2025; and Riddle, Karyn, et al. "Binge Watching and Serial Viewing: Comparing New Media Viewing Habits in 2015 and 2020." *Psychology of Popular Media*, vol. 11, no. 2, 2022, pp. 187–197.

82 One in three high school students reported persistent feelings of sadness or hopelessness, a 40% increase from 2009 to 2019 and suicide rates went up during that time by 57% among youth ages 10 to 24; the share of high school students who reported persistent feelings of sadness or hopelessness increased from 26% to 37%, and about 40% of teens reported persistent feelings of sadness/hopelessness in 2023, still significantly higher than 30% in 2013. See U.S. Department of Health and Human Services. "Protecting Youth Mental Health: The U.S. Surgeon General's Advisory." *Office of the Surgeon General,* 2021, www.hhs.gov/sites/default/files/surgeon-general-youth-mental-health-advisory.pdf. Accessed Sep. 2, 2025.

83 U.S. Census Bureau. "Annual Retail Trade Survey Shows Impact of Online Shopping on Retail Sales During COVID-19 Pandemic." *Census.gov*, Apr. 2022, www.census.

- Millennials have surpassed Generation Xers as the largest generation in the US labor force[84]
- Increased racial diversification; first decade in which whites declined in number and % of US population (>9% drop); whites are minorities in 27 of the 100 largest cities; less than 50% of children <16 are White[85]
- Decreasing church attendance[86]
- App dating is the most common way to meet a spouse[87]
- School/public shootings as the norm, and the advent of bulletproof school backpacks[88]
- Declining American patriotism[89]
- Increased political polarization[90]
- Rise of the celebrity, influencer, meme, TikTok[91]
- Rise of tech: IoT, 4G, 5G, AI, cloud, platform, bodycam, and self-driving car[92]

---

gov/library/stories/2022/04/ecommerce-sales-surged-during-pandemic.html. Accessed Sep. 2, 2025.

84 Fry, Richard. "Millennials Are the Largest Generation in the U.S. Labor Force." *Pew Research Center*, 11 Apr. 2018, www.pewresearch.org/short-reads/2018/04/11/millennials-largest-generation-us-labor-force/. Accessed Sep. 2, 2025.

85 "New Census Data Shows the Nation Is Diversifying Even Faster than Predicted." *Brookings*, brookings.edu/research/new-census-data-shows-the-nation-is-diversifying-even-faster-than-predicted/. Accessed Aug. 25, 2025.

86 Newport, Frank. "Church Attendance Has Declined in Most U.S. Religious Groups." *Gallup*, 26 Mar. 2024, news.gallup.com/poll/642548/church-attendance-declined-religious-groups.aspx. Accessed Sep. 2, 2025.

87 "Online Dating Now Dominates Relationship Building, Study Shows." *Global Dating Insights*, 15 Oct. 2024, www.globaldatinginsights.com/featured/online-dating-now-dominates-relationship-building-study-shows/. Accessed Sep. 2, 2025.

88 "A Majority of U.S. Teens Fear a Shooting Could Happen at Their School – and Most Parents Share Their Concern." *Pew Research Center*, Apr. 18, 2018; and Conkling, Gary. "'Monetizing Fear' by Bulletproofing Schools." *Oregon Capital Chronicle*, 27 Sep. 2024.

89 In 2001, approximately 90% of U.S. adults said they were "extremely" or "very proud" to be American. 2025 data shows that dropped to just 58%, marking the lowest level in over two decades. Meanwhile, among younger Americans, only around 40% of Gen Z and 60% of Millennials report strong national pride, compared to over 70% of older generations. "American Pride Slips to New Low." *Gallup Poll*, Jun. 30, 2025.

90 DeSilver, Drew. "The Polarization in Today's Congress Has Roots That Go Back Decades." *Pew Research Center*, 10 Mar. 2022.

91 Chayka, Kyle. "The Year Creators Took Over." *The New Yorker*, Dec. 11, 2024.

92 IoT Analytics. "Number of Connected IoT Devices Growing 13 % to 18.8 Billion Globally." *State of IoT Summer 2024*, Sep. 3, 2024.

- Decline of privacy[93] and marriage[94]
- Increase in debt at the personal and national levels[95]
- Increase in anxiety[96] and depression[97]

And of course, the backlash against the list above has played out in the most recent Presidential elections.

These vast, sweeping cultural changes profoundly influence people's attitudes. For the first time in US history, most Americans predict an American decline over the next few decades in several key areas, and American pride has been at an all-time low for several years in a row.[98]

While the American culture is increasingly polarized, the content and messaging from each "side" is so shockingly similar, one can't tell who is making the point:

- *"The American way of life is being dismantled by a group of bitter, deceitful, vengeful activists who have never built anything in their lives."*[99]

---

93 Concern for government use of personal data rose from 64% in 2019 to 71% in 2023; similarly, 73% of Americans say they have little to no control over how companies use their data, and 79% feel the same about government use: Pew Research Center. "How Americans View Data Privacy." *Pew Research Center*, Oct. 18, 2023, www.pewresearch.org/internet/2023/10/18/how-americans-view-data-privacy/. Accessed Nov. 4, 2025.

94 Pew Research recently (2024) found that 1 in 4 40-year-old American adults have never been married. They calculated that in 2019, 38% of adults aged 25–55 were unpartnered, compared to 29% in 1990. See Pew Research Center. "Share of 40-year-olds in US who have never married reaches new high." *Pew Research Center*, Jun. 28, 2023, pewresearch.org/short-reads/2023/06/28/a-record-high-share-of-40-year-olds-in-the-us-have-never-been-married/. Accessed Apr. 13, 2025; and Schweizer, Valerie J. "Marriage: More than a Century of Change, 1900–2022." *National Center for Family & Marriage Research*, bgsu.edu/ncfmr/resources/data/family-profiles/FP-24-10.html. Accessed Aug. 27, 2024.

95 2015: personal debt at $43,580 per citizen and national debt of $56,572 per citizen. Total: $100,152. 2025 Per-Citizen Debt Household = $52,962 per citizen, and national = $106,362 per citizen. Total combined debt per citizen: $159,324. 10-Year Increase: $59,172 (a 59% increase).

96 American Psychiatric Association. "American Adults Express Increasing Anxiousness in Annual Poll; Stress and Sleep are Key Factors Imp." *Psychiatry.org*, May 4, 2024. psychiatry.org/news-room/news-releases/annual-poll-adults-express-increasing-anxiousness. Accessed Nov. 17, 2025.

97 Gallup. "U.S. Depression Rates Reach New Highs." *Gallup News*, Mar. 26, 2025, news.gallup.com/poll/505745/depression-rates-reach-new-highs.aspx. Accessed Nov. 5, 2025.

98 "Public Sees America's Future in Decline on Many Fronts." *Pew Research Center Social & Demographic Trends Project*, Mar. 21, 2019, pewresearch.org/social-trends/2019/03/21/public-sees-americas-future-in-decline-on-many-fronts/. Accessed Aug.1, 2025.

99 "Factbox: Quotes from Trump Supporters at U.S. Republican Convention." *Reuters*, Aug. 25, 2020. reuters.com/article/usa-election-convention-quotes/factbox-quotes-from-trump-supporters-at-u-s-republican-convention-idINKBN25L08K. Accessed Aug. 25, 2025.

– *"If traditional values are shrinking, what values are taking their place? ... are these changing values a problem we can do something about ... and should we?"*[100]
– *"They will fundamentally change the nature of this country... they want to destroy this country, steal your liberty, your freedom, they want to control what you see and think so that they can control how you live. You will not recognize this country."*[101]
– *"For many, the American dream has become a nightmare."*[102]

This section makes the case that culturally derived values and goals, which take decades, centuries, and even millennia to develop, are being overtaken by massive changes on nearly every front. Unconscious cultural nodes of thoughts that were adaptive or accepted in the past are no longer adaptive. The overwhelming consequence of all this cultural change is a sense of loss, a rudderlessness, which bakes into our culture's psyche. The result of too much change too soon is a backlash, and an increased yearning for times when "things were great" (or at least understood) by those whose values have been overturned, and a feeling that their power and influence are waning. Ironically, this effect is felt across all socioeconomic, ethnic, and other groups, not just the traditionally privileged or power elites.

And Technology accelerates these changes.

As Herbert Marcuse reflected[103] and an idea the Supreme Court appeared to agree:[104] that *where* these changes are occurring is material and changing: *"Minds are not changed in streets and parks as they once were. To an increasing degree, the more significant interchanges of ideas and shaping of public consciousness occur in mass and electronic media."*

100 Liberal leaning HeraldNet newspaper: "Should We Be Alarmed about Changing American Values?" *HeraldNet*, heraldnet.com/business/should-we-be-alarmed-about-changing-american-values/. Accessed Aug. 25, 2025.

101 Kimberly Guilfoyle, Republican campaign operative: "RNC 2020 Day 1." *CNN Politics*, cnn.com/politics/live-news/rnc-2020-day-1/. Accessed Aug. 25, 2025.

102 Senator Bernie Sanders: "American Dream or American Nightmare? Depends Whom You Ask." *KWTX*, kwtx.com/content/news/American-Dream-or-American-nightmare–Depends-whom-you-ask-558040921.html. Accessed Aug. 25, 2025.

103 See Marcuse, Herbert. *Repressive Tolerance*. 1965. Reprinted in *A Critique of Pure Tolerance*, by Robert Paul Wolff, Barrington Moore Jr., and Herbert Marcuse, Beacon Press, 1969, pp. 81–123.

104 See *Denver Area Educational Telecommunications Consortium, Inc. v. Federal Communications Commission*. U.S. Supreme Court, 1996.

3 Disrupted Values in Technology

Cultures regularly use technologies, like Marcuse's electronic media mentioned in the paragraph above, to advance their needs. That's been demonstrated with pivotal human technologies like the printing press, steam engines, trains, electricity, photography, and telegraphs, but also through the often-overlooked innovations such as cement, Archimedes screw, optical glass, the moldboard plow, sanitation, invention of time (yes, that happened), vaccinations, penicillin, refrigeration, the pill, and Borlaug's solutions to world hunger.

If genetics and thinking via culture can't evolve fast enough to enable humans to adapt to the current self-created whirlwind of change, can technology? Or is technology evolving too fast for us to effectively grasp how to recruit it effectively?

Enthusiastic technologists tend toward the metaphor of exponential change, as represented by the meme of the second half of the chessboard. It's classically told by Ray Kurzweil as a fable about the inventor of the game of chess.

The emperor offered the inventor anything as a reward, and he said simply, "Some rice." "How much?" said the king. The inventor asked for one grain on the first of the 64 spaces, then doubled on the second, and so on. At first, it went well, with some spoonfuls, then bowls, then barrels.

> By the end of the first half of the chess board, the inventor had accumulated one large field's worth (4 billion grains), and the emperor did start to take notice. It was as they progressed through the second half of the chessboard that the situation quickly deteriorated .... the 63 doublings ultimately totaled 18 million trillion grains of rice. At ten grains of rice per square inch, this requires rice fields covering twice the surface area of the Earth, oceans included.[105] (Kurzweil, 2004)

More pragmatically, a few years ago, I was at a business meeting with top members of the US government, and a Chief Technology Officer (CTO) of one of the world's largest technology companies, who mentioned that technology was evolving very fast due to "*Moore's Law and the doubling of human knowledge every 6 months.*"[106]

---

105 Kurzweil, Ray. "The Law of Accelerating Returns." *Alan Turing: Life and Legacy of a Great Thinker*, Springer, 2004, pp. 381–416.

106 Personal discussion, covered under NDA and national security interests.

Both views above grossly underestimate the pace of technological change in human systems. What others view as overly enthusiastic or exaggerated, I view as a soporific analysis that lulls people into believing they have a workable understanding of their own environment. The sheer scale and complexity of the speed of technological acceleration is already beyond the understanding of any one human or human organization.

Let me explain.

3.1 Beyond Exponential: The Algorithmic Reality

Technology is not just expanding in a simple model of exponential rate against one variable, such as Moore's Law (the doubling in cost performance of computer processing every one to two years).[107] Mathematically, you'd express this as $M(t) = M_0 \times 2\^(t/2)$, where M is the variable, lower case t is time, and it keeps doubling. Let's call this function T1.

In the real "wild" world, it is moving forward not at an exponential rate, but rather an algorithmic rate that is a cumulative function of the combination of many, many variables, *each* of which is expanding at an exponential rate, all of which is additionally compounded through the ratcheting up of each other through bootstrapping feedback loops. This actual rate is changing so fast as to be invisible and incomprehensible to humans, whose cognitive limitations default back to the perceived confidence of their understanding of "modeled" complexity.

To understand this, let us start with some forces shaping the acceleration of technology, as it is generally understood.

Tech executives, like the CTO above, often refer to Moore's law as the exponent driving exponential change. And they are right – Moore's Law could result in the problem of the second half of the chessboard.

But innovation occurs within other aspects of the technology industry. Processors run software, and advances in software capability are captured in a concept known as Nathan's Laws.[108] Processors are embedded within, and software runs in computer architectures; the revolutionary growth of those architectures is described by Bell's Law.[109] Computing platforms are increasingly networked, as we mentioned briefly earlier, and the power of a network

107 Swanson, Bret. "Moore's Law at 50" *American Enterprise Institute,* Nov. 10, 2015, https://www.aei.org/research-products/report/moores-law-at-50-the-performance-and-prospects-of-the-exponential-economy/. Accessed on Nov. 19, 2025.

108 Myhrvold, Nathan. "The Next Fifty Years of Software." *ACM97 Conference*, 1997.

109 Bell, C. Gordon., et al. "The Effect of Technology on Near-Term Computer Structures." *Computer*, vol. 5, no. 2, Mar.-Apr. 1972, pp. 29–38.

goes up logarithmically with the number of nodes in that network, as laid out by Metcalfe's Law.[110] Computers process information, and as we discussed earlier, 90% of the world's current information was created in just the last few years. Buckminster Fuller described something he called the Knowledge Doubling Curve. Writing in 1981, Fuller noted that human knowledge doubled about every 12–13 months.[111] A standard estimate today is that the total data in the world is doubling about every 2–3 years,[112] but at least one estimate is that in 2020, it doubled not every 6 months, but every 12 hours.[113] We increasingly produce something scholars and the industry call "data exhaust."[114] Information isn't always knowledge because, as Machlup pointed out, there is a "knowledge half-life," after which some of our data becomes archaic.[115] So let's call the combined effects of Fuller and Machlup, a "modified Fuller's curve" or, for consistency's sake, Fuller's Law. And as fast and capable as our networks are today of handling all that data, they continue to double in reach and capacity about every 18 months, as described in Edholm's Law of Bandwidth.

The cumulative growth rate of Technology is not just Moore's Law, but is better described by the formula:

> Technology acceleration = Moore's Law × Nathan's Laws × Bell's Law × Metcalfe's Law × modified Fuller's curve × Edholm's Law

3.2 The Six Forces Multiplying Change

This is an algorithm, composed of six variables, each of which is expanding at least exponentially, if not logarithmically, and all of which amplify or multiply the other variables. You could model Change (C) as an algorithm that compounds these forces: $\mathbf{C=(T_1 \times T_2 \times T_3 \times T_4 \times T_5 \times T_6)}$ (and so on as new forces and technologies emerge).

---

110 Shapiro, Carl, and Hal R. Varian. *Information Rules*. Harvard Business Press, 1999.

111 Fuller, R. Buckminster. *Critical Path*. St. Martin's Press, 1981.

112 Kumar, Naveen. "Big Data Statistics 2025 (Growth & Market Data)." *DemandSage*, Jun. 24, 2025, demandsage.com/big-data-statistics/. Accessed Nov. 19, 2025.

113 Schilling, David. R. "Knowledge Doubling Every 12 Months, Soon to be Every 12 Hours." http://www.industrytap.com/knowledge-doubling-every-12-months-soon-to-be-every-12-hours/3950. Accessed Dec. 1, 2020.

114 Harford, Tim. "Big Data: a Big Mistake?" *Significance*, vol. 11, no. 5, 2014, pp. 14–19.

115 Machlup, Fritz. *The Production and Distribution of Knowledge in the United States*. Princeton University Press, 1962.

The complexity of articulating and understanding this is literally unfathomable. That is why one of the technology industry's leading executives refers to technology expansion more simply as Moore's law and Fuller's curve. And one of the field's most lauded visionaries, Kurzweil, provides a simple analogy of exponential rather than algorithmic accumulation. Are the CTO and Kurzweil deliberately simplifying the growth rate, or do they not understand just how complex and fast it is? Maybe both, but even within the technology industry, these popularly accepted exponential acceleration descriptions pale compared to the actual growth rate.

Regardless, the second half of the chessboard analogy has been influential. In their important book *The Second Machine Age*, Brynjolfsson and McAfee point out how much will happen so fast in this model: *"We're not claiming that no other technology has ever improved exponentially… But the exponents were relatively small, so it only went through about three or four doublings in efficiency during that period. It would take a millennium to reach the second half of the chessboard at that rate. In the second machine age, the doublings happen much faster and exponential growth is much more salient."*[116]

Critics point out we've heard this all before – every generation decries the latest technology as evolving too fast. Socrates feared reading would ruin memory and the oral tradition. Dickens wrote *"It was the best of times, it was the worst of times, It was the age of wisdom, it was the age of foolishness, It was the epoch of belief, it was the epoch of incredulity …"* in 1859.[117]

A 2010 article in Slate summarized some key historical technology scares:[118] Gessner wrote about the "*confusing and harmful abundance of books*" in 1545.[119] In 1889, a medical journal complained that schools "*exhaust the children's brains and nervous systems with complex and multiple studies, and ruin their bodies by protracted imprisonment.*" In 1936, the tech was radio, ruining the brains of children who "*developed the habit of dividing attention between the humdrum preparation of their school assignments and the compelling excitement of the loudspeaker*". Television would "*might hurt radio, conversation,*

---

116 Brynjolfsson, Erik, and Andrew McAfee. *The Second Machine Age: Work, Progress, and Prosperity in a Time of Brilliant Technologies*. W. W. Norton & Company, 2014, p. 47.

117 Dickens, Charles. *A Tale of Two Cities*. 1859. Penguin Classics, 2003, p. 1.

118 Bell, Vaughan. "A History of Media Technology Scares, from the Printing Press to Facebook." *Slate*, Feb. 15, 2010, https://slate.com/technology/2010/02/a-history-of-media-technology-scares-from-the-printing-press-to-facebook.html. Accessed Aug. 25, 2025.

119 Gessner, Conrad. *Bibliotheca Universalis*. Zurich, 1545, sig. *3V.

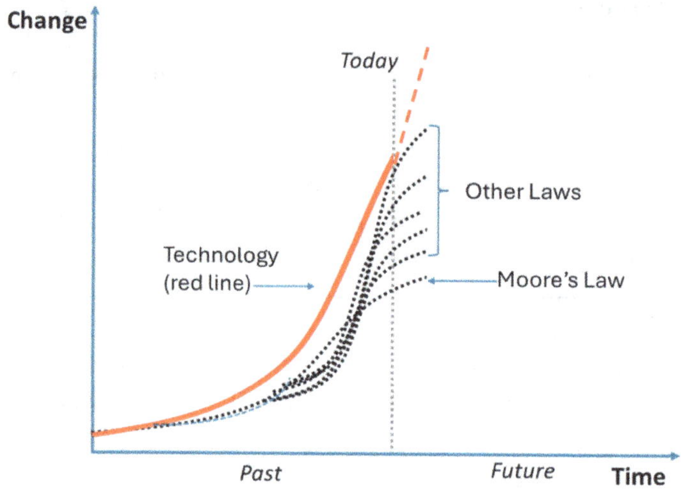

FIGURE 15.1 Two-dimensional graph of Technology's accumulated rate of change, as sum of several underlying factors

reading, and the patterns of family living and result in the further vulgarization of American culture", while according to Wired in 2005, email "hurts IQ more than pot."[120]

But the challenge is that we are now deep into the chessboard, and each step forward is an amplification many orders of magnitude more powerful than that which has come before.

Some scholars and many lay people object to Moore's law, and view Kurzweil's claims as being hyperbolic; not underestimates, but overestimates. Smil, for example, points out that there is no such thing as exponential growth in the universe; everything follows a sigmoidal function, accelerating up, and then inflecting and slowing its growth to meet the environmental limitations. It's not a Kurzweilian hockey stick, but more of a sideways S-curve.[121] One might visualize these stacked S-curves as in Figure 15.1, above. Others see recent signs of a potential slowdown in Moore's Law as evidence supporting this sigmoidal limitation.[122] It's probable that other lay people are overwhelmed by the possibilities of Kurzweil's forecast, and the massive change in a short period, forecast by Brynjolfsson & McAfee, and just reject it viscer-

120 Wallace, Lewis. "Don't Bogart That E-Mail." *Wired*, May 5, 2005.

121 Smil, Vaclav. *Growth: From Microorganisms to Megacities*. MIT Press, 2019.

122 Theis, Thomas N., and H-S. Philip Wong. "The End of Moore's Law: a New Beginning for Information Technology." *Computing in Science & Engineering*, vol. 19, no. 2, 2017, pp. 41–50.

ally, due to their own cognitive limitations, and to resolve cognitive dissonance.

However, new approaches, including quantum technology, could still give Moore's law, legs for years to come.[123] And I'd point out that we are not necessarily relying on Moore's Law, nor do I reject Smil's sigmoidal universality: I am layering sigmoidal curves on top of each other to achieve our algorithmic growth. Here is a simplified graphical representation of the technology growth rate:

But the above section only tells part of the story of how fast *digital* technology is changing, the kind most people think about when someone says the word "technology." Some definitions will be helpful here. I define:

- **"technology"**: (lower case) to be the processes, conscious or otherwise, in which techniques get adopted, combined, accumulated, and diffused within a social group.[124]
- **"technique":** the ensemble of practices by which one uses available resources to achieve certain valued ends.[125]
- **"Technology"**: (as a capitalized term) the accumulated technologies of human systems, which are increasingly digital[126]

While our world is increasingly electronic, digital, and reliant on the computing technologies we associate with Silicon Valley, Technology is not just computing, nor has it ever really been. To understand the values affected by Technology, we must accurately capture the scope of the Technology acceleration across all sectors.

This is not just a semantic distinction – it includes understanding how profound new technologies are, in part because of how rapidly they can diffuse; diffusion accelerates other adjacent technologies, spawning *multiple, accelerating,* positive-feedback loops, between more human domains.

To illustrate these dynamics, consider the Flynn Effect again, which showed that in the US and developed countries around the world, average IQ scores

123 Shankland, Stephen. "IBM's Quantum Computing Ambitions Get Exponential, Like Moore's Law." *CNET*, Mar. 3, 2019, cnet.com/news/ibms-quantum-computing-ambitions-get-exponential-like-moores-law/. Accessed Aug. 25, 2025.

124 This is from the traditions of: Ellul, Jacques. *The Technological Society*. Translated by John Wilkinson, Vintage Books, 1964; and Mumford, Lewis. *The Myth of the Machine: Technics and Human Development.* Vol. 1. Harcourt Brace Jovanovich, 1967; and Giedion, Sigfried. *Mechanization Takes Command: a Contribution to Anonymous History*. Oxford University Press, 1948.

125 Adapted from Lasswell: Lasswell, Harold D. Propaganda Technique in the World War. Ravenio Books, 1927.

126 This is identical to the definition that Kevin Kelly gives to the word "technium," in – Kelly, Kevin. What Technology Wants. Penguin, 2010.

improved by about 3 points each decade in the 20th century.[127] Flynn attributes this to improved education, nutrition, health care, and even increased social complexity. If you represented this mathematically, it would look like this: $IQ(t) = IQ_0 + (n \times wn \times t) + (e \times we \times t) + (h \times wh \times t) + (s \times ws \times t) + \varepsilon$ where:

n = nutrition factor
e = education factor
h = healthcare factor
s = societal complexity factor
w[subscript] = respective weights.

As such, the Flynn effect results from improvements in several technologies, even if it wasn't specifically driven by computer chips. Plus, there's one other force at play – emergent rate of change as the positive feedback loops between each improves the acceleration in the others. Better education leads to more Borlaugs, which leads to better nutrition, and so on. Let's call this Other factor O1.

We've come to associate computer chips with technology and Technology so much we have lost the perspective to understand how many underlying technologies scaffold each other, as technology has worked in the past, and is at work now. This inseparability caused Kevin Kelly to coin the term "technium." He explains –

> The ordinary pen you use every day seems very simple but it probably took 100 different technologies to make this pen technology, technologies of plastic, ink, ball bearing, metal, and each of those different technologies probably themselves required another 100 sub-technologies to support it and, of course, there's kind of a circular way in which pens might be necessary to make a ball bearing in the same way that electricity is necessary to make a generator, and a generator may be necessary to make the wires of an electrical system. ...so there is a sense in which all of this is very recursive and that there is a network of different supporting technologies, and that the whole web of all these things I call the technium. The technium is that largest network of all the technologies working together to support each other.[128] (Kelly, 2020)

---

127 Flynn, James R. *What Is Intelligence? Beyond the Flynn Effect.* Cambridge University Press, 2007.

128 Kelly, Kevin. "The Technium." *Edge*, edge.org/conversation/kevin_kelly-the-technium. Accessed Dec. 2, 2020.

Now, extrapolate from that example to your smartphone, its transistors, sensors, batteries, and the networks, platforms, cloud, AI, and other supporting tech. And then think forward about how those phones are influencing work, science, medicine, culture, and even the development of other Technology.

Here, I use the term Technology (capital T) specifically and explicitly in place of Kelly's technium, because a) that is how most people use the word "technology," without realizing it – as a single composite thing, and b) technium seems wonky and jargon.

With this perspective, let's now revisit the graph of our algorithm. The first graph is two-dimensional and gives us the sense that we have some thing, and we are getting more of it, at some rate.

3.3 The Third Dimension: Technology's Expanding Breadth

But there is a third dimension here we aren't seeing in 2D; we don't have growth in one thing – let's call it digital processing power, going vertically up – but we are having growth in many things, expanding both vertically and expanding sideways. The modern era can be characterized by a remarkable expansion in almost every field of human endeavor. Fundamental physics, medicine, genetics, every engineering field, agriculture, synthetic biology, material science, education, chemistry, and energy production are all being overhauled just as surely as our digital computing power technologies.[129] Yes, maybe increases in digital computing power have unlocked breakthroughs in these other fields.

For example, a recently announced discovery of neutrinos, produced in the CNO fusion cycle of the Sun,[130] is being described as one of the greatest discoveries in physics this millennium.[131] This discovery is a *keystone in astrophysics* with wide technological and cultural spillover. Its direct scientific impact is strongest in physics and astrophysics, but through instrumentation, data science, and international collaboration, it will continue to shape medicine, energy, security, and even philosophical perspectives on humanity's cosmic context. And yes, increasingly powerful digital computing has been a key

---

129 There is a body of evidence that science is actually slowing down, but I believe it overlooks the complexity of the feedback loops between fields, and the scale of the scientific progress that stems from this. And that was before the General Purpose potential of AI.

130 Borexino Collaboration. "Experimental Evidence of Neutrinos Produced in the CNO Fusion Cycle in the Sun." *Nature*, vol. 587, no. 7835, 2020, pp. 577–582.

131 "For First Time, Scientists Detect 'Ghostly' Signal That Reveals Engine of Universe." *NBC News*, Nov. 25, 2020, nbcnews.com/science/space/first-time-scientists-detect-ghostly-signal-reveals-engine-universe-n1248982. Accessed Aug. 25, 2025.

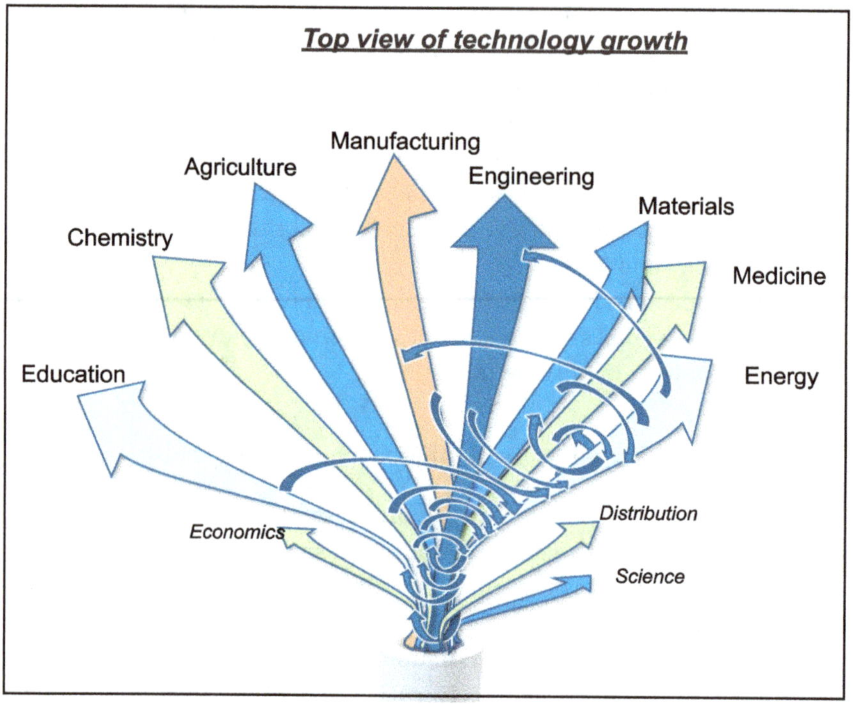

FIGURE 15.2 Conceptual representation of Technology's rapid "third dimension" expansion

component of this lab, but so has sophisticated mechanical engineering and advances in materials science.[132] While this is a highly abstract concept, Figure 15.2, above, illustrates it, conceptually.

As each field advances, it has positive feedback loops contributing to potential advances in almost every other field. While most of our attention has been on how digital computing power has influenced other industries, much less attention has been paid to how each of those other fields uplifts and bootstraps the others, including digital computing power. This creates an accelerating "tailwind" for digital computing that is not represented in our "simple" algorithm graph above. A revised, "top" view of that graph would show that as it climbs, it is also broadening; we are accelerating the growth of digital processing power AND expanding human achievement in nearly every field, with the interlocking feedback loops in all aspects of our model.

An accessible approximation of this dimension would look something like this:

---

132 "Search For High Energy 5.5 Mev Solar Axions with the Complete Borexino Dataset." *INFN Laboratori Nazionali del Gran Sasso*, https://borex.lngs.infn.it/papers/articles/search-for-high-energy-solar-axions-with-complete-borexino-dataset/ Accessed Aug. 25, 2025.

Today, conversations about accelerating tech usually focus on AI. Consider how most people discussing the dangers of AI usually frame it in a static way, based on their first- or second-hand knowledge of AI *today*. Fewer consider what happens when we apply likely advances in computing processing/architecture, and have a much more powerful AI in the *future*. A rare cohort considers what happens when you apply AI to improve all of the interrelated components of the technology that scaffolds our daily experience. (For example, just take that simple pen you are holding, and Kevin Kelly's point that it consists of over 100 discrete other technologies, and think what would happen if you applied some future advanced AI to improve every step of that. Now expand that into every facet of your life, and the grossly extended supply chains and research efforts feeding into that). And it is a rare few that consider what happens when the dramatic leap forward in computing power is achieved, when we achieve quantum supremacy[133] – when quantum computers can perform tasks in seconds that would take conventional supercomputers thousands of years.[134] THAT bootstrapping effect will radically fuel even more advanced AI, and THAT super-powered AI then has an overwhelming expanding effect on all of those facets of our lives.

But AI is taking up all the oxygen in the room in these conversations, and it causes us to lose sight of the other pieces and how they all fit together. So, let's step away from AI for a second and tease out the direct and indirect influence of just one other tech field – biotech – and unpack its influence on adjacent industries.

3.4 Force Multipliers: How Acceleration in One Technology Compounds Others

If I asked you to think about the economic sectors that benefit from biotech, you might respond: Healthcare and Pharmaceuticals, Agriculture, Energy, Food and Beverage, Chemical Manufacturing, Textiles, and Environmental Services. That's the right list of industries that will directly benefit, but biotechnology is not just transforming the industries it touches directly – it's accelerating change across a broader economic and geopolitical landscape, including in less-obvious, adjacent sectors. These ripple effects – both intended and unintended – can create entirely new markets, shift labor dynamics, and reshape global supply chains.

133 Arute, Frank, et al. "Quantum Supremacy Using a Programmable Superconducting Processor." *Nature*, vol. 574, no. 7779, 2019, pp. 505–510.

134 See Cho, Adrian. "Ordinary Computers Can Beat Google's Quantum Computer After All." *Science*, Aug. 2, 2022, science.org/content/article/ordinary-computers-can-beat-google-s-quantum-computer-after-all. Accessed Aug. 25, 2025.

Some examples:
- Real Estate & Urban Planning industries will experience shifts in infrastructure demand due to decentralized bio-manufacturing, urban vertical farms, lab-based agriculture, and bioreactors. Urban vertical farms and bioreactors for lab-grown meat reduce the need for rural land and cold-chain logistics. Unintended effects include changing land values, migration, and investment needs.
- Supply Chain & Logistics will see fundamental changes in what is moved, where, and how often. On-site or localized production of biological goods (e.g., vaccines, food, enzymes) cuts global dependencies, while legacy logistics providers and maritime trade routes get disrupted.
- Insurance & Actuarial Services will need to be repriced due to changing risks, biological enhancements, precision medicine, and new environmental realities.
- Education & Workforce Development will experience rapid obsolescence of traditional curricula and an increased need for convergence training in biology, coding, and systems thinking.
- Finance & Investment sectors must revalue assets and business models in biotech-adjacent sectors, with massive capital flows into synthetic biology and precision agriculture, and possible devaluation of oil, meat production, and traditional pharma portfolios
- Intellectual Property & Legal Systems will experience an explosion of patentable life forms, genetic code, and bio-software. New, stronger IP regimes for synthetic organisms and genetic data will emerge, but the field must overcome a quagmire of legal bottlenecks and patent wars.
- Manufacturing industries like textiles and fashion will face dramatic innovations as biological materials enable sustainable alternatives and new aesthetics in the form of new materials like mycelium leather, spider silk proteins, and biodegradable fabrics, which reduce environmental impact, but disrupt business and labor markets.
- National Security & Intelligence transform as bio-manufacturing becomes a strategic asset, akin to semiconductors or rare earths. Sovereign bioproduction capabilities reduce reliance on fragile supply chains, while the relevance of nation-states changes as innovations enable a shift to non-nation-state actors.

Setting aside these "adjacent" sectors, and looking at how synthetic biology can interact with other technologies, like AI, we can see the immediate direct impact, happening at breakneck speed.[135]

---

135 Thanks to Alexander Titus, who sent the following article out to several of us in his email newsletter. What follows is a summary of: Titus, Alexander. "*Ep 48 – The Model That*

Synthetic biology stands at a pivotal moment where AI's evolution from brute-force scaling to specialized precision is unlocking unprecedented capabilities in biological engineering. While the AI world celebrated ever-larger models like GPT-4's trillion parameters, biology demanded a different approach – one focused on the intricate specifics of protein folds and cellular mechanisms rather than raw computational power.[136] This shift (from scale-based general AI to function-specific intelligence – which emphasize cognitive precision over parameter count) materialized dramatically with GPT-4b micro, a domain-specific model trained exclusively on protein sequences and biological structures that has achieved what decades of traditional methods could not.[137]

The transformation becomes starkest when examining the Yamanaka factor problem. Yamanaka's research identified four proteins capable of reprogramming adult cells into pluripotent stem cells, winning him a Nobel.[138] However, Yamanaka himself acknowledges how poor the process is, with devastating inefficiency rates of below 0.1% conversion.[139]

After fifteen years of incremental improvements that changed only handfuls of amino acids, researchers enlisted GPT-4b micro to support the quest. The AI proceeded to rewrite entire proteins wholesale, creating variants like RetroSOX and RetroKLF that differ by over a hundred amino acids while maintaining function. The results were immediate and profound: where traditional directed evolution achieved less than 10% hit rates, the AI-designed proteins succeeded over 30% of the time, with late pluripotency markers appearing days earlier and mesenchymal stromal cells from donors over fifty showing remarkable 30% conversion rates within a week. This represents not merely accelerated discovery but a fundamental reshaping of what's possible

Reprograms Cell Identity." The Connected Ideas Project, Aug. 26, 2025, https://www.connectedideasproject.com/p/ep-48-the-model-that-reprograms-cells. Accessed Dec. 3, 2025.

136 Alvaro, D. "Reprogramming the Rules: How AI Is Transforming De Novo Protein Design in Drug Development." *Pharma's Almanac*, [date n.d.]. *Pharma's Almanac*, www.pharmasalmanac.com/articles/reprogramming-the-rules-how-ai-is-transforming-de-novo-protein-design-in-drug-development. Accessed Sep. 6, 2025.

137 OpenAI. "Accelerating Life Sciences Research with Retro Biosciences." OpenAI, Jul. 30, 2024, www.openai.com/index/accelerating-life-sciences-research-with-retro-biosciences. Accessed Sep. 7, 2025.

138 Takahashi, Kazutoshi, and Shinya Yamanaka. "Induction of Pluripotent Stem Cells from Mouse Embryonic and Adult Fibroblast Cultures by Defined Factors." *Cell*, vol. 126, no. 4, 2006, pp. 663–676.

139 Yamanaka, Shinya. "Induced Pluripotent Stem Cells: Past, Present, and Future." *Cell Stem Cell*, vol. 10, no. 6, 2012, pp. 678–684.

in synthetic biology, where AI specialists are redefining the boundaries of biological engineering itself.[140]

The bottom line is that a single technology field like biotech/synthetic biology is not a silo – it's a horizontal accelerator, rewiring how sectors interact.[141] The convergence of life sciences with data, automation, and climate tech creates a compounding effect.

As we have unwoven some of the technology strands above, such as AI or biotech, for illustration purposes, it gives the perception that they are diverging as they grow ever outward. Actually, they are converging through many interdependent mechanisms, in powerful ways.

3.5 The Figure Skater Effect: Convergence Accelerating Itself

Susan Hockfield showcases some of these feedback loops in her book *The Age of Living Machines* when she said,

> It's hard to overstate how profoundly Convergence 1.0 [of physics and engineering] transformed our world. The digital technologies and computation it made possible are now such an essential part of our lives that we take them for granted…we stand on the verge of another convergence with engineering that promises to revolutionize our lives yet again… But none of this is inevitable. Fostering Convergence 2.0 [of biology and engineering] will require the financial commitment, interdisciplinary collaboration, and political will that made Convergence 1.0 so successful.[142] (Hockfield, 2019)

We see convergence in discrete products, like when a single product, such as a mobile phone, replaces many other devices, such as a phone, camera, GPS, and music player. But more broadly, convergence refers to the integration and coupling between technologies that were originally seemingly unrelated. It is generally viewed as a cultural transformation in which diverse domains

---

140 OpenAI, and Retro Biosciences. "Accelerating Life Sciences Research with Retro Biosciences." *OpenAI*, Aug. 22, 2025, openai.com/index/accelerating-life-sciences-research-with-retro-biosciences. Accessed Sep. 7, 2025.

141 I highly recommend this 2025 Charting the Future of Biotechnology, by the National Security Commission on Biotechnology, as a great introduction/primer to the field of biotech/synthetic biology – "Charting the Future of Biotechnology – an Action Plan for American Security and Prosperity". *National Security Commission on Emerging Biotechnology*. https://www.biotech.senate.gov/final-report/chapters/. Accessed Nov. 17, 2025; and thank Alexander Titus for bringing this too, to my attention.

142 Hockfield, Susan. *The Age of Living Machines: How Biology Will Build the Next Technology Revolution*. WW Norton & Company, 2019.

of human activity become unified within a single interdependent and self-feeding platform. If you want to visualize the effects of the more tightly coupled feedback loops between the previously loosely coupled facets of humanity – picture the figure skater twirling languidly on the ice, with her arms outstretched gracefully; slowly, she draws her arms and legs into her core, and as she does, she begins to accelerate, until finally she is one tight compact core spinning so fast she becomes a blur. This is convergence.

This deeper look at technological history and our current context gives us the ability to appreciate that Technology is advancing faster and more profoundly than what Brynjolfsson and McAfee breathlessly describe. At first blush, one could make statements such as the "Internet has had a more profound impact, faster, on human potential than any other technology in human history."

What I found, however, when I look at the system holistically, is we have created a Technological niche that is so powerful that even modest breakthroughs in key fields, such as 3D printing, 5G telecommunications, quantum computing, and AI, might now *each* have a more powerful long-term effect, *much* faster, than any previous seminal human technologies such as fire, language, writing, wheel, printing press, steam engines, telegram, phone, radio, or electricity.

A mathematical model of Change (C) as a function of many technologies (T) and Other factors (O) looks like this:

$$C = (T_1 \times T_2 \times T_3 \times T_4 \times T_5 \times T_6) \times (O_1 \times O_2 \times O_3 \times O_4 \times O_5 \times O_6)$$

In the real world, Technology is not accelerating exponentially – it is, in fact, speeding up MUCH faster than that, via an algorithm, with many variables expanding simultaneously, and compounding where each increase speeds up all the others.

The consequence of this complexity and the powerful potential of what we have created is that it is currently incomprehensible to even the most advanced experts in each field. Even if they understood the acceleration of their own field (e.g., I make the case that Kurzweil and the CTO above don't do this in their own field, digital computing), they are not experts in the adjacent domains, which doubtlessly influence their areas, and so they can't see fully those effects.

And yet, even from the Kurzweilian perspective, he makes statements like this: *"We won't experience one hundred years of technological advance in the twenty-first century; we will witness on the order of twenty thousand years of progress (again, when measured by today's rate of progress), or about one thousand times greater than what was achieved in the twentieth century."*[143]

As Technology changes, so too do our goals for its use and, as does the values it seeks to extract from those who engage with it.

Let's shift now from the *rate of change* and look at how *actors and structures control that rate and respond* to it. With this dramatic change, the structure in which this innovation occurs undergoes changes in how it is controlled and responded to by the governing entities.

3.6 When Speed Wins: The New Power Hierarchy

People have perennially sought to control technological innovation and diffusion. Control and/or adaptation was done by national systems, through laws, policy, investments; and by industry coordination, investment, and self-regulation; and by social structures, such as orthodox sects like the Amish, Orthodox Jews, and Daoism; and by powerful individuals, who built empires, as well as the everyday person would make choices on technology, based on their own values and perception. That power structure typically flowed from big to little: a King or President used to have more power to affect technology diffusion than an individual, social group, or business sector. And while they may not have fully grasped the implications of technological innovation, one could say they understood the basics of what was happening, just not the long-term consequences, and they definitely tried to control it.[144]

Consider, for example, the power struggles in Europe prior to World War I. At the turn of the 20th century, world leaders gathered at two Hague Peace Conferences (1899 and 1907) with the explicit goal of constraining technological innovation in military affairs before it could unleash unprecedented destruction.[145] Count Mikhail Nikolayevich Muravyov, the foreign minister of Tsar Nicholas II, proposed specific topics for consideration: "(1) a limitation on the expansion of armed forces and a reduction in the deployment of new

---

143 Kurzweil, Ray. *The Singularity Is Near: When Humans Transcend Biology*. Penguin, 2006, p. 11.

144 See for example, the direction King Louis 15 gave to Malesherbes, his Director of the Press, to try to control the spread of newspapers in France, and Malesherbes's detailed response in 1759 on the complexity of that effort, as told in Popkin, Jeremy D. News and Politics in the Age of Revolution: Jean Luzac's "Gazette de Leyde". Cornell University Press, 2016. P36–43.

145 Davis, Calvin DeArmond. *The United States and the Second Hague Peace Conference: American Diplomacy and International Organization, 1899–1914*. Duke University Press, 1975.

armaments"[146] alongside other humanitarian concerns. The language was deliberately forward-looking – the conferences aimed not merely to regulate existing weapons but to prevent "the progressive development of the present armaments" and halt innovation itself.

The motivations behind these unprecedented efforts to limit technological acceleration were both humanitarian and pragmatic. Russia called the 1899 Conference at a time of its own intensive military expenditure and expansion, with some viewing it not primarily as an overture toward peace but rather as Russia's attempt to remove the burden of keeping pace with the armaments build-up occurring in Great Britain and Germany. It was clear by 1898 that Russia, already lagging behind the West in economic and industrial development, was unable to compete in the armaments race, with the financial burden alone being too great.[147] Tsar Nicholas II hoped the 1899 conference would ban the rapid-fire artillery being developed by Austria-Hungary, as Russia could neither develop nor purchase such weapons except at great expense. The conference agenda directly targeted technological innovation, including proposals such as the freezing of military expenditure, a ban on the use of submarine torpedoes and the throwing of explosive projectiles from balloons or other aircraft, and a ban on the development of new weapons with greater firepower than existing ones.

The conferences did produce formal agreements prohibiting specific emerging technologies and established important humanitarian principles. However, these efforts proved tragically futile. Many of the rules laid down at the Hague Conventions were violated in World War I.[148] For example, poison gas was developed, introduced, and used by all major belligerents throughout the war, in violation of the Declaration (IV, 2) of 1899 and Convention (IV) of 1907, which explicitly forbade the use of "poison or poisoned weapons."[149] The very technologies the conferences had sought to prohibit or limit – machine guns, poison gas, aircraft, and submarines – were developed at a feverish pace by all participants, and the end result defined the unprecedented carnage of World

146 "Hague Convention." *Encyclopedia Britannica*, Jul. 20, 1998, www.britannica.com/event/Hague-Conventions. Accessed Mar 16, 2024.

147 Greising, Jack H. and John Quinn. "First Hague Peace Conference." *EBSCO Research Starters*, www.ebsco.com/research-starters/history/first-hague-peace-conference, 2023. Accessed Dec 3, 2025.

148 "Hague Conventions of 1899 and 1907." *Wikipedia, Wikimedia Foundation*, last edited Sep. 25, 2025, en.wikipedia.org/wiki/Hague_Conventions_of_1899_and_1907.

149 "Hague Conventions of 1899 and 1907." *Wikipedia, Wikimedia Foundation*, last edited Sep. 25, 2025, en.wikipedia.org/wiki/Hague_Conventions_of_1899_and_1907.

War I, demonstrating how rapidly technological innovation could outpace international agreements.[150]

The parallels to today's artificial intelligence governance efforts are both striking and sobering. Modern AI safety advocates echo the concerns of early peace activists while facing identical structural constraints – the challenge of international cooperation in a competitive system where first-mover advantages could determine geopolitical dominance for decades. The Hague conferences demonstrated that even explicit efforts to halt technological innovation, backed by genuine humanitarian concern and supported by detailed treaty language, ultimately fail when nations perceive existential competitive disadvantages. Just as the conferences focused on specific technical prohibitions that became obsolete as warfare evolved, contemporary AI governance concentrates on particular applications, while artificial general intelligence development proceeds across broader fronts. The fundamental challenge remains unchanged: in an anarchic international system, states and corporations struggle to surrender potential advantages, whether in military technology a century ago or artificial intelligence today.

Today, however, we have two major twists to the power structures, which we've used over the last few thousand years, and demonstrated in the Hague example.

The first is the one described earlier – that technology is accelerating much faster than anyone acknowledges, with the result being that its complexity already exceeds the understanding of any one human or human organization.[151]

Today's technological acceleration isn't just faster – it's fundamentally different. Unlike earlier industrial revolutions that automated physical labor but left human judgment and creativity intact, artificial intelligence now automates cognition itself. As one commentator puts it, with AI "*there is nowhere left to retreat.*"[152] This shift reflects not only speed but also a discontinuity in kind: the tools of thought are being reengineered by the very sys-

150 Hull, Isabel V. *A Scrap of Paper: Breaking and Making International Law during the Great War*. Cornell University Press, 2014.

151 This is not a new consequence of course, but the scope and scale of it across our lives is profoundly more prominent than in the past. For the literature on this, see Simon, Herbert A. *The Sciences of the Artificial*. 3rd ed., MIT Press, 1996., and Gell-Mann, Murray. *The Quark and the Jaguar: Adventures in the Simple and the Complex*. W.H. Freeman, 1994.

152 Btl. "The Discontinuity Thesis: Why This Time Really Is Different." *Medium*, June 2025, medium.com/@btl101/the-discontinuity-thesis-why-this-time-really-is-different-3a63fc-5bc018. Accessed Sep. 7, 2025.

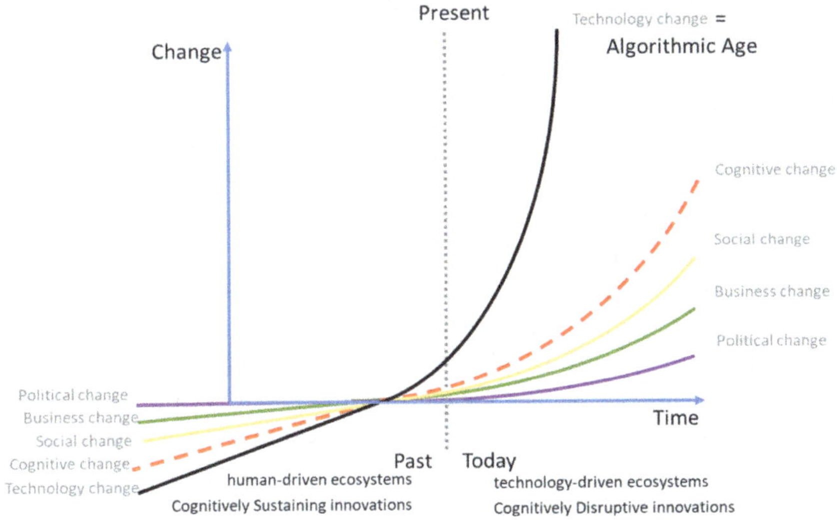

FIGURE 15.3 Differential ability to respond to change = changing power

tems they once merely supported. The specter of the technological singularity reinforces this point, describing a moment when technological progress becomes "*uncontrollable and irreversible, culminating in profound and unpredictable changes to human civilization.*"[153] Taken together, and when viewed across the landscape of ALL Technology, not just AI, these perspectives suggest that today's acceleration is already beyond the grasp of any one human or organization, making this time truly different. Figure 15.3, above, is a qualitative conceptual rendering of this shift.

Think back to the complexity of that simple Nest thermostat: even a lawyer/Ph.D. would have difficulty grasping the complexity of the web of agreements and information sharing surrounding that simple device, largely because most of those 1,000 pages can be unilaterally changed without notice to the user. They are not only composed of an immense number of pieces that are intricately interrelated – they are also all continually changing. This increased regulatory complexity (of the simple relationship between you and your thermostat) foreshadows the technological complexity: where is the data going, how is it being used? Even if one took the time to tease out that complexity, it may change by the time you've figured out what was in it originally. And that's just in one simple consumer device. Now consider the web of feedback loops that span out from the spawning of generational AI into all corners of our

153 Mucci, Tim. "What Is the Technological Singularity?" *IBM*, Jun. 72024, www.ibm.com/think/topics/technological-singularity. Accessed Sep. 7, 2025.

technological world – what's the level of complexity and ability to understand the consequences of that?

The second twist is that larger organizations typically adapt much more slowly than smaller ones. Since control is a function of understanding, and understanding is a function of adaptation to change, we see an inversion in the power to control technology, with power devolving to the individuals and social movements that can respond more nimbly and accurately to changes in technology. A rough conceptual illustration of that is:[154]

Implied in the graph above is that human cognitive potential will increase as Technology does, just as it has in the last 100 years. We discussed the mechanism for that in Chapter 11. People may not grow as fast as technology, but there's a potential for advancing our cognitive abilities far beyond what we do today. But that may be overly optimistic if we don't address the issues laid out in this book.

3.7 Summary: Changing Technology Values

What happens when we rely on technology to augment our cognition, and we adapt our thinking to the "values" embedded within that system, and those external cognitive nodes change in rapidly accelerating and incomprehensible ways?

It makes it hard to develop consistent thinking; it causes cognitive confusion, relearning, and cognitive dissonance. It makes it harder to think clearly, and at a meta-level, about the important things. We see this frustrated adoption struggle when Facebook introduces new symbols/emojis with which we can respond to content, or LinkedIn changes an algorithm, or people are forced by circumstances to work or attend school remotely from home. It gets even worse when those Technology ecosystems are influenced purposefully by outside actors for their own purposes.

If we only used technologies for small and simple tasks, such as occasional entertainment, our thoughts might not be disrupted. However, Technology is ubiquitous, and we are dependent on a web of technology for most of our everyday lives.

As technology continues to both accelerate and broaden, a key challenge exists:

154 Adapted (heavily) from Downes, Larry. *The Laws of Disruption: Harnessing the New Forces That Govern Life and Business in the Digital Age*. Basic Books, 2009.

> We need to find out how we adapt the human mind to better interface, interact, integrate, team, and effectively coexist with technology in ways that will enable more positive future human outcomes, in the face of rapidly changing technologies.

Technology is not climbing along some two-dimensional hockey stick curve, nor expanding in three dimensions as I've simplified above. It hypercatalyzes humanity, in a form best framed as an impossible-to-visualize multidimensional tesseract – accelerating not just in speed or scope, but through dimensions of change we cannot fully comprehend.[155]

4 Chapter Summary of Changing Values, across All Levels

4.1 The Compass Crisis: When All Your Navigation Systems Fail

How do you assess your portfolio of needs when the structural foundations of your cognitive world shift beneath your feet? Traditional models of thought might offer straightforward answers and concise prescriptions for adaptation. But our Three Streams Model reveals how we actually think today – and the picture isn't pretty.

4.2 Three Broken Compasses

You navigate life using three distinct compasses, each pointing you toward different goals. Like any traveler, when uncertainty strikes, you check your compasses for direction. But what happens when all three compasses malfunction simultaneously?

Your biological compass carries ancient programming – evolutionary imperatives that once ensured survival but now often mislead. This compass moves slowly, creating an illusion of stability that breeds false confidence. Your ancestors needed to hoard calories and seek immediate gratification;

155 A tesseract is a four-dimensional cube – imagine extending a regular cube into a fourth spatial dimension we can't physically see. Just as a square becomes a cube when extended into 3D, a cube becomes a tesseract when extended into 4D. See Ciechanowski, Bartosz. "Tesseract." Bartosz Ciechanowski, 2024, ciechanow.ski/tesseract/. Accessed Jul. 13, 2025. This explains – "*Our human brains are hardwired to understand things in three dimensions – reasoning about the fourth one requires some amount of imagination or even delusion. We can try to understand how things work in 4D, but I don't think we can fully experience it.*"

you live in a world of relative abundance where these impulses sabotage long-term health and happiness.

Your social compass spins wildly as cultural constructs reshape themselves at breakneck speed. Yesterday's social norms become today's taboos. The direction that felt right last decade now leads nowhere useful. You chase moving targets, trying to orient yourself by reference points that keep relocating.

Your technological compass spins fastest of all. The values embedded in your apps, devices, and platforms – created by distant engineers and executives – pull you in countless directions simultaneously. These digital architects embed their priorities into your thinking tools, and their arrows blur into an incomprehensible whir of competing demands.

4.3 The Disorientation Effect

Without steady values to pursue, you lose your bearings entirely. This disorientation triggers a cascade of problems: How can you measure progress toward unclear goals? How can you allocate resources effectively when you can't identify the right objectives? The questions multiply: Am I doing enough? Am I doing the right things?

4.4 Emergent Chaos

The interactions between these failing navigation systems create new complications. Individuals, movements, and corporations adapt faster than governments. Technology enables instant global connection, allowing people to organize around shared values rather than geography. Nation-states lose influence as power structures realign around digital networks and ideological communities.

4.5 Two Possible Futures

This transformation offers both promise and peril. The optimistic view: You might adapt faster than larger, slower institutions, gaining unprecedented individual agency. The pessimistic view: You depend on external systems for thinking and valuing, and these systems are effectively breaking down simultaneously. While some individuals will successfully pivot and thrive, many will experience the changes as overwhelming loss.

4.6 The Bottom Line

The net conclusion from this section is that thought in its current state, and across all three planes, is organized around a portfolio of largely unconscious or unexamined values that are often outdated and/or are tailored to circumstances that are radically changing. At best, thought is not optimally suited to our current human niche. But, technology as a magic wand, offers us the ability to step outside traditional values, and craft a portfolio of previously unheard of goals and wants. But first, you must recognize that your navigation systems have failed, and charting a new course requires conscious, deliberate effort rather than autopilot reliance on broken compasses.

CHAPTER 16

Our Thoughts Are Experiencing Varying Rates of Acceleration

> "Well, in our country," said Alice, still panting a little, "you'd generally get to somewhere else – if you run very fast for a long time, as we've been doing."
>
> "A slow sort of country!" said the Queen. "Now, here, you see, it takes all the running you can do, to keep in the same place. If you want to get somewhere else, you must run at least twice as fast as that!"[1] (Lewis Carroll)

Change, change, change; it's everywhere, and it's accelerating. It's been a theme throughout the book, and we plumbed those depths in the last chapter, illustrating how fast what we want has changed within each of the Three Streams of our thinking. Yet, some of my biggest-thinking friends are the first to tell me this is nothing new. It's been around since the dawn of humanity. But again, I'd say, this time is different.

The goal of this chapter is to highlight the unprecedented situation we find ourselves in. If you've already bought this tale I've spun, then feel free to skip this one. If you don't, perhaps your heels are dug in, and nothing I can say will change your mind. But for the curious, open-minded people who aren't quite sure and would like to really understand this point better, I've laid out my case in a little more detail in this chapter, as we look at how each stream changes at a different rate.

Consider the historical Global Gross Domestic Product chart, for example, which measures the total output of the world economy. The graph comes from Our World in Data and combines three historical inflation-adjusted datasets – the Maddison Database (before 1820), the Maddison Project Database (1820–1989), and the World Bank (1990 onward) – to chart the last 12,000 years' collective human output. Viewed across its 12,000-year scale, it looks completely flat until about Y2K, when it makes a 90-degree upward turn and climbs straight like a rocket.[2] Zoom into the modern era and you can see that the flat line is say under 3 trillion or so, and that continues until

1 Carroll, Lewis. *Through the Looking-Glass, and What Alice Found There*. Macmillan, 1871, p. 42.

2 "Global GDP over the Long Run." *Our World in Data*, ourworldindata.org/grapher/global-gdp-over-the-long-run. Accessed Aug. 25, 2025.

about 1870 or so, when the line starts to climb. This era is often referred to as the Second Industrial Revolution, a time when the use of electricity in manufacturing and huge deposits of coal and oil became economically feasible, unleashing new sources of energy, and industries saw revolutions in communications and transportation. The line begins to inflect here, tripling from $3 trillion to just over 10 trillion about 50 years later, and then it has continually doubled every 20 years since. Examined at this scale, of just a century-and-a-half, the graph looks like the familiar hockey stick growth we are used to in the tech sector. For the entire recorded human history, it was essentially flat, until it saw a sudden and unprecedented global explosion of productivity. In just the last 150 years.

One might argue that this is driven by exploding populations, but the World Bank has modeled GDP per capita since 1960, and it provides a similar story.[3] The World Bank has tracked GDP per capita worldwide since 1960, and we see it had a >5x increase in the years 1960 to 1980 (from $452/per year to $2,580), and it has about doubled every 2 decades since, reaching $13,736 in 2020.

Note that this picture is an average of every person on the globe. Given the huge disparities in economies, many countries have experienced significantly more change than this, on a per capita basis. The Western developed countries, including the US, were already at $17,000 in 1960, and as a group, now realize >$50,000 GDP per year per capita.

So, while I'm a big fan of putting change in the broadest possible frame, and realizing every generation goes through change and thinks they are unique and unprecedented, I think the data above shows that our pace of change is clearly unprecedented when viewed through the lens of deep human economic history.

Here, economic value serves as a proxy or a visual catchall metaphor, providing a starting reference point for the more abstract changes we are facing, such as the streams of our model.

There's an additional factor that is implied so far that should now be made explicit: while the factors influencing the effectiveness of each of our layers of thought are changing, they are changing at different rates, and they each have a different speed at which they can adapt.

We are not the first to think about this. Pinker has said, for example,

> I don't think that our man-made environment is necessarily running away from us and it's going to be a matter of how the genes are going to keep up. ... Instead we've designed computers so that THEY mesh with OUR way of thinking. We

3 "GDP per Capita (Current US$)." *World Bank*, data.worldbank.org/indicator/NY.GDP.PCAP.CD. Accessed Aug. 25, 2025.

> have designed elaborate graphic interfaces that translate quite abstract information into representations of physical objects, in a particular location in space, that can be moved in a particular way, because that's how human intuition works. So I think the answer is: our minds are going to shape the environment in ways that we can cognitively deal with.[4] (Pinker, 2016)

Here, he is basically saying it's okay if our brain can't keep up, because we can leverage our brain to change the technology and manage the interfaces and adaptations, with the caveat that we would be responsible if we didn't. He's not explicitly using the Three Streams Model, but his statements are consistent with it.[5]

I think Pinker has set forth an optimistic and even idealistic approach. One, by the way that isn't yet happening, at least not in a coordinated fashion, at scale. Before we examine whether it is feasible, let's briefly review how quickly each of the layers can adjust:

Hutchins, in his book *Cognition In the Wild* – "*the collection through time of partial solutions to frequently encountered problems is what culture does for us*",[6] a process that takes centuries, according to him.[7] Then he adds: *"If we ascribe to individual minds in isolation the properties of systems that are actually composed of individuals manipulating systems of cultural artifacts, then we have attributed to individual minds a process that they do not necessarily have, and we have failed to ask about the processes they actually must have in order to manipulate the artifacts"*.[8]

Hutchins' work is considered a landmark achievement in helping us to understand cognition "in the wild" (but only with respect to those tasks focused on specific occupational contexts), and was written relatively recently, in 1995.

But in noting that tasks are accumulated over centuries, Hutchins has unwittingly documented a key flaw in human thought; one that as recently as 1995 wasn't yet a problem. Culture cannot adapt fast enough to optimize, genes most certainly are not, and technology is changing so fast that it is now part of the problem, not part of the solution.

---

4 Pinker, Steven, and Michael Shermer. "Progress and Its Discontents: The Edge Annual Question 2016." *Edge*, edge.org/3rd_culture/pinker_rose/part2/part2_p2.html. Accessed Aug. 25, 2025.

5 See also Pinker, Steven. *How the Mind Works*. W. W. Norton & Company, 1997.

6 Hutchins, Edwin. *Cognition in the Wild*. MIT Press, 1995, pp. 168–169.

7 Hutchins, Edwin. *Cognition in the Wild*. MIT Press, 1995, p. 168.

8 Hutchins, Edwin. *Cognition in the Wild*. MIT Press, 1995, p. 173.

TABLE 16.1 Three streams, three speeds

Layer	Speed of change	Comments
Technology 	Seconds to months to years	Technology can launch in days, weeks, months, or years, but due to its layered nature, it appears to be undergoing significant changes constantly. Feedback loops exist between this and other layers of thought.
Culture 	Seconds to decades	Culture (1:1) can change with every new person you meet; however, due to synchronization, it can take years or decades for 51% of a population to change and thus it appears to be changing slower than technology. Feedback loops exist between this and other layers of thought.
Brain & experience 	Seconds to months to years	The brain undergoes microscopic changes with every new experience. The adult brain can be materially modified after 3–4 months of repeated exposure, and significantly, with years of exposure (think 10,000-hour rule; London cabbies' hippocampus). Feedback loops exist between this and other layers of thought.

One more factor we need to consider is the quote at the beginning of this chapter. As many of my bosses have told me over the course of my career, there's a difference between activity and results. Lewis Carroll creatively illustrates this point with the Red Queen, who says that since everyone else is running, we need to run twice as fast to achieve any progress.

1 The Red Queen Effect: Why Standing Still Means Falling Behind

In 1973, evolutionary biologist Leigh Van Valen[9] adopted Carroll's theme to explain a pattern he was seeing in the fossil record: species and ecological niches appeared to be remarkably linear and stable. This is surprising because of the concept of evolution; Van Valen would have expected significant changes in populations and species due to evolution creating more fit species, better able to compete in their niche.

He concluded that while species adapted, so did the other species in their niche, resulting essentially in a standoff. Like improving your running time to beat your rival, you get better, but so do they, so you keep ending your races in a tie. Van Valen used the term the Red Queen Hypothesis to describe this dynamic.[10] Sure, some species went extinct, but since all species have an equal probability of going extinct across an ecological niche, you again have a zero-sum game. This creates unexpected stability over time. Even if species A becomes more competitive and causes species B to go extinct, then the niche stabilizes, as species A is now deadlocked against a new competitor, species C:

As two species ebb and flow over time in their constant battle for supremacy in their niche, within each species, you would have different populations rising and falling, according to their ability to fit and survive in the new niche.

How is that relevant to our brief discussion about change, and specifically the varying rates of acceleration?

Matt Ridley applied this concept to our modern world when he wrote his 1994 book, *The Red Queen*, in which he describes The Red Queen Effect – you must constantly adapt and improve to stay in harmony with your constantly evolving environment.[11] The faster your niche evolves and the more competition you face, the faster you must adapt to stay alive. Ridley brings this up in the context of sex, but at the end of his book, he notes that this selection process shapes human intelligence as well.

Since his writing, numerous people and organizations have adopted this approach to look at organizations and economies.

Consider a practical application of this: network connectivity.

---

9 According to my research, no relation to Eddie Van Halen.

10 Van Valen, Leigh. "A New Evolutionary Law (1973)." *Foundations of Macroecology; University of Chicago Press*, 2014, pp. 284–314.

11 Ridley, Matt. *The Red Queen: Sex and the Evolution of Human Nature*. Macmillan, 1994.

Remember, my background as a technologist is grounded in helping people, organizations, and countries adapt to evolving network technologies. I've done this since the first browser hit the World Wide Web in the early nineties, and then through successive waves of cellular connectivity: 2G, 3G, 4G, and now 5G (and the beginnings of the next one, too).

Each of those cycles begins with an overhyped set of pronouncements of how this wave will change the world. Even today, I'm torn between expert communities that think 5G is a nothingburger, and those that think it will be a General Purpose Technology that transforms the world. The truth is they are both right, but you must look at things from multiple perspectives and against different timescales to see that. If your expectations are that 5G will result in a self-driving ambulance that comes to your house and drops off a robot to operate on you, you are going to be disappointed, especially if you think that's going to happen any time soon.

On the other hand, if you open the aperture a little wider and leave it open a little longer, you will see a different picture. First, if you wait long enough, the scenario I just described will probably happen, and most likely in my lifetime.

But secondly, look at 5G as an accelerant, the same way that we applied the effects of AI on Kelly's pen in the last chapter, or synthetic biology's indirect effects on far-flung sectors. With 5G, connectivity will be more ubiquitous, cheaper, and faster. More things will be online, and more people. Disruption and intermediation will happen. You say you want a pizza, and a drone drops it off on your porch; you don't have to talk to anyone or spend time manipulating apps on your phone. Frictions are removed.

And that brings us back to the Red Queen Effect. The more people and things are connected to an ecosystem, the faster innovation will occur for all successful entities within that ecosystem. If Metcalfe's Law describes the power of a network (network power expands exponentially with the size of the network), the Red Queen Effect describes the acceleration of the pace of innovation necessary to survive in that network as a function of network growth.

Of course, part 2 of that corollary is that as networks grow and the pace of necessary innovation accelerates, the number of failing entities will increase, because not everyone will be able to keep up. Put simply, this model predicts that larger networks lead to exponentially more failed entities within them.

You can see that demonstrated in the American economy. McKinsey conducted a study on the average life span of companies in the S&P 500, and they found an accelerating rate at which companies disappear (or go extinct, from a Van Valen perspective).[12] In 1958, the average life of an S&P 500 was 61 years, and now it is 18. McKinsey predicts that by 2027, 75% of our current companies will disappear. Where do they go? They get bought, like Yahoo and my old firm, merged, like AOL and Time Warner, or go bankrupt, like Enron and Lehman Brothers. Although some companies are old, many of the most valuable firms today are relatively young, such as Apple, Alphabet, Microsoft, and Amazon.

I see the Red Queen Effect at work in this context, particularly with 5G and other technologies. In Part 4 of this book, I will make the case that the Red Queen Effect is at work in people and culture. Cultural landmark events like Black Lives Matter, MeToo, and January 6th are not necessarily reflecting an increase in the number of bad things that happen to segments of our society, but they do reflect that these events are happening in a world that is increasingly connected. While driven by unique causes, the rapid diffusion and amplified visibility of these events reflect Red Queen dynamics in highly connected networks.

Accelerating change won't be ubiquitous and consistent across networks, of course. Consider again Van Valen's models: some species will ebb and flow, like Amazon and Walmart. Some species will go extinct, like Blockbuster. Even within a species, some populations will rise and fall – consider how Google's search business competes within the company against their own products like Google Nest, while concepts like Google Loon may rise, spin out, and then eventually fall.

It is a system of systems. The pace of change within health care will be different than the pace of change in nail salons, retail, or construction.

Now consider the material we covered in Chapter 14 – the multiple feedback loops within and across the various fields of human endeavor, including technology, and the various subcomponents of technology, like computer chips, software, and computer architectures.

If you consider 5G in this broad context, it's like jet fuel and oxygen being applied to a fire: it accelerates everything, so much that the fire will create its own climate, which further accelerates the process.

Bringing this back to thought, you can see how the varying rates of change combine with the feedback loops within the model and the complex interconnectedness of the various fields of human endeavor. Now you'll begin to see

---

12 McKinsey & Company. "McKinsey on Finance" *McKinsey & Company*, no. 84, Dec. 2023, www.mckinsey.com/~/media/mckinsey/business%20functions/strategy%20and%20corporate%20finance/our%20insights/mckinsey%20on-finance%20number 84/mckinsey-on-finance-number 84.pdf. Accessed Sep. 7, 2025.

thought as I've slowly begun to view it over the last few years – as an amazingly complex integrated Red Queen skater, spinning ever faster on the ice, as all of the expanding pieces integrate into a self-reinforcing, all-powerful quickening.

Just remember, as Megginson wrote in his paraphrase of Darwin, "*It is not the strongest of the species that survives, nor the most intelligent, but the one most responsive to change.*"[13]

13 Megginson, Leon C. "Lessons from Europe for American Business." *Southwestern Social Science Quarterly*, vol. 44, no. 1, 1963, pp. 3–13, at p. 4.

CHAPTER 17

Our Thoughts Are Controlled by a Complex Web of Actors

> There is a Zen story about a man riding a horse that is galloping very quickly. Another man, standing alongside the road, yells at him, "Where are you going" and the man on the horse yells back, "I don't know. Ask the horse."
>
> I think that is our situation. We are riding many horses that we cannot control.[1] (Thích Nhất Hạnh)

Let's talk about that quote from Thích Nhất Hạnh – how are we to control our horses, our thoughts, given the factors discussed so far?

1 Control over Thinking in the Brain

Let's start with the basics: at least we control our own thoughts, right?

Not according to psychologist Jonathan Haidt who says we are like the rider on the back of an elephant, and our conscious reasoning part of the mind has only limited control of what the mind does.[2]

Haidt notes the flaws in the traditional view of our control of our own thoughts, starting by pointing out that we don't control our own bodies, using the independence of the penis as a prime example. Then he introduces Gazzaniga's work in the 1960s on patients who had the hemispheres of their brain separated (to treat cases like severe epilepsy). Gazzaniga determined that separate intelligences operated in each half of the typical brain.[3] Later studies showed these hemispheric modules could even be in conflict with each other, as we've noted with McGilchrist's work. Haidt summarizes this work as suggesting that the mind is "a confederation of modules" capable of operating independently and sometimes at cross purposes.[4]

---

1 Hạnh, Thích Nhất. *Being Peace*. Parallax Press, 1987, p. 42.

2 Haidt, Jonathan. *The Happiness Hypothesis: Finding Modern Truth in Ancient Wisdom*. Basic Books, 2006, p. 1.

3 Gazzaniga, Michael S. "The Split Brain in Man." *Scientific American*, vol. 217, no. 2, 1967, pp. 24–29.

4 Haidt, Jonathan. *The Happiness Hypothesis: Finding Modern Truth in Ancient Wisdom*. Basic Books, 2006, p. 6.

 | HTTPS://DOI.ORG/10.1163/9789004739161-022

Unconscious emotions, too, are required for rationality. Drawing on Damasio, Haidt argues that reason and emotion must work together to produce intelligent behavior, but that "*emotion ... does most of the work.*"[5] As we covered earlier, Lisa Feldman Barrett, too, would have our thinking shaped largely by emotions, but she would say those are controlled by sensations we get from our body.

John-Dylan Haynes built on work done by Libet and others when, in 2007 (and subsequently), he put people in an MRI machine and asked them to perform various tasks, like selecting a letter. Like Libet, Haynes demonstrated that up to 7 seconds before conscious decision-making kicks in, the unconscious brain was gearing up for the decision.[6] In subsequent work, Haynes and colleagues used imaging studies to accurately predict which conscious decision would ultimately be made, well before conscious processes kicked in.[7]

Now, consider again the ideas of Kaplan and Minsky – that our minds are really just an unknown number of agents fighting it out amongst themselves to try to get what they want. Or as Dennett put it in his Multiple Drafts Model,[8] our mind is just a "virtual machine" that operates, as Oliver Selfridge suggested, in a pandemonium[9] of unconscious demons.[10]

Merlin Donald dismisses this, as he says these approaches may work for those like Minsky who are developing AI, but these "AI hardliners" are just replacing the little person in our head (homunculus) with several unconscious little people (homunculi), without necessarily proving they exist. He suggested instead his "hybrid model," in which consciousness develops out of a combination of forces within the brain, interwoven and tightly coupled with human culture as a distributed cognitive network.[11]

These scholars fundamentally challenge the notion of conscious control over a unified self. We are not autonomous agents in control of our thoughts

---

5 Haidt, Jonathan. *The Happiness Hypothesis: Finding Modern Truth in Ancient Wisdom.* Basic Books, 2006, p. 13.

6 Haynes, John-Dylan, et al. "Reading Hidden Intentions in the Human Brain Using High-Resolution fMRI." *Current Biology*, vol. 17, no. 4, 2007, pp. 323–328.

7 Soon, Chun Siong, et al. "Unconscious Determinants of Free Decisions in the Human Brain." *Nature Neuroscience*, vol. 11, no. 5, 2008, pp. 543–545.

8 Dennett, Daniel C. *Consciousness Explained.* Little, Brown and Company, 1991.

9 It is worth noting that Pandemonium was Milton's word for Hell in his epic poem Paradise Lost.

10 Selfridge, Oliver G. "Pandemonium: A Paradigm for Learning." *Symposium on the Mechanization of Thought Processes*, HM Stationery Office, 1959.

11 Donald, Merlin. *A Mind So Rare: The Evolution of Human Consciousness.* W. W. Norton & Company, 2001.

and actions. Instead, brain structures, unconscious emotions, social pressures, and cultural forces shape our behavior and create an illusion of agency we rarely recognize. What we experience as conscious control represents, in Wegner's analysis, an illusion – a post-hoc narrative constructed to explain decisions already determined by unconscious processes:[12] you may be under the delusion that you have a conscious experience under your control.

2 Control over Thinking with Culture

You have far less control over your Thinking with Culture, of course.

Picture this: You look across a crowded room, and see a good friend on the phone, with pain flashing across their face: your pre-wired empathy circuits cause you to feel pain as well, plus other feelings trying to understand what caused your friend to feel this way, and why, and what you can do about it. Partly to distract yourself from these uncomfortable feelings, you glance down at your phone, scroll through your social media feed, and see, ever so briefly, content related to dozens of people, curated according to your interests and habits (and ultimately in a way that maximizes the profit of those companies, of course), further spurring on your unconscious thoughts and emotions.

If Donald is right, and you buy the story we laid out in Chapter 14 about the complexity of Thinking with Culture, your cognition is really a function of vast numbers of people, living and dead, as well as corporate and other organizational interests, media (entertainment & journalism), science, government, religion, and even, increasingly, artificial entities passing themselves off as people.

As Sterelny says, *"Indeed, our most important cognitive tools – language, well designed notation systems and the like – are the multi-generational product of many minds....Notebooks, experiments, programs and papers are more often than not the result of many hands and minds. The same is true of decision and action in many commercial and administrative organisations."*[13]

In the 1950s, psychologist Solomon Asch conducted his now-classic experiments on conformity. Participants were asked to judge which of several comparison lines matched the length of a target line. While the task was

12 Wegner, Daniel M. *The Illusion of Conscious Will*. MIT Press, 2002.

13 Sterelny, Kim. "Externalism, Epistemic Artefacts and the Extended Mind." *The Externalist Challenge*, 2004, pp. 239–254.

simple and the correct answer obvious, the results were striking when participants were surrounded by others (confederates) who had been instructed to give incorrect answers. Faced with unanimous opposition, about one-third of participants conformed and gave the wrong answer. Across trials, 75 percent of participants conformed at least once, despite clear evidence from their own eyes.

Asch concluded that individuals will often subordinate their own perceptions to group pressure, even when the group is clearly wrong. His experiments reveal the powerful role of social influence in shaping thought and decision-making. The lesson is clear – our thinking is not only vulnerable to unconscious drives within the brain but also deeply shaped, and sometimes overridden, by cultural and social forces around us.

Asch concluded, "*That reasonably intelligent, well-meaning, young people are willing to call white black is a matter of concern.*"[14]

Lipsitz's *The Possessive Investment in Whiteness: How White People Profit from Identity Politics* illustrates how cultural values shape modern American society through systematic racial advantage. He demonstrates that public policy and private institutions collaborate – often unconsciously – to assign monetary value to whiteness, creating entrenched racial hierarchies.[15] Both White and Black Americans navigate opportunities through insider networks that perpetuate discrimination across housing, education, employment, and other cultural domains.

The system encourages White Americans to maintain an identity that provides structural advantages. As Lipsitz argues: "*We do not choose our color, but we do choose our commitments. We do not choose our parents, but we do choose our politics. Yet we do not make these decisions in a vacuum, they occur within a social structure that gives value to whiteness and offers rewards for racism.*"[16] Writing two decades before 70 million Americans (predominantly white) supported Donald Trump in 2020, Lipsitz reveals how unconscious cultural forces shape our thinking more powerfully than we recognize or understand.

14 Asch, Solomon E. "Opinions and Social Pressure." *Scientific American*, vol. 193, no. 5, Nov. 1955, pp. 31–35.

15 Lipsitz, George. *The Possessive Investment in Whiteness: How White People Profit from Identity Politics*. Revised and expanded ed., Temple University Press, 2006.

16 Lipsitz, George. *The Possessive Investment in Whiteness: How White People Profit from Identity Politics*. Revised and expanded ed., Temple University Press, 2006, p. 3.

3 Control over Thinking with Technology

We might assume our devices remain under our control – turning on the TV and selecting a movie in our homes represents the pinnacle of personal autonomy.[17]

Except, the available options emerge as cultural and technical byproducts. Netflix algorithms generate your feed by balancing the platform's corporate interests against your viewing history and data from millions of other users.

3.1 Unknowable Systems: The Impossibility of Understanding Tech

Corporate interests and technological features control your television viewing, filtering content and imposing their priorities through barely regulated systems that claim to serve your preferences. While these companies assert they prioritize user interests, their fundamental mission involves generating shareholder profits by manufacturing market demand.

Consider the ubiquitous iPhone with its average of 90 applications. In 2019, *New York Times* reporter Kevin Litman-Navarro analyzed 150 privacy policies from common apps.[18] These policies required 15–20 minutes to read on average and, like our earlier Nest example, often exceeded the cognitive capabilities of PhDs and lawyers. Facebook's policy matched the average length of other applications but rivaled Kant's *Critique of Pure Reason* in linguistic complexity and conceptual intricacy.

No wonder, then, that Pew Research found that 36% of adults *never* read them, and 38% of the rest read them only occasionally.[19] Other research shows this might be conservative. A 2017 Deloitte survey of 2,000 U.S. consumers found that 91% of people consent to terms of service without reading them; the rates were highest among younger adults, 97% of whom accepted without reading.[20] In 2017, Purple, a Wi-Fi provider, tried to draw attention to this

17 Ellul literally uses the phrase *chacun chez soi* to refer to our ability to control the things that happen in our own houses, in Ellul, Jacques, "The Autonomy of Technique," *Controlling Technology: Contemporary Issues*, Katz, Eric, Light, Andrew, and Thompson, William B., Eds., Prometheus Books, 2003.

18 "We Read 150 Privacy Policies. They Were an Incomprehensible Disaster." *The New York Times*, Jun. 12 2019, nytimes.com/interactive/2019/06/12/opinion/facebook-google-privacy-policies.html. Accessed Aug. 25, 2025.

19 "Americans' Attitudes and Experiences with Privacy Policies and Laws." *Pew Research Center*, Nov. 15, 2019, pewresearch.org/internet/2019/11/15/americans-attitudes-and-experiences-with-privacy-policies-and-laws/. Accessed Aug. 25, 2025.

20 "2017 Global Mobile Consumer Survey: Executive Summary". *Deloitte*, 2017, www2.deloitte.com/content/dam/Deloitte/us/Documents/technology-media-telecommunica

trend when they put in place a Terms and Conditions acceptance that required users of the free service to commit to 1,000 hours of public services – cleaning toilets, scraping gum off the sidewalk, and "manually relieving sewer blockages."[21] 22,000 people signed up during the two-week experiment. Only one person read the Terms and Conditions closely enough to win a prize embedded within it.

So – what's the big deal?

Let's remember what this phone is: it's as powerful in some ways as the world's best supercomputer in 1998; it can track anything you put into it or do with the apps on the phone, plus it knows the exact location, altitude, and movements of itself. On board are microphones and cameras that can record everything within range, including the presence, name, and other details of other networked devices; and behind these sensors are cloud-based AI and ML connected to the phone by 4G/5G and Wi-Fi networks, which can increasingly track and analyze your written, spoken and visual signals deeply, and accurately determine everything from emotions and lying to diagnosis of key diseases.

So again – what's the big deal; not a bad deal for $50 bucks a month, right?

In 2019, the *Washington Post* tracked an average smartphone and discovered that 5,400 hidden app trackers transmitted user data to third parties through the device's connection. These applications used 1.5 gigabytes of monthly data, equivalent to a typical AT&T subscriber's own monthly data use at the time. DoorDash is a great example – they'd employ at least nine trackers that recorded device specifications (name, model, memory size), behavioral data (accelerometer readings), advertising identifiers, and personal information, including delivery addresses, names, emails, and cellular carriers, to measure advertising effectiveness. The company shares this data with trading partners such as Facebook and Google Ad Services while disclaiming responsibility for these partners' privacy practices deep within their privacy policy: "*DoorDash is not responsible for their privacy policies or practices*."[22]

And your phone is just one of many connected devices in your world, each of which is trying to understand you better, so they can serve (and shape) your needs. Companies use this information, and other information they farm from around the world about you (in a process known as data aggregation and using

---

tions/us-tmt-2017-global-mobile-consumer-survey-executive-summary.pdf. Accessed Aug. 25, 2025.

21 "Purple Community Service." *Purple.ai*, purple.ai/blogs/purple-community-service/. Accessed Aug. 25, 2025.

22 "Privacy Policy." *DoorDash*, help.doordash.com/consumers/s/privacy-policy-us. Accessed Dec. 9, 2020.

tactics like de-anonymization and re-identification),[23] to shape the experience it offers to you, and/or to tell the advertising they present to you, which is itself designed to influence and control your behavior.

3.2 Surveillance Capitalism: When You Become the Product

The 2020 docudrama The Social Dilemma describes how online platforms closely watch, track, measure, and track users, and leverage this data to increase engagement, growth, and advertising revenue through various manipulation techniques.[24] These tactics include sophisticated psychological research. These platforms are highly effective at shaping your behavior, and when your behavior conforms to their needs, the result is that you behave very much like an addict. Technology itself crafts ever more narrowly targeted ads, which are highly personalized for effectiveness (e.g., influence on your behaviors) that constantly evolve in response to your actions and non-actions.[25] And remember, the technology you interact with is often free, because it is ad-supported; you are exposed to advertising throughout your day, to the tune of 5,000–10,000 times a day. As technology platforms try to exert their needs on you, so, too, do the advertisers who pay for the platform. The industry saying is "if the platform is free, then *you* are the product."

Shoshana Zuboff coined the term "surveillance capitalism" to describe this dynamic in her powerful 2019 book *The Age of Surveillance Capitalism*. Surveillance capitalism harnesses human experience as the raw material from which new precision products are made, in anticipation of what you will need, according to Zuboff. She found "*it is no longer enough to automate information flows about us; the goal now is to automate us.*"[26]

Marshall McLuhan, writing in 1964, saw this coming. He said, "*Archimedes once said, 'Give me a place to stand and I will move the world.' Today he would have pointed to our electric media and said, 'I will stand on your eyes, your ears, your nerves, and your brain, and the world will move in any tempo or pattern I*

23 Rocher, Luc, et al. "Estimating the Success of Re-Identifications in Incomplete Datasets Using Generative Models." *Nature Communications*, vol. 10, no. 1, 2019, pp. 1–9.

24 *The Social Dilemma*. Directed by Jeff Orlowski, performances by Tristan Harris, Shoshana Zuboff, Jaron Lanier, and others, Exposure Labs, Netflix, 2020.

25 Athey, Susan, and Joshua S. Gans. "The Impact of Targeting Technology on Advertising Markets and Media Competition." *American Economic Review*, vol. 100, no. 2, 2010, pp. 608–613.

26 Zuboff, Shoshana. *The Age of Surveillance Capitalism: The Fight for a Human Future at the New Frontier of Power*. Profile Books, 2019, p. 339.

choose.' We have leased these "places to stand" to a myriad of private corporations."[27]

It should be no surprise, then, to learn that the inventor of the Internet's pop-up advertisement now expresses regret over his creation. Ethan Zuckerman says, "*I wrote the code to launch the window and run an ad in it. I'm sorry. Our intentions were good,*" and then adds: "*I have come to believe that advertising is the original sin of the web. The fallen state of our internet is a direct, if unintentional, consequence of choosing advertising as the default model to support online content and services.*"[28]

Zuboff, McCarthy-Jones, and others place control squarely in the hands of tech companies – sophisticated actors deliberately manipulating our behavior for profit. This framing dominates contemporary tech criticism. But is it the full story? Before accepting that companies like Facebook and Google wield ultimate control, we need to examine a deeper question.

3.3 Who's in Charge?

Maybe technology even controls itself?

In his book *What Technology Wants*, Kevin Kelly paints a picture of technology as the seventh kingdom of life on earth. He says the technium is "*the organism of ideas,*"[29] and "*it's the greater, global massively interconnected system of technology vibrating around us.*"[30]

Just as Donald saw consciousness as a tight coupling between our brains and our culture, Kelly's technium blends technology and culture, and is an outgrowth of the human mind, as it – "*as it extends beyond hardware to incorporate culture, art, social institutions, and intellectual creations of all types.*"[31]

Kelly ascribes autonomy and intent to his technium – "*The technium also wants what every living system wants: to perpetuate itself, to keep itself going. And as it grows, those inherent wants are gaining in complexity and force*" and "*We don't have everything that the technium demands, but we can learn to work with this force rather than against it.*"[32]

Kelly is not alone. Katz, White, and Thompson published *Controlling Technology*, containing 34 essays debating "Do we control technology or does tech-

27 McLuhan, Marshall. *Understanding Media: The Extensions of Man*. McGraw-Hill, 1964, p. 68.
28 Zuckerman, Ethan. "The Internet's Original Sin." *The Atlantic*, Aug. 14, 2014.
29 Kelly, Kevin. *What Technology Wants*. Penguin, 2010, p. 11.
30 Kelly, Kevin. *What Technology Wants*. Penguin, 2010, p. 12.
31 Kelly, Kevin. *What Technology Wants*. Penguin, 2010, p. 11.
32 Kelly, Kevin. *What Technology Wants*. Penguin, 2010, p. 45.

nology control us?"[33] They cite Heidegger and Ellul's arguments that "*modern technology, although it may seem to be just a more efficient means of doing what humans have always done, confronts humanity with issues that go to the very core of who we are and how we live.*"

Ellul uses the much narrower term "technique" and has argued that technique is autonomous from economics and politics, and it shapes social, political, and economic change. Pitt cautions that treating technology as a "*thing*" and "*attributing causal powers to it*" or intentions of its own can lead us "*down blind alleys.*"[34] He argues that whether technology is autonomous depends on definition, noting that if autonomy means being free from influence in development and use, then "*technology can never be autonomous*" because it is created to serve specific goals.[35] He seems to reluctantly side with Charlton Heston, who said, "*Guns don't kill people, people do.*"[36] We shouldn't claim technology is the culprit, but ourselves – we have met the enemy, and they are us. Pitt says, "*Tools by themselves do nothing.*"[37]

3.4 Collective Agency: You Get What You Ask For

Does technology control us, or do we control it? The answer may actually cut against both prevailing narratives. Technology isn't controlled (or even really understood, as we've argued earlier) by corporate executives, nor has it achieved autonomous agency. Instead, technology serves the market – which means it reflects our collective demands. And that means WE are in charge of the asylum. Let's turn to economist Carl Menger and gluten-free bread to help explain how and why.

---

33 Katz, Eric, et al., editors. *Controlling Technology: Contemporary Issues.* Prometheus Books, 2003, p. 3.

34 Pitt, Joseph C. "Thinking About Technology." *Controlling Technology: Ethics and the Responsible Engineer,* edited by John D. Katz, Linda White, and Charles Thompson, Temple UP, 1994, p. 31.

35 Pitt, Joseph C. "Thinking About Technology." *Controlling Technology: Ethics and the Responsible Engineer,* edited by John D. Katz, Linda White, and Charles Thompson, Temple UP, 1994, p. 32.

36 Heston, Charlton. *"From My Cold, Dead Hands."* Speech at the NRA Annual Meeting, Charlotte, NC, 20 May 2000. Transcript available via C-SPAN, www.c-span.org/video/?155368-1/nra-annual-meeting. Accessed Sep. 7, 2025.

37 Pitt, Joseph C. "Thinking About Technology." *Controlling Technology: Ethics and the Responsible Engineer,* edited by John D. Katz, Linda White, and Charles Thompson, Temple UP, 1994, p. 29.

Menger was famous in part for making the distinction in the role of the human actor in economic markets.[38] Prices reflect human wants, and causal processes are purposefully designed to satisfy those wants. The complex economy can be wholly explained as the result of the driving force of human needs.[39] Kelly, in *What Technology Wants*, quotes Gordon Moore himself as saying "*Moore's Law is really about economics,*" and iconic American engineer Carver Mead, who said that Moore's Law is "is *really about people's belief system, it's not a law of physics, it's about human belief, and when people believe in something, they'll put energy behind it to make it come to pass.*" Control is exercised by the market; it gives us what we demand.

I've argued that technology's invisible acceleration proves difficult or impossible to perceive due to both complexity and dematerialization. AI, cloud computing, 5G networks, and social media appear to exist merely as applications and data on our phones – as these technologies expand exponentially, our devices remain the same size, creating an illusion of technological stasis.

Yet I live in Loudoun County, Virginia, known as "Data Center Alley," where this dematerialization reveals its physical reality. Over 200 massive data centers currently operate here, with another 100 in development. Industry experts regularly claim that most global internet traffic flows through these windowless behemoths – each roughly the size of a Walmart Supercenter – that now dominate our formerly rural landscape.[40] Driving the county's back roads, forests of construction cranes and looming concrete structures in fresh industrial paint make "digital" transformation impossible to ignore.

What drives this explosive growth? According to David Cappuccio, a managing vice president and chief of research at Gartner, the answer lies in human behavior: "*That's what's driving the massive growth – the end-user expectation of anything, anytime, anywhere. We're what's causing the problem.*"[41]

Academically, this approach – that people drive the evolution and adoption of technology – is further developed within an area of scholarship known as

38 He was also the founder of the Austrian School of economics; though Menger died before Nobel Prizes were awarded in Economics, his students mentored Friedrich Hayek, who was awarded a Nobel in 1974.

39 Schumpeter, Joseph A. "Carl Menger." *Ten Great Economists: From Marx to Keynes.* Routledge, 1969.

40 Loudoun County, Virginia. *Loudoun County, VA – Data Centers and Virtual Infrastructure.* Press release, Loudoun County Government, accessed Sep. 7, 2025.

41 Glanz, James. "Power, Pollution and the Internet." *The New York Times*, Sep. 22, 2012, www.nytimes.com/2012/09/23/technology/data-centers-waste-vast-amounts-of-energy-belying-industry-image.html. Accessed Jul 13, 2021.

the Social Construction of Technology (SCOT),[42] itself a branch of the sociology field of Science and Technology Studies (STS).

So, we control technology, and it isn't controlling us? How does this agree with Pitt?

Not so fast. Just because technology in its various forms and features is designed to meet *human* needs and wants, doesn't mean it was suited to meet *your* specific needs, and that you have control of the technology in your life, including most of the technology to which you are outsourcing your thoughts.

You do not control Technology if it is shaped by a combination of corporate interests and mob rule.

It might be democratic control, as discussed by Bertrand Russell, who warns that democracy can slide into apathy, allowing corrupt or unaccountable leaders to go unchecked.[43] Has the aggregate cultural influence of what the rest of the people in society desire created such demands for technology that you have no influence, and a dour future is foreordained? Have we slipped into a popular apathy? If so, can it be reversed?

I find hope in gluten-free bread.

30 years ago (and maybe even 10), a person who wanted to buy gluten-free products in their store would be treated like an alien from another planet. There was no gluten-free bread available to meet their needs, as not enough people were asking for it. But now, even the average store has many offerings: enough people demanded something that was healthier for them, so the markets evolved to meet those demands. Alone, we may not control the new innovations of technology, but together we can. Can this apply to thought?

42 See Pinch, Trevor, and Wiebe Bijker. "The Social Construction of Facts and Artifacts: Or How the Sociology of Science and the Sociology of Technology Might Benefit Each Other." *The Social Construction of Technological Systems: New Directions in the Sociology and History of Technology*, edited by Wiebe Bijker et al., MIT Press, 1987, pp. 17–50.; and Pinch, Trevor. "The Social Construction of Technology (SCOT): The Old, the New, and the Nonhuman." *Material Culture and Technology in Everyday Life: Ethnographic Approaches*, 2009, pp. 45–58.

43 Russell, Bertrand. *Proposed Roads to Freedom: Socialism, Anarchism and Syndicalism.* George Allen & Unwin, 1918.

3.5 Mental Autonomy under Siege

Let's adopt Metzinger's definition of mental autonomy as the ability to control one's own mental functions, including attention, memory, planning, and decision-making.[44]

In a provocative and important 2019 paper, The Autonomous Mind: The Right to Freedom of Thought in the Twenty-First Century, Simon McCarthy-Jones argues that twenty-first-century technological advances pose new threats to our mental autonomy – our freedom of thought. He notes the research on how the data stored at technology companies like Facebook and Google includes *"facial expressions, actions, possessions, purchases, musical preferences, websites visited, words used in Facebook posts, and 'likes' registered on social media."*[45] He further warns that advances in machine learning, such as systems that can infer sexuality from facial images, pose a "threat to the privacy and safety of gay men and women."[46]

McCarthy-Jones draws on work by Zuboff and others to show how behavioral data are monetized through micro-targeting designed to influence user behavior. He notes that governments can also use such systems to "nudge" citizens – for example, into joining organ donor registers, increasing fruit and vegetable consumption, or improving tax compliance.[47] Beyond commercial and policy uses, similar techniques are deployed by state actors to influence elections, and in systems such as China's social credit framework to shape citizen behavior.

All the while, technology studies its own effectiveness, in meta-optimization, to meet its ultimate goals. Facebook's first president, Sean Parker, explains what they are looking for – *"All about how do we consume as much of your time and conscious attention as possible?" "by exploiting a vulnerability in human psychology... The inventors, creators, it's me, it's Mark [Zuckerberg]... understood this consciously. And we did it anyway."*[48]

---

44 Metzinger, Thomas. *The Ego Tunnel: The Science of the Mind and the Myth of the Self.* Basic Books, 2009, p. 222.

45 McCarthy-Jones, Simon. "The Autonomous Mind: The Right to Freedom of Thought in the Twenty-First Century." *Frontiers in Artificial Intelligence*, vol. 2, 2019, article 19, p. 3.

46 McCarthy-Jones, Simon. "The Autonomous Mind: The Right to Freedom of Thought in the Twenty-First Century." *Frontiers in Artificial Intelligence*, vol. 2, 2019, article 19, p. 4.

47 McCarthy-Jones, Simon. "The Autonomous Mind: The Right to Freedom of Thought in the Twenty-First Century." *Frontiers in Artificial Intelligence*, vol. 2, 2019, pp. 4–5.

48 Pandey, Erica. "Sean Parker: Facebook Was Designed to Exploit Human 'Vulnerability.'" *Axios*, 9 Nov. 2017, axios.com/sean-parker-facebook-was-designed-to-exploit-human-vulnerability-1513306782-6d18fa32-5438-4e60-af71-13d126b58e41.html. Accessed Aug. 25, 2025.

4 My Take: Does Technology Control Us, or Do We Control It?

Technology is a direct function of human needs, as expressed in the market through prices, purchases, and attention. To this extent, technology is controlled by us, through our actions.[49] However, these are collective needs; technologies are often adapted to mass markets and so are geared to the common denominator. These needs are increasingly derived or synthetic because they are a function of powerful technology, which learns and adapts to be a powerful influence on people's needs and behaviors, artificially shaping needs and amplifying others. They are often the result of manufactured demand. Technologies are geared to the needs that drive monetization, regardless of their adaptability to individuals or populations. The result, as articulated by Abbas and colleagues, is that we found ourselves in "enslavement by design" through Technology's manipulation of human biological and psychological responses.[50]

The typical American individual has ineffective control over Technology today, to the extent that control means they use it consciously to meet their specific needs in a way that leaves them better off, subjectively.

At the collective level, technology operates beyond human control precisely because it serves the market of our least common denominator desires. However, crowd values are often outdated, leading us to unconsciously desire the wrong things. This makes us subject to manipulation by elites, such as sophisticated marketers or political actors seeking to influence our beliefs, and by the foibles of the rest of society. And our own.

We don't invest the effort required to understand and consciously direct these impulses, largely due to their overwhelming complexity. This book maps these dynamics systematically.

We now turn to Part 4, which examines the model's sixth insight: *human cognition functions poorly in our contemporary environment.*

This conclusion is an important one, which, if accurate, demands that we find and implement solutions. Thus, I've grouped this insight together with the concluding chapters, in which we will explore potential solutions that might generate more promising and widespread futures.

---

49 I highly recommend the classic book *The Cluetrain Manifesto*, whose first two theses are: 1. Markets are conversations. 2. Markets consist of human beings, not demographic sectors. Levine, Rick, et al. *The Cluetrain Manifesto*. Basic Books, 2009.

50 Abbas, R., et al. "Machine Learning, Convergence Digitalization, and the Concentration of Power: Enslavement by Design Using Techno-Biological Behaviors." *IEEE Transactions on Technology and Society*, vol. 3, no. 2, June 2022, pp. 76–88.

5 Summary of Part 3

This section introduced six key insights gained from using the Three Streams Model. We have developed five insights in some depth, revealing a picture of cognition that diverges sharply from conventional understanding.

Popular view: We navigate life via a simple natural process of making conscious, rational decisions based on individual self-interest via processes we control within our minds – processes evolution has optimized over millions of years for maximum effectiveness in our ecological niche. We are evolution's pinnacle achievement.

Three streams view: Cognition operates as a highly unconscious, complex, and infinitely distributed process over which we exercise limited control. It serves numerous discrete needs and values held by multiple selves within each individual – many competing with one another – as well as countless external entities. The pace of change across the model's three streams struggles to match the evolving contexts of modern life. Each stream adapts at different rates, operating with varying degrees of anachronism and responsiveness to contemporary human needs.

PART 4

The Path Forward

1 Purpose of This Part of the Book

Parts 1 through 3 diagnosed the disruption, built the model, and extracted its insights. This part asks the obvious next question: what do we do about it? I begin by evaluating how well our current cognitive environment serves human flourishing and where it falls dangerously short. From there, I work through practical pathways at three levels: individual agency, community norms, and the broader social and economic systems that shape them.

Rather than offering simplistic fixes, I provide frameworks and levers for conscious participation in shaping our shared cognitive future. The goal is not to retreat from technology but to align it intentionally with autonomy, competence, relatedness, and meaning, and to move from adapting to disruption toward designing what comes next.

 | HTTPS://DOI.ORG/10.1163/9789004739161-023

CHAPTER 18

The Disruption of Thought Is Not Going Well for Us, So Far

> The applications of science have built man a well-supplied house, and are teaching him to live healthily therein. They have enabled him to throw masses of people against another with cruel weapons. They may yet allow him truly to encompass the great record and to grow in the wisdom of race experience. He may perish in conflict before he learns to wield that record for his true good. Yet, in the application of science to the needs and desires of man, it would seem to be a singularly unfortunate stage at which to terminate the process, or to lose hope as to the outcome.[1] (Vannevar Bush)

> You take the blue pill, the story ends, you wake up in your bed and believe whatever you want to believe. You take the red pill, you stay in Wonderland, and I show you how deep the rabbit hole goes. Remember: all I'm offering is the truth, nothing more.[2] (Morpheus, The Matrix)

> I think the next century will be the century of complexity.[3] (Stephen Hawking)

But before we build, we must confront what we're building from. Construction requires honest assessment of the foundation. So let's be clear: the disruption of thought is not serving us well. Understanding why – and for whom – sharpens our architectural vision for what comes next.

Why should you, the reader, care about this disruption? What's the big deal, especially if it doesn't appear to affect you personally now? Perhaps, it's even benefiting you – why invest time understanding it?

Even if you buy my story in chapter 5 that this is the most important issue facing the world, why should you allocate your scarce cognitive resources to this? After all, the world is facing so many other problems.

Our Sun will burn out in 5 billion years, which is a big deal, but there's not much we can or should do about it now, with everything else going on. One

1 Concluding paragraph to his iconic essay: Bush, Vannevar. "As We May Think." *The Atlantic Monthly*, vol. 176, no. 1, July 1945, pp. 101–108.

2 *The Matrix*. Directed by the Wachowskis, Warner Bros., 1999.

3 Hawking, Stephen. Interview. *San Jose Mercury News*, Jan. 23, 2000.

would think that a reasonably aware person would already be stressed out enough by climate change, the next pandemic, upcoming elections, various shooting wars breaking out across the world with their attendant atrocities, and the continued threat of plastic straws.[4] Why should we carve out some attention cycles to consider the disruption of thought?

In Chapter 5, I made the sweeping statement that the disruption of thought was a root cause for many of America's pressing social challenges today, without providing the reasoning for that. Here, I will lay out the case that Technology isn't, on net, serving most of us well, and why this becomes a negative driver of social disorder. This is a headline summary of the case made out in great depth and detail in my other work.

You should care as an individual because we can all take steps to improve our lives and our relationship with Technology. Even if Technology is serving you well today, it can serve you better. And if you are concerned about the challenges above or your current quality of life, then Technology can be a mighty force for good, if we embrace it correctly. I've taken a cut at what that might look like in the next few chapters, but mostly I would encourage you to take the time, reflect, and consider your own lived experience with tech.

---

4 Lest you think I'm making light of the world's existential threats, keep in mind that I'm Irish, and culturally we live by the MO that if we weren't laughing at something, we'd be crying. Which is why I'm hoping there are a great many people laughing at my numerous faults during my funeral, ideally several decades from now.

I spend an inordinate amount of time trying to understand humanity's challenges, and am fascinated by our tendency to ignore preventable crises, something Wucker explored at length in *The Gray Rhino.*

In this one case, however, I'm still mystified by the PR against plastic straws. I recognize that at one time, the US used an estimated 500 million straws every day, contributing to ocean pollution that annually kills a hundred thousand marine mammals like sea turtles. That's bad. But then I watch *Seaspiracy*, which shows that most ocean plastic pollution comes from the fishing industry. Fishing, through practices like bottom-trawling, "bycatch," and lost fishing gear, may be on the order of magnitudes more detrimental to ocean life than plastic straws. So, if I really cared, I should be far more upset about people eating fish than the existence of plastic straws. But by avoiding the use of my plastic straw in my plastic-cupped venti pumpkin cream cold brew and in my plastic bottled water, I've greenwashed my conscience, made me feel good about myself, and I stopped worrying about those poor sea turtles, as I bite into my tuna sandwich. And I go back to doomscrolling my phone. Straws represent classic left-brain solutioning, which is good, but absent the countervailing right-brain understanding of the whole context; tactics like this actually erode our ability to affect the desired outcome because they take the resources and attention off the main objective, and funnel them to some small part, thereby hijacking overall progress. I can't handle the complexity of the whole, but I want it to get better, so I'll do a small piece and feel like I'm helping. That's ok, if, as we'll discuss in the next chapter, there are folks connecting all the dots together into an integrated larger whole (see Hawken).

That may be the most important takeaway from the book. I have faith in human ingenuity and resilience, and I believe many of you will come up with new and better solutions that I haven't even considered. In my own life, I follow or pursue much of what I recommend in the next chapters, and I know I'm better off for it. My potential for positive impact on the world is greatly (miraculously) amplified by Technology, as is my optimism and belief that technology can be a tool that can be used to foster broader and equitable prosperity.

Once you understand what's going on, I'm sure you will almost immediately begin living with technology better in your own way, maybe using some of the ideas I'll sketch out, or your own, or others' solutions. We don't need to wait for Silicon Valley billionaires to provide us with new or better technologies. Or for a miracle, such as Congress or other government efforts to somehow have the foresight, expertise, will, and consensus to pass better laws. We can be the change we want to see in the world. And while this is a cliché, we *must be*. Otherwise, we will be the victim of our own undoing. And the operative word there is "our," as in plural.

I have a brilliant friend, who is both a technologist and a talented college professor, who intuitively understands most of the dynamics I cover in this work, and some of the possible negative outcomes. He is very comfortable with the idea that our world could come burning down at any time, in large part due to our own actions, or inactions. His solution? He has an off-grid cabin where he can unplug and be self-sufficient, even if the world continues its current course. I respect that.

Another friend is also a brilliant, multi-talented technologist and entrepreneur who believes that Technology has done nothing but uplift his life, family, and his various high-tech businesses. This includes technological innovation in his family's rural farm. I think that's great.

This is classic rugged American individualism – pulled up by our own bootstraps. Cutting through all the noise and trouble to succeed (or at least, try to survive) and taking responsibility for oneself, family, and destiny, using the tools at hand. One of my favorite sayings that Teddy Roosevelt popularized is "You do what you can, with what you have, where you are."[5] The operative words here are 'you' and 'self'. The individual.

Think back to the beginning of the book though and that quote from Daniel Humm: once we understand something, we have a responsibility to share that with others. That's because very few of us stand alone. In the acknowledgment

5 Roosevelt, Theodore. *The Autobiography of Theodore Roosevelt: with Illustrations*. Charles Scribner's Sons, 1913.

section, I note we are all buttressed by the communities we live in. Philosophers, including Aristotle,[6] Locke,[7] Rousseau,[8] and up to modern prophets like Hannah Arendt,[9] point out that being part of a community includes responsibilities to that community. That requires going beyond sharing with your tribe: it asks for action, which is what Humm did.

It's asking a lot of people who are struggling through their own hectic lives to take on that responsibility to do something on behalf of others. I know. But I've found this research journey compels me to do just that. Using the right side of my brain to deeply understand an intricately entangled world, I can first tell others, then perhaps engage my left brain to take actions toward the prescriptions outlined in the next chapters.

It gives me a purpose. When you use technology to make your life better, you flourish. When you use technology to serve your community, it is likely to flourish, but you'll find that, counterintuitively, you will, too.

But we are jumping ahead.

This chapter examines how poorly technology is serving us today, rather than how it might serve us in the future. And through that lens, we see additional reasons to motivate us to lean into this issue.

As I alluded to earlier, I'm also currently finishing another book project that digs deeply into this question. It's called *The Great Irony of Technology*, and its premise is that Technology has objectively left most of us better off compared to our grandparents and great-grandparents, but paradoxically, we feel worse off subjectively. That book is much longer than this one, and it was initially a VERY long section within the initial drafts of this manuscript. You can imagine what an unmanageable tome that was – over 600 pages. It is unsurprising, then, that my initial reviewers suggested cleaving them into (at least) two titles. While I accepted that great advice, to me, these books are all shoots from the same tree, and specifically, all part of a set of answers to my initial research questions. In this case: how *well* does technology serve us today?

It turns out to be a very hard and thorny question. What categories of progress or valued outcomes are we looking at? For whom, specifically? What yardsticks or metrics do we use? Is it a direct measure or a proxy? Where is our data, and how was it collected? How can it be rigorously and consistently measured across populations or time?

---

6 Aristotle. *The Politics*. Translated by Carnes Lord, 2nd ed., University of Chicago Press, 2013.

7 Locke, John. *Two Treatises of Government*. Edited by Peter Laslett, Cambridge University Press, 1988.

8 Rousseau, Jean-Jacques. *The Social Contract*. Translated by Maurice Cranston, Penguin Classics, 2004.

9 Arendt, Hannah. *The Human Condition*. 2nd ed., University of Chicago Press, 1998.

The question is rife with methodological challenges and biases.

I chose to look mainly at the United States to keep focus. But I look across gender, age, affluence, geography, immigration status, sexual preference, education, and other factors. I wrestle with the yardsticks of "progress" and "flourishing," with a deep review of the pros and cons of the available tools. My research relies on hundreds of data points from many sources and fields.

All of that will be condensed down into just a few pages here. If you'd like to "see the work" behind the findings, you might pick up that other book.

1 The Great Irony of Technology

Earlier, I talked about techno-optimists and tech-dystopians. Why can't there be a middle ground? Not one where each side is a little right and they cancel each other out, but one that looks across the domains and finds that the groups are both right, but simply talking past each other?

Yes, according to techno-optimists like Pinker, we have leveraged all kinds of technologies to double human life expectancy, and in our back pockets, we all have something that resembles some of the performance of 1998's world's best supercomputer. We can use that tool to arrange a hookup on Tinder and to hail an Uber to that tryst. And so much more. We have amazing foods from all over the world, and miraculous medicine. Even my poorest and most underprivileged friends facing food insecurity usually have mobile phones and flatscreens, because our society has invested in the social technologies that make up the current US civic safety nets (and asks us all to be the best consumer we can be). The average person in America today has luxuries that the Robber Barons of just 150 years ago, like Cornelius Vanderbilt, Andrew Carnegie, and John D. Rockefeller, would have envied. Life is amazing; yay!

Yet, politicians of all stripes can and do claim the world is "going to hell."[10] How can this be true in the face of the litany of facts that people like Pinker lay out about the amazing leap forward humanity has made over the last century?[11]

10 Scannell, Patrick. *The Great Irony of Technology*. Brill. Forthcoming.

11 While I could formally cite his book, *Enlightenment Now*, let me instead encourage you to watch his 2018 TED talk, which has gotten over a million views, and in 18 minutes, summarizes the thrust of his point. Regardless of where you stand on this question, it is a useful reference: https://www.youtube.com/watch?v=yCm9NgobbEQ. Pinker, Steven, "Is the World Getting Better or Worse? A Look at the Numbers". *YouTube*. https://www.youtube.com/watch?v=yCm9NgobbEQ. Accessed Nov. 17, 2025. And let's be real: if

The Great Irony of Technology tackles these points, and bridges these two camps by separating objective gains from subjective assessments.

1.1 Dying of Despair: When Privilege Becomes Poison

It starts with America's "privileged" – the whites who have traditionally been the ruling majority. Most Whites do not have a college degree. By looking at the state of this demographic – Whites without college degrees – we are starting with the largest group in the largest demographic in America, to develop a baseline, albeit one that is skewed by historic privilege.

But that privilege doesn't seem to manifest here, now. This demographic is actually experiencing *declining* life expectancies, which is a reversal of a 150-year trend of growing longevity.[12] The premature deaths are due largely to a growing epidemic of "deaths of despair"[13] (suicides, overdoses, and other similar factors), as well as from lifestyle diseases, many of which stem from another epidemic: obesity. Their privileged lives are either killing them, or they are killing themselves to get away from it.

Technology, writ large, plays a significant role here. Technological unemployment, driven by innovations such as factory automation, offshoring, and AI, eliminates jobs or lowers pay in occupations that have been in families for generations.[14] Today, that includes mass elimination of jobs in manufacturing, coal, trucking, and agricultural sectors. These transformations hollow out the communities built around these industries.[15] And AI and other tech threaten to do the same to our service sectors, including fast food workers, taxi drivers,

---

you are the average American, you are more likely to watch an 18-minute video, perhaps at 1.5 speed, than you are to read a book.

12 Leaving aside WWI: United Nations Department of Economic and Social Affairs (UN DESA), and Gapminder. "Annual Life Expectancy at Birth in the United States, from 1850 to 2023, with Projections until 2100." *Statista*, Statista Inc., Jul. 11,2024, www.statista.com/statistics/1040079/life-expectancy-united-states-all-time/. Accessed Nov. 17, 2025.

13 See Case, Anne, and Angus Deaton. *Deaths of Despair and the Future of Capitalism*. Princeton: Princeton University Press, 2020. And Case, Anne, and Angus Deaton. "Rising Morbidity and Mortality in Midlife among White Non-Hispanic Americans in the 21st Century." *Proceedings of the National Academy of Sciences* 112, no. 49 (Dec. 8, 2015, pp. 15078–15083).

14 Brynjolfsson, Erik, and Andrew McAfee. *Race Against the Machine: How the Digital Revolution Is Accelerating Innovation, Driving Productivity, and Irreversibly Transforming Employment and the Economy*. Digital Frontier Press, 2011.

15 Yang, Andrew. *The War on Normal People: the Truth About America's Disappearing Jobs and Why Universal Basic Income Is Our Future*. Hachette Books, 2018.

cashiers, warehouse workers, and customer service representatives.[16] Other technologies, such as social media and television, combined with the Internet and mobile phones, also play an important role in shaping and amplifying the information this group uses to try to make sense of the world.

Changes in the "technology" of US social systems (laws, mores, values) have changed the social order and norms that were the traditions and heritage of many communities. Social innovations in the form of changes to immigration policy and enforcement bring millions of people to the USA, often in jobs that US Citizens don't want.[17] Other policies that provide a growing set of rights and privileges to non-heteronormative Whites land in this population as a chaos of "wokeness."[18] These efforts are not just changing who's in power; they influence what's being taught to their children. One plus one no longer equals two, when accepted bedrock ideas such as fixed male and female genders are now fluid, muddying everything from parental notice to school bathrooms and sports.[19] From this perspective, it's no wonder that many in this group are angry and want to return to a time when they remember things as being great (or, at least for them, and better and more in line with the values, opportunities, and influence this group had in the mid-20th century). The subjective well-being of this demographic is in clear decline.

1.2 The Affluent Paradox: Success Often Means Fat, Broke, and Busy

What about the American elites who are living the successful American Dream? This affluent and typically highly educated group of many races (though still disproportionately White) has an unequal share of power and resources, including, often, college education. At least *they* must be thriving, right? Important work is going on across many aspects of society to ensure

16 Knowledge workers are not immune either, as even advanced skill positions like lawyers, doctors, and software developers are and will be replaced too, but here we are looking at those roles typically held by those with only high-school degrees.

17 Mukherjee, Sahana, and Jens Manuel Krogstad. *"Most U.S. Voters Say Immigrants – No Matter Their Legal Status – Mostly Take Jobs Citizens Don't Want." Pew Research Center*, Oct. 21, 2024.

18 Stallworth, Charles. "The Elites Have Stopped Hiding Their Hatred of the Working Class." *Newsweek*, Oct. 12, 2022.

19 Please keep in mind that I'm writing these words not as indictments or as my personal position, but in the voice of the segment I'm discussing, after spending many years researching their issues, problems, and personal viewpoints.

others have a fair shot at this lifestyle. So why assess it? What can be wrong in this Shangri-La, other than some whining from "entitled snowflakes"?[20]

Life in rich America presents its own distinct challenges. Remarkably, families earning three to four times the average American household income often report negative and deteriorating assessments of their present circumstances and prospects.[21] Their children often exhibit poorer mental health outcomes than youth from lower socioeconomic backgrounds.[22] These privileged families remain susceptible to lifestyle diseases while experiencing significant pressure to maintain social status and pursue the elusive American Dream.[23] Many Americans – even those with advanced degrees and large incomes – find themselves "fat, broke, and busy,"[24] while simultaneously battling stress, isolation, psychological distress, and mounting personal debt. These individuals, apparently beneficiaries of American prosperity, instead navigate lives resembling a Winslow Homer seascape – their vessels dismasted and rudderless, being tossed about by the Sturm und Drang[25] of the modern lives we have built for ourselves.

When we look at people in the 80, 90, and 99% income levels, we find a common theme of people living on a hamster wheel, maybe best described as a hedonic treadmill.[26] Where are they trying to go? Materialism and "keeping up" loom large here. Compared to lower-income Americans, the rich are far more likely to live in suburban enclaves, where the cost of living is signifi-

20 Camlin, Thaddeus. "Addiction and Society: Snowflakes and a Culture of Outrage." Practical Recovery, Jun.23, 2017.

21 Picard, Joe. "They're in the Top 10% of Earners. They Still Don't Feel Rich." *The Wall Street Journal*, Jun. 30, 2025. (In this article, individuals earning in the top 10% – typically over $250,000 annually – report feeling financially insecure due to high housing costs, college tuition, and lifestyle obligations. Over a quarter of those earning $200,000–$300,000 say they feel financially unsatisfied).

22 Luthar, Suniya S. "Children of the Affluent: Challenges to Well-Being." *Current Directions in Psychological Science*, vol. 12, no. 2, 2003, pp. 49–53.

23 Luthar, Suniya S. "The Culture of Affluence: Psychological Costs of Material Wealth." *Child Development*, vol. 74, no. 6, 2003, pp. 1581–1593.

24 Gaines, Jeff, and Sorensen, Jim. *Fat, Broke, and Busy: Escaping America's "Never Enough" Culture*. Jul. 25, 2013.

25 I encourage the reader to spend a few minutes becoming familiar with the origins of this phrase. It's used very consciously here, as a nod to the late 18th-century German literary and artistic movement that emphasized raw emotion, individualism, and a rejection of Enlightenment rationalism. I believe it's an appropriate precedent and analogy to our current American lives. See, for example: Chantler, Abigail. "The" Sturm und Drang" style revisited." *International Review of the Aesthetics and Sociology of Music*, 2003, pp. 17–31.

26 Diener, Ed, Richard E. Lucas, and Christie Napa Scollon. "Beyond the Hedonic Treadmill: Revising the Adaptation Theory of Well-Being." *American Psychologist* 61, no. 4, 2006, 305.

cantly higher. Relative to the consumer-driven economic expansion of the 1950s, they typically have larger homes, more sophisticated cars, and more complex and intense shopping behaviors than their parents and grandparents.[27] Meanwhile, the price they pay for these goods is, on average, and after inflation adjustments, much higher than in earlier generations. So too are the costs for vital needs such as health care and college education.[28]

Sure, they are getting smart homes, miraculous medicines, self-driving cars, and Amazon Prime deliveries, but it's taking a bigger bite out of their wallet than earlier generations. Relative to earlier generations, they have more, but at higher costs. Meanwhile, social media, television, and all advertising create the illusion that everyone else's lives are better.[29] Parents work harder and longer to keep up, spending less time with their children,[30] who absorb both the parental stress and pressure to excel. The result: crushing anxiety across generations from both the real and perceived threats to their future.[31] To cope, family members retreat to separate screens and binge-watching – activities that worsen mental health.[32] The family unit that might be strengthened in response to working together to tackle these problems instead fragments under the very pressures it could help address, leaving isolated individuals in their McMansions to face mounting stress alone.

Lewis Mumford, in his classic *Technics and Civilization* (1934), examined the profound relationship between technology and civilization – including the

27 Hall, Chris. "The More This Rolex Costs, the More You Want It. Here's Why." *Wired*, Sep. 6, 2024, https://www.wired.com/story/the-more-this-rolex-costs-the-more-you-want-it-this-is-why/. Accessed Nov. 19, 2025. See also: Veblen, Thorstein. *The Theory of the Leisure Class*. Oxford University Press, 2007, for the foundational theory on this topic; while it is > 100 years old, it is conceptually sound, and insightful.

28 "Average Cost of College Over Time: Yearly Tuition Since 1970." *EducationData.org*, Sep. 9, 2024, educationdata.org/average-cost-of-college-by-year. Accessed Nov. 17, 2025.

29 Sabatini, Fabio, and Francesco Sarracino. "Keeping Up with the E-Joneses: Do Online Social Networks Raise Social Comparisons?" *Economics Discussion Papers, No 2018–43*. Kiel Institute for the World Economy, *2018*. http://www.economics-ejournal.org/economics/discussionpapers/2018-43. Accessed Nov. 5, 2024.

30 "Time Poverty." *Wikipedia*, https://en.wikipedia.org/wiki/Time_poverty. Accessed Sep. 16, 2025.

31 Luthar, Suniya S., and Shawn J. Latendresse. "Children of the Affluent: Challenges to Well-Being." Current Directions in Psychological Science, vol. 14, no. 1, 2005, pp. 49–53.

32 American Psychological Association. "Screen Time and Emotional Problems in Kids: A Vicious Circle?" *APA News Releases*, Jun. 9, 2025, https://www.apa.org/news/press/releases/2025/06/screen-time-problems-children. Accessed Nov 17, 2025.

paradox that abundance can itself become a burden.[33] Returning to this theme decades later, Mumford argued that technological overprovision produces passivity and a loss of meaning, as human beings grow dependent on systems that supply everything while demanding nothing. He illustrates the point with an Egyptian story that: "*... reveals the emptiness of a Pharaoh's life, in which every desire was too easily satisfied, and time hung with unbearable heaviness on his hands. Desperate, he appealed to his counselors for some relief from his boredom; and one of them put forth a classic suggestion: that he fill a boat with thinly veiled, almost naked girls, who would paddle over the water and sing songs for him. For the hour, the Pharaoh's dreadful tedium, to his great delight, was overcome.*"[34]

After climbing to the top of society, armed with an outsized and unequal share of wealth and opportunity, many rich Americans find themselves trapped in a frantic lifestyle they need to kill themselves to acquire, pay for, and maintain, which is itself ephemeral, in that once they achieve something, they find it is never enough. They only "achieve" the emptiness that comes from having, as Mumford said, "*every desire was too easily satisfied.*"

Keynes, too, foreshadowed this in his 1930 essay, Economic Possibilities for our Grandchildren,[35] "*Yet there is no country and no people, I think, who can look forward to the age of leisure and of abundance without a dread. For we have been trained too long to strive and not to enjoy. It is a fearful problem for the ordinary person, with no special talents, to occupy himself, especially if he no longer has roots in the soil or in custom or in the beloved conventions of a traditional society.*"[36] With too much time on our hands and with fraying roots in custom or beloved conventions, we become discontent with what we have and want more, leading us to work harder to gain that abundance and even more.

That is not progress.[37] It is quite the opposite.

---

33 Lewis Mumford. *Technics and Civilization*. Harcourt, Brace & Company, 1934.

34 Mumford, Lewis. *The Myth of the Machine: Technics and Human Development*. Vol 1. Secker & Warburg, 1967, p. 206.

35 It's poignant to read this essay now; Keynes had no children, but married in 1926, at the age of 43, and he and his wife had a miscarriage the next year, when he was 44. If that child had lived and become a parent themselves at age 44, that offspring (Keynes' grandchild) would be my age now. So, in a sense, Keynes is looking into the future and sending this note to my generation.

36 Keynes, John Maynard. "Economic Possibilities for our Grandchildren." *Essays in Persuasion*, Harcourt, Brace & Company, 1931, p. 369. https://roar-assets-auto.rbl.ms/documents/43407/Intro_Session1.pdf. Accessed Nov. 12, 2025.

37 It reminds one of the proverbs: For what shall it profit a man, if he shall gain the whole world, and lose his own soul? (Mark. 8 Verses 36).

Where is the fairness uplifting the rest of America into this "prosperous" state if, after achieving it, people feel like they are going backwards in their quality of life? The wheel turns ever faster, but the hamster never gets the cheese.

To create lasting, widespread positive changes, the first step is understanding the perceived quality of life of all the people in our society before we set out to ensure everyone has the same opportunity, or what the Australians refer to as a "fair go."[38] While everyone deserves their fair chance at whatever American Dream they choose to pursue, it seems imprudent to blindly follow and copy today's pinnacle achievers, and maybe a different aspiration should be crafted.

Turning now toward the traditionally oppressed, we find very different problems.

1.3 Progress Invisible, Injustice Amplified

Blacks face disproportionate incarceration rates,[39] poverty, disease, employment, healthcare discrimination, high rates of single parenting, and are more likely than Whites and other races to be the victims of certain crimes such as homicides.[40] Social technologies of slavery, racism, and practices like redlining[41] set the stage for these structural problems, while others, such as civil rights legislation and changes to education and employment practices, have made some modest progress against these injustices.[42] But any improvements in their quality of life over the last 50 years are overshadowed by both a) the outsized gains experienced by Whites during this period and b) the ongoing

38 "Equity" and "inclusion" have become divisive triggers to Americans these days, so I include "fair go" here, and encourage readers of all political persuasions to read up on this, as I believe this concept is something most of us can rally behind. And don't we need to be building common ground now? See – Howard, Cosmo. "What Did a 'Fair Go' Originally Mean to Australians?" *Australian Journal of Political Science* 58, no. 2, 2023. pp. 210–229.

39 "Criminal Justice Fact Sheet." *NAACP*, Nov. 4, 2022, naacp.org/resources/criminal-justice-fact-sheet. Accessed Jun. 3, 2025.

40 "One in Five: Disparities in Crime and Policing." The Sentencing Project, Jan. 16, 2024, www.sentencingproject.org/reports/one-in-five-disparities-in-crime-and-policing/. Accessed Jul. 17, 2025.

41 Jacoby, Sarah F., et al. "Historic Redlining, Structural Racism, and Firearm Violence: A Structural Equation Modeling Approach." *Preventive Medicine*, vol. 153, 2021.

42 "Racial Inequality in the United States." *U.S. Department of the Treasury*, 21 July 2022, home.treasury.gov/news/featured-stories/racial-inequality-in-the-united-states. Accessed Mar. 15, 2024.

trauma of injustices they still face. Blacks are still getting pulled out of their cars and shot by police for "driving while black."[43] This is not progress.

How does Technology affect this? Historical data is not consistently and rigorously available, but it's fair to say that despite the reality of cases like Ahmaud Arbery, Breonna Taylor, and George Floyd, it's likely the *rate* of injustices that blacks face is far lower now than 50, 75, or 100 years ago.[44] However, ubiquitous cellphone videos and social media platforms rightly capture, highlight, and amplify these events when they happen. The net result is that while Technology seems to show how life is improving for Whites (who are prominent among those presenting an artificial "best version" of their life on social media), this portrayal contrasts strongly with the direct struggling and lived experience of being black, which is then deepened by the amplified signals (such as viral bodycam and bystander videos) that other Blacks' quality of life, opportunity, and safety are going backwards.

1.4 The Pattern Repeats: Women, Gen Z, Rural America

I won't try to summarize each demographic of the other book here. In general, electronic media of all types (TV/streaming, social media, web), artificially amplify to every demographic that a) others are doing better than they are, b) the world is going to hell, and always, due to the nature of algorithms, c) their own community is bearing the brunt of all bad things. Additionally, special circumstances apply to each population. E.g.:

- Women are now not only working three shifts[45] – day job, cooking and cleaning, and parenting – but now face a fourth shift of carrying an unequal share of caring for the aging population that has dramatically longer lifespans;[46]

43 Singer, Jenny. "What to Know About Daunte Wright, Who Was Fatally Shot by Police at a Traffic Stop." *Glamour*, Apr. 13, 2021.

44 I'd love to show authoritative citations here to support this, but the data is highly problematic, across time and geographies. I'll leave this flagged as speculation on my part, but I hope this gets researched better and a baseline develops. Apparently, there is a lot of …latitude that comes up when writing up records of these events, if they get written up at all.

45 Gerstel, Naomi. "The Third Shift: Gender and Care Work outside the Home." *Qualitative Sociology*, vol. 23, no. 4, 2000, pp. 467–483.

46 Jokinen, Eeva, et al. "The Fourth Shift: on the Borders of Home, Labour and Affects." Research plan submitted to the Academy of Finland, 31 Jan. 2007, 4kaanne.wordpress.com/2009/02/14/neljas-kaanne-tutkimusprojekti/. Accessed Aug. 26, 2025.

- Gen Z (across all income levels) has the highest rates of poor mental health of any previous generation, with crushing rates of anxiety, stress, loneliness, and future foreboding;[47]
- Rural America finds that instead of growing food (they've been supplanted by industrialized farming and under the thumb of increasingly consolidated large transnational or foreign-owned corporations like Archer Daniels Midland, Bayer, Bunge, Cargill, and Smithfield Foods), they are increasingly growing "deserts of opportunity."[48]

These are just a few examples, but it is fair to say that across America, with few exceptions, subjective well-being, agency, life satisfaction, outlook, perceived progress, and sense of connectedness to others are all experiencing declines in large swaths.

Technology (capital T) in its various forms plays a critical role in all these declines.

Samuel D. James has written an interesting contemporary assessment of Technology on modern Christian life called *Digital Liturgies*.[49] He points to the case of the transgender female swimmer Lia Thomas. Lia Thomas, who later won the NCAA women's 500-yard freestyle, had mid-pack men's rankings prior to transition. After a yearlong transition, she won an NCAA women's title. James notes the national controversy around this example, one of many cultural transformations going on in America. While there was a large visceral critical outcry against this situation from many, the decision to transition and compete as a woman was respected by many other groups and people, who cheered Thomas, according to James.

"What does this have to do with technology?" you say.

Let's set aside the medical innovations that enabled Thomas's transition, and the technological (little t) cultural evolution and innovation context in which this occurred. Instead, James says bluntly: *"Digital technology has recalibrated our worldview and reshaped our consciences…This isn't merely a problem of content; it is a process of form. In other words, it's not simply that on social media and the web that we read sentences that devalue the physical."*[50]

47 "Generation Z and Mental Health." *Annie E. Casey Foundation*, Mar. 3, 2021, https://www.aecf.org/blog/generation-z-and-mental-health. Accessed Nov. 9, 2025.

48 "Measuring America: 10 Years and Counting." *Measure of America*, Social Science Research Council, 2018, measureofamerica.org/10yearsandcounting/. Accessed on Aug. 26, 2025.

49 Samuel D. James. *Digital Liturgies: Rediscovering Christian Wisdom in an Online Age*. Crossway, 2023.

50 Samuel D. James. *Digital Liturgies: Rediscovering Christian Wisdom in an Online Age*. Crossway, 2023. pp. 30–31.

This quote echoes the point that Baudrillard made, and that which Justice Kennedy put in *Packingham v. North Carolina* (2017): social media has supplanted "real life" to become "the modern public square" – they are the forums in which today's public life is conducted.[51] Technology's power to distort reality through selective amplification reveals itself starkly in recent events: despite fewer than 10 transgender college athletes nationwide, the topic generated thousands of news stories consumed on our devices, creating the illusion of a widespread crisis.[52] This exemplifies how technology both drives and responds to cultural transformation – the "recalibration" James referenced above directly connects Technology and our wellbeing (and agency). It also touches on a broader White Christian frustration and the "woke" anger I alluded to earlier.

Similar erosion of subjective well-being is ubiquitous, and occurs across all demographics, each with distinct narratives but shared underlying technologically fueled acceleration of anxiety and division.

I'm only representing one side of the coin in this chapter. I recommend you watch Pinker's TED talk to understand the amazing advances in quality of life that essentially all these groups have experienced, objectively, over the last century.[53] I've also recommended several other books in the suggested reading section at the end of this book that give me, personally, an awareness of the miracles of modern technology and a balanced and informed optimism as I look around now, and think about the future.

With full recognition of those miracles and progress, I've come to *realize those gains don't matter* if people don't *sense* them. If you are moving forward but looking backwards or at data that shows you are moving backwards, *that's* what matters. When people tell you that you should be grateful for how much better off you are than your grandparents, you don't feel enlightened, but instead develop a growing sense of resentment that your lived/experienced pain and struggle are being ignored.[54]

This disconnect between objective progress and subjective experience points to something deeper than mere ingratitude or misperception. The

---

51 Kennedy, Anthony M. *Packingham v. North Carolina*. 582 U.S. 98. Supreme Court of the United States, 2017.

52 Patel, Shiwali. "With Fewer Than 10 Transgender NCAA Athletes, the Outcry Is Out of Proportion." *The Washington Post*, Jul. 5, 2025.

53 The one exception I found was Native Americans, who've had essentially no progress in quality of life over the last century, at least none that I've been able to find. I literally broke down crying several times during the month of research in this area.

54 And I can share from personal direct experience that telling them they are wrong to feel this way, and they'd feel better and understand more if they just read several long expensive books written by affluent White highly educated men (and 1 woman). That doesn't work.

very nature of how we relate to the technologies that surround us has fundamentally shifted. To understand why this matters for our collective well-being, we need to examine not just what technology does for us, but how our relationship with technology shapes our understanding of the world itself.

One of the most profound shifts is how we comprehend – or fail to comprehend – the systems we depend on.

2 The Return of Magic

I call this shift "The Return of Magic" – a transformation in how we comprehend and interact with systems that increasingly govern our daily lives. This shift has profound implications for democracy, community, and human agency that extend far beyond any individual's personal relationship with their smartphone or social media feed.

I've given talks on this subject to global technology and think tank events, the European Association of Archaeologists, and to an audience of world-leading scientists and astrophysicists. What one topic would be relevant to such varied audiences?

Typically, I start lightly, with a short video of a hidden camera magician in a big box store demonstrating a new smart fridge to a consumer. You order on the front of it, and the food downloads "from the cloud," appearing in the fridge seconds later. "This is the best thing I've ever seen; I've got to get one!" says the shopper. (This, even though during the demonstration, there's the inevitable tech bug – a network error halfway through the download. The demonstrator opens the fridge to display a bunch of bananas, neatly sliced in half.)

My point – technology has become so entangled, convoluted, and advanced that none of us really understand how it all works anymore.

We are rewarded for specializing, and so we know our little part. However, what is known across so many fields is expanding faster than our ability to keep up. It is becoming increasingly specialized and opaque to both us and the end user.

All of us – even sophisticated consumers and technologists see wondrous emerging technologies and say, 'It's magic!'

So what?

It's more than a cute TikTok meme or observation. Humans' unique period of progress comes from advances in science, technology, education, and medicine, particularly over the last few centuries. Since the Enlightenment, these realms have grown at the expense of magic – the light of knowledge drives out the dark shadows of ignorance.[55] The archaeologists loved this part of the discussion. Myths, superstitions, folk medicine, folk wisdom, and even religion have been swept aside after open and sometimes violent confrontations between magic and the dark.

Before this time, people would have laughed at the idea of perpetual human progress. As science fiction writer Cixin Liu writes,[56]

> Medieval Europe was materially impoverished compared to the Classical Rome of a thousand years earlier, and was more intellectually repressed. In China, the lives of the people were worse during the Wei, Jin, and Southern and Northern Dynasties compared to earlier Han Dynasty, and the Yuan and Ming Dynasties were much worse than the earlier Tang and Song Dynasties. But after the Industrial Revolution, progress became a constant feature of society, and humanity's faith in the future grew. (Liu, 2016).

But that Icarus-like flight we've enjoyed since the Enlightenment, which climbed to dizzying heights during the Industrial Revolution and Information Age, may have stalled, its wax wings melting, and we might be starting to tumble.

We are now seeing a "re-enchantment" of our world. How a thing works is not to be understood – it is to be accepted without question. How does your self-driving car, AI, or robotic surgical arm work? 'Gee, we don't know, but it sure is great; I need that!'.[57]

---

55 Outram, Dorinda. *The Enlightenment*. 3rd ed., Cambridge University Press, 2019.

56 Liu, Cixin. *Death's End*. Translated by Ken Liu, Macmillan, 2016, p. 72.

57 At this point many of my technology and complexity science folks jump up and argue. "Pat, that's an increasingly complicated system, where expert knowledge is being deployed; it is not a complex one where emergent and unforeseeable consequences arise", and/or "*I* know how that works!". But these folks miss two important considerations. The first point is that we can observe how increasingly complicated things become complexified in ways that are beyond the expert's knowledge. We see this in mishaps like the Bhopal Gas Tragedy, the Toyota Camry runaway acceleration recall, the Challenger explosion, the Boeing 737 Max crashes, and the recent collapse of the Francis Scott Key Bridge. In this last, a mismarked electrical wire was the ultimate root cause of what turned into a multi-billion economic disaster. Events like this happen inevitably as systems get more complex. Magic enters, both good and bad. Similarly, the technologist might understand how some specific feature or technology works in great detail, but they don't understand the whole; they almost always lack the expertise and incentive to really understand how it will be used and to what effect across all end users and use cases. Which brings me to the second and most important factor my friends may miss – you, the developer, may know how it works, but increasingly, the consumer is left to wonder, mouth open, at this new magic.

There's a downside to this re-enchantment, though. Fairy tales consistently show us that there is a cost to using magic.[58] These stories are not simplistic fantasy, but rather subversive narratives often portraying magic as morally complex and consequential. The genie may grant the wish you asked for, but not the one you meant to ask for. Using a magic wand may come at the expense of great pain, or it may not work the way you'd expect (you turn the frog into a pen, instead of a prince). It takes a lot of work to learn how to use magic safely and responsibly, and even then, you use it rarely and carefully, because of its cost and destructive potential.

In my discussions with technologists and scientists, I emphasized that they were complicit in building these "magical" things, and there were consequences and responsibilities that became apparent when we understood the magification we are collectively performing.

Our present circumstances were wonderfully foreshadowed by von Goethe's 1797 poem "The Sorcerer's Apprentice."[59] In it, the sorcerer leaves his workshop in the care of his apprentice. Tiring of his chores and emboldened by the little he has learned so far, the apprentice casts a spell he had overheard once to have the broom go and fetch the water for him. It works! The broom comes to life and gets to work. But the broom spills water everywhere. Wanting to prevent a bigger mess, the apprentice realizes he doesn't know the magic to turn it back into a normal broom. He tries other things he has learned, but only succeeds in speeding up the broom and makes things worse. The sorcerer returns, stops all the misguided magic and mayhem, and admonishes: the magic should only respond to masters who know how to use it.

In this metaphor, we are both the sorcerer and the apprentice. As masters of various domains, we have created the tech and unleashed it into a world full of apprentices. Worse, we are also increasingly the apprentices, using Technology without knowing what it will do, or how to stop it, and then sending that out into the world for others to experience. But we, as the sorcerers, don't even know the incantations to control the technology we've unleashed. We don't know how to use it.

Also, while magic is respected, it is also to be feared. A growing reliance on the technology interwoven into our lives also leaves us with heightened fear and distrust of the things that are around us. It's not just because we don't know how they work, which scares us; it's because we don't know how they

58 Zipes, Jack. *Fairy Tales and the Art of Subversion: the Classical Genre for Children and the Process of Civilization*. Routledge, 1983.

59 Goethe, Johann Wolfgang von. "The Sorcerer's Apprentice." Translated by Katrin Gygax, *Gygatext*, 2013, www.gygatext.ch/english_translations_zurich_sorcerers_apprentice.html . Accessed Nov. 12, 2025.

can be used against us by others, whether that is the people giving us this magic, or the bad actors who hack that tech and use it against us.

Most importantly, in a reversal of trends that are a few centuries old, the Return of Magic is also accompanied by a fading support and faith in the foundations of civilization, with growing distrust in science, education, government, and medicine. We're progressively conscious of the amazing things that technologies like AI can do for us, but unconsciously, we're harboring a burgeoning fear of the tech that is our world.

But it need not be so. Once we become aware of this and the incentives of the social systems that we inhabit, we can learn to be better people and to build and adopt better technology and social systems.

3 The Hidden Burden of Ubiquitous Technology

Technology carries its own burden in the mundane real world – a tax we pay in time, attention, and mental energy that compounds daily. Setting up devices, maintaining them in working order, and navigating their constant demands has become a second job we never applied for. Network outages leave us stranded mid-task. Error codes demand we become amateur diagnosticians. Malware intrusions force us into cybersecurity roles we're unqualified for. Even simple low-battery alerts interrupt our thoughts with urgent demands for attention.

When you spend more time maintaining your Roomba – connecting it to your network, setting up the password and new account, installing the app on your phone, then training, unclogging, charging, and prepping rooms so it won't terrorize the cat, knock things over, or devour power cords – than you would have spent simply sweeping, the promise of convenience has become its own burden.

Brod coined the term "technostress" in 1984 to capture the "disease of adapting" to new technologies,[60] and Tarafdar and colleagues extended it in 2011, operationalizing the concept into measurable stressors: techno-overload, techno-complexity, techno-insecurity, techno-uncertainty, and techno-invasion.[61]

---

60 Brod, Craig. *Technostress: The Human Cost of the Computer Revolution.* Addison-Wesley, 1984.

61 Tarafdar, Monideepa, et al. "Impact of Technostress on End-User Satisfaction and Performance." Journal of Management Information Systems, vol. 27, no. 3, 2011, pp. 303–334.

Consider the password paradox: we've traded remembering our children's phone numbers for managing dozens of unique login credentials across countless portals. Checking a water bill now requires navigating two-factor authentication. Refilling a prescription means remembering which of three healthcare portals holds your account, then solving digital puzzles to prove you're human. Signing up for a yoga class – an activity meant to reduce stress – becomes a multi-step authentication dance.

This represents a fundamental shift in human cognitive burden. Our ancestors dealt with complexity too, but it was the complexity of weather patterns, seasonal cycles, and social relationships – systems that, once learned, remained relatively stable. Today's digital complexity is deliberately unstable. Software updates constantly change interfaces we've just mastered. Security requirements multiply annually. Apps we rely on disappear overnight, taking our data with them.

The accumulating taxation quietly exhausts us, creating what researchers call "continuous partial attention"[62] – a state where we're always monitoring multiple technological demands simultaneously. We're not just users of technology; we've become its unpaid maintenance staff, trapped in an endless cycle of updates, hacks, troubleshooting, and relearning. And managing the cybersecurity of all of that is by itself its own special layer of hell. As *Wired* succinctly puts it, every internet-connected device – from smart thermostats to security cameras – adds to our personal **attack surface**, multiplying entry points for cyber intrusion and escalating the burden of digital housekeeping.[63]

The tools that promised to simplify our lives have instead created new categories of work that consume the very time and mental space they were supposed to free.

This isn't progress – it's a hidden form of technological debt that we're all paying, one password reset at a time.

4 When Complexity Backfires

If calling technology "magic" sounds too abstract, here's a concrete reason to pay attention: our increasingly sophisticated world is becoming dangerously fragile.

62 Stone, Linda. "Beyond Simple Multi-Tasking: Continuous Partial Attention." *Linda Stone*, Mar. 8, 2014, lindastone.net/2009/11/30/beyond-simple-multi-tasking-continuous-partial-attention/. Accessed Sep. 13, 2022.

63 Newman, Lily Hay. "Hacker Lexicon: What Is an Attack Surface?" *Wired*, Mar. 30, 2017, https://www.wired.com/2017/03/hacker-lexicon-attack-surface/. Accessed Nov. 19, 2025.

Think about it. Your smartphone contains billions of transistors. Your car runs on dozens of interconnected computer systems. The Internet relies on millions of servers, routers, and cables spanning the globe. Each marvel of modern life depends on countless tiny, interdependent parts working in perfect harmony.

Here's the problem: the more complex these systems become, the more ways they can break. And when they do break, they don't fail gracefully. One small malfunction – a single line of code, a loose connection, a miscommunication – can trigger a cascade that brings down the entire system.

This isn't just theoretical. It's the reality of living in a world where everything is connected to everything else.

Charles Perrow is the seminal thinker in the field of accident research. When he set out to understand the Three Mile Island nuclear accident, he discovered something unsettling: the catastrophe wasn't caused by human error or equipment failure alone, but by the system's staggering complexity itself.

In systems with thousands of interconnected components – what Perrow calls "*tightly coupled*" with "*interactive complexity*" – it becomes impossible to predict how all the pieces will behave together.[64] Small, unexpected failures don't just stay small. They ricochet through the system, triggering cascades that quickly spiral into major disasters. These weren't freak accidents, Perrow realized. They were inevitable – so inevitable he called them "*normal accidents.*"

History proved him right. Bhopal. Chernobyl. The Challenger explosion. Each followed the same pattern.

Consider the 2003 blackout that plunged 45 million Americans into darkness. It started with something absurdly minor: a software glitch that prevented a single alarm from sounding when power lines got overloaded in Ohio.[65] Then a tree brushed against one of those overloaded lines. Within seconds, that local hiccup cascaded across the interconnected Eastern grid. Suddenly, tens of thousands of New Yorkers were trapped in elevators and dark subway tunnels. Hundreds of thousands more walked home through

---

64 Perrow, Charles. *Normal Accidents: Living with High-Risk Technologies.* Updated ed., Princeton University Press, 1999, p. 5.

65 Minkel, J. R. "The 2003 Northeast Blackout–Five Years Later." *Scientific American*, Aug. 13, 2008, www.scientificamerican.com/article/2003-blackout-five-years-later/. Accessed Aug. 26, 2025.

gridlocked streets. Detroit lost water pressure for four days. Sewage spilled into waterways from Cleveland to New York.[66]

The COVID pandemic offered another stark lesson in how our complex, interconnected world can unravel. A virus jumping from animals to humans in one corner of the globe brought the entire planet's supply chains – and our daily lives – to a grinding halt.

But here's the thing: Perrow focused on what happens when complex systems break. What about when they don't break?

That's where anthropologist Joseph Tainter comes in. Studying collapsed civilizations, he discovered a troubling pattern: complexity doesn't just create the risk of catastrophic failure. Even when everything works perfectly, complexity eventually devours itself.

The math is brutal. As systems grow more complex, they demand ever-increasing resources to maintain, expand, and innovate. Eventually, those costs overwhelm the benefits. The system collapses under its own weight.

Tainter uses the Roman Empire as a classic example. Its vaunted road system, bureaucracy, and military organization reached throughout the world, providing order and increased quality of life for ordinary Roman Citizens during the centuries of the Pax Romana. Due to this expansion, however, it faced various military, economic, and administrative difficulties. Rome invested heavily in maintaining its vast territorial holdings and supporting the complex apparatus of governance and military infrastructure.[67]

Eventually, the returns on these investments began to diminish, and Rome became needy and addressed this burgeoning costs by conquering neighboring people, and using that wealth, accrued over decades, centuries, or longer, to fuel Rome's operating costs for a few years. But this, too, is a dead-end road, as the conquering itself begets more complexity, and even smaller marginal returns.[68] Rome grew unwieldy, frail, and collapsed.

---

66 "Sewage Spill During the Blackout Exposed a Lingering City Problem." *The New York Times*, Aug. 28, 2003, www.nytimes.com/2003/08/28/nyregion/sewage-spill-during-the-blackout-exposed-a-lingering-city-problem.html. Accessed Aug. 3, 2025.

67 Cartwright, Mark. "Pax Romana." World History Encyclopedia, Dec. 8, 2015, www.worldhistory.org/Pax_Romana/. Accessed Nov. 19, 2025.

68 Keynes provides another example, noting that Britain's own accumulation of capital began to inflect upwards in 1580, when Drake returned home on the Golden Hind, bearing the treasure he looted from Spain. Queen Elizabeth was a significant investor in the venture, and she and her empire benefitted enormously – "Out of her share she paid off the whole of England's foreign debt, balanced her Budget, and found herself with about £40,000 in hand."

To bring Tainter back to our own modern individual experience, think back two pages to when I mentioned the toll that technology demands on us to acquire, install, and effectively maintain. And how hard you have to work to pay for all that, and the technology of debt that allows you to make it all happen. That combined toll is but one of the many technologically driven factors that contribute to the average American finding themselves fat, broke, alienated, depressed, busy, stressed out, unhappy, and increasingly in some circles – angry. That is not progress. Tainter might be the first to say it is a mistake to apply his nation-state level theories to our individual lived experiences, but you can scale up our collective experience and see the parallels.

This is why the two tech camps talk past each other. Yes, you may be way better off in some ways than your great-grandparents. After all, you have miraculous technologies like clean indoor running water, flushing toilets, and a Roomba. But if you feel worse off, it doesn't matter.

And this is but a micro-scale representation of the dynamics that Perrow warns of, and Tainter sees playing out historically at the civilizational scale. This is the state of the water in which we find ourselves swimming: We've built an amazing world in such a way that disaster lurks in its very sophistication.

5 Our Thinking Has Shaped Our World in Ways That Don't Serve Our Thinking

Here's an unsettling thought: human consciousness hasn't just discovered the world – it has actively created the world we now inhabit. Our particular way of thinking has shaped our institutions, technologies, cities, and culture so thoroughly that we now mistake this human-made environment for reality itself.

This isn't immediately obvious. When you look around, you see buildings, roads, smartphones, and social media platforms that feel like they "just are" – like natural features of the landscape. But every single element of the modern world emerged from human minds operating in specific ways, with particular biases and blind spots.

Iain McGilchrist helps us understand this uniquely human feedback loop – and why it's becoming dangerous.

You'll remember McGilchrist from earlier chapters: the psychiatrist-philosopher who spent decades researching how the brain's two hemispheres create fundamentally different ways of experiencing reality. The left hemisphere excels at focused analysis, categorization, and manipulation of the world. It's detail-oriented and mechanistic. It views everything as a problem to be solved or a resource to be exploited. The right hemisphere, by contrast, sees the bigger picture, grasps context and meaning, and understands that we're embedded in a living world of relationships.[69]

Here's McGilchrist's crucial insight: our modern civilization has been built predominantly by left-hemisphere thinking – breaking things down, quantifying everything, optimizing systems, and treating the world as a machine to be controlled. This bias hasn't just influenced how we think; it has literally constructed the reality we inhabit.

The feedback loop works like this: Left-brain thinking creates institutions and technologies that reinforce left-brain thinking, which creates more left-brain-designed systems, which further reshape how we experience the world, and so on. We've built a world that mirrors the left hemisphere's particular way of seeing – and now that world is shaping us in return.

According to McGilchrist, this explains why the modern world often feels mechanical,[70] fragmented, and alienating despite our unprecedented technological capabilities. We've created a civilization primarily favoring only half our brains. You can see it in these ways:

5.1 The Left-Brain Lens: How Modern Thinking Works

- **Emphasis on Logic and Reason:** Modern education and intellectual pursuits often prioritize logical reasoning and analytical thinking, which are left-brain activities. This can lead to undervaluing intuition, creativity, and holistic understanding, which are associated with the right brain.
- **Focus on Abstraction and Analysis:** Left-brain thinking emphasizes breaking down complex phenomena into smaller parts, leading to an emphasis on abstract reasoning and analysis over holistic understanding.
- **Emphasis on Quantification and Measurement:** There is a strong tendency to value what can be quantified and measured, often at the expense of qualitative and experiential aspects of life. This can be seen in the dominance of metrics and statistics in various fields.
- **Linear and Sequential Thinking:** The left hemisphere's preference for linear, sequential processes has influenced how we structure time, education, and even narratives, favoring clear beginnings, middles, and ends.

---

69 Again, the pop-science around this is problematic, and even some serious contemporary neuroscientists fault McGilchrist for not respecting the mutual interconnectedness and synergistic integration of the two sides of the brain. Play along with me as I attempt to thread the simplification paradox enough for both of us to learn something, and use the 2-sided brain as a helpful metaphor.

70 I highly recommend: Giedion, Sigfried. *Mechanization Takes Command: A Contribution to Anonymous History*. Oxford University Press, 1948.

- **Control and Manipulation:** Left-brain dominance is associated with a desire to control and manipulate the environment, leading to technological advancements and industrialization. However, this can also result in an exploitative relationship with nature.
- **Specialization, Compartmentalization, and Fragmentation:** The left hemisphere's preference for focusing on details and specific tasks has led to a world where specialization is highly valued. This results in compartmentalized knowledge and professions, often at the expense of interdisciplinary approaches and broader perspectives. These fragmented perceptions of the world contribute to social polarization.

5.2 The Left-Brain World: What We've Built

- **Technological and Scientific Dominance:** The scientific and technological advancements that shape much of contemporary life are driven by left-brain thinking. This includes a focus on empirical data, quantifiable results, and mechanical processes, often sidelining more humanistic and qualitative aspects.
- **Bureaucratic and Procedural Systems:** Modern institutions, including governments, corporations, and educational systems, are often structured in a hierarchical and procedural manner. These systems prioritize efficiency, order, and predictability, aligning with left-brain attributes but sometimes stifling creativity and adaptability.
- **Reductionism in Science and Medicine:** A left-brain approach in science and medicine usually breaks down complex systems into simpler parts. While this has led to significant advancements, it can also overlook the importance of context and the interconnectedness of systems, leading to a fragmented understanding of health and the environment.
- **Economic and Capitalist Structures:** The global economy, especially in its capitalist form, is heavily influenced by left-brain thinking. This includes a focus on competition, quantifiable success (such as profits and growth), and efficiency. These priorities can overshadow considerations of social welfare, environmental sustainability, and ethical concerns.
- **Literal Interpretation of Language:** The left hemisphere's focus on literal and explicit meaning affects how language is used and interpreted. This can result in a preference for straightforward, unambiguous communication (which we noted earlier as common in "low-context cultures"), often neglecting the subtleties, metaphors, and emotional nuances conveyed by the right brain (as seen in "high-context cultures").

- **Urban Planning and Architecture:** Even modern urban planning and architecture often reflect left-brain biases through their emphasis on geometric shapes, functional design, and systematic layouts. This can create environments that are efficient and orderly but may lack the organic, human-centered qualities that foster well-being and community.
- **Legal and Political Systems:** The structure and functioning of legal and political systems often rely on left-brain principles of logic, consistency, and codification. While these systems aim to ensure fairness and order, they can also become rigid and detached from the lived experiences and complexities of human life.

5.3 The Cost: What We've Lost

The bottom line is that through these means and others, left-brain thinking literally manifests[71] the world we live in – physically, culturally, and virtually. The loss of context, understanding, and meaning makes it difficult to appreciate the increasing interconnectedness of various aspects of life.

McGilchrist argues that these left-brain biases can lead to a disconnection from the more holistic, integrative, and empathetic ways of understanding and interacting with the world. He advocates for a rebalance, where right-brain attributes are given more recognition and integrated into our ways of thinking and organizing society.

The way we choose to think, and the models we adopt to understand something, shape how we see the world, according to McGilchrist:

> Our talent for division, for seeing the parts, is of staggering importance – second only to our capacity to transcend it, in order to see the whole. These gifts of the left hemisphere have helped us achieve nothing less than civilisation itself, with all that that means. Even if we could abandon them, which of course we can't, we would be fools to do so, and would come off infinitely the poorer. There are siren voices that call us to do exactly that, certainly to abandon clarity and precision (which, in any case, importantly depend on both hemispheres), and I want to emphasise that I am passionately opposed to them. We need the ability to make fine discriminations, and to use reason appropriately. But these contributions need to be made in the service of something else, that

71 Remember, the word manifest is rooted in the Latin "manus", meaning hand, and "infestus" – meaning struck or hit. IE to bring something about using the hand. Now, consider that 90% of people are right-handed. That means that the vast majority of things in your world were likely brought about through the right and left hands. It literally manifests our world.

> only the right hemisphere can bring. Alone they are destructive. And right now they may be bringing us close to forfeiting the civilisation they helped to create. (McGilchrist, 2019)[72]

Like Baudrillard's precession of simulacra (representations preceding and shaping what we take to be reality), McGilchrist points out that when we experience the world through the left-brain-dominated lens of Technology, we don't see it as it is.

McGilchrist argues that we come to "experience our representation of the world" rather than the world itself, such that reality is increasingly "re-presented" as a kind of virtual construct. In this state, a self-referential, fragmented, and decontextualized worldview can emerge, marked by both unwarranted optimism and underlying anxiety

While McGilchrist acknowledges we can't abandon these traits (and indeed would be foolish to try), he does contend we need to move toward a better balance between the left and right hemispheres so that we can have healthier, more holistic lives. The way to do that would be to consciously focus on and develop the strengths of right-brain thinking by fostering:

- **Multidisciplinary Approaches:** McGilchrist encourages the integration of various disciplines and perspectives to address complex societal issues. By engaging in interdisciplinary dialogue, we can gain a more comprehensive understanding of the challenges we face and develop innovative solutions that consider the diverse dimensions of human existence.
- **Meaningful Human Connections**: McGilchrist highlights the significance of human relationships and face-to-face interactions. He emphasizes the importance of fostering empathy, compassion, and social connections to counter the potential isolating effects of technology. Building and maintaining strong interpersonal relationships can contribute to a sense of belonging and shared purpose.
- **Education for Deep Understanding**: McGilchrist advocates for an education system that nurtures deep understanding and fosters critical thinking. He emphasizes the importance of cultivating curiosity, encouraging interdisciplinary learning, and promoting contemplative practices to facilitate a deeper engagement with the world and foster a more holistic understanding of complex issues.

72 McGilchrist, Iain. *The Master and His Emissary: the Divided Brain and the Making of the Western World.* Expanded 2nd ed., Yale University Press, 2019, p. 429.

McGilchrist's path forward requires rebalancing our cognitive and cultural biases – a challenge that becomes urgent when we recognize how far we've drifted.

6 Technology Is Not Working Out So Well for Us, So Far

There you have it – my case for why our technology isn't serving us well. Despite miraculous innovations that would have seemed like pure magic to our great-grandparents, we find ourselves increasingly anxious, isolated, and mystified by the very systems we've created. We've built a world of such staggering complexity that managing it consumes ever more of our energy while the risk of catastrophic failure grows daily.

This isn't an accident. It's the inevitable result of constructing civilization primarily through our left brain's narrow focus – breaking everything into parts, optimizing individual components, and losing sight of the living whole. We've gained tremendous analytical power but sacrificed the right hemisphere's gift for meaning, context, and coherent understanding.

Why should you care?

Whether you see this as your problem or not, these forces are reshaping the world you inhabit. You might be thriving personally, benefiting from technological conveniences and economic opportunities.

Cough. French Revolution. Cough.

The social fabric around you is fraying. Others in your community are struggling, and that instability will eventually reach everyone.

More fundamentally, the engines of civilization itself are breaking down. Trust in science, education, medicine, and democratic institutions continues to plummet.[73] When mainstream publications like Newsweek[74] and The Nation[75] seriously discuss the possibility of American civil war, we're no longer talking about abstract social trends – we're watching the foundations of our world crack in real time.

---

73 "Americans' Trust in Scientists and Views of Science Decline in 2023." *Pew Research Center*, Nov. 14, 2023, www.pewresearch.org/science/2023/11/14/americans-trust-in-scientists-positive-views-of-science-continue-to-decline/. Accessed Nov. 3, 2025.

74 Stanton, Andrew. "Plurality of Americans Believe 'Civil War' Likely in Next Decade – Poll." *Newsweek*, Jun. 17, 2025, http://www.newsweek.com/plurality-americans-believe-civil-war-likely-poll-2086771. Accessed Aug. 26, 2025.

75 Feffer, John. "You Can't Just Vote Out Trumpism." *The Nation*, Jul. 2, 2020, www.thenation.com/article/politics/trump-election-2020/. Accessed Aug. 26, 2025.

W.B. Yeats, writing after the carnage of World War I, captured our moment perfectly, when he questioned if: "*the centre cannot hold.*"[76] His next lines feel prophetic today: "*Mere anarchy is loosed upon the world... The best lack all conviction, while the worst are full of passionate intensity.*"

But here's where despair transforms into possibility.

The Nobel Laureate Ilya Prigogine discovered something remarkable about complex systems pushed far from equilibrium: they don't just break down – they can also break through to entirely new forms of organization. These "*dissipative structures*" maintain stability not by resisting change, but by embracing it, constantly adapting and evolving in response to environmental pressures.[77]

This gives me hope. Yes, our current systems – which, to be clear, are the ways we think, individually and collectively – exhibit all the characteristics Prigogine described: nonlinear behavior, instability, and sudden phase transitions. But these same conditions that feel so threatening also represent the preconditions for breakthrough – for the emergence of new ways of organizing human society that we can barely imagine.

Rather than waiting to see what Yeats described as "*rough beast slouches towards Bethlehem to be born,*" we have the opportunity – perhaps the obligation – to consciously participate in shaping whatever comes next. The breakdown of the old order isn't just crisis; it's also invitation.

The question isn't whether change is coming. The question is whether we'll help guide it toward something more humane, sustainable, and wise than what we're leaving behind.

A principle spans from William James to modern neuroscience: understanding how we think enables us to think differently. One can paraphrase James[78] by saying that we may "alter our lives by altering our mental attitudes"; neuroscientist Anil Seth shows how thought processes construct our experienced reality, mostly unconsciously.[79] This book grounds that insight in material reality: the three-stream map you've just acquired isn't just diagnos-

76 Yeats, William Butler. "The Second Coming." *The Collected Poems of W. B. Yeats*. Edited by Richard J. Finneran, Scribner, 1996, p. 187.

77 See Kondepudi, Dilip, et al. "Dissipative Structures and Irreversibility in Nature: Celebrating 100th Birth Anniversary of Ilya Prigogine (1917–2003)." *Chaos: An Interdisciplinary Journal of Nonlinear Science*, vol. 27, no. 10, 2017.

78 James, William. "Is Life Worth Living?" *International Journal of Ethics*, vol. 6, no. 1, Oct. 1895, pp. 1–29. James says "Be not afraid of life. Believe that life is worth living, and your belief will help create the fact."; this is popularly styled/mis-attributed as "we can alter our lives by altering our thinking."

79 Seth, Anil. "Your Brain Hallucinates Your Conscious Reality." *TED*, Jun. 2017, www.ted.com/talks/anil_seth_your_brain_hallucinates_your_conscious_reality. Accessed Nov. 18,

tic – it's operational. Understanding how brain evolution, cultural programming, and technological acceleration shape your thought processes – how you evaluate needs, interpret your environment, allocate resources, and achieve outcomes – transforms that understanding into leverage for intentionally reconstructing new and better forms of thought.

As Alan Kay famously said, "*The best way to predict the future is to invent it.*"[80]

2025. In this talk he says, "We're all hallucinating all the time; it's just that when we agree about our hallucinations, we call it reality."

80 Kay, Alan. "Learning vs. Teaching with Educational Technologies." *EDUCOM Bulletin*, no. 3/4, Fall/Winter 1983, p. 17.

CHAPTER 19

The Purpose of Prescriptions

> It took a breakthrough....a new way of thinking ....for a much later generation to step back at last and see what had happened year after year, century after century...[1] (David Brin)

1 Where Are We Heading, and Why?

In this part of the book, we are consciously setting out to build new ways of thinking – nothing less than that.

While Brin is predominantly a science fiction writer, real-world leaders from Einstein[2] and Churchill[3] to JFK,[4] Baudrillard,[5] Toffler,[6] Ellul,[7] and Obama[8] have called on humanity to develop new modes of thought. These next chapters sketch how we can bring this about.[9]

In tackling this mission, we are effectively discussing and architecting new ways to be human.

This book didn't begin with such daunting and auspicious aspirations. However, when one asks, "how does technology affect how we think?," it's hard not to conclude that Technology is fundamentally overhauling thought.

This book has catalogued those ways and the conditioning factors that have brought us to this point. A myriad of intentional and unintentional organizing

---

1 David Brin. "More 'Prophetic'(?) Extracts from Earth..." *Contrary Brin*, Nov. 2023, https://davidbrin.blogspot.com/2023/11/more-extracts-from-earth.html. Accessed Mar. 20, 2026.

2 Einstein, Albert. *The World As I See It.* Citadel Press, 1993.

3 Churchill, Winston. *The Sinews of Peace.* Westminster College, Fulton, Missouri, Mar. 5, 1946.

4 Kennedy, John F. *Commencement Address at American University.* Jun. 10,1963. John F. Kennedy Presidential Library and Museum.

5 Baudrillard, Jean. *Simulacra and Simulation.* Translated by Sheila Faria Glaser, University of Michigan Press, 1994.

6 Toffler, Alvin. *Future Shock.* Random House, 1970.

7 Ellul, Jacques. *The Technological Society.* Translated by John Wilkinson, Vintage Books, 1964.

8 Obama, Barack. *Remarks by the President at the Acceptance of the Nobel Peace Prize.* Oslo City Hall, Oslo, Norway, Dec. 10, 2009." *The White House,* https://obamawhitehouse.archives.gov/the-press-office/remarks-president-acceptance-nobel-peace-prize. Accessed Aug. 3, 2025.

9 A special thanks to my good friend Ed Kirschner for helping me to think through and articulate the considerations in this chapter. To the extent that they are muddled and/or not helpful to you, it is his fault 😊.

 | HTTPS://DOI.ORG/10.1163/9789004739161-025

forces are breaking down and discarding old forms of thought, and, among other reactions, new processes are flooding in to fill the gaps, ad hoc.

So why should it be considered audacious to begin thinking holistically and proactively about how to think in new and better ways? The urgency is clear: this work is vital. Few endeavors could be more ambitious – yet none is more necessary right now.

It is helpful here to re-state my definition of thought: the processes and tools a person uses (consciously or unconsciously) to:

- evaluate their needs and goals
- make sense of their environment (including other people and themselves)
- consider how to use available resources to achieve valued outcomes (goals)
- control processes used to manage resources to achieve those valued outcomes

In Chapter 3, we explored how technology provides enhanced cognitive scaffolding for modern jet pilots through the Rockwell Collins helmet, and how smart glasses could offer a similar experience in a few years as you navigate social events.

In Chapter 7, we noted that if the cognitive processing occurs within a technology outside of your body, it still counts as thought, if "*were it done in the head, we would have no hesitation in recognizing as part of the cognitive process.*"[10]

The fighter jet helmet illustrated how thought will be greatly extended beyond the limited sensors we were born with. But the use cases throughout this book don't just give us *more* information; better thinking isn't likely to feature more information – after all, we're likely already suffering from information overload.

These use cases feature Technology that is also parsing and curating data, filtering out information you don't need, while highlighting others. Based on your preferences, history, and settings you can adjust, the combined suite of technologies can even spot opportunities and navigate you towards unexpected behaviors. Order this food, not that. Talk to this person and bring up trout fishing. Etc.

You might not know why your technologies are steering you in certain ways (it's magic!), and so all of this happens unconsciously to you, but it is definitely thought, as it is helping you evaluate your needs, make sense of

10 Clark, Andy, and David J. Chalmers. "The Extended Mind." *Analysis*, vol. 58, no. 1, 1998, pp. 7–19.

your environment, allocate resources, and manage processes and resources to achieve valued outcomes.

People are already doing this via technology in other parts of your life in ways that benefit themselves, so why can't we craft or seek technologies that let us do it to ourselves, for our own benefit?

It's a scary proposition, to be sure, because it is uncharted territory. We are born with a small set of imperfect sensors that, together with our memory, constitute our reality and the things we hold most precious. We navigate each day using our imperfect senses, memory, and the ability to process the world's challenges.

When we change that, we change the reality of the world as we know it. It might replace some normal functions (like when your phone serves as your long-term memory for your family's phone numbers) or augment your natural senses (such as the augmented reality glasses use case).

But it could also perform functions your brain can't natively handle today. Think back to Chapter 9, and our near-future parenting morning. Imagine that suite of smart home technologies using more real-time data than a person could possibly track. A system diagnoses a child's illness and, immediately and proactively reshapes the mother's day around that sudden development.

In his book *The Man Who Mistook His Wife For A Hat*, Oliver Sacks explores how various neurological conditions affect the perception, cognition, and behavior of individuals, including some unusual and bizarre ones.[11] This includes a chapter on a patient featured in the book's title with prosopagnosia. That man cannot recognize faces, including his wife's, and instead, he perceived his wife's head as if it were a hat. Other cases discuss loss of memory, or phantom pains emanating from limbs which have long since been amputated. What Sacks does in the book is rigorously and humanely show us that the brain adapts to these conditions. Despite its enormous complexity, the brain compensates to even radically shifted visions of reality.

Similarly, as we discussed in Chapter 3, Michael Chorost shows us how the brain can react and learn to interpret even new physical hardware inserted into it; in his case, artificial cochlear implants that replace malfunctioning natural sensory pathways. While the new signals are initially confusing and disorienting, as they are not what the brain is accustomed to processing, over time, the brain learns, adapts, and establishes a new ability to incorporate the technology into its new augmented reality.

---

11 Sacks, Oliver. *The Man Who Mistook His Wife for a Hat: And Other Clinical Tales*. Picador, 1986.

That gives me hope when I think about human thought in the near future. No matter who is causing thought to be disrupted, humans can adapt. But the question is, where do we want to go, and which path do we want to take to get there? One option is to wait patiently to see how we adapt to the many disruptive forces from outside agents that affect our cognitive ability. But I prefer the one where we set out consciously and intentionally to improve it ourselves.

That is the purpose of the next few chapters.

Within that charter, there's still a range of dimensions and options. "Better" forms of thought will mean different things to each of us because we all have different experiences and values. And the goal(s) will probably even vary across time within our own lives as our respective circumstances, wants, and needs evolve. Certain people might choose to have technology provide us *more* information and control, so we can process that information consciously. Others may opt to have technologies handle higher-level functions, such as deciding what we want, thereby reducing the need for information and mental processing.

2 A Vision for a Better Future: The Human Operating System

Consider how we'll manage technological complexity through the metaphor of a future car. This vehicle will serve as much more than transportation – it will also be one of the gateways to edge computing, AI networks, and the interconnected systems that will define our technological lives.

They will be capable of extraordinary things – navigating autonomously, predicting hazards, or even recommending destinations. It will monitor us, observing our attention and even our health, entertain and inform us, all while moving us from point A to point B.

And it will shape how we monitor the outside world, providing heads-up displays on the windows, or notifications via sound or haptics, performing functions like enhanced safety information (e.g., identifying black ice on the road or alerting the driver to objects on a possible collision course). The driver could set preferences to dial up or dial back the information and could select the content they want. Perhaps the driver would like to see the prices of homes displayed under each house they pass, or to learn key historical or natural facts about the landscape they are passing through.[12]

12 This was the focus of a technology platform I actually worked on for a year with a Fortune 100 CTO and his team.

But no driver could ever manage the complexity undergirding all of that, directly. What makes the system usable is the car's operating system, which condenses thousands of technologies and data streams into something comprehensible and actionable. Yet, despite its staggering capabilities, you'll interact with all this complexity through a familiar interface: the dashboard.

This operating system and the dashboard will be radically customizable. Information-seeking drivers might configure it to display real-time safety data, property values of passing homes, or historical facts about the landscape. They want granular control and comprehensive feedback. Others will prefer minimal information – perhaps just destination and arrival time – trusting the car's autonomous systems to handle the complexity invisibly. Still others might simply say "take me somewhere fun" and delegate even destination choice to the vehicle's AI.

The key insight is that the same underlying technological complexity can serve vastly different user preferences through interface design. The car's operating system manages thousands of processes – sensor fusion, route optimization, safety monitoring – but presents only the information each user wants to see based upon their goals and preferences, and the level of control they want to access.

Now extend this model beyond transportation to modern or near-future lives. What if we could develop a personal operating system that manages the technological complexity infiltrating every domain of our existence? Like the car's OS, this system would have three essential components:

- **Deep customization beneath the surface:** Just as the car's AI handles navigation, safety, and efficiency automatically, your personal OS could manage information filtering, decision support, and resource allocation according to your values and preferences.
- **Autonomous functions that hide complexity:** Rather than drowning you in data streams and notifications, the system would handle routine cognitive tasks in the background – scheduling, prioritizing, even suggesting behaviors – while keeping you informed at the level you prefer.
- **A customizable dashboard for conscious control:** Visual, audio, and haptic interfaces would let you monitor your life's various domains – health, relationships, work, learning – and adjust the underlying systems when needed, and shaped by your preferences.

Some people will want detailed metrics and fine-grained control over every process. Others will prefer high-level outcomes with minimal cognitive overhead. The same technological infrastructure could serve both preferences.

This kind of human operating system could help us navigate an increasingly tangled world, to the places we want to go – translating technological overload into experiences that feel comprehensible, intentional, and even empowering.

This isn't science fiction – it's an architectural approach to organizing the technological tools that are already reshaping how we think, decide, and act.

3 Summing Up Our Approach to Drafting Prescriptions

In these pages, we've seen how technology is not only reshaping thought but also extending, fragmenting, and sometimes overwhelming it.

We've established three foundational principles for the work ahead:

1. Technology is already reshaping human thought, whether we guide that process or not.
2. Humans possess a remarkable capacity to integrate new cognitive tools.
3. Better ways of living with technology are possible – if we choose to design these relationships intentionally.

Grounded in this shared understanding, we can now turn to the urgent task ahead: shaping concrete solutions that make these possibilities not only comprehensible, but livable. The solutions in the following chapters aren't abstract theories; they're actionable strategies for building these systems in your own life, starting with the tools and technologies available today.

CHAPTER 20

What Is to Be Done? Designing Solutions for a Better World Is Hard

> You never change things by fighting the existing reality. To change something, build a new model that makes the existing model obsolete.[1] (Buckminster Fuller)

> If you don't cannibalize yourself, someone else will.[2] (Steve Jobs)

I'm wary of introducing solutions. This wariness stems from the situation we find ourselves in and the complexity and immaturity of our understanding of it.

In my experience, proposing fixes too early invites easy dismissal – and lets people reject the crisis itself.[3] Some problems simply don't yet have workable solutions. Others resist decomposition: focus on one part, and you miss the emergent whole that inflicts the real damage.

Many of my most successful colleagues deliver results in hard conditions by breaking problems into parts, acting effectively, and moving to the next part.[4] You eat an elephant one bite at a time. You focus and specialize. When that's your background, it becomes the default mode.

But what happens when the problem cannot (yet) be decomposed? Focusing on one part, you overlook the emergent whole that has a greater impact than the pieces you see. What happens, for example – and strictly hypothetically – if the problem isn't just with social media and unethical companies, but with a complex set of interactions and interdependencies between humans and all the programmable technologies we've built to scaffold us in this self-created niche?

What about problems that have arrived quickly, for which there aren't yet solutions, because the world hasn't digested them, gestated the solutions, or figured out how to manipulate them? Could the makers of seemingly trivial

1 Fuller, R. Buckminster. *Critical Path*. St. Martin's Press, 1981.

2 Jobs, Steve, as quoted in Isaacson, Walter. *Steve Jobs*. Simon & Schuster, 2011, p. 408.

3 Critiquing and dismissing unworkable fixes are valid and necessary, especially in the early stage. I'm not bemoaning that, rather I'm stepping back from that instinctive process to ask us to sit with the problem for a while, let it soak, and move through a process that may not initially require or have immediate workable solutions.

4 I thank my dear colleague Carl Pabo for helping me flesh out my thinking here.

IoT devices – like webcams or smart plugs – anticipate they'd be conscripted into Mirai-style botnets during real cyber conflicts?[5] What are the downsides to Apple's new Vision Pro, and how will we address them?

Your successful peers may be the only ones equipped with the vision, experience, and curiosity to grasp your big ideas or complex problems. Yet in unfortunate irony, your harmonization with them often stalls precisely when they pivot toward problem-solving mode, responding exactly as their training demands.

Recall the preface where I discussed the inability to discuss carbon credit trading schemes in the 1970s because there wasn't clear and broad awareness and acceptance of the problem and its causes.

1 The Timing of Ideas: Two Critical Frameworks

In the public policy space, we have a name that describes the timing of ideas – the Overton Window. Based on work done in the 1990s by Joseph P. Overton, this concept describes the range of ideas that are politically viable or acceptable within a given context. His focus included the boundaries of acceptable political discourse and how certain policy ideas become viable. Ideas can't be broken down and acted on when they are not politically feasible. They don't emerge ready to be implemented. They follow a progression.

1.1 The Overton Window: When Ideas Become Possible

Schopenhauer says: "*To truth only a brief celebration of victory is allowed between the two long periods during which it is condemned as paradoxical or disparaged as trivial.*"[6] Today that is often paraphrased as – All truth passes through three stages: First, it is ridiculed. Second, it is violently opposed. Third, it is accepted as self-evident. A more rigorous way to look at this is an Overton Window. This is a framework that Joshua Treviño formalized based on Joseph Overton's framework, after Overton's death in 2003. An Overton Window can be viewed as a spectrum of six categories: Unthinkable, Radical, Acceptable, Sensible, Popular, and Status Quo Policy.[7] The spectrum

5 Antonakakis, Manos, et al. "Understanding the Mirai Botnet." *26th USENIX Security Symposium (USENIX Security 17)*, USENIX Association, 2017, pp. 1093–1110.

6 Schopenhauer, Arthur. *The World as Will and Representation.* Translated by E. F. J. Payne, Dover, 1966, vol. 1, p. xvii.

7 Treviño, Joshua. "The Overton Window." *Swords Crossed* (blog), 29 Apr. 2006, web.archive.org/web/20060716000532/http://www.swordscrossed.org/?p=50. Accessed 23 Aug. 2025.

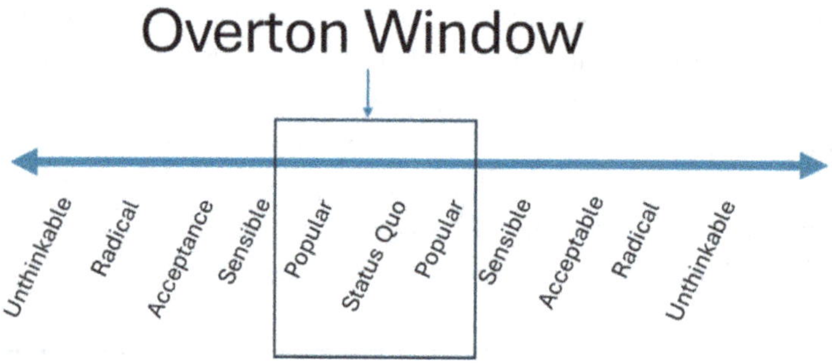

FIGURE 20.1 Overton Window

can run in different directions, based on the perspectives of the stakeholders. In a two-sided debate, it might look like this:

Consider the gun control debate: one side might stake out the position that only police officers should have guns in the USA (unthinkable to many American citizens), while the other extreme argues that anyone should be able to buy any firearm, anytime, anywhere (equally unthinkable to many, though to a different group). Policymakers tend to operate within the range of the politically feasible – the Overton Window – rather than at the extremes. Importantly, the window can shift through framing, events, and advocacy.

When I discuss solutions, people often assume I'm proposing sensible, popularly accepted policy, which skips several steps. However, my role more likely involves moving an idea from Unthinkable to Acceptable (or at least to Radical).

Note that the window doesn't adjudicate which position is *correct* – only which positions are *discussable*. Policymakers work within the feasible range not because it's morally optimal, but because it's politically survivable.

1.2 The Collingridge Dilemma: When Control Becomes Impossible

Designing solutions amid rapid technological acceleration faces another challenge. Here, another concept helps us understand our circumstances – we face a Collingridge Dilemma.[8] Philosopher David Collingridge outlined this phenomenon as the special difficulties present in planning for emerging and complex technologies.

8 Thanks to my colleague Jim Giordano for bringing this concept to my attention during our discussion on his podcast; I've thought a lot about it and the other nuggets that pour out of Jim's brilliant and unique mind.

In the early stages of these technologies, predicting and controlling their potential negative consequences proves impossible. We don't know the problems they may cause, so we can't regulate or shape them optimally. We lack control. Later in the technology's maturation, as it becomes more established and widely adopted, the problems become more apparent. But by this time, it has become embedded in societal structures and practices. We can see the problem, but there is a "lock-in" – technological, economic, social, and institutional – where various interests, incentives, and norms prevent any change, however well-intended.

Collingridge articulated it succinctly: "*When change is easy, the need for it cannot be foreseen; when the need for change is apparent, change has become expensive, difficult, and time-consuming.*"[9]

Even good ideas must pass through the Overton Window – from unthinkable to policy. And with technology, the Collingridge Dilemma creates a double bind: early on, harm is hard to foresee; later, the system locks in and becomes costly to change. We tend to work where problems are both legible and tractable, leaving the largest, entangled ones to fester.

Here's a simplified representation of that:[10]

There's a third constraint worth noting. Biologist Stuart Kauffman's concept of the "adjacent possible" describes how innovation spaces expand: you can only reach possibilities that are one recombination away from existing elements.[11] You cannot skip steps – even brilliant ideas depend on accumulated prior work. This means some solutions simply don't exist yet, not because we can't imagine them, but because the prerequisite innovations haven't emerged.

Together, these frameworks[12] suggest change operates in a constrained three-dimensional space:

---

9 Collingridge, David. *The Social Control of Technology*. St. Martin's Press, 1980, p. 18.

10 Thanks to Per-Olof Hedvall, https://www.peohedvall.com/free-resources/.

11 Björneborn, Lennart. "Adjacent Possible." In *The Palgrave Encyclopedia of the Possible*, edited by Vlad Glaveanu, Palgrave Macmillan, 2020.

12 There's others I've considered and left off here, for conciseness. If this is an area you are interested in, you might also consider:
- Path Dependence / Lock-In: (Brian Arthur, Paul David) *Inertia from success; hard to switch.*
- Technology Readiness Levels (TRLs, pp. (NASA) *Steps from idea to deployment.*
- Gartner Hype Cycle: *Attention rises, crashes, stabilizes.*
- Punctuated Equilibrium: (Eldredge & Gould, later by Tushman & Anderson) *Long stasis, sudden leaps.*
- Diffusion of Innovations: (Everett Rogers) *Adoption spreads unevenly across groups.*
- Technological Momentum: (Thomas Hughes) *Mature systems defend themselves.*

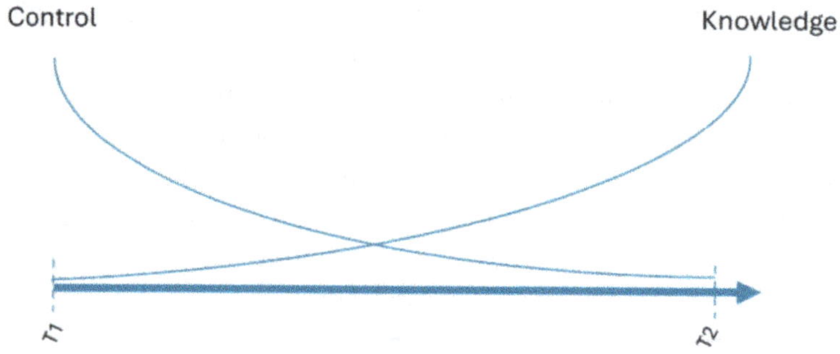

FIGURE 20.2 Collingridge dilemma

- Technical feasibility (adjacent possible): What combinations are currently achievable?
- Social acceptability (Overton window): What changes fall within legitimate discourse?
- Temporal leverage (Collingridge dilemma): What moment in a technology's lifecycle are we in?

Notice what's missing: moral or existential imperative. These constraints describe what we *can* do and when we *can* act – but say nothing about what we *should* do. The urgent and the right often lie outside this viable space.

2 Humanity's Problem-Selection Dilemma

These frameworks reveal our problem-selection dilemma: we spend time where there's a viable political window (Overton), where we can still exert control (Collingridge), and where solutions are structurally reachable (adjacent possible). That leaves a class of mega-problems – often the most existentially or morally pressing ones – that we can't easily sense, segment, or act on, so we ignore them.

This blind spot may be Humanity's Achilles' heel.

- Cynefin Framework: (Dave Snowden) *Match approach to system complexity.*
- Shifting Baseline Syndrome: (Daniel Pauly) *Each generation resets "normal."*
- Innovation S-Curve: *Rapid gains, then plateau; new curve begins.*
- Requisite Variety (Ross Ashby's Law, pp. *Complexity must match environment.*
- Valley of Death: *Funding gap between lab and market.*

Dator's Four Futures: (Jim Dator) *Growth, Collapse, Discipline, Or Transformation.*

These thoughts were going through my head when I listened to my colleague Jim Rutt interview Peter Turchin on his podcast.[13] Peter Turchin – professor emeritus at the University of Connecticut and researcher at the Complexity Science Hub Vienna – argues in *End Times* that US instability follows long waves. Indicators like "*popular immiseration*," elite overproduction, and fiscal stress began worsening in the 1970s; his models flagged the 2020s as a likely turbulence spike. When asked what to do, he's candid: his science can't yet prescribe remedies.

As a reader, this statement frankly frustrates me. You can't tell me about a big problem without giving me some guidance on what I can do about it. But as a writer, knowing what I know now, I can see how accurate and honest this assessment is.

His metaphor captures our predicament: a ball rolling down a valley toward a cusp; small nudges can matter enormously, but knowing where to push is the hard part.[14]

So here I am, operating in the first stage, prior to an Overton Window, waiting until I'm in the early-middle phase of a Collingridge Dilemma, hoping for a chance to know where I can make tiny influences that might take an almost certain negative outcome and make it somewhat benign.

Maybe I should stick with problem description and accept that my role is to make the invisible visible (but initially ridiculous) and foster that thing through violent opposition. Or, in Overton's language, shepherd the unthinkable into maybe acceptable or sensible. Then others may solve them.

You can see why I get frustrated. I say to the few others who think deeply about this, "*OK, but we can't just talk about it – we have a responsibility and an opportunity to fix it; what should we do*?" Most people aren't aware of these problems, and the majority of those who are, treat them as Michele Wucker's Gray Rhinos – highly likely, negative events that are predictable (and indeed, predicted) but ignored.

---

13 "Rutt, Jim. "EP 190 Peter Turchin on Cliodynamics and End Times." *YouTube, uploaded by IBHA International Big History Association,* https://www.youtube.com/watch?v=-UKSm39YJGw. Accessed Nov.18, 2025.

14 Note, I'm NOT sharing the premise of End Times as accepted fact; several of my peers vociferously disagree with a cyclical view of human systems. My own work says we are entering uncharted waters, full of emergent factors that cannot be forecast, and paradigms informed by the past may be ineffective. Our future may be unpredictable, which might be good, because as bad as things are, we might still pull it off: it is, after all, unpredictable. The point of invoking Prof. Turchin here is in looking at how he addresses solutions to the complex problems he senses, not to judge the validity of his conclusions.

But there are a few people I know who will stand in the breach and solve these mega problems, accepting the inherent challenges I've described above. People who focus not primarily on ideas (which they use as means to an end) but instead on outcomes.

3 The Paradox of Optimism

This brings me to an apparent contradiction. How can I catalog these many serious systemic challenges and structural impediments to change while maintaining genuine optimism about our future? The answer lies in recognizing that the same systems creating our problems also contain the seeds of our solutions.

Before we turn to specific tactics, let me share a couple of general themes I consider when people ask me, "How can you still be optimistic?" I respond simply: people and technology.

4 People: Why I'm Still Optimistic

After a decade-plus deep dive into human evolution and systems, with ever-growing awareness and acknowledgment of humanity's flaws and anachronisms, and despite the current oppressive deluge of negative headlines, I've developed a new, much deeper, and overwhelming appreciation for humans and human nature.

To understand why, you might read Rebecca Solnit's *A Paradise Built in Hell*, where she describes how diverse coalitions of people, even traditional enemies, come together in the face of disasters to save themselves and others. Or you might look away from your screens for a minute and just watch the people around you. You might see your transgender friend hold a door for an elderly man wearing a MAGA cap, or an apparently affluent person in the bathroom who calls the janitor 'Sir'. Or the person in that old rusty pickup truck who stops on the rainy highway to help the driver of a new Mercedes change a tire.

Or you might look back at your screen, past the "if it bleeds, it leads" content and algorithms, and use that technology to find positive and much more accurate views of our world. Perhaps view the YouTube clip about how 20,000 motorcyclists came together to drive past the home of a young boy with terminal cancer, whose dying wish was to watch motorcycles go past his house.[15]

Consider the asymmetry of the world's good actors to bad actors: 19 hijackers perpetrated 9/11; approximately 400 emergency responders died running toward the crisis,[16] tens of thousands of others helped on site,[17] while hundreds of thousands donated blood.[18] The spontaneous evacuation of Manhattan that day by water saw 150 fishing boats, ferries, water taxis, and other volunteers rescuing over 500,000 people, overshadowing the legendary Dunkirk evacuation, which rescued 300,000 trapped British soldiers over nine days in World War II.[19]

Look beyond the "Missing Girl Feared Kidnapped in the Outback" headlines[20] to see that police arrived on scene within minutes. Within an hour-and-a-half, a drone was in the air, and shortly after, the road to the area was closed with an expanding police presence, augmented by dozens of volunteers scouring the terrain. The next day or two saw these ranks swell to hundreds of people, supported by multiple helicopters, more drones, a military jet, as well as coastal marine search and rescue and volunteer fishing boats. A single perpetrator had kidnapped the young girl from the campground, and he was caught.

If you don't look too deeply, the message that lingers is that we live in a world where our children are not safe. Instead, with the scales falling from my eyes, I see the more accurate picture: child kidnapping is so rare that the media must bring me a story from over 9,000 miles away to titillate our primal circuits. People are the reason to have optimism, not fear.

---

15 "Humanity at Its Finest, 20,000 Bikers Respond to 6-Yr-Old Kilian..." YouTube, http://www.youtube.com/watch?v=MZ85aFEivGo. Accessed Dec. 3, 2024.

16 "12th Annual Blood, Sweat and Stairs: Honoring 9/11 First Responders." *Goodfellow Air Force Base Newsroom*, Sep. 6, 2024, http://www.goodfellow.af.mil/Newsroom/Article-Display/Article/3907701/12th-annual-blood-sweat-and-stairs-honoring-911-first-responders/. Accessed Jul. 17, 2025.

17 "Volunteers Remember 9/11." *American Red Cross Greater New York*, American Red Cross, http://www.redcross.org/local/new-york/greater-new-york/volunteer/volunteers-remember-9-11.html. Accessed Sep. 27, 2025.

18 Korcok, M. "Blood Donations Dwindle in US after Post-Sept. 11 Wastage." *PMC*, 2002, http://www.ncbi.nlm.nih.gov/pmc/articles/PMC128413/. Accessed Jul. 16, 2024.

19 Kendra, James M., and Tricia Wachtendorf. *American Dunkirk: The Waterborne Evacuation of Manhattan on 9/11*. Temple University Press, 2016.

20 See, for example – Vesoulis, Abby. "Police Looking for Kidnapped Girl Tipped Off after Suspect Bought Diapers: Reports." *Newsweek*, 12 Nov. 2021, https://www.newsweek.com/police-looking-kidnapped-girl-tipped-off-after-suspect-bought-diapers-reports-1645999. Accessed Jul. 29, 2025.

5 Technology: Build to Catch the Train

What about technology? Consider the anecdote above. The mother used a cellphone to report the missing child to police, who used that overlooked miracle of cars and roads to respond quickly. The drones, helicopters, jets, and boats operated thanks to countless technical innovations. Social media, Internet, and other technologies helped amass that volunteer pool. She was finally located using a single ping from a phone to a cell tower.

We might simply surrender, acknowledging that constantly accelerating technology has already escaped our grasp – that we can no longer understand or control it. We're sprinting at full speed to catch a train that's rapidly pulling away from us.

But I see it differently. We can and should leverage existing technologies while developing new ones that help us address the consequences of our existing creations. If we can't run fast enough to catch the train, what can we build to help us reach it?

Yes, these tools may include those that emerge from surveillance capitalism and greedy tech firms, but they remain powerful instruments for good – if we recruit and deploy them purposefully rather than merely indulging our vestigial hardwiring mindlessly.

6 A Framework for Moving Forward

I'm now a strong believer in bottom-up approaches, driven by those inflected power curves from Chapter 15. Power once resided in elites and large organizations. Now, accelerating technology shifts power from yesterday's elites to those who adapt most effectively to change.

This new power[21] belongs to individuals and small groups, largely because they face less inertia.[22,23] Here, we'll witness both the most successful adapta-

---

21 I highly recommend the book *New Power:* Heimans, Jeremy, and Henry Timms. *New Power.* Macmillan Publishers., 2018.

22 Inertia: the concept was articulated by Galileo and formalized by Newton; the term 'inertia' entered early modern physics via Kepler, where he used to mean resistance to motion, now meaning resistance to change.

23 This is a controversial point of mine, especially in the face of rising strongman politics, spiraling inequality and common wisdom. Everyone wants to make their billions and have millions of followers so that they can change the world. But I'd recommend one take a read through the End of Old Power and the Birth of New Power section of the Suggested Readings.

tions to our emerging niches and the most spectacular failures. We can't ignore this reality: anyone seeking to harness change at scale must build tactics that account for and mitigate increased failures amid the *Sturm und Drang* of social response to new technologies.

We need systems that search for, monitor, and recruit effective adaptation within the chaos of social response to new technologies. By identifying key individuals and groups who have successfully achieved their goals, we can discover solutions to the futures we aim to create.

At the individual level, using thoughtful, comprehensive approaches, we can achieve the pace we need.

However, while individuals may adopt faster, our solutions need to scale. For that, we need collective approaches.

While power shifts, organizations and movements still wield enormous influence. Technologies develop to meet demand – there must be sufficient demand for newer technologies before we see them emerge. That's how you change Silicon Valley and Wall Street: not just by shaming them, but by creating market pressure. Hope remains for government as well. At collective levels, we achieve the scale we need.

The complex interplay between our thinking and the environment we have created means that numerous levers exist for us to pull to improve outcomes. These levers operate at multiple levels – individual, community, institutional, and systemic. Some require technological innovation; others demand social innovation. Some work through markets; others through culture or policy. What follows in the next chapters represents a survey of my own modest view of some of those levers, but there will be many more.

One last point is worth highlighting: perhaps we don't have to do anything. I don't suggest this in a lazy, defeatist manner.

Michael Sacasas warns against the "Borg Complex": the lazy assumption that technological change is irresistible.[24] In their book *Our Next Reality*, Graylin and Rosenberg discuss solutions for "AI Alignment," especially facing possible Artificial Super Intelligence (ASI) and its intersection with AR/XR/metaverse. Drawing on Daoist *wu wei*, they add their own caution: hurried actions

---

24 Sacasas, Michael. "Borg Complex: A Primer." *The Frailest Thing*, Mar. 1, 2013, thefrailestthing.com/2013/03/01/borg-complex-a-primer/. Accessed Dec. 21, 2023. In discussing the rhetoric of technological inevitability, Sacasas coined the phrase Borg Complex, describing it as a stance "exhibited by technologists, writers, and pundits who explicitly assert or implicitly assume that resistance to technology is futile."

often destroy more than they solve.[25] Both perspectives argue for purposeful – but not panicked – intervention.

With that advice in mind, let's proceed deliberately. The next two chapters begin where we have leverage: first, tools for individuals; then, tactics for communities and institutions. We'll explore how to navigate the Overton Window and Collingridge Dilemma, not by avoiding them but by understanding how to work within and around their constraints, turning the very frameworks that create our challenges into guides for effective action.

25 Graylin, Alvin W., and Louis Rosenberg. *Our Next Reality: How the AI-Powered Metaverse Will Reshape the World.* Foreword by Neal Stephenson, Nicholas Brealey Publishing, 2024.

CHAPTER 21

Taking Control: How Individuals Can Navigate Technological Disruption

About 85 per cent of my 'thinking' time was spent getting into a position to think, to make a decision, to learn something I needed to know.[1] (J.C.R. Licklider)

I'm starting with the man in the mirror / I'm asking him to change his ways[2] (Michael Jackson)

1 The Agency We Didn't Know We Had

Technology shapes how we think – this insight reaches back at least to the ancient Egyptians, thousands of years ago, as they worshipped Thoth (credited with creating the art of writing, inventing the calendar, and controlling space and time).[3] We saw in Chapter 3 that folks like Bush, Licklider, and Engelbart foresaw the positive potential of technology many decades ago. Contemporary scholars study facets of the negative consequences of this dynamic, focusing on aspects such as screen time,[4] social media,[5] digital surveillance,[6] surveillance capitalism,[7] misinfo/disinfo,[8] and myriad other moving parts. While

1 Licklider, Joseph Carl Robnett. "Man–Computer Symbiosis." *IRE Transactions on Human Factors in Electronics*, vol. HFE-1, Mar. 1960, pp. 4–11, at p. 4.

2 Glenn, Siedah Garrett, and Michael Jackson. "Man in the Mirror." *Bad*, performed by Michael Jackson, Epic Records, 1987.

3 Budge, E.A. Wallis. *Thoth, the Hermes of Egypt: a Study of Some Aspects of Theological Thought in Ancient Egypt*. Kegan Paul, Trench, Trubner & Co., 1904.

4 Twenge, Jean M., and W. Keith Campbell. "Associations Between Screen Time and Lower Psychological Well-Being Among Children and Adolescents: Evidence from a Longitudinal Study." *Preventive Medicine*, vol. 116, 2018, pp. 135–142.

5 Masri-Zada, Tariq, et al. "The Impact of Social Media & Technology on Child and Adolescent Mental Health." *National Institutes of Health*, 2025.

6 Michael, Katina. "Uberveillance: Microchipping People and the Assault on Privacy." *Quadrant*, vol. LIII, no. 3, 2009, pp. 85–89.

7 Zuboff, Shoshana. *The Age of Surveillance Capitalism: The Fight for a Human Future at the New Frontier of Power*. PublicAffairs, 2019.

8 Breuer, Pablo, and S.J. Terp. "DISARM: A Framework for Analysis of Disinformation Campaigns." *Proceedings of the 2022 IEEE International Conference on Cognitive Security and Information Management (COGSIMA)*, IEEE, 2022, pp. 1–8.

 | HTTPS://DOI.ORG/10.1163/9789004739161-027

some approaches provide guidance for individuals, many place root cause on elites, claiming technology companies exploit us while political leaders fail to protect society.[9]

They do exploit us. Leaders do fail to protect us. But throughout this book, especially in Chapters 4, 15, and 18, we've seen that the root cause of constant technological acceleration is us. Companies give us what we demand, and leaders struggle within complex systems they can barely understand, much less control. They function like modern parents trying to understand how technology shapes their children's lives – managing to nudge small changes like "no phones at dinner table," while the rest of the world stands ready to serve those desires.

This seems to place enormous blame on billions of people who appear to be innocent victims of complex technological machinations, who don't understand AI or sometimes even the newest iPhone, much less how to code or fabricate these technologies.

But I'm not assigning blame – I'm identifying agency. The people deploying AI, building iPhones, and creating disruptive technologies don't fully understand how their creations will affect us either. Looking to these leaders to build better technologies or more effective policies won't work unless we all operate from a shared understanding. Within that framework, we can begin making changes, large and small, working together to create better systems and lives.

This is a positive story because it means we have *agency*. We don't have to wait for elites to fix problems they/we are creating. If we are the driving force behind it, then that means we can control it. Like discovering you're in quicksand and learning to stop struggling – this reorientation alone constitutes a tremendously powerful and optimistic step.

But where do we start?

2 Awareness: Seeing the Water We Swim In

David Foster Wallace warned graduating students at Kenyon College's 2005 commencement, *"The immediate point of the fish story,"* he said, *"is that the most obvious, ubiquitous, important realities are often the ones that are hardest to see and talk about."*[10]

---

9 See, for example: Schaake, Marietje. *The Tech Coup: How to Save Democracy from Silicon Valley*. Stanford University Press, 2024.

10 I recommend watching the whole speech. It's among the best and most meaningful commencement messages I've ever seen. There are numerous links to it on YouTube, and will be 22 minutes well spent. Here's one: "This is Water by David Foster Wallace

Later in that talk, he expanded: "*Learning how to think really means learning how to exercise some control over how and what you think. It means being conscious and aware enough to choose what you pay attention to and to choose how you construct meaning from experience. Because if you cannot exercise this kind of choice in adult life, you will be totally hosed.*"

Awareness begins with recognizing several uncomfortable truths:

Our thinking is anachronistic. Much of our cognitive apparatus evolved over hundreds of thousands of years under completely different selection pressures. We shouldn't assume these mental tools remain adaptive in our current environment.

Modern thought results from uncoordinated creations. Both social and technological innovations influence us outside conscious awareness. These inventions emerged and spread without coordination, creating chaotic and disordered effects on our cognitive abilities.

Technology is accelerating algorithmically. Not just exponentially, but through compounded effects of multiple exponential factors with nearly infinite reinforcing feedback loops, creating additional emergent accelerations.

The great irony of technology: Despite receiving untold benefits from technology, most people report feeling worse off. The fit between our adaptations and our current niche remains poor.

We hold a magic wand. Today's technological capability can give us almost anything we ask for. However, we must think deeply about what we truly want or need, considering technology's potential to provide it.

This represents transformational opportunity. Just as writing and external representations freed cognitive resources for civilization development, we face a similar moment. We need only recognize the opportunity and seize it.

This book succeeds when it causes you to think about these points in daily interactions with Technology, as Vint says it did for him. As Ken Wilber suggests, growth involves expanding awareness to include more perspectives, capacities, and dimensions of reality.[11]

(Full Transcript and Audio)" *FarnamStreet*, https://fs.blog/2012/04/david-foster-wallace-this-is-water/. Accessed Aug. 19, 2025.

11 Wilber, Ken. *A Brief History of Everything*. Shambhala, 1996.

3 Planning: The Magic Wand Mindset

In the streaming series *Halo*, the antagonist reflects:[12] "*We are on a journey. We are born, we live and die according to the rules of blind and unguided evolution. As a result, our species is simply not equipped to survive what comes next. It is time for us to take control over our evolution.*"

What would our plan for "taking control" look like?

It starts with awareness, then builds a new mindset. Consider this thought experiment: If I lost both legs in an accident, rather than immediately pursuing regrowth (assuming technology allowed it), I would pause and examine my assumptions. Why do I value legs? What functions do they serve? What possibilities might emerge if I replaced them with something else entirely – or embraced their absence?

This isn't merely abstract. Future astronauts permanently in space, in zero gravity, might find that legs become evolutionary baggage. Legs consume precious calories, occupy valuable real estate in cramped vessels, and serve no locomotion purpose. Clinging nostalgically to terrestrial anatomy – trying to "make legs great again" – would miss the opportunity to optimize for a new environment.

My own experience reinforces this mindset. A military accident crushed my right foot, derailing planned athletic careers. Multiple surgeries followed, but chronic pain persisted. I contemplated asking surgeons whether amputation might be better than endless agony. Because I had discovered Hugh Herr's story.

In 1982, frostbite forced surgeons to amputate both of Hugh's legs below the knee.[13] After recovering, he shocked his medical team by returning to climbing using prosthetics, and with greater skill than before. Eventually, he created several interchangeable and adjustable lower legs, with stiff toes (for climbing on dime-thin rock edges), and titanium spikes (for climbing steep and overhanging ice). He could even adjust them to change his height, from five feet tall to a towering, unnatural height of eight feet! Through drive and prototyping, he became the first major amputee athlete who could compete and win against able-bodied elite athletes.

Today's prosthetic technology evolves at breakneck speed. Advances in design, materials, and adaptability reframe what amputation might mean, transforming yesterday's last resort into today's viable strategic choice.

---

12 "Reckoning." *Halo*, created by Kyle Killen and Steven Kane, season 1, episode 9, Paramount+, May 12, 2022.

13 Herr, Hugh. "The Intersection of Man and Machine." *NIH Catalyst*, vol. 29, no. 5, 2012.

This experience taught me to flip the script – transform apparent problems into strategic advantages. If faced with similar situations today, I would contact friends to help. I know world-class technologists and fabricators. I've started makerspaces and visited dozens of innovation facilities worldwide. If I reached out to these communities and challenged them to help me replace missing limbs with something better, we would get excited and work through structured processes: storming, norming, and forming. We'd discuss requirements, reconcile perspectives, assess constraints, and develop iterative products shaped by user feedback.

Just as advanced prosthetics turn loss into possibility, we can design new ways of thinking that don't merely replace what disruption takes from us but create capacities we never had before.

4 Why Can't We Do the Same for Thought?

What makes thought a sacred cow that shouldn't be tackled the same way as upgrading missing body parts? Others already change how we think – that was the whole point of Chapter 16. Why can't we set out consciously to do it ourselves?

Here's how I visualize the opportunity space:

Most of us inhabit the bottom left quadrant: muddling through a technology-infused world and suffering net negative outcomes. True, exceptions exist – people blissfully doing well with little thought. But most are not.

A typical prescription suggests unplugging to avoid bad effects. Limit screen time. We see this in movements like Wellness 2.0 and digital detoxes, [14] and in books like Cal Newport's *Deep Work*[15] and even more so in his *Digital Minimalism.*[16] But completely unplugging abandons technology's power to elevate your life. You drift into mediocrity – neither worse off nor better.

Instead, absorb insights from this analysis and deliberately chart a course toward intentional technology consumption with positive well-being outcomes. This journey demands experimentation, and experiments fail. Some people pursuing better technology habits will crash into the bottom right

14 Hawthorn Center. "The Deeper Benefits of a Digital Detox." *Hawthorn Center*, 22 May 2025, https://www.hawthorncenter.com/the-deeper-benefits-of-a-digital-detox/. Access ed 19 Mar. 2025.

15 Newport, Cal. *Deep Work: Rules for Focused Success in a Distracted World.* Hachette UK, 2016.

16 Newport, Cal. *Digital Minimalism: Choosing a Focused Life in a Noisy World.* Penguin, 2019.

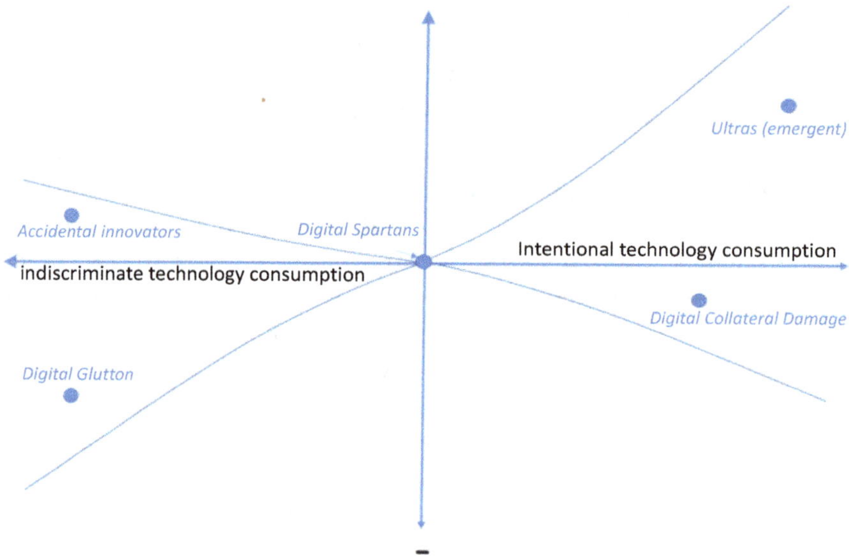

FIGURE 21.1 Conceptual diagram of the intersection of technology consumption and well-being

quadrant. But if you're already struggling and sliding downward, what do you risk by plotting a different path? Failure leaves you no worse off than before, while success transforms your trajectory entirely.

Consider the Figure 21.1, above. Success places us in the top right quadrant – better off, feeling better, more effective, with cognitive capabilities that constitute unfair advantages over our earlier selves.

The first step in this process involves reflecting deeply on your situation and your potential goals. I've described our opportunity here this way:

- *If I gave you a hammer and told you to build something, you'd approach the task with a certain mindset and goals.*
- *If instead I gave you a large bag of money, you'd see the problems, opportunities, and potential actions differently. Maybe even the goals you could accomplish.*
- *But how would you think about that task if I gave you a magic wand?*

Technology is a magic wand. You might think of it as some oppressive thing that is out of your control, that victimizes you, but you need to shift your hammer/money mindset to the fact you are now wielding a magic wand. One way to conceive of that magic wand is that it gives you a Framework for Individual Action; spells you can cast to help yourself.

Jaco Hamman, a theologian, looks at how we are adapting to our modern world in his book *Growing Down*.[17] In a world focused on "growing up" – reaching forward, accomplishing, and progressing (left brain attributes, I'd point out) – Hamman invites us to consider "growing down": finding meaning and developing roots (i.e., McGilchrist's right brain functions). These roots sustain and nourish us. A "grown-down" person sidesteps the anxieties, distractions, and compulsion of normal tech, to have a measured, reflective, and purposeful relation with our tools. Being grown-down does not mean using technology to effect change externally; it is diagnosing ourselves, revealing parts of our own personal, interpersonal, and spiritual development. That means intentional and mindful use of technology is also prognostic.

We need to, as Lester Brown summed up: *"First we need to decide what needs to be done. Then we do it. And then we ask if it is possible."*[18]

Implied in this planning is that each of us must take ownership of this task and do the necessary headwork for ourselves. That's the bad news. The good news: we have agency.

5 The Framework for Individual Action

5.1 Reclaim Your Purpose

Technology functions as a magic wand. You might perceive it as an oppressive force victimizing you, but shift the perspective: you're wielding the wand, not suffering under it. Abandon the hammer-and-money mindset that society programmed into you.

Ask yourself: What would you create if you truly held magic in your hands? Forget what others taught you to want. What dreams would you chase if limitations disappeared?

This requires confronting fundamental questions philosophers have wrestled with for millennia: What is my purpose? Who am I? How do I relate to others?

One approach suggests we need no reflection – simply pick up signals from our social system and conform. Thoreau criticized this when he observed, *"The mass of men serve the state thus, not as men mainly, but as machines, with their*

17 Hamman, Jaco J. *Growing Down: Theology and Human Nature in the Virtual Age*. Baylor University Press, 2017.

18 Brown, Lester R. *Plan B 4.0: Mobilizing to Save Civilization*. W. W. Norton & Company, 2009, p. XIV.

bodies".[19] More recently, former child-soldier turned author Emmanuel Jal observes how we abdicate our self in the process, "*Whoever owns your mind owns you and everything you create.*"[20]

Every culture establishes mainstream views and pressures us to conform.[21] We instinctively crave belonging. Technology – evolving from print to radio, television, Internet, and connected platforms – amplifies these cultural norms at unprecedented scale. This medium doesn't just reflect who we are; it actively shapes who we aspire to become.

McLuhan, in *Understanding Media*, captures the gravitational pull of mass belonging: "The pleasure of being among the masses is the sense of the joy in the multiplication of numbers.[22] Media brought us together into a global village with a shared cultural space, fostering a sense of collective identity and interconnectedness. But he also mused that some media, like television, encouraged passive consumption of media by the masses, and he worried about its potential to create a "tribal" society with a "hive mind." McLuhan felt we should be aware of the *structural* media effects like this, in addition to the influence of media's *content*.[23]

German philosopher Martin Heidegger explored this dynamic in *Being and Time*, using "das Man" (often translated as "the They") to describe everyday anonymous social conformity.[24] He argued that pervasive societal expectations create pressure that dictates how individuals should think, feel, and behave. The result: we become alienated from our true selves, trading authenticity for belonging and anonymity.

To be authentic, in contrast to the They, we must accept our mortality and take responsibility for our choices and actions.

---

19 Thoreau, Henry David. *Civil Disobedience.* Dover Publications, 1993, p. 3.

20 Jal, Emmanual. *My Life is Art*, Counterpoint, 2023, p. 4.

21 See Zheng, Yongnian. "Harmonious Society and Harmonious World: China's Policy Discourse under Hu Jintao." *China Policy Institute Briefing*, no. 26, 2007, https://www.nottingham.ac.uk/iaps/documents/cpi/briefings/briefing-26-harmonious-society-and-harmonious-world.pdf. Accessed Jul. 8, 2025.; Whybrow, Peter C. *American Mania: When More Is Not Enough.* W.W. Norton & Company, 2005; and Bazzi, Samuel. "The Roots and Persistence of 'Rugged Individualism' in the United States." *Frontier Culture*, 2017, https://www.bu.edu/econ/files/2017/03/Frontier-Culture-The-Roots-and-Persistence-of-%E2%80%9CRugged-Individualism%E2%80%9D-in-the-United-States-2.pdf. Accessed Nov. 13, 2025.

22 *McLuhan, Marshall. Understanding Media: The Extensions of Man. McGraw-Hill, 1964,* p. 107.

23 McLuhan has much more to say here, and I'm tempted to dwell on him more, but instead let me point you to the suggested readings under Media, Information, and Knowledge.

24 Martin Heidegger. *Being and Time.* Translated by John Macquarrie and Edward Robinson, Harper & Row, 1962, pp. 164–168.

Aristotle distinguished between two paths to flourishing: eudaimonia (comprehensive well-being through virtue and realizing potential) versus hedonic pleasure (temporary emotional satisfaction). Unlike hedonic happiness's fleeting spikes, eudaimonia creates lasting fulfillment through character development and moral growth.[25]

Technology can give us almost anything we conceive. Using it shapes who we are and creates new future opportunities. Shouldn't we use this as an opportunity to examine our goals consciously?

Living well with technology might mean turning away from the hamster wheel of consumerism and materialism – from conforming behaviors currently encouraged by our technology adoption habits. These conclusions emerge empirically through experience, not dogmatic principles.

The good news: we have agency. The bad news: we must take responsibility for carving our own way, developing newly possible selves. While immediate gratification and transient joy have their place, deeper meaning and flourishing come from pursuing authentic goals we consciously choose – whether hedonic, eudaimonic, or something else entirely.

5.2 Cultivate Intellectual Humility

Since we're examining philosophical foundations, consider the ancient Skeptics who emerged after Aristotle.[26] Beginning with Pyrrho, they tried to achieve mental tranquility (ataraxia) by refraining from forming firm opinions. Rather than judge situations dogmatically, they highlighted that what seems true to one person or culture may not be perceived similarly by others.[27]

They favored curious, peaceful, but critical attitudes toward knowledge and willingness to engage in reasoned inquiry. They weren't stressed about being right or spending energy defending positions; they pursued peaceful journeys to learn and understand.

This approach isn't unique to ancient Greeks. The Jesuits have a central tradition of "discernment" with three key parts: being attentive, being reflective, and being loving, leading to better decision-making.

---

25 Crisp, Roger, ed. *Aristotle: nicomachean ethics*. Cambridge University Press, 2014. (esp. Books I and X).

26 "Epistemology." *Encyclopædia Britannica*, Aug. 2, 2025, https://www.britannica.com/topic/epistemology. Accessed Sep. 17, 2025.

27 Vogt, Kathleen. "Ancient Skepticism." *Stanford Encyclopedia of Philosophy*, Stanford University, Nov. 4, 1997, https://plato.stanford.edu/archives/fall2007/entries/skepticism-ancient/. Accessed Jun. 8, 2025.

Tania Tetlow, President of Fordham University, explains discernment's relevance to modern discourse:[28] "*It is basically the opposite of social media. Discernment means taking time to consider big decisions, not jumping to conclusions. It means being open and curious. It means assuming good intentions of those you disagree with. And it means being self-aware enough of your own biases that you realize what prevents you from seeing truth.*"

Both ancient Skepticism and Ignatian discernment demonstrate that agency emerges through intellectual humility and reflective inquiry. These traditions establish frameworks for evaluating information – systematic approaches for determining justified, rational, warranted knowledge.

As Alvin Toffler observed (quoting Herbert Gerjuoy), "T*omorrow's illiterate will not be the man who can't read; he will be the man who has not learned how to learn.*"[29] It's from Gerjuoy's formulation, as Toffler transmitted it that we get today's meme that today's illiterate will not be those who can't read, but those who cannot learn, unlearn, and relearn.

Humility brings with it some additional requirements. At the top of that list is compassion, an overlooked skillset. We need it for those who are resilient and those who are not; for those who endure change and those who resist. Sometimes, we need it most for ourselves when we don't handle change perfectly or even well. Community is important to the humble, as we awaken to the fact that few survive change alone. Cultivate tribe and community(ies); know your commonwealth. Being a humble thinker also means you need to become interdisciplinary – build up many strengths and skills; learn to work and collaborate with others, with some that complement ours (and to be compassionate that those with other skills often think, act, and believe differently than you; that's good, but takes compassion). Finally, we all need to prioritize effective communication. My pursuit of answers to my research questions has changed me, and at every turn, I've needed to learn to communicate better. It will be an increasingly important expertise in the future. Effective collaborations require effective communication. This involves communicating with people from other fields, at various levels of experience, across diverse functional expertise, through different channels.

---

28 Dubner, Stephen J. "If We're All in It for Ourselves, Who Are We?" *Freakonomics Radio*, Aug. 7, 2024, https://freakonomics.com/podcast/if-were-all-in-it-for-ourselves-who-are-we/. Accessed Jun. 13, 2025.

29 Toffler, Alvin. *Future Shock*. Random House, 1970, p. 414.

5.3 Balance Unplugging and Plugging In

5.3.1 Strategic Unplugging

While embracing technology's benefits, deliberately reduce thoughtless tech consumption. This manifests in several ways:

Consider minimal social media presence, accepting that you'll miss some updates but avoiding accompanying noise. When connecting with friends, call them or meet for lunch – direct engagement yields richer exchanges than digital scrolling.

Practice what the Japanese call *shinrin-yoku* (forest-bathing):[30] regular immersion in nature for physical, mental, and emotional well-being. Take walks in woods, cycle rural roads, embrace natural tranquility.

Practice deliberate boredom.[31] Rather than reflexively reaching for phones while waiting, simply wait. In cars, occasionally turn off radios and embrace silence. While technology offers easy solutions to boredom, philosopher Lars Svendsen argues that deliberate boredom catalyzes creativity, self-reflection, imagination, and personal growth.[32]

5.3.2 Intentional Plugging In

Simultaneously, harness technology to advance yourself. Only technology can match algorithmically accelerating tech. We need tools that help us catch the speeding train before it leaves us behind.

When I reflect on my watching habits and consider ideas like unplugging or boredom, my thoughts often turn to Hans Selye, who pioneered stress research.

He found that too much stress (distress) is bad, but also that lack of stress is bad too. The key was in finding the right amount of stress, which Selye calls "eustress." Selye's work has been critical in sports physiology, in developing the idea of periodization (systematically varying training volume and intensity over different phases of a training cycle), and the importance of recovery periods. It's an unwritten rule in training that your muscles don't grow while

30 Li, Qing. *Forest Bathing: How Trees Can Help You Find Health and Happiness*. Viking, 2018.

31 Or what the folks on TikTok call raw-dogging: "Raw Dogging: The New Travel Trend." *CNN Travel*, Jun. 28, 2024, https://www.cnn.com/2024/06/28/travel/raw-dogging-travel-trend-explainer-intl-hnk/index.html. Accessed Dec 9, 2025.

32 Svendsen, Lars Fr. H. *A Philosophy of Boredom*. Translated by John Irons, Reaktion Books, 2005.

you are working out; they do when you stop stressing them, and recover and adapt to the previous stress.

I need to balance the time I spend staring at screens with consciously constructed periods of boredom and natural tranquility. This has been very helpful to me. And we can use Technology to monitor that stress, and other important metrics from our bodies and thoughts.

This follows advice from mid-20th-century visionaries like Vannevar Bush (who envisioned "memex" technology for storing and linking information),[33] Douglas Engelbart (who demonstrated early prototypes of technologies underlying our current world),[34] and J.C.R. Licklider (who advocated for global computer networks).

Engelbart's goals were the same as Bush's: harnessing tech to augment human potential. At about this same time (1960s), Licklider advocated for the development of a global network of computers, which he termed the "Intergalactic Network,"[35] envisioning a future where computers would be interconnected and accessible to all. It was the siren call, and the people who responded (including, notably, Vint) built what was eventually the Internet. All three shared a vision of technology that enhances human capabilities, whether through augmenting intellect, fostering collaboration, or democratizing access to information.

Leverage available tools purposefully. Use search engines not just for quick answers but for deep research. Employ AI for summarizing complex topics or finding patterns in large datasets – while maintaining skeptical awareness of biases and limitations.

Look forward to emerging technologies. Wearable devices can monitor vital signs reflecting internal processes you're unconscious of. Smart devices can track sleep quality and duration. Eventually, we might develop integrated "human operating systems" that span all dependent technologies, simplifying interaction through intuitive dashboards while keeping most processes unconscious until you need awareness.

---

33 Bush, Vannevar. "As We May Think." *The Atlantic Monthly*, vol. 176, no. 1, July 1945, pp. 101–108.

34 Engelbart, Douglas C. "Augmenting Human Intellect: A Conceptual Framework." *Stanford Research Institute*, Oct. 1962, https://www.dougengelbart.org/pubs/papers/scanned/Doug_Engelbart-AugmentingHumanIntellect.pdf. Accessed Jul. 7, 2025.

35 Licklider, Joseph C.R. "Memorandum for Members and Affiliates of the Intergalactic Computer Network." *Advanced Research Projects Agency*, Apr. 23, 1963, https://psmag.com/social-justice/the-social-life-of-gense-64616/the_Intergalactic_Computer_Network.pdf. Accessed Sep. 23, 2024.

5.4 Curate Your Information Diet

UCLA psychologist Steve Cole observes: "*Your experiences today will influence the molecular composition of your body for the next two to three months, or perhaps for the rest of your life. Plan your day accordingly.*"[36]

McGilchrist argues that attention is how our world comes into being for us,[37] and that the way we attend carries profound moral significance.[38]

Marcus Aurelius noted similar ideas 2,000 years ago: "*Our life is what our thoughts make it,*"[39] and "*The happiness of your life depends upon the quality of your thoughts.*"[40]

Ideas like those above are one of the reasons I love the idea of an "information diet." This idea is explored in Clay Johnson's *The Information Diet: A Case for Conscious Consumption*, published in 2012. Johnson points out that just as our physical health can be influenced by what we eat, our mental well-being and cognitive abilities can be affected by the quality and quantity of information we consume. Spend time considering what you attend to. Just as physical health can be influenced by what we eat, mental well-being and cognitive abilities are affected by information quality and quantity we consume.

When coming home from stressful days, avoid content that subconsciously increases stress and anxiety. Consider how media artificially distorts worldview negativity – think about how many people you've seen killed on TV versus real life.

Avoid TV news, which only reports events because they're "new" (rare). Rare bad events from around the world, condensed into daily shows, make us think these unusual occurrences are ordinary.

Recognize that most media operates commercially. Media doesn't just tell; it sells. If products or services are free, you're the product. Advertising shapes programming, content gets tailored to deliver large audiences to advertisers, and product placements influence storylines.

---

36 Cole, Steven. "The Social Life of Genes." *Pacific Standard*, in *Stories – The Social Life of Genes – Showcase*, Apr. 6, 2025, https://psmag.com/social-justice/the-social-life-of-genes-64616/. Accessed Nov. 27, 2025.

37 McGilchrist, Iain. *The Matter with Things: Our Brains, Our Delusions, and the Unmaking of the World*. Perspectiva Press, 2021, p. 27.

38 McGilchrist, Iain. *The Matter with Things: Our Brains, Our Delusions, and the Unmaking of the World*. Perspectiva Press, 2021, p. 30.

39 Aurelius, Marcus. *Meditations*. Penguin, 2015, p. 35.

40 Aurelius, Marcus. *Meditations*. Penguin, 2015, p. 15.

Be conscious of commercial influence. Prefer ad-free content when possible. Seek out alternate perspectives deliberately, including positions you disagree with. Study viewpoints from people in society you don't understand well.

5.5 Practice Conscious Consumption

Think carefully about what you buy. Consider whether purchases align with your consciously chosen values rather than culturally imposed expectations.

Support products whose features reflect your values and emerging needs. When we buy things that are better for us, companies make more of those products. Vote with your wallet for the future you want to see.

As Paul Hawken notes: "*Business is the only mechanism on the planet today powerful enough to produce the changes necessary to reverse global environmental and social degradation.*"[41]

This applies to navigating technological disruption – if we don't demand the right things, businesses can't offer them.

Again, we can find Aurelius's ancient wisdom useful on this topic: "*Enjoy material comforts if they're there, but don't miss them if they aren't there. Treat the things you don't have as nonexistent. Look at what you have and think how much you'd want them if you didn't have them.*"[42]

6 Developing New Forms of Thought

Einstein warned we need new ways of thinking to survive and evolve as a species, while Obama noted that technological advancement without corresponding advancement in human institutions endangers us.

Crafting new forms of thought represents one of our greatest tasks. This isn't distinct from the items outlined above; rather, it results from practicing many of them.

Traditional approaches resolve complexity by breaking it into chunks and developing hyper-specialization. But as David Weinberger[43] demonstrates –

41 Hawken, Paul. *The Ecology of Commerce: a Declaration of Sustainability*. HarperBusiness, 1993.

42 Aurelius, Marcus. *Meditations*. Translated by Gregory Hays, Modern Library, 2002.

43 See Weinberger, David. *Too Big To Know: Rethinking Knowledge Now That the Facts Aren T the Facts, Experts Are Everywhere, and the Smartest Person in the Room Is the Room*. Basic Books, 2011, and Weinberger, David. *Everyday Chaos: Technology, Complexity, and How We're Thriving In New World of Possibility*. Harvard Business Press, 2019.

what each of us knows is expanding greatly. Yet, what there is to know expands even faster. The result is that fewer people can understand wholes or anticipate downstream consequences.

David Epstein's *Range* argues that generalists often outperform specialists in innovation and uncertainty navigation.[44] Their success comes from cultivating diverse experiences, skills, and knowledge across broad interests. They're less tied to traditional solutions and more likely to think outside established domains.

However, in standard environments, specialists remain the default and generalists aren't rewarded. Developing generalist skills requires accepting this dichotomy, seeking roles and opportunities that broaden experience and test comfort zones – perhaps at the expense of traditional success measures.

Specialization is not wrong. McGilchrist notes that "*our talent for division, for seeing the parts, is of staggering importance*," and that these capacities "*have helped us achieve nothing less than civilisation itself.*"[45]

But McGilchrist argues this talent is "second only to our capacity to transcend it, in order to see the whole." The left hemisphere craves certainty, clarity, and confidence, and it plays a crucial role. The right hemisphere, however, holds the paramount ability: it "makes it possible to hold several ambiguous possibilities in suspension without premature closure." This perspective suggests that instead of seeing ourselves as uniquely suited to solve problems, we should leverage our experience to connect with stakeholders and resources, catalyzing more effective engagement.

This thinking is curious, solution-focused, and stakeholder-oriented rather than self-oriented. The goal: uplift the room and thereby uplift others, not prove individual intelligence.

These are the characteristics of thinking needed to address complex problems at the intersection of thought and technology. Such thinking draws on diverse intellectual foundations, considers problems from multiple perspectives without premature judgment, and occasionally uncovers surprising connections that traditional experts might miss. It is curious, open-minded, humble, and focused on collective rather than individual gain. In short: right-brain thinking in a left-brain world.

---

44 Epstein, David. *Range: Why Generalists Triumph in a Specialized World*. Penguin, 2021.

45 McGilchrist, Iain. *The Master and His Emissary: the Divided Brain and the Making of the Western World*. New Expanded ed., Yale UP, 2019, p. 429.

7 Moving Forward

Research strongly supports several other habits this analysis only touches on: healthy relationships with exercise, mindfulness, meditation, nutrition, sleep, and play – all positively correlated with better thinking, mental wellness, and overall flourishing.

While many view these as antidotes to technology, the ultimate opportunities emerge through hybrid approaches: unplugging during these activities while plugging in to current and future technologies to monitor and assess their effectiveness in real-time. Let technology scaffold performance in important domains.

Rather than cover these extensively here, recognize their importance. Others address them adequately elsewhere, usually specialists dealing with sub-facets of the complex whole.

The path forward requires accepting that this journey is deeply personal. Everyone's situation, perspective, resources, and desired outcomes differ. What follows must be adapted to individual circumstances while drawing from shared frameworks for conscious technology engagement.

In short: right-brain thinking in a left-brain world. Most importantly: start somewhere. Begin with awareness, develop your own plan, take ownership of your relationship with technology, maintain intellectual humility, balance strategic unplugging with intentional plugging in, curate your information consumption, and practice conscious choice-making.

The future belongs to those who consciously navigate technological disruption rather than being swept along by it. In the end, conscious navigation means awareness, humility, balance, curation, and choice. Each small practice builds toward new forms of thought. The tools exist. The agency is yours. The question remains: How will you use your magic wand?

8 When Unprecedented Problems Require Unprecedented Methods

These next chapters survey solutions across multiple domains – from social psychology to economic theory – which would normally warrant separate books. I'm aware this creates vulnerabilities: superficial treatment of complex topics, potential contradictions between analytical frameworks, and the risk of mistaking breadth for depth.

Yet, the central premise of this work – that technological disruption of thought creates systemic challenges – demands this integrative approach.

Unprecedented problems require methods that stretch beyond traditional disciplinary boundaries. If our thinking shapes our technologies, which reshape our social structures, economic systems, and political institutions, then addressing technological disruption requires engaging the entire web of interdependencies.

Consider these chapters as reconnaissance across interconnected territories rather than detailed cartography of any single region. I'm sketching my impression of the landscape where collective solutions must operate, identifying leverage points and systemic constraints, while acknowledging that each section raises more questions than it answers.

I move now to more speculative ground, transitioning from personal empirical findings to proposing solutions for others – sometimes, at society's grandest scales. After a decade as a researcher surveying the landscape, I am now shifting to become an architect, helping us navigate these challenges together.

Consider this version 0.1 of a blueprint, offered humbly as a starting point rather than final answers. Think of me as Engelbart demonstrating his vision of computing's future – not as the engineers who built the Internet. I'm sketching possibilities, not <yet> engineering solutions.

Start by stress-testing these ideas. Workshop them. Replace what doesn't work with better solutions, then share improvements with others who will do the same. That's how we tackle complex issues: together.

CHAPTER 22

The Science of Scaling Change

> The better we get at getting better, the faster we will get better.[1] (Douglas Engelbart)

1 Building a Bridge to a Better Future

In Chapter 15 and the previous chapter, we discussed how the new power in our culture is the ability to positively adapt to change, and how technology was the primary driver for change in our society. The new luxury good is the ability to embrace technology in ways that will leave you better off.

What makes this dynamic "new" has two dimensions.

The first is the unprecedented scale and pace of change we're experiencing.

The second reality is the structural shift we are seeing in "power" in human systems, as individuals adapt faster to technological change than larger entities. This represents a democratization of opportunity. But we've also established that cultures function as markets where businesses serve collective demand rather than individual needs, and that we remain intricately interconnected and interdependent.

Therefore, individual solutions are not enough – to achieve lasting progress even for the individual, we must build a bridge to that future and bring others along with us. Positive adaptation must scale. So let's spend a few minutes reviewing the science behind scaling innovation and social change so that we can incorporate it into our blueprint for responding to the disruption of thought.

---

1 Engelbart, Douglas. "*Improving Our Ability to Improve: a Call for Investment in a New Future.*" Keynote Address, World Library Summit, Singapore, Apr. 2002. Doug Engelbart Institute, dougengelbart.org/pubs/papers/scanned-original/2002-augment-133320-Improving-Our-Ability-to-Improve-eic11.pdf. Accessed Jun. 12, 2025.

 | HTTPS://DOI.ORG/10.1163/9789004739161-028

2 Understanding How Change Spreads

The first challenge: moving from individual actions to group behaviors. After all: *True leadership is not in how far you go – but in how many you bring with you.*[2]

How can we help others become aware of new ideas, such as these, and recruit and encourage them to collaborate on solutions that lead to more widespread, durable human flourishing?

Fortunately, researchers have already charted how innovation scales.

2.1 The Adoption Curve: From 5% to Tipping Point

Everett Rogers' *Diffusion of Innovations* suggests that as little as 5% of a population needs to accept an idea before it embeds within that culture. Rogers identified distinct adoption phases: Innovators (2.5%), Early Adopters (13.5%), Early Majority (34%), Late Majority (34%), and Laggards (16%).[3] The critical moment comes when enough early adopters create self-sustaining momentum – priming the pump so water continues flowing on its own.

Rogers felt the critical event, reached with just a relatively small number of early adopters, is when enough individuals have adopted the innovation sufficiently that further adoption becomes self-sustaining. Sort of priming the pump, after which the water continues to flow all on its own. Once an idea is embedded, it is not only much harder to dislodge, but according to Rogers, it is positioned to continue to expand.

Rogers is just one of the most well-known thinkers to have studied these dynamics. Back in the 19th century, Frenchman Gabriel Tarde pioneered similar ideas (a very small percentage of people develop new ideas, which then diffuse through the whole through a network of elite influencers),[4] while Joseph Schumpeter introduced the idea of "creative destruction"[5] describing how new ideas subsume older ones.

---

2 This was expressed by Raleigh, NC, Trash truck driver Charlen Parker, who used it as detailed mismanagement and corruption in the city of Raleigh's solid waste services departmento at a City Hall meeting. I was unable to find any other earlier source to attribute this to, and in his eloquence, I find it right to credit him, at least for now: https://www.youtube.com/watch?v=Xqy73nmKovs Accessed Aug. 18, 2025.

3 Rogers, Everett M., Arvind Singhal, and Margaret M. Quinlan. "Diffusion of Innovations." *An Integrated Approach to Communication Theory and Research*, Routledge, 2014, pp. 432–448.

4 Tarde, Gabriel. *The Laws of Imitation.* Translated by Elsie Clews Parsons, Henry Holt, 1903.

5 Schumpeter, Joseph A. *Capitalism, Socialism, and Democracy.* Harper and Brothers, 1942.

More recently, scholars such as Duncan Watts ("6 degrees of separation"),[6] Mark Granovetter (the strength of "weak ties"),[7] and Ronald Burt ("structural holes" in social networks),[8] have built on and/or revised ideas within the field of innovation diffusion. But it's fair to say that no one has popularized it in recent times more than Malcolm Gladwell, and his application of the phrase "tipping point" to describe the critical event in which diffusion becomes self-sustaining.

2.2 The Vital Few: Who Makes Ideas Spread

Gladwell draws on a broad set of social sciences to suggest that little things can have outsized diffusion results. But not just any small thing, done by anyone. He says, "*The success of any kind of social epidemic is heavily dependent on the involvement of people with a particular and rare set of social gifts.*"[9]

Gladwell builds on Rogers' adoption timeline by identifying the "vital few" – the specific people who transform ideas into epidemics. He argues that scaling any innovation requires three distinct personality types beyond the Innovator:

Mavens are information brokers who accumulate marketplace knowledge and share it strategically. They spark word-of-mouth epidemics through credibility, social skills, and ability to communicate across settings. Mavens prove most valuable when ideas are complex or require trusted endorsement.

Connectors possess extraordinary talent for making friends across diverse social worlds. With networks exceeding one hundred people, they bridge different communities through curiosity, confidence, and boundless energy. Connectors become essential when you need rapid, wide-scale adoption across new markets or social circles.

Salesmen are charismatic persuaders with an indefinable quality that makes others want to agree with them. They excel at converting skeptical audiences from awareness to action through natural influence that transcends mere words.

That helps us understand the type of people who help scale ideas, but can it be done with as few as 5% of the overall? Gladwell might point out that it depends on the strengths of the initial Mavens, Connectors, Salesmen, and

---

6 Watts, Duncan J. *Six Degrees: The Science of a Connected Age*. W. W. Norton, 2003.

7 Granovetter, Mark S. "The Strength of Weak Ties." *American Journal of Sociology*, vol. 78, no. 6, May 1973, pp. 1360–1380.

8 Burt, Ronald S. *Structural Holes: The Social Structure of Competition*. Harvard University Press, 1992.

9 Gladwell, Malcolm. *The Tipping Point: How Little Things Can Make a Big Difference*. Little, Brown and Company, 2000, p. 33.

Persuaders involved in the early days, and the nature of the idea and the audience.

While Harvard professor Erica Chenoweth thinks the number could even be as low as 3.5%,[10] other modern scholars say that an idea needs to get to a tipping point of ranging from 10%[11] up to 25%,[12] before it catalyzes widespread behavioral change.

2.3 Tactics for Scale: From A.C.E. to Layered Networks

Verne Harnish wrote an influential book in 2014, tackling this topic, called *Scaling Up: How a Few Companies Make It...and Why the Rest Don't.* The book centers on four critical decision areas that companies must master: People (focus on the top 10%), Strategy (a single, action-oriented page), Execution (clear goals, accountability, and regular/effective communications rhythm), and Cash. Harnish advocates for creating organizations that can thrive independently of their founders' daily involvement.[13]

Heimans and Timms, in their book *New Power*, discuss how movements and the proliferation of ideas have evolved over time, and how technology has driven these changes.[14] They distinguish between "old power," which operates like currency (held by the few, jealously guarded, hierarchical, and leader-driven), and "new power," which works like a current (made by many, open, participatory, and peer-driven). They lay out their A.C.E. Framework: For ideas to spread in a new power world – they must be Actionable (asking

---

10 BBC Future. "It Only Takes 35% of People to Change the World." *BBC*, May 13, 2019, https://www.bbc.com/future/article/20190513-it-only-takes-35-of-people-to-change-the-world. Accessed Dec 9, 2025.

11 See Porter, Mason Alexander, Jukka-Pekka Onnela, and Peter J. Mucha. "Communities in Networks." *Notices of the American Mathematical Society*, Vol. 56, No. 9, 2009, pp. 1082–1097; and. Galam, Serge, and Taksu Cheon. "Tipping Points in Opinion Dynamics: A Universal Formula in Five Dimensions." *Frontiers in Physics* vol. 8, 2020, p. 566–580.

12 See Centola, Damon, et al. "Experimental Evidence for Tipping Points in Social Convention." *Science* vol. 360, no. 6393, 2018, pp. 1116–1119. See also Berger, Jonah. *Contagious: Why Things Catch On.* Simon & Schuster, 2013 (doesn't specify a number, but argues against Gladwell's "few"), and Aral, Sinan, and Dylan Walker. "Creating Social Contagion Through Viral Product Design: A Randomized Trial of Peer Influence in Networks." *Management Science, vol.* 57, no. 9, 2011, pp. 1623–1639 (again, doesn't specific a number, but describes robust conditions that need to exist for contagion).

13 Verne Harnish, *Scaling Up: How a Few Companies Make It...and Why the Rest Don't.* Gazelles, Inc., 2014.

14 Heimans, Jeremy, and Henry Timms. *New Power: How Power Works in Our Hyperconnected World – and How to Make It Work for You.* Doubleday, 2018.

people to take action), Connected (creating peer connections that set off network effects), and Extensible (allowing customization, remixing, and shaping by participants). They also note that new power is characterized by radical transparency, willingness to allow communities to reinvent content, shared control, and actionable ideas that people make their own rather than simply consume.

Most of these thinkers would agree, however, that network structure, social influence, and the nature and actions of the committed early minorities are factors in driving collective behavior to scale.

As a result, in an organization of say 10,000 people – such as a large firm, a medium-sized town, or a typical university – culture, mindset, behavior, and paradigm change can begin by reaching as few as 350 people, or as many as 2,500. The exact percentage will depend on the type and effectiveness of the early members of that specific diffusion in that organizational network.

But one other factor needs to be considered: Robin Dunbar's "number."

Dunbar is an anthropologist and evolutionary psychologist who has spent significant time looking at group dynamics in primate species and early human species. He's become famous for noting that our modern human brains are constrained by the evolutionary factor that our forebears' group sizes were traditionally no larger than about 150. As a result, the maximum number of people we can functionally keep track of and maintain working ties with is 150 – now known in many circles as Dunbar's Number. But even Dunbar notes that in practice, we might have only 50 close friends and family.[15]

Collectively, these thinkers provide us with a blueprint for the quantity, quality, and tactics needed to successfully launch and scale revolutions.

If thought is being disrupted, and *if* there is a significant benefit to proactively building new forms of thought and disseminating them into mainstream society, one need not boil the ocean by crafting a single tactic or message that reaches everyone in the group in the same way, in one shot.

But individual change that ignores proven scaling practices noted above rarely grows into larger solutions on its own.

The solution? A strategic approach, layered like an onion: Start with cross-disciplinary Innovators and Mavens who create mass change through layered networks. Innovators form the inner core, connecting formally and informally

---

15 We'll skip over the nuance, for now, introduced by Granovetter, about the emergence of "weak ties" and modern phenomena of large networks of social media "followers," and stick at this point to the basic limitations of our ability to functionally connect with social groups. Also, like any popular idea in science, this one has a number of critiques, and even among supporters, including Dunbar himself, the actual number is debated. We're using this here as a touchstone, not an ironclad rule/threshold.

to Mavens. These Mavens then leverage Connectors' extensive networks, including authority figures, while early participants with Salesperson and Persuader traits catalyze the process. While some participants may command large-scale "Influencer" reach, most change emerges through numerous loosely coordinated personal networks and/or events of 50–150 people each. This creates momentum that sustains itself at scale, independent of the original Innovators' ongoing efforts.

The only difference I'd point out: Gladwell's reductionist model, designed to appeal to mass audiences, overlooks crucial real-world nuances. Organizations should encourage Persuaders/ Salespeople to innovate and push Innovators to network actively. And Connectors can also be Mavens, and vice versa. In our future world, we'll all possess unique strengths, but these strengths will emerge when we step outside prescribed narrow roles and tackle whatever jobs demand attention.

The different functions remain important, but unlike Gladwell's rigid model, we can view these functions as interchangeable hats that each group member wears. This messier approach still creates layered networks, but each person becomes their own network center for different ideas and innovations while simultaneously playing various roles in others' networks.

This is the blueprint that will guide humanity through these challenges toward the opportunity of a more flourishing future. One innovator can host a small event or share their thoughts with a small group of, say, 20 people. Assume 50% of the people curate and/or connect that information to a network of 50 people each. If 50% of those do the same, and so on, you get to over 150,000 in just 3 cycles. Of course, in real life, the response/engagement rate is far lower, but at the same time, Technology platforms amplify messages now far beyond Dunbar's number, especially if you follow the best practices that Heimans and Timms shared with us.

Via this conscious and strategic approach, we can achieve, as the Dalai Lama shared, *"Just as ripples spread out when a single pebble is dropped into water, the actions of individuals can have far-reaching effects"*.[16]

The science is clear: small committed groups can catalyze system-wide change when they understand and apply proven diffusion principles. The question isn't whether individual solutions can scale – it's whether we'll organize deliberately to make it happen.

---

16 His Holiness the Dalai Lama (@DalaiLama). "Just as ripples spread out when a single pebble is dropped into water, the actions of individuals can have far-reaching effects." X (formerly Twitter), 10 May 2013, twitter.com/dalailama/status/332790603966476288. Accessed Feb. 12, 2022.

CHAPTER 23

Building Technology-Enabled Communities

> Community is not a place, a building, or an organization; nor is it an exchange of information over the Internet. Community is both a feeling and a set of relationships among people. People form communities to be with others like themselves, to belong to something greater than themselves, to do meaningful work together.[1] (Marvin Weisbord and Sandra Janoff)

> We are not going to be able to operate our Spaceship Earth successfully nor for much longer unless we see it as a whole spaceship and our fate as common. It has to be everybody or nobody.[2] (Buckminster Fuller)

1 The Foundation of Collective Prosperity

The same technologies disrupting our thinking can be redirected to forge the meaningful connections necessary for coordinated response, transforming potential isolation into purposeful collaboration.

In the last chapter, I argued it is not enough to jump a chasm to a better future, alone – we must build a bridge to bring others along. Or, as Reid Hoffman counsels, "*No matter how brilliant your mind or strategy, if you're playing a solo game, you'll always lose out to a team.*"[3] Let's sit with that a moment.

Groups require common ground to unite around activities supporting collective goals. One powerful organizing principle: behaviors that lift and sustain collective prosperity. If you earn $400,000 annually as a cloud security architect, you should care when basic necessities become unaffordable for others. We're all in the same boat.

---

1 Weisbord, Marvin R., and Sandra Janoff. *Don't Just Do Something, Stand There!: Ten Principles for Leading Meetings That Matter*. Berrett-Koehler Publishers, 2007, p. 21.

2 Fuller, R. Buckminster. *Operating Manual for Spaceship Earth*. 1969. Lars Müller Publishers, 2008, p. 83.

3 Hoffman, Reid, and Ben Casnocha. *The Start-up of You: Adapt to the Future, Invest in Yourself, and Transform Your Career*. Random House, 2012, p. 176.

 HTTPS://DOI.ORG/10.1163/9789004739161-029

Success begins with an accurate and compassionate understanding of our actual circumstances. Hans Rosling's *Factfulness*[4] highlights why we even need to talk about this. Rosling created a survey of 13 basic questions about world population, education, poverty, and societal wellness. Administered to thousands across hundreds of events, he found we systematically get wrong answers – so wrong that chimpanzees choosing randomly consistently outperform humans.

Our answers aren't randomly wrong – we see the world as far worse than it really is. We think it's poorer, less healthy, less educated, and more dangerous than it is in reality. Ironically, higher education correlates with more negative perceptions. World Economic Forum members scored below average. Nobel laureates were among the most inaccurate. Even specialized knowledge and interest didn't help. A global women's rights group had among the lowest and most negative results on female gender-related questions.

The first step in fostering collective prosperity: gratefully and accurately appreciating our situation. If we misperceive how well we're doing, working harder at improvement feels futile when people think things only worsen.

If we don't improve collective prosperity, we get Peter Turchin's concept of "popular immiseration" – situations where general population living standards decline despite overall economic growth. Contributing factors include increasing inequality, declining purchasing power, rising living costs, and limited access to social services or economic opportunities. Turchin's theory suggests popular immiseration can precede social upheaval as discontent grows, affecting everyone, even those in the highest income brackets.

I argue in *The Great Irony of Technology* that structural objective factors like this ("I can't afford McDonald's anymore!")[5] are compounded by declines in various subjective parts of daily life, including high stress levels, unhappiness, loneliness, and depression. Technology plays a key role in both the objective structural factors and in instilling and amplifying the declines in subjective well-being.

Turchin's theory suggests that popular immiseration, as described in the previous paragraph, can be a precursor to social upheaval, revolutions, or other forms of collective action and "disintegration" as discontent among the population grows. This is the challenge facing all of us, even those in the high-

4 Rosling, Hans, Ola Rosling, and Anna Rosling Rönnlund. *Factfulness: Ten Reasons We're Wrong About The World–And Why Things Are Better Than You Think.* Flatiron Books, 2018.

5 "'I can't afford McDonald's anymore': Customers slam McDonald's prices after paying almost $25 for 3 items." *Reddit*, u/SpatchcockedBonky, r/LateStageCapitalism, Aug. 30, 2023, www.reddit.com/r/LateStageCapitalism/comments/17wn5n8/i_cant_afford_mcdonalds_anymore_customers_slam/. Accessed Sep. 9, 2025.

est income brackets. But we can take heart in historian Arnold Toynbee's observation – *"Now civilizations, I believe, come to birth and proceed to grow by successfully responding to successive challenges. They break down and go to pieces if and when a challenge confronts them which they fail to meet."*[6]

Instead of ignoring this, running away from fears, or resisting this challenge, we should move toward collective aspirations through increased social cohesion and support – a critical first step toward building sustainable futures for all.

2 From Social Capital Decline to Revival

Community is a moderating influence on subjective well-being.

Robert D. Putnam famously explored how Americans are experiencing a decline of social capital and community engagement in his book *Bowling Alone*.[7] He starts by referencing De Tocqueville in the 1830s, who noted Americans' unique propensity for "free association"[8] and participation in civic life. But by 2000, Putnam found these proclivities were in decline, and despite achieving material prosperity and technological advances, Americans had become increasingly disconnected from one another, leading to many social and political challenges. Putnam specifically pointed to changes in technology and thinking as the root cause for these trends, citing changing work patterns, suburbanization, shifts in generational values, and technological "advancements" such as television and the Internet.

Fortunately, Putnam returns a few years later with a more optimistic perspective, working with co-author Lewis M. Feldstein to pen *Better Together: Restoring the American Community*.[9] This book focuses on solutions and strategies for rebuilding social capital and strengthening communities, and draws on examples from across the United States where individuals and organizations have successfully fostered social connections, revived civic engagement, and revitalized local communities. Putnam and Feldstein highlight the power

6 Toynbee, Arnold Joseph, and David Churchill Somervell. *Civilization on Trial*. Oxford University Press, 1948, p. 56.

7 Putnam, Robert D. *Bowling Alone: The Collapse and Revival of American Community*. Simon & Schuster, 2000.

8 Tocqueville, Alexis de. *Democracy in America*. Translated by Harvey C. Mansfield and Delba Winthrop, University of Chicago Press, 2000.

9 Putnam, Robert D., Lewis Feldstein, and Donald J. Cohen. *Better Together: Restoring the American Community*. Simon & Schuster, 2004.

of collective action and the potential for positive change when individuals come together to address common challenges.

In my experience, I've seen both sides: technology isolating people by replacing thick real-life friendships with thin network connections, and also tech bringing people together in unprecedented ways.

3 The Maker Movement: A Case Study in Technology-Enabled Community

Consider the maker movement – a cultural phenomenon characterized by growing communities engaging in DIY activities, tinkering, and creating. Makers use diverse tools and technologies, from traditional woodworking and welding to 3D printing, CNC machining, and Arduino programmable circuits. They typically meet in makerspaces – shared tools and facilities, like high school shops but open to communities.

I became interested in this movement in 2013, seeing potential solutions to educational system challenges and as a way of regaining US manufacturing competitiveness. After researching and visiting many makerspaces, I helped start one with several others. That organization now operates two makerspaces in different communities with hundreds of members.

What I learned during launch and growth: a makerspace isn't tools and facilities – it's primarily community. People from all walks of life unite over shared interests and projects. Sometimes they teach, sometimes learn. We help each other. Technology includes not just building tools but also Kickstarter for fundraising, Slack for collaboration, websites for discovery, email for communication, and payment platforms for operations. We harness technology to open and democratize making, empowering individual agency, creativity, and innovation; fostering collaboration and learning; and building community resilience.

The surprising takeaway: it fosters community eudaimonia by connecting people meaningfully as part of something bigger than themselves, allowing personal progress toward valued skills and projects. It improves individual and collective quality of life.

Yes, technology can isolate and estrange us, but it can also unite us. Which wins depends on the conscious choices we make.

4 The Two Wolves: Choosing Connection over Isolation

A parable captures this choice perfectly: an old chief tells his grandson about a fight inside him between two wolves – one evil (full of anger, envy, sorrow, regret, greed, arrogance, self-pity, guilt, resentment, inferiority, lies, false pride, superiority, and ego), the other good (full of joy, peace, love, hope, serenity, humility, kindness, benevolence, empathy, generosity, truth, compassion, and faith). When the grandson asks which wolf wins, the chief replies: "The one you feed."[10]

The maker movement and similar technology-enabled communities demonstrate something crucial: we can consciously choose which wolf to feed. When we approach technology with intentions of service, collaboration, and mutual flourishing rather than consumption and competition, entirely different outcomes become possible.

But individual communities, however vibrant, face limits. They operate within larger systems – economic, political, and technological – that can either support or constrain their flourishing. The same forces that enable small groups to form meaningful connections can be harnessed to create networks of communities, coalitions of purpose, and movements that shift entire systems.

The question becomes: how do we scale the insights from local community building to address challenges that span institutions, industries, and nations? How do we move from feeding the good wolf in our immediate circles to changing the larger environment in which all wolves must compete for resources?

---

10 It should be noted that this parable has become memetic. Its origins may have come from earlier preachers in the 50s and 60s, who used dogs instead of wolves, but it did NOT come from Indigenous Americans as it is widely attributed to be on the Internet. Graham admits this here – Graham, Billy. *The Holy Spirit: Activating God's Power in Your Life*. Word Books, 1978.

CHAPTER 24

From Networks to Collective Action

> Each time a man stands up for an ideal, or acts to improve the lot of others, or strikes out against injustice, he sends forth a tiny ripple of hope… those ripples build a current which can sweep down the mightiest walls of oppression and resistance.[1] (Margaret Mead)

1 The Pacing Problem and Power Shifts

As traditional institutions struggle with the pace of technological change, new collaborative networks emerge as essential vehicles for collective solutions that can match the speed and complexity of our challenges.

Complex problems require enhanced collaboration. Better collaborations become possible through technology's power, and collaborators' thinking gets shaped by their collaborative tools as their medium for shared effort.

Former Venezuelan Trade Minister Moisés Naím argues in *The End of Power*[2] that traditional power structures face disruption, with influence shifting from large hierarchical institutions toward smaller, more agile actors and networks. As traditional power brokers react to these changes by trying to consolidate power, they find that it can backfire, leaving governance and leadership ill-suited to address the complexities of the modern world. Naim calls for new approaches that emphasize adaptability, collaboration, and inclusivity. This conclusion would seem to defy how most people think of the rising power of billionaire CEOs, government leaders, and prominent cultural figures, but since the book was written from the position of an insider who is a regular participant in government efforts and global events like the World Economic Forum, it should be heeded.

I illustrated this pacing problem in Chapter 15, showing differential change rates: Technology (fastest), individual cognition, social systems, business, and politics (slowest).

This isn't just theory – it reflects lived reality for anyone trying to address technological disruption through traditional institutional channels. While

1 Kennedy, Robert F. "*Day of Affirmation Address.*" University of Cape Town, 6 June 1966.

2 Naím, Moisés. *The End of Power: from Boardrooms to Battlefields and Churches to States, Why Being In Charge Isn't What It Used to Be*. Basic Books, 2013.

institutions debate, technology advances. While committees form, algorithms reshape thinking. While policies are crafted, new platforms emerge that render regulations obsolete.

2 New Power Frameworks for Collective Response

Jeremy Heimans and Henry Timms' *New Power* provides frameworks for understanding this transformation.[3] They distinguish "old power" (operates like currency – held by few, jealously guarded, hierarchical, leader-driven) from "new power" (works like current – made by many, open, participatory, peer-driven).

New power is characterized by radical transparency, willingness to allow communities to reinvent content, shared control, and actionable ideas that people make their own rather than simply consume. Examples include crowdfunding platforms (Kickstarter, Indiegogo), social media movements (#MeToo, Black Lives Matter, Arab Spring), open-source communities (Linux, Wikipedia), peer-to-peer platforms (Airbnb, Uber), and civic engagement tools (Change.org, Avaaz). They argue digital-era leaders need new power principles – empowerment, collaboration, and responsiveness – to navigate hyperconnected complexity and achieve meaningful change.

The authors argue that in the digital era, old power leaders need to embrace new power principles such as empowerment, collaboration, and responsiveness to better navigate the complexities of our hyperconnected world and achieve meaningful change.

According to John Alexander and Ariane Conrad, shifting power presents both an opportunity and a responsibility.[4] In their book *Citizens*, they urge individuals to become active, engaged participants in society. This involvement not only enables people to effect positive change but also fosters a sense of belonging and provides avenues for personal growth.

This mirrors exactly what I found in the maker movement and shows how we can create social movements that overcome the pacing problem facing large traditional power players, through new strategies, tactics, and technologies.

---

3 Heimans, Jeremy, and Henry Timms. *New Power: How Power Works in Our Hyperconnected World – and How to Make It Work for You*. Doubleday, 2018.

4 Alexander, Jon, and Ariane Conrad. *Citizens: Why the Key to Fixing Everything Is All of Us*. Canbury Press, 2022.

3 Historical Precedents: Small Groups, Large Changes

J.C.R. Licklider's 1963 memo calling for what became the Internet demonstrates how visionary collaboration works.[5] He suggested exploring "*possibilities for mutual advantage*" to determine "*who is dependent upon whom for what and who may achieve a bonus benefit from activities of other group members.*" This was necessary because manifesting his vision required many people working in silos to subsume efforts into new, holistic, emergent collaboration.

Many examples throughout this book follow that pattern. Paul Hawken's *Blessed Unrest* describes the environmental movement as the "largest movement in the world" – not through traditional top-down organization but through diverse, decentralized efforts tackling causes from human rights to economic equity.[6] Hawken argues these share common purpose and desire to address global challenges like poverty, climate change, and social injustice. Technology catalyzes these collaborations by facilitating communication, information sharing, crowdfunding, and providing advocacy tools.

Historically, small groups of idealists have successfully overturned established systems many times:

- **American Civil Rights Movement** traces back to a few strategists like W.E.B. Du Bois (NAACP founder) and later figures like Bayard Rustin and Ella Baker who developed nonviolent resistance philosophy and tactics.[7]
- **Women's Suffrage Movement** emerged from small gatherings like the 1848 Seneca Falls Convention, organized by Elizabeth Cady Stanton, and Lucretia Mott. Stanton's "Declaration of Sentiments" provided intellectual foundation that secured voting rights across the Western world.[8]
- **Environmental Movement** began with thinkers like John Muir, Aldo Leopold, and Rachel Carson. Carson's *Silent Spring* and Leopold's "Land Ethic"

5 Licklider, Joseph C.R. "*Memorandum for Members and Affiliates of the Intergalactic Computer Network.*" Advanced Research Projects Agency, Apr. 23, 1963, https://worrydream.com/refs/Licklider_1963_-_Members_and_Affiliates_of_the_Intergalactic_Computer_Network.pdf. Accessed Sep. 23, 2024.

6 Hawken, Paul. *Blessed Unrest: How the Largest Movement in the World Came into Being, and Why No One Saw It Coming*. Viking, 2007.

7 Lewis, David Levering. *W.E.B. Du Bois: Biography of a Race, 1868–1919*. Henry Holt, 1993; and Jonas, Gilbert. *Freedom's Sword: The NAACP and the Struggle Against Racism in America, 1909–1969*. Routledge, 2005.

8 Baker, Jean H. *Votes for Women: The Struggle for Suffrage Revisited.* Oxford University Press, 2002.

provided frameworks that led to the EPA, Clean Air Act, and modern environmental movement.[9]

The pattern repeats: small groups develop new frameworks for understanding problems, create tools and strategies for action, build networks that amplify their efforts, and eventually shift entire systems through persistent, coordinated effort.

4 Practical Integration: Where Individual and Collective Converge

These frameworks converge on clear imperatives for organizing the collective response that technological disruption demands, using the very tools that created the challenge to build solutions at the scale required.

Given these dynamics, how do we move from individual awareness to collective transformation? Several practical approaches emerge:

4.1 Create Local Networks

Start with your existing community. Host small gatherings (20–50 people) focused on conscious technology use, applying diffusion principles by identifying and engaging local Mavens, Connectors, and natural Salespeople. Share insights from individual practice while learning from others' experiences. Use technology platforms strategically to coordinate and amplify rather than consume passively.

4.2 Build Cross-Sector Collaboration

Connect diverse stakeholders around shared interests in human flourishing amid technological change. Bring together educators, technologists, health professionals, artists, and community leaders. Different perspectives and skills create richer solutions than any single expertise could generate, leveraging the "smartest person in the room is the room" principle.

9 Carson, Rachel. *Silent Spring*. Houghton Mifflin, 1962; and Leopold, Aldo. *A Sand County Almanac: And Sketches Here and There*. Oxford University Press, 1949; and Dunlap, Thomas R. *DDT: Scientists, Citizens, and Public Policy*. Princeton University Press, 1981.

4.3 Demand Better Technology Through Collective Consumer Power

I previously recommended opting out of rabid consumerism, or at the very least, to practice conscious consumerism. But if only one person changes, it makes no difference. Together, we can shift markets. Money drives the world.

If we all stopped clicking on links to polarizing content, tech platforms would make less money and automatically shift their content algorithms. That applies not just to social media and other online content, but all the Technology in our lives. That's how we shape businesses, and it's at the scale of business and industry that we can shape our social structure and future. If we asked the genie for better things, the genie *might* give us better things.[10]

I'm not alone; conscious Technology consumerism for better tech is underway. It may take place in the fringes, such as Hapifork,[11] and the Kolibree Smart Toothbrush,[12] but as more of us ask for it, it becomes more mainstream – in products such as the Oura Ring[13] and the several features in the latest Apple Watch. To achieve impact and scale, better Technology will emerge if we ask for it.

4.4 Foster Institutional Innovation

Work within existing institutions – schools, workplaces, and community organizations – to pilot new approaches to technology integration. Document what works and share findings with similar institutions facing comparable challenges, creating networks of institutional learning.

4.5 Support Policy Development

Engage in policy discussions about technology's role in society. While individual change matters enormously, systemic challenges also require institutional responses. Support policies that protect human agency while enabling beneficial technological development.

10 But remember, we must be VERY careful of what we ask for, and how we ask it. Genies are notoriously tricky.

11 SmartFeed USA. https://smartfeedusa.com/ *SmartFeed USA,* Accessed Aug. 18, 2025.

12 Kolibree. *Kolibree.* https://www.kolibree.com/en. Accessed 18 Aug. 2025.

13 Oura Health Ltd. *Oura Ring.* https://ouraring.com. Accessed 18 Aug. 2025.

5 Up Next: Understanding Systemic Constraints

The deeper question emerges: if our individual practices and collaborative networks conflict with the basic logic of the systems they operate within, which will ultimately prevail?

Put simply: we don't exist in a vacuum. We are fish in water. How we respond to the disruption of thought is bound to our relationship to that environment. Understanding this tension requires examining the economic and energy foundations that currently shape technological development and social organization – starting with the economic systems that provide the fundamental organizing logic for how we allocate resources, set priorities, and define success.

CHAPTER 25

New Economic Alternatives

Money is the answer. What was the question?[1] (Richard Shockey)

I conclude this discussion of solutions with my boldest and most provocative assertion: the current Western economic system – how we handle and pursue monetary capital and its underlying premise of growth-oriented consumption – may not be ideal for our technological future.

Why does a book on technology and thought need to address economic strategy? Because economic systems are organizing principles of our social systems, quietly but powerfully shaping the environmental niche in which thought happens. Economics provides answers to many societal questions and creates much of the water we swim in today.

Churchill observed that, "*it has been said democracy is the worst form of government, except for all those others that have been tried.*"[2] Similarly, while capitalism may not be ideal, it has been better in many ways than the traditional alternatives, like socialism, communism, strongman feudalism, or even tribalism.

But perhaps now we can create better alternatives using technology to get there. That transition phase is critical – we can't simply propose new social systems and adopt them. If they differ radically from today's systems, they must navigate the Overton Window. New alternatives need not only to work better on paper but also be achievable from our current position.

1 Problems with Growth-Dependent Economics

The current system faces two fundamental problems. First, it's predicated on constant growth, which isn't sustainable. At some point, we hit the globe's carrying capacity.

1 Gray, Terry. "Some of My Favorite Quotations." *University of Washington*, 2000.

2 Churchill, Winston S. Speech on the Parliament Bill. *House of Commons Official Report* (Hansard), 5th ser., vol. 444, 11 Nov. 1947, col. 206–207.Accessed Aug. 26, 2025.

The USA has maintained impressive growth – in roughly 70 years since WWII, we've had only 10 years of negative GDP growth, averaging over 3% annual expansion.[3]

Why do we believe we need to grow every year? Part of it might be the factors that Tainter pointed out – that as systems get increasingly complex, each technological innovation or bureaucratic expansion yields a smaller marginal return, so we must grow to keep pace with these smaller returns. Or part of it might be to pay for the growing debt. As Tainter showed us, Rome sought new neighbors to conquer, and used their gold and accumulated wealth to pay for the burgeoning cost of running the empire. Now, in very much a Red Queen Effect, we use economic growth to fund the growing costs of past decisions. And some of that growth comes from extracting in years or decades "surpluses" of resources that accreted over millions of years.

Paul Hawken captures this unsustainability: "*At present we are stealing the future, selling it in the present, and calling it gross domestic product.*"[4]

Second, using GDP as a proxy for wellness overlooks broader human flourishing concepts. The COVID response illustrates this: the government spent over $5 trillion, but most went to economic stimulus rather than healthcare.[5] The message was clear – getting people back to buying stuff mattered more than health outcomes. This economic structure treats us as consumers rather than citizens.

2 When the Metric Eats the Mission

COVID is a mini-case study, but a telling one. Politicians in both parties have maintained the primacy of economic growth, and championed policy accordingly. Reagan famously stated, "*The best social program is a job,*"[6] underscoring his belief that economic growth was essential for addressing social issues.

3 U.S. Bureau of Economic Analysis. *Real Gross Domestic Product (GDPC1)*. Federal Reserve Bank of St. Louis, https://fred.stlouisfed.org/series/GDPC1. Accessed Aug. 26, 2025.

4 Hawken, Paul. "You Are Brilliant, and the Earth Is Hiring." *The NAMTA Journal*, vol. 38, no. 1, Winter 2013, pp. 269–272, files.eric.ed.gov/fulltext/EJ1078017.pdf. Accessed 13 Oct. 2025. Originally delivered as a commencement address, University of Portland, 3 May 2009.

5 Washington Post Staff. "Where Did the COVID Aid Money Go?" *The Washington Post*, Sep. 8, 2022, https://www.washingtonpost.com/business/interactive/2022/covid-money-trail-investigation-explained/. Accessed Aug. 26, 2025.

6 Reagan, Ronald. "Address to the Nation on the Economy." 5 Feb. 1983. *Public Papers of the Presidents of the United States: Ronald Reagan, 1983*, U.S. Government Printing Office.

In the 90s, Clinton advisor James Carville coined the phrase best describing the path to getting elected: "it's the economy, stupid."[7] Seeing the juggernaut of US economic prosperity, it is no surprise to find other countries adopting the same playbook, with governments as diverse as China, India, the European Union, and Japan all pursuing economic growth as their central strategic pillar.[8]

Many economists and scholars have advocated economic growth as a primary economic goal, especially when promoting prosperity, job creation, and improving living standards. This includes Kuznets,[9] Solow,[10] Romer,[11] Schumpeter,[12] Boserup,[13] Friedman,[14] and North[15] (with 5 Nobels between them), all focused on the fundamental importance of economic growth and its positive effects on society.

3 Alternative Economic Thinking

But a growing number of scholars are making the case that economic growth occurs via trade-offs, particularly in the areas of social justice and environmental sustainability. Partly in response to that, several prominent economists and others have explored the idea that capitalism can adapt to different conditions, including steady-state economies or periods of economic contraction.

7 "It's the Economy, Stupid." *Political Dictionary*, https://politicaldictionary.com/words/its-the-economy-stupid/. Accessed Aug. 26, 2025.

8 See OECD. "Economic Policy Reforms 2021." *OECD Publishing*, 2021, https://www.oecd.org/en/publications/economic-policy-reforms-2021_3c796721-en.html. AccessedAug. 26, 2025; and OECD. "Economic Policy Reforms 2023." *OECD Publishing*, 2023, https://www.oecd.org/en/publications/economic-policy-reforms-2023_9953de23-en.html. Accessed Aug. 26, 2025.

9 Kuznets, Simon. *Modern Economic Growth: Rate, Structure and Spread.* Yale University Press, 1966.

10 Solow, Robert M. "A Contribution to the Theory of Economic Growth." *Quarterly Journal of Economics, vol.* 70, no. 1, 1956, pp. 65–94.

11 Romer, Paul M. "Endogenous Technological Change." *Journal of Political Economy* vol. 98, no. 5, 199, pp. S71–S102.

12 Schumpeter, Joseph A. *The Theory of Economic Development: an Inquiry into Profits, Capital, Credit, Interest, and the Business Cycle.* Harvard University Press, 1934. Originally published 1911.

13 Boserup, Ester. *Population and Technological Change: a Study of Long-Term Trends.* University of Chicago Press, 1981.

14 Friedman, Milton. *Capitalism and Freedom.* University of Chicago Press, 1962.

15 North, Douglass C. *Structure and Change in Economic History.* W. W. Norton, 1981.

The highly influential John Maynard Keynes helped us to understand how market economies can adapt to periods of economic contraction.[16] Economist and mathematician Nicholas Georgescu-Roegen contributed to the idea of a steady-state economy and challenged perpetual growth.[17] Ecological economists Herman Daly and Robert Costanza[18] go further, advocating steady-state economics, arguing that infinite economic growth is not sustainable on a finite planet.

E.F. Schumacher, in his book *Small Is Beautiful: Economics as if People Mattered*[19] discusses appropriate-scale technology and sustainable development, and proposes a shift toward smaller, decentralized economies that focus on human well-being and environmental sustainability over continuous growth. Nobel Laureate Amartya Sen emphasizes the importance of enhancing people's capabilities and well-being rather than focusing only on GDP growth.[20] His recommended approach would consider a broader set of indicators, including education, health, and personal freedoms. Kate Raworth, in her book *Doughnut Economics,* proposes a model that aims to meet the needs of all within the means of the planet.[21]

Taken together, this work challenges the assumption baked into the current system that growth is required for prosperity and should be the primary organizing function of an economy and society.

---

16 See Keynes, John Maynard. *The General Theory of Employment, Interest and Money.* London: Macmillan, 1936; and the IMF analysis (Jahan, Sarwat, Ahmed Saber Mahmud, and Chris Papageorgiou. "What Is Keynesian Economics?" *Finance & Development* vol. 51, no. 3, 2014, pp. 53–54) in which they find that he "spearheaded a revolution in economic thinking that overturned the then-prevailing idea that free markets would automatically provide full employment" and "asserted that free markets have no self-balancing mechanisms that lead to full employment."

17 Georgescu-Roegen, Nicholas. *The Entropy Law and the Economic Process.* Harvard University Press, 1971.

18 Costanza, Robert, et al. *An Introduction to Ecological Economics.* St. Lucie Press, 1997.

19 Schumacher, E. F. *Small Is Beautiful: A Study of Economics As If People Mattered.* Blond & Briggs, 1973.

20 Sen, Amartya. *Development as Freedom.* Knopf, 1999.

21 Raworth, Kate. *Doughnut Economics: Seven Ways to Think Like a 21st-Century Economist.* Chelsea Green Publishing, 2017. (The "doughnut" represents a balance between social foundations and ecological ceilings, offering a framework for a sustainable and just economy).

4 Emerging Alternatives

I've pointed out earlier that in trying to sense the future, there are tactics you can use. Gibson's "*the future is already here, just not evenly distributed*"[22] means that if a particular future will play out, we should see signs of it today. If we fail to find any, it's possible that particular future may not be in the cards. Alternatively, we might be too early, in an Overton sense. If we find evidence of the future we are thinking about, we should see where it exists, and why, and see what it might take to scale and gain lock-in.

Leaving aside the traditional alternatives to capitalism, scouting the fringes reveals interesting experiments. Bhutan explicitly rejected GDP for the Gross National Happiness Index, taking a holistic development approach considering social, environmental, cultural, and spiritual well-being dimensions.[23] While intriguing, this approach emerged from unique circumstances and may face implementation challenges elsewhere.

Scandinavian countries – Denmark, Norway, Sweden, Finland – consistently rank among the world's happiest populations.[24] They blend capitalism with social democracy, balancing economic prosperity with social equity. Their success stems from strong social welfare, high social trust, work-life balance, low corruption, gender equality, and quality public services. Sweden transformed from an agrarian society to one of the world's richest through sustained investments in education, innovation, social welfare, and sustainable development.

Many of these factors take time to enculture though and can't be easily picked up and grafted into other mature societies.

5 Game B: A Transition Framework

Jim Rutt champions "Game B" as a proactive alternative arising out of response to "Game A" – the current dominant societal paradigm in the West. He characterized Game A as competitive, zero-sum thinking, hierarchical

22 Gibson, William. Interview. *NPR*, 30 Nov. 2003.

23 Thinley, Jigmi Y., and Janette Hartz-Karp. "National Progress, Sustainability and Higher Goals: The Case of Bhutan's Gross National Happiness." *Sustainable Earth* 2, no. 1, 2019, p. 11.

24 Tenhunen, Marja-Liisa. "Why Finland Is the Happiest Country in the World?." *Cogito-Multidisciplinary Research Journal, vol.* 3, 2020, pp. 18–28.

structures, and short-term incentives.[25] Game A is about 300 years old, but according to Rutt, it's showing signs of instability. If Game A fails, people will move on to another option. Looking in the wings, the current options include neofeudalism, neofascism (both currently experiencing resurgence, globally), and even a neo-Dark Ages.

Game B is more than theory; they have assembled a worldwide community working to create better societal alternatives, aspiring to leverage people's authentic long-term interests toward a healthier, more cooperative society.

They argue that a non-hierarchical, bottom-up organization can work because it was humanity's default mode historically and can be again with new technologies. Game B envisions moving from centralized, extractive, goods-based economics to decentralized, regenerative, services-based systems, emphasizing access and sharing over ownership.

They are not just talking about "tuning out and dropping out" of society in some kind of "back to the land" movement – they have working groups in economics, currency and monetary theory, the future of cities, political metamodernism and nation-states, permaculture, parenting, and intentional communities.

It starts with creating a better sense of values, away from "he who dies with the most toys wins" and money-on-money returns, and towards real sustainability, improved quality of life, and self-actualization. They aim to move from current economic features (centralized, linear material economy, depleting and extractionary, fixed goods, possession and ownership, natural and technological divides, commodity-based, and a human labor workforce) to a new sustainable future (decentralized and distributed, closed-loop materials economy, regenerative, modular and adaptable, services-based, access and sharing structures, biomimicry, technology-focused, and with automated labor).

5.1 Transition B: Building the Bridge

Critically, they propose "Transition B" – a system interfacing with current economics. These must be mutually beneficial systems that outcompete exploitative alternatives through manifest appeal and voluntary participation. They would become "islands of coherence" that grow into attractors – ways of living that succeed because they're more resilient, sustainable, and effective.

25 Gilliland, Mike, and Euvie Ivanova, hosts. "FTP133: Jim Rutt – How to Transition to a GameB Society." *Future Thinkers*, Dec. 8, 2020, https://futurethinkers.org/jim-rutt-gameb/. Accessed Aug. 26, 2025.

Game B acknowledges requiring two to three generations and hard collaborative work on institutional problems. But building alternatives proactively seems more rational than waiting to see how current systems fare. After all, hope is not a strategy.

6 It's Been Done Before

Rutt's not alone in re-envisioning our civic pillars. Writing back in the revolutionary 1970s, Ivan Illich advocated dismantling centralized institutions – schools, hospitals, transportation systems – that had grown beyond human scale and replacing them with "convivial tools" that individuals could control directly. He proposed creating decentralized "learning webs" where people would access educational resources, exchange skills, and connect with peers without formal schooling structures. Illich envisioned tools and institutions that would "*give people tools that guarantee their right to work with high independent efficiency*" rather than creating dependency on professional experts and bureaucratic systems.[26]

Several other seminal thinkers have advocated reimagining civic institutions. This includes Jane Jacobs (urban planning – challenged conventional urban planning by advocating for neighborhood-scale, organic development),[27] E.F Schumacher (advocated for human-scale technologies and decentralized economic institutions),[28] Elinor Ostrom (first woman to win the Nobel Prize in Economics, by documenting successful community-governed alternatives to markets or state control),[29] among many others.

Each of these thinkers, like Illich, identified fundamental problems with existing institutional structures and proposed alternative frameworks centered on human needs, local knowledge, and community autonomy.

Still other writers tackled the intersection of technology and civic norms and institutions.

Kevin Kelly, whose pen we've touched on throughout the book and who is a former editor of the influential tech magazine Wired, demonstrates that technological systems create opportunities for decentralized governance and self-

26 Illich, Ivan. *Tools for Conviviality*. Harper & Row, 1973, p. 11.

27 Jacobs, Jane. *The Death and Life of Great American Cities*. Random House, 1961.

28 Schumacher, E. F. *Small Is Beautiful: Economics as if People Mattered*. Harper & Row, 1973.

29 Ostrom, Elinor. *Governing the Commons: the Evolution of Institutions for Collective Action*. Cambridge University Press, 1990.

organizing communities.[30] Yochai Benkler builds on Ostrom's "commons," analyzing how digital technologies enable new forms of collaborative production that challenge traditional economic institutions and property regimes.[31] Langdon Winner explores how technologies have inherent political qualities and can be designed to support different social arrangements,[32] while technologist and author Jaron Lanier argues for humanistic technology design and alternative economic models for data.[33]

And Arbib and Seba, in the book *Stellar* that we touched on previously, arrive at pretty much the same conclusions as Rutt's Game B.

Keep in mind, visionary idealists CAN have impact. In addition to the civil rights, suffrage, and environmental catalysts we covered just a few pages, consider Illich's impact.

While his full vision was never achieved, Illich's influence extends across several educational movements today. Homeschooling advocates frequently cite his work as a philosophical justification for education outside traditional systems.[34] Unschooling practices, which emphasize child-directed learning, draw directly from his belief that natural curiosity drives authentic education.[35] Adult education theorists have incorporated his emphasis on self-directed learning and the recognition of informal learning experiences.[36]

Visionaries can and have shaped our institutions, but they work within a construct and a paradigm. And our current paradigm has borders inscribed by energy, which we'll turn to next.

---

30 Kelly, Kevin. *Out of Control: the New Biology of Machines, Social Systems, and the Economic World.* Addison-Wesley, 1994; and Kelly, Kevin. *The Inevitable: Understanding the 12 Technological Forces That Will Shape Our Future.* Viking, 2016.

31 Benkler, Yochai. *The Wealth of Networks: How Social Production Transforms Markets and Freedom.* Yale University Press, 2006.

32 Winner, Langdon. *The Whale and the Reactor: a Search for Limits in an Age of High Technology.* University of Chicago Press, 1986.

33 Lanier, Jaron. *Who Owns the Future?* Simon & Schuster, 2013.

34 See Bartlett, Lesley, and Daniel Schugurensky. "Deschooling Society 50 Years Later: Revisiting Ivan Illich in the Era of COVID-19." *Otras Voces en Educación* 1, no. 1, 2020, pp. 1–19, and Farenga, Patrick. "Homeschooling: Creating Alternatives to Education." *Bulletin of Science, Technology & Society* 18, no. 2, 1998, pp. 127–133.

35 See McDonald, Kerry, and Peter Gray. *Unschooled: Raising Curious, Well-Educated Children Outside the Conventional Classroom.* Chicago Review Press, 2019.

36 Ogawa, Masahiro. "The Theories of Ivan Illich and Paulo Freire and Environmental Education: Re-thinking from the Standpoint of Adult and Community Education." *Japanese Journal of Environmental Education* 19, no. 3, 2010, pp. 3–29.

7 When Energy Shapes Thought

The last 150 years' economic expansion was largely driven by the extractionary exploitation of fossil fuels. Transitioning away from fossil fuels – necessary for climate stability – will likely mean some period of economic contraction. Contractions typically self-reinforce through debt defaults, job losses, and foreclosures.

This would be softened if renewable energy had the same cost per energy unit as fossil fuels. Fossil fuels achieve those cost advantages largely because of the economies of scale they enjoy from existing institutionalized processing, distribution, and use advantages.

Replacement energies can achieve the same return to us, and perhaps so much so that people will be as resistant to paying for energy as they are to paying to use a Wi-Fi hotspot now.

But that is in the future. For now, we must get there; we must somehow cross this "valley of death."

James Arbib and Tony Seba wrote *Stellar*, a very compelling 2024 book proposing tactics for making that transition.[37] I will leave it to them and others to work out the how. But here I want to draw the line connecting the dots and showing the interdependencies between thought, Technology, and energy.

The brain is an expensive organ, energy-wise, using a disproportionate amount of the calories we consume. Technology is, too, even if you don't see it. The new cloud architectures and AI require massive facilities, typically built where electricity is abundant and cheap.

I'm not requiring these datacenters to use renewable energy or power companies to provide carbon-neutral electricity for our homes and devices.

I'm saying Technology depends on energy, so the ethics of our tools are inseparable from the economics of our power. And if weaning off fossil fuels means a period in which energy costs will increase, we need to begin thinking about that. This allows us a unique opportunity to reconsider basic assumptions. Fueled by advances in Technology, where micro-generation (solar, wind, water, and other) and local storage in our homes and other edge nodes of our electrical grids become more common, the grid itself can be rethought. Transmission and distribution systems can be made less interdependent and more resilient.

This will undoubtedly shape our human systems, Technology, and how we think. The technical pathways are well-covered elsewhere; our task is awareness, alignment, and choice.

---

37 Arbib, James and Seba, Tony. *Stellar: a World Beyond Limits, and How to Get There*, 2024.

8 Turning the Page

These last few chapters map the terrain where solutions must operate – technology, community, economics, power, and energy reveal why isolated fixes inevitably fail. Each domain connects to others in complex webs of causation and consequence.

But surveying the landscape raises a deeper question: if technological disruption of thought operates across all these systems simultaneously, what does conscious response actually require? How do we move from recognizing problems to directing solutions? From understanding disruption to steering evolution?

The next chapter steps back from tactical recommendations to examining the strategic choice we face as a species. Having explored both individual practices and collective mechanisms, we can now see the full scope of what's at stake – and what's possible when we choose to become conscious architects of our own cognitive future rather than passive recipients of change designed by others.

CHAPTER 26

Conclusion: Steering the Next Revolution

To observe the scope and rapid pace of change of the last 150 years, go to Oregon in the western United States and take Route 242 to the summit of McKenzie Pass. It may have been first traversed by a human over 20,000 years ago. For millennia, this route was a Native American trail, up until as recently as the 1860s, when it saw its first white man come through the pass on horseback. 10 years later, the pass saw its first wheel when the first covered wagon came, and by 1910, the first automobile struggled over. Just 50 years later, Walter Cunningham tripped on the rough lava that makes up the ground of the pass, ripping his prototype spacesuit while training for his Apollo moon mission. Today, if you look west down the pass, you can see the 100-foot microwave tower that CenturyLink uses to transport high-speed wireless data over the mountain range.

In the last century and a half, the ancient hardy windswept hemlock trees growing on the pass witnessed a human culture, which was relatively stable for thousands of years, disrupted in a rapid succession of technological disruptions: domestication of animals, wheeled vehicles, the industrial age, space age, and now the information age.

We entered the period with, as Mumford says, "this smug Victorian belief in the inevitable improvement of all other institutions through command of the machine" and we fast forward to now, when the messages going over Qwest's microwave tower increasingly signal the world's bewildering complexity and our perception of numerous existential threats to us as a species.

All in the evolutionary nanosecond. How to navigate? (Pat Scannell, The History of Thought, Book 2)[1]

The future cannot be predicted, but futures can be invented. (Dennis Gabor (and various others))[2]

We began this book with some relatively simple questions: How do we think? How has modern technology affected our thinking? How will we think in the

1 Scannell, Patrick. *The History of Thought, Book 2: The Co-evolution of Technology and Cognition in the Human Species*, Forthcoming.

2 Gabor, Dennis. *Inventing the Future*. Secker & Warburg, 1963, p. 184.

 | HTTPS://DOI.ORG/10.1163/9789004739161-032

near future, especially in the face of rapidly accelerating technology? And, crucially, how well is our thinking serving us now – and where is that trajectory taking us?

Since then, they have expanded, slightly to include inquiries such as:

- How will human cognition change as technology rapidly accelerates?
- What new cognitive capabilities can/ should technology unlock for us?
- As we outsource thinking to technology, what should we do with our freed mental capacity?
- What should we value when technology can increasingly fulfill most of what we wish for?
- Will technology ultimately save or destroy humanity?
- How do we thrive when few, if any, fully comprehend the pace of change or the nature of the technologies we depend on?

After a decade of research spanning fields as diverse as neuroscience, anthropology, philosophy, economics, and technology, I don't arrive at simple answers. Instead, I find myself at a new, distant place that feels more like a beginning than a conclusion – with a more profound question:

Can we learn to consciously direct our own cognitive evolution before unaligned systems – technological, corporate, cultural, algorithmic, and our evolutionary legacy – reshape us according to their needs and momentum?

1 The Three Streams Revolution

The Three Streams Model reveals something remarkable about this moment in human history. Consider this metaphor: for 300,000 years, we built the human version of the standalone PC – complete with a biological operating system and cultural software. Early tools and tribal knowledge were equivalent to the era when we relied solely on floppy disks (or in my case, cassette tapes) to transfer data into or out of our early computers, before the advent of the Internet. But in the last 50 years, since the Internet's birth, and especially the last 20 years as broadband became ubiquitous, our standalone PC transformed into a networked computer. Like our computers, our cognitive abilities have gained the potential to become infinitely more powerful as part of a global network.

This isn't just a metaphor. We literally think across three interconnected streams now:

- **Our embodied brains** – the 86 billion neurons processing almost all of our cognition unconsciously, shaped by millions of years of evolution for archaic environments.

- **Our cultural networks** – the social constructs, languages, and shared beliefs that span our relationships to other people, connecting us to ideas that stretch across generations but change now at unprecedented speed.
- **Our technological extensions** – the devices, algorithms, and connected systems that have become literal extensions of our cognitive abilities, processing information and making decisions at scales no individual brain could handle.

Each stream operates at different speeds of change, subject to control by numerous actors, pursuing their own goals.

But here's what makes this moment unique: unlike the printing press or steam engine, we're not just adopting new tools. We're becoming hybrid cognitive systems where the boundaries between internal and external thinking blur beyond recognition.

If we want to know how to steer the next decade, we should consult the only long-run experiment we've ever run: deep time. Evolution is an unforgiving lab where minds, bodies, tools, and niches co-evolve under pressure. The patterns that decided winners and losers then might still govern us now. With that lens, consider the world 2 million years ago.

2 From Niche Partitioning to the Fourth Cognitive Revolution

Our species is not the pinnacle achievement of some linear progression; we are simply the last branch left on a large tree where all the other branches died. At least some of those extinctions happened because our relatives couldn't adapt fast enough to the changes they faced. And we came close to being wiped out a couple of times too.[3]

About 2 million years ago, multiple human species shared the same landscapes. The robust, small-brained, heavily-muscled *Paranthropus boisei* lived alongside our more gracile (slender) and slightly larger-brained[4] ancestors *Homo habilis* and *Homo erectus*. Each developed distinct survival strategies

3 See, for example Gee, Henry. *The Decline and Fall of the Human Empire: Why Our Species Is on the Edge of Extinction*. First US edition, St. Martin's Press, 2025.

4 This is a very simplified statement: a) we don't actually have any brains from these species, so we go off cranial volume, b) they have different body sizes, which means we have to bring in a measure called encephalization quotient (EQ), and c) there are a range of specimens within each species, with correspondingly varying brain and body sizes. That said, with all that taken into account, the statement above is easily supported.

through what biologists call "niche partitioning." They could live alongside each other without necessarily competing for resources.

Boisei evolved into specialists – powerful chewing machines with wide cheekbones and massive molars that could process the toughest plant materials. Meanwhile, *erectus* and *habilis* became ecological generalists, using their liberated hands for something revolutionary: tools.

When climates shifted and tough vegetation waned, the specialists starved while the generalists adapted with tools, broader diets, and more flexible brains.

Across deep time, our own cognition changed in stacked revolutions, that roughly summarized, can be described as:

- **Hands and tools** (~4–2 million years ago): Bipedalism set us up for manual dexterity, which unlocked tool use, scaffolding new circuitry for manipulation and planning, setting the stage for language.
- **Language and shared intentionality** (~50–40 thousand years ago): We recruit other minds at scale – coordinating goals, teaching, and transmitting norms – turning isolated problem-solvers into scalable distributed systems.
- **Writing and external memory** (last several millennia): Ideas persist beyond any brain; knowledge *ratchets* across time and distance, accumulating rapidly, and unlocking civilization and progress.

Each step shifted cognitive load: we remembered less *content* but more about *where and how* to find it. The result wasn't "more of the same" intelligence, but differently organized intelligence. And each step set the stage for the next.

That arc drops us into the present: our niche is now self-constructed and accelerating. We live inside vast technological, social, and informational systems that we must maintain, steer, and – crucially – learn to align with positive human outcomes.

3 The Megafauna Mirror (or Why I Don't Want Us to End Up Like the Mastodon)

As waves of *Homo* spread across the globe, other species had to adapt to us, and that process provides us with another relevant insight.

These species included megafauna – massive creatures like woolly mammoths and giant ground sloths, which had evolved size as their primary survival strategy. Biologists name these traits "K-selected."[5] These species thrived in stability but lived perpetually near their environment's carrying capacity (the largest population size of a species that an environment can support over the long term, given the available resources, without damaging the environment or exhausting those resources). Their slow reproduction made them exquisitely vulnerable to environmental disruption.

This approach contrasts sharply with "r-selected" species like frogs, dandelions, and mice, which reproduce rapidly and produce countless offspring. R-selected populations succeed in periods of uncertainty and change because their large numbers maintain vast behavioral and genetic diversity – like a portfolio of survival strategies. When conditions shift unpredictably, hopefully a lucky few variants will possess traits suited to the new reality, and survive.

When humans arrived with new technologies – spears, fire, and coordinated hunting – we changed the carrying capacity of ecosystems. Megafauna's previously successful adaptations became liabilities (they became attractive targets for hungry humans), and their limited populations and long lifecycles, including extended childhood/maturity, meant they could not respond.

Wherever we went around the world, megafauna went extinct. Whether we were the smoking gun or the final straw, the lesson endures: when environments transform quickly, specialization breaks and flexibility bends.

Here's the hard truth in this mirror: modern humans have made ourselves a K-selected species, pressed against the limits of a carrying capacity we've artificially inflated with technology.

We've extracted vast accumulated resources that took eons to develop – tapping millions of years of fossil fuels to support 200 years of our energy needs, and stripping remote islands of thousands of years of bird droppings for fertilizer to satisfy 150 years of our growing appetites.

We've redesigned plants and animals to extract more nutrition from our world. We've eliminated our natural population controls – killing off predators and conquering diseases that literally plagued our ancestors as recently as a century ago.

Today, massive satellite fleets cocoon the planet while fiber-optic networks, electrical grids, and supply chains weave across the land. Wireless communications saturate the air. As a species, we now bear the enormous overhead of maintaining this technological infrastructure.

Like the megafauna before us, we've specialized – but our specialization is technological complexity. We require decades of education to function in our

5 Pianka, Eric R. "On r- and K-Selection." *The American Naturalist*, vol. 104, no. 940, 1970, pp. 592–597.

niche. Our survival strategy depends on interdependent systems spanning the globe.

This is our mastodon moment: COVID exposed how finely balanced – and therefore brittle – our technologically scaffolded niche is; at carrying capacity, even one disruption can send the whole global system over the edge.

We've spun the dials on nearly every factor that shapes human experience – climate, information flow, social structures, and the basic cognitive demands of daily life.

4 The Unconscious Revolution Already Underway

The Three Streams Model reveals something else that most people don't recognize: the cognitive revolution isn't coming – it's already here, happening largely outside our awareness.

Most cognitive processing across all three streams occurs unconsciously. In your brain, parallel processing handles everything from visual recognition to emotional responses without conscious involvement. In culture, you absorb and transmit social cues and behavioral norms automatically. In technology, algorithms curate your information environment and shape your decisions through processes you never see.

This means the disruption of thought operates primarily below the threshold of conscious awareness. You experience the results – the feeling that you can't focus like you used to, the sense that the world is changing faster than you can understand – but the underlying mechanisms remain invisible.

The complexity is staggering. Your thinking now depends on interactions between that complicated brain, connecting to millions of humans (living and dead, real and fictional) through cultural transmission, and engagement with billions of connected devices processing exabytes of data. This system exhibits complex, nonlinear dynamics where small changes can cascade into transformations.

Most troubling, the values and goals embedded in each stream are becoming misaligned. Your biological inheritance optimizes for conditions that no longer exist. Your cultural environment changes too rapidly for traditional adaptation mechanisms. Your technological tools pursue objectives determined by their market forces rather than your human flourishing.

5 The Alignment Challenge (or Why I Don't Want Us to End Up Like My Dog)

In artificial intelligence research, we call this the "alignment problem" – ensuring that AI systems' goals align with human values.

But we face a larger alignment challenge: the accumulated technologies affecting all three streams of cognition.

Let me illustrate what awaits us with a personal example.

My dog enjoys having me open the door and letting him out to play in the yard, riding in the truck with the windows down, and having treats (who doesn't, right?). Maybe he knows how the door works, but just can't operate it. But he doesn't understand how to operate the truck's power windows or why I sometimes keep them closed. To him, these things are just magic and mystery. He's frustrated and confused when it's too cold to lower the windows or when he can't have treats because the vet says he needs to lose weight. I am his benevolent overlord, making decisions based on knowledge he cannot comprehend – or sometimes just because I don't feel like doing what he wants.

I don't want human life to occupy that same position relative to our technology. If we do, the *best-case* scenario is that the technology we've created becomes our benevolent overlord. The worst-case scenario is considerably darker.

Long before ChatGPT's launch, I warned that soon our technology would be smarter than us. Writing now in 2025, this saying isn't as controversial as it was back then; in fact, it is more likely to strike you as boringly self-evident. The question isn't whether this will happen, but how we prepare for it.

6 The Conscious Path Forward

As McGilchrist concludes, "*We don't need a lot more quick fixes. We need a change in the paradigm.*"[6]

The solution isn't to unplug or return to some imagined golden age. Instead, we must become conscious architects of our own cognitive evolution.

Each previous revolution forced us to rewire our cognition. When we developed complex social thinking, we had to remember who possessed that knowledge and master the skills to extract it from them. Writing demanded differ-

6 McGilchrist, Iain. *The Master and His Emissary: The Divided Brain and the Making of the Western World.* Yale University Press, 2009, p. xxi.

ent adaptations – we memorized less but learned to encode (write) and retrieve knowledge (read) from external sources.

The pattern is clear: some brain regions worked less while others worked harder. Neural pathways are strengthened in some areas and pruned in others. We learned to think differently – not worse, but differently, in ways suited to success in the new environment.

Now we stand at the threshold of the Fourth Revolution, one that will dwarf all previous transformations in scale, scope, and speed. Unlike previous revolutions that have occurred over millennia, this one can happen through our conscious choice and design.

We begin by asking critical questions:

- What capabilities should we develop as cognitive tasks become automated (what should we be thinking about when Technology can think better than us in every domain)?
- What should we focus on as our ability to outsource routine thinking expands exponentially?
- What new meta-skills should we cultivate to maintain human agency, and to promote widespread human flourishing?
- Most critically: What should we want, value, and pursue when technology can deliver everything we've previously dreamed of?

Technology-driven disruption of how you think is already underway through the unplanned actions of others. We routinely build better transportation, energy, and medicine. What's wrong with consciously applying that logic to how we think?

If the response is, "We can't, because it's too hard," then I think we are drastically underestimating what our minds can solve, given the technology available to us today.

7 The Choice before Us

Humanity stands at a crossroads unlike any we've faced in our past. The same adaptive intelligence that carried us through previous transformations can help us build a better world – or at least position ourselves to survive amid the changes rushing toward us. Our ancestors navigated similar pivotal moments (or at least some of them did; let's face it, all of them went extinct, except for us, and even we found ourselves hanging on by our fingertips at least a couple of times), but those positive outcomes to earlier transitions unfolded across millennia.

Today's transformation compresses into mere decades.

We can become conscious architects of the next phase of human cognitive evolution. We can design systems that enable widespread human flourishing rather than standing by while market forces and algorithmic optimization determine our species' trajectory.

The tools we design and build will lower cognitive noise and raise signal by default. They will help us:

- Support those disoriented by rapid change without condescension.
- Accept that multiple social operating systems may need to coexist in parallel rather than forcing one brittle monoculture.
- Build governance that is as adaptive as the technologies it steers.

We are both the players and the game's makers. Uncontrolled innovation turns technology from ally to adversary; controlled, value-aware innovation can broaden the human flourishing frontier.

8 Why I Wrote This for You

We opened with fish without a conception of water. Along the way, we met the tuna – a creature that moves faster than its muscles alone would allow, by exploiting the eddies they create.

That's my vision for the future of thought: becoming conscious of our technological environment and proactively learning to exploit it rather than merely being reactively and passively shaped by it. Just as our ancestors learned to think with fire, tools, language, and writing, we can learn to think with artificial intelligence, global networks, and cognitive augmentation systems. We can use our heads and our tech to be ahead of this curve, rather than behind it.

This doesn't mean becoming dependent on systems we don't understand. It means actively designing the cognitive technologies we want, building the cultural frameworks that support human flourishing, and developing practices that keep our brains healthy and adaptable.

The alternative is to accept the position of my dog – subject to decisions made by entities, whose goals may not align with our deepest values. Or we accept that we are the mastodon.

Now that we've traveled this path together, I hope you see your world more clearly – both as it exists, and as it could become.

More importantly, I hope you see your role not as a passive recipient of technological change, but as an active architect of conscious cognitive evolution.

That transformation begins the moment you become aware of the need for it, and accelerates when you decide to pursue it.

That's exactly why I wrote this for you.

> ...the test of a first-rate intelligence is the ability to hold two opposed ideas in the mind at the same time, and still retain the ability to function. One should, for example, be able to see that things are hopeless and yet be determined to make them otherwise.[7] (F. Scott Fitzgerald.)

---

7 Fitzgerald, F. Scott. "The Crack-Up." *The Crack-Up*, edited by Edmund Wilson, New Directions, 1945, p. 69. Originally published in *Esquire*, Feb. 1936.

APPENDIX 1

On Disruption

> Ever since *The Innovator's Dilemma,* everyone is either disrupting or being disrupted. There are disruption consultants, disruption conferences, and disruption seminars.[1] (Jill Lepore)

1 Introduction

I've written this chapter for people who might ask, "What do you mean – specifically – when you say thought is being 'disrupted'?" Originally, it was at the beginning of the book, but reviewers said that it stood in the way of getting into the crux of the book, and I agree.

However, I find it critical when evaluating someone's points to clearly and explicitly know what they mean when they invoke certain words. Especially, common words or those that could have many meanings, to different people.

"Disruption" fits both categories, so I will unpack my take on the word here.

This is also important because the fact that thought is going through disruption, in the way that I mean it, is both provocative and profound. My use of the word "disruption" is central to my argument. To understand my views on what is happening to thought, I must lay out specifically what I mean when I invoke the word "disruption."

I'm not making the case for how all people should use the word; this is simply my explanation of how I'm using it.

2 Disruption in Common Use: A Brief Literature Review

The term "disruption" is everywhere now and used in so many ways that it can be somewhat meaningless. It is, at a minimum, vague; often, it is a commonly used proxy for any change.[2]

---

1 Lepore, Jill. "The Disruption Machine". The New Yorker, Jun. 16, 2014. https://www.newyorker.com/magazine/2014/06/23/the-disruption-machine. Accessed Dec. 12. 2025.

2 Gobble, MaryAnn The Case Against Disruptive Innovation. Research-Technology Management, vol. 58(1), 2015, pp. 59–63.

HTTPS://DOI.ORG/10.1163/9789004739161-033

Our obsession with the word continues with a trend that is almost as old as humanity. It carries on in the tradition of Socrates' fear of reading, because he felt it would ruin memory and the oral tradition, and Gessner's worry about the "confusing and harmful abundance of books" in the 1500s. It is likely that every new technology of the past, such as stone tool use and metal working, was feared by the people who saw it happening, just as trains, electricity, the telephone, radio, television, and even VHS machines were met with critical concerns and resistance.

I'll begin by reviewing what other key people mean when they say disruption, and some other relevant work on similar phrases.

The word *disruption* derives from the Latin *disrumpere*, meaning "to break apart."[3] Its modern usage spans two senses: a general sense of "disturbance or problems that interrupt an event, activity, or process," and a business/technology sense of "a radical change to an existing industry or market due to technological innovation."[4]

But in today's digital marketplace, all "disruption" roads lead quickly to Clayton Christensen, who says those definitions are wrong. Christensen defines disruptive innovation as the creation of new markets through different value propositions that eventually displace established competitors.[5]

Christensen cemented this definition within a specific framework and theory of "disruptive innovation" that was introduced in a 1995 Harvard Business Review article, followed two years later by his book *The Innovator's Dilemma*. This definition and framework have been widely adopted in the technology community.

However, during and after the dot com boom, there was an accelerating use of the term disruption to describe all Silicon Valley technical innovation more broadly. Uber, for example, is often cited in the popular press as a disruptor because after only 5–6 years in business, it became worth more than the century-old company Ford and was at one time worth more than Ford and GM combined.[6]

Christensen revisited the concept in 2015, arguing that companies like Uber, while innovative, are not truly disruptive within his framework. Disrup-

---

3 "Disruption." *Online Etymology Dictionary*, etymonline.com/word/disruption. Accessed Sep. 15, 2019.

4 "Disruption." *Oxford English Dictionary*, Oxford University Press, oed.com. Accessed Jan. 3, 2020.

5 Christensen, Clayton M. *The Innovator's Dilemma: When New Technologies Cause Great Firms To Fail*. Harvard Business Review Press, 2013, pp. xv–xvi.

6 Isaac, Mike. "Uber Is Said to Be Raising Money at a $62.5 Billion Valuation." *The New York Times*, Dec. 4, 2015.

tive innovations, he explains, "begin by successfully targeting overlooked segments, gaining a foothold by delivering more-suitable functionality – frequently at a lower price," or by creating a "new-market foothold." Incumbents, focused on more profitable segments, often fail to respond, and "entrants then move upmarket, delivering the performance that incumbents' mainstream customers require. When mainstream customers start adopting the entrants' offerings in volume, disruption has occurred."[7]

Uber did not target an overlooked segment and move upmarket; rather, Uber developed a sustaining innovation that represented only an incremental improvement in the taxi marketplace. They didn't start with an overlooked segment or by creating a new marketplace, so they don't count, according to Professor Christensen.

The National Research Council defines disruptive technology as one that produces "sudden and unexpected effects."[8] This seems more in line with the popular use of the term. Why did the NAS care about disruption? Because they build models of human systems and noted *"All forecasting methodologies depend to some degree on the inspection of historical data. However, exclusive reliance on historical data inevitably leads to an overemphasis on evolutionary innovation and leaves the user vulnerable to surprise from rapid or nonlinear developments."*[9]

In this 2010 report, *Persistent Forecasting of Disruptive Technologies,* NAS added caveats about disruption – *"A methodology that can forecast disruptive technologies must overcome the evolutionary bias and be capable of identifying unprecedented change. A disruptive event often arrives abruptly and infrequently and is* therefore particularly hard to predict using an evolutionary approach. The technology that precipitates the event may have existed for many years before it has its effect, and the effect may be cascading, nonlinear, and difficult to anticipate."[10] (emphasis mine)

In my defense and intelligence community work, we sometimes refer to the events leading to and possibly signaling an attack or bad event as "left of boom" because of our Western technique of reading left to right. We can't react to the aftereffects of an unwanted event ("right of boom"), but we must

7 Christensen, Clayton M., Michael E. Raynor, and Rory McDonald. "What is Disruptive Innovation." *Harvard Business Review*, vol. 93.12, 2015, pp. 44–53.

8 National Research Council. *Persistent Forecasting of Disruptive Technologies.* National Academies Press, 2010, p. 14.

9 National Research Council. *Persistent Forecasting of Disruptive Technologies.* National Academies Press, 2010, p. 15.

10 National Research Council. *Persistent Forecasting of Disruptive Technologies.* National Academies Press, 2010, p. 16.

be able to pick up the signals and intelligence that might show that it could happen, and ideally prevent it before it occurs. This is really hard.

Here, NAS is discussing the ability to sense "left of boom" on something Taleb terms Black Swan events: infrequent, unpredictable based on historical records, and with complex effects. Also, NAS gives us the reason to be specific and rigorous in our use of terms here by showing the utility of the definition: we need to distinguish between the events that are informed by history, and those events that may appear in our blind spots, in which our experience and track record don't prepare us to see.

But their definition isn't a definitive literature review or a pat answer to what I mean when I say thought is being disrupted; it simply provides an authoritative alternative to Christensen, and this gap supports our motivation for pursuing an explicit definition.

Fortunately, we don't just have these three choices (Christensen, popular, or NAS); the literature on "disruption" and related terms is actually robust and diverse, spanning many disciplines.

Take related terms, for example. Lundgaard and Rosenstand adopted a nomenclature in a 2019 paper that distinguishes between:

- "transformation" – how novel technologies open up new action possibilities for entrepreneurs and consumers; as they summarize, disruptive innovations "typically result from new scientific discoveries or engineering inventions" and are based on a "new knowledge base";[11] transformations uncover new possibilities.[12][13]
- "radical innovation" – how such technologies result in new-to-the-world products and services; these contrast sharply with incremental innovations that originate from improvements to well-understood and already adopted technologies.[14] Radical innovations offer new-to-the-world value. The uniqueness and newness of radical innovations may not be a good thing;

11 Lundgaard, Stine Schmieg, and Claus Andreas Foss Rosenstand. *Investigating Disruption: a Literature Review Of Core Concepts Of Disruptive Innovation Theory*. Aalborg University Press, 2019, pp. 7–8.

12 Danneels, Erwin. "Disruptive Technology Reconsidered: a Critique And Research Agenda." *Journal of Product Innovation Management*, vol. 21.4, 2004, pp. 246–258.

13 In this definition, there is a distinction between invention (which I would say create new techniques), and the innovation, resulting from exploitation and commercialization (which I might term "technology" – the diffusion of the technique).

14 Slater, Stanley F., Jakki J. Mohr, and Sanjit Sengupta. "Radical Product Innovation Capability: Literature Review, Synthesis, And Illustrative Research Propositions." *Journal of Product Innovation Management*, vol. 31.3, 2014, pp. 552–566.

it's common for radical ideas not to gain traction and not have a disruptive impact on their markets.[15]

- "(dis)continuity" – the changes in the trajectories of markets or industries that result from the emergence of radical innovations.
- "disruptiveness" – the effects that certain kinds of innovations have on incumbent businesses.

Note that the level of focus is an important distinction to tease out; some words or definitions focus on the impact at the company, sector, or industry level, where other academics are looking at the broader level of society when they discuss disruption and related terms.

Yet another useful inquiry comes from a little-cited paper on MP3s. In their review of the major changes within the music industry in the late 90s and 00s, Kai Riemer and Robert B Johnston see two critical faults of the value in using Christensen's narrow and specific disruption model to explain the now classic disruption of the music industry.[16] The first is that Christensen's model looks narrowly at the incumbents' reactions, and it doesn't provide whole industry explanations. Second, Christensen assumes market and customer demands as outside the scope of his model (exogenous), and so any transformation of an industry caused by wholesale changes in what the market wants couldn't be explained by Christensen.

Instead, Riemer and Johnston find that Thomas Kuhn provides a better way to understand how the music industry changed in the face of technological and cultural innovation.

Kuhn challenged the way we understand science with his 1962 book *The Structure of Scientific Revolutions*. In what has become one of the most cited books of all time, Kuhn's central argument is that science proceeds for periods of time in which there is a continuous "paradigm" – "*universally recognized scientific achievements that, for a time, provide model problems and solutions for a community of practitioners.*"[17] These paradigms dictate the nature of the questions the field will pursue, how they are to be investigated and interpreted, but also, as Kuhn developed in later work, the skills, beliefs, goals,

15 Hill, Charles WL, and Frank T. Rothaermel. "The Performance of Incumbent Firms in The Face of Radical Technological Innovation." *Academy of Management Review,* vol. 28.2, 2003, pp. 257–274.

16 Riemer, Kai, and Robert B. Johnston. "Disruption As Worldview Change: A Kuhnian Analysis of the Digital Music Revolution." *Journal of Information Technology,* vol. 34.4, 2019, pp. 350–370.

17 Kuhn, Thomas S. The Structure of Scientific Revolutions, 3rd edition. University of Chicago Press, 1996, p. 10.

and how scientists make sense of the world.[18] He is careful to point out that these elements compose a holistic "worldview." They are not just the sum of some individual shared assumptions and beliefs, or, as some would say, shared activities and technological commitments;[19] these worldviews are the source of meaning for core concepts; they provide the core elements of a field and direct the attention of the people in that paradigm.

But every so often, Kuhn points out, anomalies occur, which create a crisis, and confidence is lost in the paradigm. New paradigms rise and are adopted. Critically, Kuhn points out that research done under earlier paradigms can't be compared with science done under new ones: they are "incommensurable." This was controversial because it challenged the positivist tradition's idea that science builds on a steady accumulation of knowledge and that newer ideas are better approximations of truth than older ideas. So instead of a steady linear process of "normal" science, Kuhn sees periodic revolutions.

Riemer and Johnston believe this model better explains the digital music revolution: the MP3 format and platforms like Napster were anomalies. Combined with the cultural shift among customers who transitioned from owning high-quality music to sharing moderate-quality music, this created a "crisis" in the industry. Kuhn explained that in a paradigm crisis, the participants can't understand what's going on because their old model doesn't explain it. These revolutions can be exciting opportunities for change for some, but for those rooted and trained in the old paradigms, they are existential crises and even feel like personal failures. Because the two schools of thought are incompatible, they don't build on each other; the new one replaces the old. In Kuhn's words, this is a "paradigm shift," which is "a reconstruction of the field from new fundamentals."[20] This makes more sense to Riemer and Johnston in the context of the digital revolution in the music industry than what Christensen's model helps us to understand.

Specifically, they point out:

- Post-hoc explanations are often told through the lens of the new paradigm; because of incommensurability, many actions done under and justified by the old paradigm are often left out, so post-hoc explanations are often incomplete.

18 Kuhn, Thomas S. "Second Thoughts on Paradigms". In: Suppe, Frederick (ed.) *The Essential Tension.* University of Chicago Press, 1977, pp. 293–319.

19 Dosi, Giovanni. "Technological Paradigms and Technological Trajectories: a Suggested Interpretation Of The Determinants and Directions Of Technical Change." *Research Policy*, vol. 11.3, 1982, pp. 147–162.

20 Kuhn, 1962, p. 85.

- Incumbents can't sense the beginning of the crisis; even if they could, it would only raise confusion, since they have no way of dealing with it. So, *solutions to resolve crises must come from outside the established paradigm.*
- During a crisis, old and new compete, but people operating under old models can't understand the fight. One is not better than another; it's just that one is doing business as usual, while the new model plays by a different set of rules, one that the old model doesn't understand.

Building on Kuhn and later disruptive innovation scholarship by Dosi and others, Riemer and Johnston propose three points to overcome the challenges above:

1. **Disruption results from a change in worldview:** Any attempt to understand discontinuous innovation must consider the changing worldview that goes along with the disruption. (They note that most disruptive theories, including Christensen's, fail to do this).
2. **Disruption as worldview change is unpredictable:** Because of the incommensurability, existing paradigms can't sense and predict pending changes. "*The best we can achieve is to try to theorize the nature of technology-enabled industry crises and how incumbents might cope with them tactically.*"[21]
3. **Disruption becomes retold in hindsight:** Post-hoc descriptions cannot be relied on.

Helpfully, Riemer and Johnston provide practical implications for those potentially facing disruption:

- **Understand paradigm shift:** By understanding the mechanisms of potential disruption, incumbents are better positioned to understand it as it occurs.
- **Scout the fringes:** Since a new paradigm forms outside of traditional technologies or business principles, incumbents are best served by staying actively engaged with projects that would otherwise seem "nonsense" or "crazy."
- **Be experimental:** By creating projects and businesses independently budgeted that may "break the rules" and experiment with new technologies without having to justify themselves to the old paradigm, incumbents can begin to sense the potential of new technologies.
- **Give novices a voice:** During a crisis, incumbents should actively recruit voices from outside the established paradigm, who might imagine alternative ways of doing business that incorporate emerging new technologies.

---

21 Gans, Joshua. "Disruptive Innovation and the Theory of the Firm." *MIT Sloan Management Review*, vol. 57, no. 3, 2016, pp. 71–78.

Christensen's prominence in the innovation literature causes him to draw criticisms from many corners. In a 2016 Techcrunch article, Alex Moazed and Nicholas Johnson point out that Christensen's work focused on companies like Hitachi, Sony, and Quantum Corporation.[22] But Uber isn't a company, as much as it is a platform, according to Moazed and Johnson. While they don't explicitly say Christensen is trapped within a company paradigm, and technology has moved onto the platform paradigm. Christensen's focus is on why companies fail. His unit of analysis is existing companies, and this biases his conclusions. Christensen himself seems to confirm this when he famously said, *"the prediction of the theory would be that Apple won't succeed with the iPhone,"* adding, *"History speaks pretty loudly on that."*[23]

Lepore's 2014 article takes Christensen to task on this. *"Disruptive innovation is a theory about why businesses fail. It's not more than that. It doesn't explain change."*[24] She points out that while he intended it to be predictive, *"Christensen's Disruptive innovation can reliably be seen only after the fact.. Transfixed by change, it's blind to continuity. It makes a very poor prophet."* Riemer and Johnston have pointed out above that post-hoc explanations are incomplete and faulty. Perhaps this is why Christensen's $3.8-million Disruptive Growth Fund was quietly liquidated after less than a year in business, losing 64 cents on the dollar.[25] Not all disruption is constructive.[26]

While companies seek sustaining innovations, platforms create new markets differently, according to Moazed and Johnson. Companies like Airbnb, Alibaba, Etsy, Facebook, Google, Instagram, Pinterest, WeChat, and YouTube launch *"disruptive innovations that build new networks, communities and marketplaces."*[27] Once they have established a strong network around their original service, these companies *"can easily tap into that network to unlock new*

---

22 Moazed, Alex and Nicholas L. Johnson. "Why Clayton Christensen Is Wrong About Uber and Disruptive Innovation", *Techcrunch*, Feb. 27, 2016, https://techcrunch.com/2016/02/27/why-clayton-christensen-is-wrong-about-uber-and-disruptive-innovation/. Accessed Dec. 13, 2025.

23 Mcgregor, Jena. "Clayton Chistenson's Innovation Brain". *Bloomberg*, Jun. 18, 2007. https://www.bloomberg.com/news/articles/2007-06-15/clayton-christensens-innovation-brainbusinessweek-business-news-stock-market-and-financial-advice. Accessed Dec. 13, 2025.

24 Lepore, Jill. "The Disruption Machine". The New Yorker, Jun. 16, 2014. https://www.newyorker.com/magazine/2014/06/23/the-disruption-machine. Accessed Dec. 12, 2025.

25 "Don't Judge a Fund By the Book". *Bloomberg*, Jun. 25, 2001, https://www.bloomberg.com/news/articles/2001-06-25/dont-judge-a-fund-by-the-book. Accessed Dec 13, 2025.

26 Thanks to Vint Cerf for this observation on this section.

27 Christensen, Clayton M. *The Innovator's Dilemma: When New Technologies Cause Great Firms to Fail.* Harvard Business School Press, 1997, p. xviii.

customer groups and create new markets. Networks are extensible in a way that traditional supply chains are not."[28]

Christensen's original research was based on linear business (where value and information flow primarily in one direction – from the supplier to the consumer, according to Moazed and Johnson), such as disk-drive manufacturers and steel mills. Perhaps trapped in the company paradigm, Christensen couldn't see how a platform like Uber could leverage a disruptive innovation in the taxi market as a foundation for nonlinear growth by creating and expanding new markets in food delivery services or other sectors. Platforms differ from traditional linear businesses in that they "don't own or control supply the way a linear business does" and instead "have at least two distinct customer groups: producers and consumers."[29]

Thales Teixeira, a former Harvard Business School professor, and his co-author Greg Piechota point out that disruption is not always a new technology or market, but often a new business model. He suggests that startup businesses can succeed not just with new technologies, but by targeting one part of the customer value chain, decoupling it from the rest, and doing it better than existing firms in an industry. Teixeira and Piechota note that discussions of disruption often focus on technologies that are exciting or attention-grabbing, rather than on how customer value is actually created and captured.[30]

Sephora, for example, is a retailer in the cosmetic industry. The customer value chain for cosmetics involves the customer visiting the store, sampling products, selecting the right one, making a purchase, bringing it home, using it, and then replenishing it. Birchbox is an online firm disrupting this model by focusing solely on the sampling aspect: for a monthly subscription fee, it sends you a variety of samples every month. You can buy it anywhere you want. Teixeira and Piechota point out that many incumbent firms feel they are being "disrupted" by new technologies or by startup companies, but the disruption is actually just customers' changing behavior. They say, "*The changing needs and wants and behaviors of customers are actually the root cause of*

28 Parker, Geoffrey G., Marshall W. Van Alstyne, and Sangeet Paul Choudary. *Platform Revolution: How Networked Markets Are Transforming the Economy.* W. W. Norton & Company, 2016, p. 21.

29 Moazed, Alex, and Nicholas L. Johnson. *Modern Monopolies: What It Takes to Dominate the 21st Century Economy.* St. Martin's Press, 2016, p. 31.

30 Teixeira, Thales S., and Greg Piechota. *Unlocking the Customer Value Chain: How Decoupling Drives Consumer Disruption.* Currency, 2019.

this huge shift away from large retailers into other startups and other online retailers."[31]

Long before Christensen, Teixeira, and the recent research reviewed above, Dick Foster wrote a book called *Innovation: The Attacker's Advantage* in 1986. In it, he described a phenomenon he called "*an age of discontinuity,*"[32] a phrase he adopted from Peter Drucker. Leaning heavily on Kuhn, Foster predicted that established firms face "*technological discontinuities*"[33] where the market abruptly shifts, and existing competitive advantages will become worthless. He cites the example of how electronic cash registers overtook traditional cash registers, increasing their market share from 10% to 90% over the four years between 1972 and 1976. NCR, the market leader for traditional cash registers, didn't see the disruption coming, and instead of growing their business, they experienced major losses and layoffs. "*I will forever rue the day I didn't call it 'disruption'*" he has said.[34] Among Foster's tips for firms facing disruption: just being aware of the possibility of a technological shift.

So far, we've been discussing the term "disruption" from the perspective of business and Kuhn's review of science. But the field of biology offers us a similar model: punctuated equilibrium.

The theory of punctuated equilibrium was first put forth in 1972 by paleontologists Niles Eldredge and Stephen Jay Gould.[35] They were looking into speciation – the divergence between species, through effects like geographic isolation or separation. They concluded that the prevailing interpretation of Darwin's work was wrong: species do not adapt gradually. For most biological species, life is characterized by remarkable stability across time, often despite extreme climatic changes such as glacial cycling.[36] Gradual changes are rare, according to Eldredge and Gould. But the relatively sudden appearance of new species had been documented as far back as 1954 by Ernst Mayr.[37] Mayr felt

31 Teixeira, Thales S., and Greg Piechota. *Unlocking the Customer Value Chain: How Decoupling Drives Consumer Disruption.* Currency, 2019, p. 4.

32 Foster, Richard N. *Innovation: The Attacker's Advantage.* Summit Books, 1986, p. 36.

33 Foster, Richard N. *Innovation: The Attacker's Advantage.* Summit Books, 1986, p. 36.

34 Fox, Justin. "The Disruption Myth" *The Atlantic*, Oct. 2014, https://www.theatlantic.com/magazine/archive/2014/10/the-disruption-myth/379348/. Accessed Dec. 13, 2025.

35 Eldredge, Niles and Stephen J. Gould "Punctuated Equilibria: an Alternative To Phyletic Gradualism", 1972, In T.J.M. Schopf, ed., *Models in Paleobiology.* Freeman Cooper. pp. 82–115. Reprinted in N. Eldredge Time Frames. Princeton University Press, 1985, pp. 193–223.

36 Cronin, Thomas M. "Speciation and Stasis in Marine Ostracoda: Climatic Modulation of Evolution." *Science*, vol.227.4682, 1985, pp. 60–63.

37 Mayr, Ernst "Change Of Genetic Environment and Evolution" In J. Huxley, A. C. Hardy and E. B. Ford. Evolution as a Process. Allen and Unwin, 1954, pp. 157–180.

that large populations diluted out mutations and adaptations, even the beneficial ones. He concluded that *"Rapidly evolving peripherally isolated populations may be the place of origin of many evolutionary novelties."* Building on this, Eldredge and Gould make the case that what seems to be a gradual evolution over time is the result of several pulses of rapid change averaged together.

The theory has been contentious. A 2006 paper in the prestigious journal Science, for example, examined the DNA of 122 species across a wide range of taxonomies (e.g., fungi, plants, animals), and concluded that punctuated equilibrium events or pulses accounts for a significant number of speciation events (22%) but the majority (78%) presumably occurred over more extended or gradual timescales.[38]

The concept of punctuated equilibrium has been adopted outside of biology, into other fields. In 2012, Policy Studies Journal, for example, dedicated a whole issue to Punctuated Equilibrium Theory (PET). In it, scholars discussed the dynamics of apparently stable politics ("gridlock"), incremental policy change, and a common occurrence of major policy changes.[39] Drawing on fields like complexity theory, models were developed that saw *"policymaking as a continual struggle between the forces of balance and equilibrium, dominated by negative feedback processes, and the forces of destabilization and contagion, governed by positive feedback processes. In the former, a disturbance is met with countervailing actions in a thermostatic-type process."*[40] In other fields, PET is applied to technology diffusion,[41] congressional budgeting,[42] and organizational transformation.[43]

---

38 Pagel, Mark, Chris Venditti, and Andrew Meade. "Large Punctuational Contribution of Speciation To Evolutionary Divergence At The Molecular Level." *Science,* vol. 314.5796, 2006, pp. 119–121.

39 Jones, Bryan D., and Frank R. Baumgartner. "From There To Here: Punctuated Equilibrium To The General Punctuation Thesis To A Theory Of Government Information Processing." *Policy Studies Journal, vol.* 40.1, 2012, pp. 1–20.

40 Baumgartner, Frank R., Bryan D. Jones, and John Wilkerson. "Studying Policy Dynamics." *Policy Dynamics,* ed. Frank R. Baumgartner, and Bryan D. Jones. University of Chicago Press, 2002, pp. 29–46.

41 Loch, Christoph H., and Bernardo A. Huberman. "A Punctuated-Equilibrium Model Of Technology Diffusion." *Management Science,* vol.45.2, 1999, pp. 160–177.

42 Robinson, Scott E., and Floun'say R. Caver. "Punctuated Equilibrium And Congressional Budgeting." *Political Research Quarterly,* vol. 59.1, 2006, pp. 161–166.

43 Romanelli, Elaine, and Michael L. Tushman. "Organizational Transformation As Punctuated Equilibrium: An Empirical Test." *Academy of Management Journal,* vol. 37.5, 1994, pp. 1141–1166.

A provocative 2009 article in the Harvard Business Review challenges punctuated equilibrium from a different angle. The paper was not written by academics, but by technology executives John Hagel III (a longtime Silicon Valley icon, who had leading roles at Atari, McKinsey, and Deloitte), John Seeley Brown (former Chief Scientist of Xerox Corporation and the director of its Palo Alto Research Center), and Lang Davison (partner and leader at McKinsey, and author of several technology industry books). They ask the stark question *"what if the historical pattern – disruption followed by stabilization – has itself been disrupted?"*[44]

They point out that economies typically stabilize after the introduction of new technologies. Carlota Perez documented these stabilizations in her book *Technological Revolutions and Financial Capital*, presenting them as a function of infrastructure. Steam engines, electricity, and telephone systems require massive infrastructures; once those infrastructures are built, companies and models adapt to the new platforms.

But Hagel, Brown, and Davis, say:[45]

> We now face something entirely different. Today's core technologies – computing, storage, and bandwidth – are not stabilizing. They continue to evolve at an exponential rate. And because the underlying technologies don't stabilize, the social and business practices that coalesce into our new digital infrastructure aren't stabilizing either. The economic disruptions which in the past were concentrated around the infrequent deployment of new infrastructures now erupt on a continuing basis, driven by the rapidly evolving capabilities of our digital infrastructure. This instability has been further magnified by a long-term global trend towards liberalization of economic activity, systematically removing regulatory barriers to entry and barriers to movement. (Hagel, Brown, and Davis, 2025).

The article concludes that punctuated equilibrium is being disrupted by multiple cycles of accelerating change, to which we never fully adapt before the next cycle kicks in. They call it *"the mother of all disruptions, a big shift into a world without equilibrium, one that will continue to shift rapidly. A world in which extreme events, such as the ongoing financial turmoil across global mar-*

44 Hagel III, John, et al. *"The New Reality: Constant Disruption." Harvard Business Review*, Jan. 17, 2009, hbr.org/2009/01/the-new-reality-constant-disruption. Accessed Dec. 5, 2025.

45 Perez, Carlota. *Technological Revolutions and Financial Capital*. Edward Elgar Publishing, 2003.

kets, become increasingly likely. A world of shifting product economics, and increasing volatility in brand equity, share values, and commodity prices."[46]

3 My Take on the Term "Disruption"

The disruption I see is a massive change in how we think, which we've discussed throughout this book. It is non-linear and abrupt and can't be predicted based on historical precedents. My use of the term disruption is equivalent to a Kuhnian paradigm shift. It's a stronger use than the National Academy of Science definition, because it is not simply something that is difficult to forecast and that "*triggers sudden and unexpected effects.*"

A key difference is that it can't be understood with our existing worldviews, or from within our traditional specialized perspectives.

Riemer and Johnston's recent but little-cited paper on the music industry offers profound insights into how we might begin to deal with the disruption of thought.

The difference between understanding thought in the last century and thought in the next will be at least as profound (and far more complex) than the differences between seeing Apple as a platform versus Apple as a company.

Disruption will occur across the whole of thought (macro level), as well as within different focus areas of thought (micro level), just as BirchBox decoupled cosmetic samples from the cosmetic customer value chain.

This disruption will not result from any single technology, market, model, culture shift, or change in our behavior; rather, it will occur in response to major changes in all of these areas. And Hagel et al. make an important distinction in terms of how the dynamics of our current technological self-constructed niche are far different than what has come before. Rather than some complex model of feedback processes and countervailing forces fighting it out to a draw in some closed environment governed by thermostatic-type stabilizing forces, we are entering a period of accelerating chaos and highly open systems that make the organization of human systems over the last century look like child's play. We will no longer have a period in which we adapt to each change; the changes themselves will evolve before we can do that.

---

46 Hagel III, John, et al. "*The New Reality: Constant Disruption.*" *Harvard Business Review*, Jan. 17, 2009, hbr.org/2009/01/the-new-reality-constant-disruption. Accessed Dec. 5, 2025.

This disruption is not an academic one. When Kuhn is talking about paradigm shifts in science, he is referring to major changes to how we *see* and *understand* the world. He is not implying that the world itself changes. My research leads me to believe that changes in our thinking will literally affect the world, just as McGilchrist said we've created a society biased toward activity centered on the left hemisphere of our brain, and that, in turn, has created a lock-in that promotes incentives and behavior that shapes our society even further in this direction.

Earlier, I mentioned "massive"; in our highly scaled and connected world, "massive" can be a matter of perspective. A better descriptor may be that this disruption will be "existentially profound" – where punctuated equilibrium posits that relatively sudden events might change the trajectory of a species, the disruption that I envision can shape the future of all living things on the planet.

While we can't predict it using traditional specialized tools from any one discipline, if we use multiple lenses and combine them differently, in ways that might seem crazy or too complicated, we can begin to sense the change and see the changing fundamentals. We can heed Foster's advice *to just* being aware of the possibility of a technological shift, and note his concern that during these shifts, existing competitive advantages will become worthless. We may not begin to describe the specific consequences of this transformation, which NAS said may be cascading, nonlinear, and difficult to anticipate. However, we might "scout the fringes," experiment, and begin to cast about for new voices or ideas that can begin to help us *understand* the change.

This change will not occur in pockets of isolated populations – precisely because of our highly connected societies, it will propagate widely.

For these reasons and more, the disruption that cognition will undergo meets Lundgaard and Rosenstand's definition of "transformation" because it will create new possibilities for people, from a new knowledge base; it will be "radical" because it will result in new-to-the-world products and services; and it will be discontinuous because it will change the trajectories of markets or industries.

It is in this sense that I use the term "disruption" in this book.

APPENDIX 2

Suggested Reading List

This collection maps intellectual landscapes that I've found relevant, rather than establishing a definitive/exhaustive canon. I've focused primarily on books to keep the list concise. Soon, I plan to expand this into an online resource where others can contribute and curate other relevant articles, papers, essays, and additional materials.

These writings reveal (to me) that technology is not just mere tools but is a dynamic force reshaping human experience – rewiring our cognition, social structures, and collective potential.

Think of this as an interconnected tapestry, though I've organized it into digestible threads for easier exploration. Each work may span multiple categories, reflecting the complex and transdisciplinary web of technological thought, but I've only listed (with some exceptions) each within a single category for conciseness. Keep that in mind if you are looking for material within a single topic.

This is a big patch, and it's speeding up, immeasurably, across the whole front.

The breadth of these authors is represented within this initial eclectic first list of thinkers – these are the ones I refer back to at least monthly, if not daily, across this huge swathe of topics:

1 Most impactful to me, across all categories

- *The Technological Society* – Jacques Ellul (technological systems and social control)
- *The Human Evolutionary Transition* – Magnus Enquist, Stefano Ghirlanda (human species development)
- *Master and His Emissary* – Iain McGilchrist (brain hemispheres and cognitive processing)
- *Simulacra and Simulations* – Jean Baudrillard (media theory and representation)
- *Factfulness* – Hans Rosling (global progress and data-driven perspective)
- *The Myth of the Machine* – Lewis Mumford (technology's impact on human civilization)
- *Technics and Civilization* – Lewis Mumford (philosophical implications of technology)

 HTTPS://DOI.ORG/10.1163/9789004739161-034

- *Mechanization Takes Command* – Siegfried Giedion (historical analysis of technological systems)
- "Playground" – Richard Powers (amazing novel on AI, friendship, environment)
- *50 Years Hence* – essay, Winston Churchill (future technological predictions)
- *Too Big to Know* – David Weinberger (networked knowledge systems)
- *Analogia* – George Dyson (emergent technological consequences)
- *Future Shock* – Alvin Toffler (societal adaptation to technological change)
- *The Age of Surveillance Capitalism* – Shoshana Zuboff (technology and power structures)
- *The Singularity Is Near* – Ray Kurzweil (technological singularity and future predictions)
- *Infinite Jest* – David Foster Wallace (postmodern exploration of media and culture)
- *Growth* – Vaclav Smil (technological and economic development)
- *Energy* – Vaclav Smil (global energy systems and transformation)
- *The Society of Mind* – Marvin Minsky (computational theory of consciousness)
- *How the Mind Works* – Steven Pinker (cognitive science and computational theory)
- *The Modularity of Mind* – Jerry Fodor (cognitive architecture and mental processes)
- *The Architecture of the Mind* – Peter Carruthers (cognitive structure and processes)
- "Overflowing Brain" – Torkel Klingberg (explores limits of working memory in face of information overload)
- *Behave* – Robert Sapolsky (biological foundations of human behavior)
- *Gödel, Escher, Bach: an Eternal Golden Braid* – Douglas Hofstadter (recursion in complex systems and cognition)
- "How Emotions are Made" – Lisa Feldman Barrett (emotional basis for cognition)
- *The Prehistory of the Mind* – Steven Mithen (evolutionary origins of cognition)
- *A Mind So Rare* – Merlin Donald (consciousness and cultural evolution)
- *The Camel and the Wheel* – Richard Bulliet (technological adaptation in ancient societies)
- *Catching Fire* – Richard Wrangham (human technological and dietary evolution)

- *The Mind in the Cave* – David Lewis-Williams (cognitive evolution of human consciousness)
- *How to Think Like a Neanderthal* – Thomas Wynn, Frederick Coolidge (prehistoric cognitive processes)

2 Foundational Thinkers on Technology and Society

- *As We May Think* – Vannevar Bush (visionary essay on information organization)
- *Deschooling Society* – Ivan Illich (critique of institutionalized education)
- *Technology and the Character of Contemporary Life* – Albert Borgmann (philosophical analysis of modern technology)
- *The Question Concerning Technology* – Martin Heidegger (fundamental philosophical examination of technological essence)
- *Ethics in the Real World* – Peter Singer (technological ethics)
- *Autonomous Technology* – Langdon Winner (technics out of control in political thought)
- *What Things Do* – Peter-Paul Verbeek (philosophical reflections on technology and design)
- *The Atom Bomb and the Future of Man* – Karl Jaspers (technological existential implications)
- *We Have Never Been Modern* – Bruno Latour (critical analysis of technological modernity)
- *Technics and Praxis* – Don Ihde (philosophical exploration of technology)
- *Sapiens* – Yuval Noah Harari (historical human development)
- *The Technological Bluff* – Jacques Ellul (critique of technological illusions)

3 Cognitive Science and the Mind

- *The Construction of Reality* – Peter Berger, Thomas Luckmann (sociology of knowledge)
- *The Mind Doesn't Work That Way* – Jerry Fodor (critique of computational mind models)
- *Consciousness Explained* – Daniel Dennett (philosophy of consciousness)
- *The Emotion Machine* – Marvin Minsky (AI and human emotion)
- *The Sensory Order* – F.A. Hayek (theory of perception and cognition)
- *The Brain that Changes Itself* – Norman Doidge (neuroplasticity)
- *Incognito* – David Eagleman (unconscious brain processes)

- *Self Comes to Mind* – Antonio Damasio (consciousness and neurological systems)
- "The Embodied Mind" – Francisco J. Varela, Evan Thompson, and Eleanor Rosch (seminal book on embodied cognition)
- "How Things Shape the Mind" – Colin Renfrew and Lambros Malafouris (treatise on embodied, extended and distributed cognition)
- "A Whole New Mind" – Daniel Pink (exploration of the need for more right-brain creative thinkers)
- "The Spike" – Mark Humphries (follows journey of 1 single spike within the brain)
- "Thinking Fast and Slow "– Daniel Kahneman (seminal book on how the brain processes)
- "The Future of the Brain" – Gary Marcus, Jeremy Freeman (eds) (explores frontiers of cognitive sciences with fields leaders)
- "Brain and Culture" – Bruce E. Wexler (explores neurological effects of cultural factors)
- "Brain Sense" – Faith Hickman Brynie (interaction of the brain and senses)
- "Subliminal" – Leonard Mlodinow (unconscious cognitive processes)
- "The Compass of Pleasure" – David J. Linden (how brain responds to pleasure)
- "Cognition in the Wild" – Edwin Hutchins (distributed cognition in task/occupational settings)
- "Quantifying Human Information Processing" – Dennis K. McBride, Dylan Schmorrow (eds)
- *The Feeling of What Happens* – Antonio Damasio (consciousness and emotion)
- *The Conscious Mind* – David Chalmers (philosophical exploration of consciousness)
- *Embodied Cognition* – Lawrence Shapiro (cognitive theory)
- *Being There* – Andy Clark (cognitive philosophy)
- *Predictably Irrational* – Dan Ariely (cognitive biases)
- *Noise: A Flaw in Human Judgment* – Daniel Kahneman, Olivier Sibony, Cass Sunstein (decision-making processes)
- *Surfing Uncertainty* by Andy Clark – Predictive processing and intelligence
- On Intelligence – Jeffrey Hawkins
- *A Thousand Brains* – Jeffrey Hawkins
- *Social Cognition* – Shelley E. Taylor and Susan Fiske (survey on field of social cognition)

4 Evolution and Human Cognition

- *Evolving Ourselves* – Juan Enriquez, Steve Gullans (biotechnological impact on evolution)
- *Thought in a Hostile World* – Kim Sterelny (cognition as a threat response)
- *Cultural Origins of Human Cognition* – Michael Tomasello (bridges evolutionary psychology and cultural psychology)
- *The Case against Reality* – Donald D. Hoffman (distorted view of reality as perceived by brain)
- *The Human Age* – Diane Ackerman (examines the anthropocene)
- *The Prehistory of the Mind* – Steven Mithen (bridges evolutionary psychology and cognitive archaeology)
- *The Construction* of Reality – Michael Arbib, Mary Hesse (how the brain constructs reality through social and physical interactions)
- *Darwin Comes to Town* – Menno Schilthuizen (rapid biological evolution)
- *How Things Shape the Mind* – Lambros Malafouris (material engagement theory)
- *The Ape That Spoke* – John McCrone (language evolution)
- *Tools, Language & Cognition in Human Evolution* – Kathleen Gibson, Tim Ingold (tool use and cognitive development)
- *Not by Genes Alone* – Peter J. Richerson, Robert Boyd (cultural evolution)

5 The Digital Age and Its Cognitive Impact

- *Reclaiming Conversation* – Sherry Turkle (technology's impact on human interaction)
- *Born Digital* – John Palfrey, Urs Gasser (digital natives and technological transformation)
- *Alone Together* – Sherry Turkle (technology and human relationships)
- *Stolen Focus* – Johann Hari (how technology erodes attention spans)
- *The Hacking of the American Mind* – Robert Lustig (dopamine, social media, and tech addiction)
- *Irresistible* – Adam Alter (psychology of habit-forming technology)
- *The Great Disruption* – Francis Fukuyama (technological and social change)
- *This Will Change Everything* – John Brockman (ed) (radical ideas coming to humanity
- *Rapt* – Winifred Gallagher (your attention dictates your cognition)
- *Mind Change* – Susan Greenfield (how tech like video games rewires our brains)

- *Growing* Up Digital – Don Tapscott (effects of tech on youth)
- *Controlling Technology* – Andrew Light, Eric Katz, William B. Thompson (do we control tech, or does it control us?)
- *Technological Nature* – Peter H. Kahn (how our relationship with nature is increasingly mediated by technology)
- *Technological Slavery* – Theodore Kaczynski (critique of modern technology/society)
- *Turning Away from Technology* – Stephanie Mills (macro effects of technology on culture)
- *iBrain* – Gary Small (effects of tech on youth brains)
- *The 24 hour Brain* – Rosalind Cartwright (relationship between sleep and mental processes)
- *The Attention Merchants* – Tim Wu (attention economy)
- *Distracted* – Maggie Jackson (technology and human attention)
- *Distracted: A Philosophy of Cars and Phones* – Robert Rosenberger (technology and perception)
- *The Tell-Tale Brain* – V.S. Ramachandran (neuroscience and cognitive understanding)
- *The Man Who Mistook His Wife for a Hat* – Oliver Sacks (neurological case studies)
- *The Design of Everyday Things* – Donald Norman (design and cognitive interaction)

6 Emerging Technological Paradigms

- *The Inevitable* – Kevin Kelly (technological trend analysis)
- *The Second Machine Age* – Erik Brynjolfsson (computational and technological transformation)
- *Faster* – James Gleick (acceleration of technological change)
- *End Times* – Peter Turchin (systemic analysis of societal transitions)
- *Dawn of the New Everything* – Jaron Lanier (technological innovation)
- *The Coming Convergence* – Stanley Schmidt (technological integration)
- *The Age of Speed* – Vince Poscente (technological acceleration)
- *Present at the Future* – Ira Flatow (technological change)
- *The Paradox of Choice* – Barry Schwartz (critically examines abundance of choices)
- *Technology as Magic* – Peter M.R. Stirk (blurring of technology and magic)
- *Technology vs. Humanity* – Gerd Leonhard (how do we embrace technology without becoming it)

- *Is Future Given?"* – Ilya Prigogine (questions if human catastrophe is certain)
- *The Coming Convergence* – Stanley Schmidt (human technology convergence)
- The Age of Missing Information – Bill McKibben (cultural critique)
- *The Code* – Margaret O'Mara (birth of Silicon Valley and Big Tech)
- *Beyond the Obvious* – Phil McKinney (innovation in a rapidly changing world)
- *I Live in the Future* – Nick Bilton (cultural critique of rapid tech change)
- *Blur* – Bill Kovach and Tom Rosenstiel (age of information overload)

7 Ecological and Planetary Systems

- *The Sixth Extinction* – Elizabeth Kolbert (technological impact on global ecosystems)
- *Speed & Scale* – John Doerr (technological solutions to climate change)
- *The Wizard and the Prophet* – Charles Mann (technological versus ecological approaches)
- *Drawdown* – Paul Hawken (technological and systemic environmental solutions)
- *The Uninhabitable Earth* – David Wallace-Wells (climate crisis scenarios)
- *The Overstory* – Richard Powers (human–forest relationships)
- *Under a White Sky* – Elizabeth Kolbert (human interventions in natural systems)
- *The Sustainability Revolution* – David Orr (ecological challenges)
- *Hot, Flat, and Crowded* – Thomas Friedman (global environmental challenges)
- *This Changes Everything* – Naomi Klein (capitalism and climate change)
- *The Story of More* – Hope Jahren (climate change and human consumption)
- *Regeneration* – Paul Hawken (ecological renewal strategies)
- *Origin* – Jennifer Newsom (Systemic approaches to planetary challenges)
- *The Planet Remade* – Oliver Morton (technological interventions in global ecosystems)
- *A Short History of Progress* – Ronald Wright (civilizational ecological consequences)
- *Scale* – Geoffrey West (mathematical principles of complex systems)

8 On the Virtues of Unplugging and Digital Minimalism

- *Digital Minimalism: Choosing a Focused Life in a Noisy World* – Cal Newport (intentional technology use)
- *Deep Work: Rules for Focused Success in a Distracted World* – Cal Newport (focused work strategies)
- *How to Do Nothing: Resisting the Attention Economy* – Jenny Odell (reclaiming attention from capitalism)
- *Irresistible: The Rise of Addictive Technology and the Business of Keeping Us Hooked* – Adam Alter (tech addiction mechanics)
- *Dopamine Nation: Finding Balance in the Age of Indulgence* – Anna Lembke (neurochemical addiction recovery)
- *Indistractable: How to Control Your Attention and Choose Your Life* – Nir Eyal (attention control techniques)
- *Ten Arguments for Deleting Your Social Media Accounts Right Now* – Jaron Lanier (complete social media rejection)
- *Reclaiming Conversation: The Power of Talk in a Digital Age* – Sherry Turkle (face-to-face communication revival)
- *The World Beyond Your Head: On Becoming an Individual in an Age of Distraction* – Matthew Crawford (philosophical attention defense)
- *World Without Mind: The Existential Threat of Big Tech* – Franklin Foer (tech monopoly resistance)
- *Radical Acceptance: Embracing Your Life with the Heart of a Buddha* – Tara Brach (mindful digital living)
- *How to Break Up with Your Phone: The 30-Day Plan to Take Back Your Life* – Catherine Price (phone addiction recovery)
- *The Sleep Revolution: Transforming Your Life, One Night at a Time* – Arianna Huffington (tech-free sleep habits)
- *24/6: The Power of Unplugging One Day a Week* – Tiffany Shlain (weekly digital sabbath)
- *The Shallows: What the Internet Is Doing to Our Brains* – Nicholas Carr (cognitive decline research)

9 Artificial Intelligence and the Future of Humanity

- *Superintelligence* – Nick Bostrom (AI risks and potential futures)
- *Our Final Invention* – James Barrat (threats posed – AI development)
- *The Singularity Is Nearer* – Ray Kurzweil (technological singularity)
- *Life 3.0* – Max Tegmark (AI and intelligence evolution)

- *The Future of the Mind* – Michio Kaku (neuroscience and AI potential)
- *The Age of AI* – Daniel Huttenlocher, Eric Schmidt, Henry Kissinger (AI's human future)
- *AI Superpowers* – Kai-Fu Lee (global AI development)
- *The Alignment Problem* – Brian Christian (aligning AI with human values)
- *The Sentiment Machine* – Simon Beck (AI's emotional and social implications)
- *To Be a Machine* – Mark O'Connell (transhumanism and technology)
- *Machine, Platform, Crowd* – Andrew McAfee and Erik Brynjolfsson (AI's economic and social transformations)
- *AI 2041* – Kai-Fu Lee and Chen Qiufan (Speculative fiction exploring AI futures)
- *The Book of Why* – Judea Pearl (causality and machine learning)
- *Prediction Machines* – Ajay Agrawal, Joshua Gans, Avi Goldfarb (AI's economic implications)
- *The Master Algorithm* – Pedro Domingos (machine learning approaches)
- *Artificial Intelligence: A Guide for Thinking Humans* – Melanie Mitchell (Accessible overview of AI capabilities and limitations
- *AI Superpowers: China, Silicon Valley, and the New World Order* – Kai-Fu Lee (Global AI development)
- *The Centered Mind* – Peter Carruthers (Consciousness and AI)
- *The Shape of Thought* – Richard Harper (Reasoning in the age of AI)
- *The Line* – James Boyle (AI and the Future of Personhood)
- *Generation AI* – Matt Britton (AI and Gen A: 15–25 year olds)

10 Technological and Societal Futures

- *The Fourth Industrial Revolution* – Klaus Schwab (emerging technologies)
- *The Third Wave* – Alvin Toffler (technology waves shaping history)
- *The End of Certainty* – Ilya Prigogine (science, complexity, and unpredictability)
- *The End of Work* – Jeremy Rifkin (automation and post-labor economy)
- *A World without Work* – Daniel Susskind (AI and workforce transformation)
- *The Wealth of Networks* – Yochai Benkler (open-source economies)
- *Who Owns the Future?* – Jaron Lanier (digital labor and ownership)
- *Our Next Reality* – Alvin Graylin, Louis Rosenberg (VR/AR/MR and AI future)
- *21 Lessons for the 21st Century* – Yuval Noah Harari (contemporary technological challenges)

11 Media, Information, and Knowledge in the Digital Age

- *Too Big to Know* – David Weinberger (how digital networks redefine knowledge)
- *Weapons of Math Destruction* – Cathy O'Neil (algorithmic bias and societal impact)
- *Understanding Media* – Marshall McLuhan (media theory and communication technology)
- *Present at the Future* – Ira Flatow (scientific advances and their implications)
- *Why Things Bite Back* – Edward Tenner (unintended consequences of technology)
- *The Essential Herman Kahn: In Defense of Thinking* – Paul Aligica, Kenneth Weinstein (technological thinking)
- *Proust and the Squid* – Maryanne Wolf (reading and brain evolution)
- *Musicology and Neuroscience* – Robert Zatorre (music and cognitive processes)
- *This Is Your Brain on Music* – Daniel Levitin (music and neurological experience)
- *Reading in the Brain* – Stanislas Dehaene (neuroscience of reading)
- *Sound* – Mike Senior (cognitive aspects of sound perception)
- *The Universal Sense* – Seth Horowitz (hearing and perception)
- *Sensation* – Thalma Lobel (sensory experiences)
- *The Information* – James Gleick (history and theory of information)
- *Linked* – Albert-László Barabási (network science and complexity)
- *Reinventing Knowledge* – Ian McNeely (knowledge systems' evolution)
- *Media Virus* – Douglas Rushkoff (information propagation)
- *Convergence Culture* – Henry Jenkins (digital media interactions)
- *Algorithms of Oppression* – Safiya Noble (technological discrimination)
- *The Tyranny of Metrics* – Jerry Muller (quantification of knowledge)
- *Big Data* – Viktor Mayer-Schönberger (data's societal implications)
- *Net Smart* – Howard Rheingold (mindful digital engagement)
- *The Unintended Consequences of Technology* – Nick Bostrom (systemic technological risks)

12 Optimistic Perspectives on Human Progress (Candles in the Dark)

- *Enlightenment Now* – Steven Pinker (data-driven argument for human progress)
- *Paradise Built in Hell,* Rebecca Solnit (how disasters spark extraordinary human solidarity)
- *Abundance* – Peter Diamandis, Steven Kotler (technological solutions for global challenges)
- *Factfulness* – Hans Rosling (misconceptions about global progress)
- *The Rational Optimist* – Matt Ridley (human potential and collaborative progress)
- *Progress* – Johan Norberg (human advancement narrative)
- *Why the World Is Getting Better* – Gregg Easterbrook (optimistic global perspective)
- *The Future Is Better Than You Think* – Peter Diamandis (technological optimism)
- *Humankind: A Hopeful History* – Rutger Bregman (positive view of human nature)
- *The Empathetic Civilization* – Jeremy Rifkin (human cooperation potential)
- *The Future We Choose* – Christiana Figueres, Tom Rivett-Carnac (optimistic climate action)
- *What Are You Optimistic About?* – John Brockman (ed) (optimistic views from leading thinkers)

13 Complexity, Systems Thinking, and Network Effects

- *The Web of Life* – Fritjof Capra (systems thinking in biological and technological evolution)
- *Scale* – Geoffrey West (mathematical principles of growth)
- *The Limits to Growth* – Donella Meadows et al. (technological and ecological limits)
- *The Nature of Technology* – W. Brian Arthur (technology's complex evolution)
- *Linked* – Albert-László Barabási (network science and complexity)
- *Complexity* – M. Mitchell Waldrop (complex adaptive systems)
- *Over-Complicated* – Sam Arbesman (systemic complexity)
- *Thinking in Systems* – Donella Meadows (systems thinking approach)

- *Chaos: Making a New Science* – James Gleick (complexity and unpredictability)
- *Superforecasting* – Philip Tetlock, Dan Gardner (improving predictions in complex worlds)

14 Cybernetics, Artificial Life, and Self-Organizing Systems

- *Cybernetics* – Norbert Wiener (control and communication in animals and machines)
- *The Human Use of Human Beings* – Norbert Wiener (philosophical exploration of cybernetic principles)
- *Design for a Brain* – W. Ross Ashby (Early exploration of adaptive systems)
- *An Introduction to Cybernetics* – W. Ross Ashby (system dynamics and control)
- *Out of Control* – Kevin Kelly (cybernetics and self-organizing systems)
- *Vehicles: Experiments in Synthetic Psychology* – Valentino Braitenberg (simple rules and complex behaviors)
- *The Recursive Mind* – Michael Corballis (recursion in thought)
- *Natural-Born Cyborgs* – Andy Clark (human–technology integration)
- *The Emotion Machine* – Marvin Minsky (cognitive architecture)
- *The Extended Mind* – Andy Clark (cognition beyond individual brain)
- *The Age of Em* – Robin Hanson (brain emulation and AI economics)
- *Cybernetic Revolutionaries* – Eden Medina (cybernetics in historical contexts)
- *The Cybernetic Brain* – Andrew Pickering (cybernetics' intellectual impact)
- *Autopoiesis and Cognition* – Humberto Maturana, Francisco Varela (biological cybernetic thinking)
- *A Cyborg Manifesto* essay – Donna Haraway (vision for feminism, socialism, and materialism in a digital age)
- *The Cyborg Handbook* – Donna Haraway's (examines cyborg society and technologies)

15 End of Old Power and the Birth of New Power

- *New Power* – Jeremy Heimans and Henry Timms (unexpected ways power is changing)
- *Collaborative Intelligence* – Dawna Markova (about new and powerful ways to collaborate)

- *Blessed Unrest* – Paul Hawken (origins/structure of world's largest movement – environmentalism)
- *The End of Power* – Moisés Naím (declining power in established leaders and institutions)
- Citizens – Jon Alexander (new tools /frameworks for citizen impact)
- *The Wisdom of Crowds* – James Surowiecki (collective intelligence)
- *Mutual Aid* – Peter Kropotkin (cooperative human potential)
- *SuperCooperators* – Martin Nowak (cooperation in evolution)
- *Team of Teams* – General Stanley McChrystal (organizational adaptability)

16 Human Adaptation to Extreme Technological Change

- *The Psychology of the Future* – Stanislas Dehaene (neuroscience and future learning)
- *Hyperfocus* – Chris Bailey (mastering attention in a digital world)
- *The Extended Mind* – Annie Murphy Paul (offloading cognition to environment)
- *Deep Work* – Cal Newport (focus strategies in a distracted world)
- *How to Break Up with Your Phone* – Catherine Price (managing digital addiction)
- *Bored and Brilliant* – Manoush Zomorodi (reclaiming attention in digital age)
- *The Organized Mind* – Daniel Levitin (cognitive strategies for information overload)
- *Resilient* – Rick Hanson (growing an unshakable core of calm and strength)
- *Atomic Habits* – James Clear (building adaptive personal systems)
- *The Distraction Addiction* – Alex Soojung-Kim Pang (managing technological overwhelm)
- *Thinking in Bets* – Annie Duke (decision-making under uncertainty)
- *Alone Together* – Sherry Turkle (technology's emotional impact)
- *Reclaiming Conversation* – Sherry Turkle (human connection in digital age)
- *Net Smart* – Howard Rheingold (mindful internet engagement)
- *Antifragile* – Nassim Nicholas Taleb (thriving in chaos and uncertainty)
- *Whiplash* – Joi Ito (learning to adapt to rapid change)
- *Reinventing Organizations* – Frederic Laloux (adaptive organizational structures)
- *Emotional Intelligence 2.0* – Travis Bradberry (navigating technological social landscapes)
- *Range* – David Epstein (value of broad thinking in specialized world)

- *How to Do Nothing* – Jenny Odell (resisting attention economy)
- *Stolen Focus* – Johann Hari (reclaiming deep thinking in digital world)
- *How to Stay Smart in a Smart World* – Gerd Gigerenzer (human intelligence versus algorithms)
- *The Hacking of the American Mind* – Robert Lustig (technology's dopamine manipulation)
- *Peak Mind* – Amishi Jha (strengthening cognitive resilience)
- *Ultralearning* – Scott Young (adapting to new knowledge quickly)
- *The Art of Noticing* – Rob Walker (cultivating awareness in age of distraction)
- *A World without Work* – Daniel Susskind (automation's impact on employment)
- *The Adaptation Advantage* – Heather McGowan, Chris Shipley (developing adaptability as a career skill)
- *Who Owns the Future?* – Jaron Lanier (digital platforms and individual agency)
- *Radical Technologies* – Adam Greenfield (critically engaging with emerging technologies)
- *Postcapitalism* – Paul Mason (technology reshaping economic systems)
- *You Are Not a Gadget* – Jaron Lanier (resisting dehumanizing digital trends)
- *The Age of Surveillance Capitalism* – Shoshana Zuboff (digital privacy and agency)
- *Weapons of Math Destruction* – Cathy O'Neil (algorithmic decision-making risks)
- *The Paradox of Choice* – Barry Schwartz (managing decision overload)
- *Make Time* – Jake Knapp, John Zeratsky (focusing on what matters)
- *Smarter Than You Think* – Clive Thompson (technology's positive cognitive impact)

17 Existential and Long-Term Perspectives on Humanity's Future

- *The Precipice* – Toby Ord (existential risks and humanity's future)
- *Life 3.0* – Max Tegmark (being human in the age of artificial intelligence)
- *Homo Deus* – Yuval Noah Harari (possible future evolution of humanity)
- *The Age of Em* – Robin Hanson (speculative look at brain emulation)
- *Superintelligence* – Nick Bostrom (managing advanced AI risks)
- *500 Year Delta* – Jim Taylor, Watts Wacker (long-term future perspectives)

- *The Extreme Future* – James Canton (technological and societal projections)
- *Our Final Invention* – James Barrat (artificial intelligence's ultimate impact)
- *Re-engineering Humanity* – Brett Frischmann, Evan Selinger (technology's transformative potential)
- Who Owns the Future? – Jaron Lanier (concentration of money and power in digital networks)
- The Extreme Future – James M. Canton (about a world experiencing fast, radical, and complex change)

18 Science Fiction that is Not Completely Dystopian

The ability to tell stories about the future comes with a certain responsibility because what we think about manifests itself in how we perceive the world, behave, and become.

I have deep respect for the science fiction authors who shoulder that burden, and particularly those who shy away from the low-hanging fruit and ever-present dystopian views of the future. What follows is just a short list of some of those I keep returning to most often:

- *Uplift* (series), Earth, Kiln People – David Brin
- *Stealing Worlds* – Karl Shroeder
- *Infinity Born* – Doug E. Richard
- *The Culture Series* – Iain M Banks
- *Neuromancer, the Peripheral, Agency* – William Gibson
- *Snow Crash, Diamond Age* – Neal Stephenson
- *I, Robot* – Isaac Asimov
- *The Three-Body Problem* – Liu Cixin
- *A Fire Upon the Deep* – Vernor Vinge
- *Down and Out in the Magic Kingdom* – Cory Doctorow
- *Daemon* – Daniel Suarez

Ready Player One – Ernest Cline

Bibliography

12th Annual Blood, Sweat and Stairs: Honoring 9/11 First Responders. *Goodfellow Air Force Base Newsroom*, Sep. 6, 2024, http://www.goodfellow.af.mil/Newsroom/Article-Display/Article/3907701/12th-annual-blood-sweat-and-stairs-honoring-911-first-responders/. Accessed Jul. 17, 2025.

1504 Hunt-Lenox Globe, https://en.wikipedia.org/wiki/Hunt%E2%80%93Lenox_Globe. Accessed Aug. 23, 2025.

2017 Global Mobile Consumer Survey: Executive Summary. *Deloitte*, 2017, www2.deloitte.com/content/dam/Deloitte/us/Documents/technology-media-telecommunications/us-tmt-2017-global-mobile-consumer-survey-executive-summary.pdf. Accessed Aug. 25, 2025.

Abbas, Roba, et al. "Machine Learning, Convergence Digitalization, and the Concentration of Power: Enslavement by Design Using Techno-Biological Behaviors." *IEEE Transactions on Technology and Society*, vol. 3, no. 2, Jun. 2022, pp. 76–88.

"A Brain Chemical Changes Locusts from Harmless Grasshoppers to Swarming Pests." *University of Cambridge*, www.cam.ac.uk/research/news/a-brain-chemical-changes-locusts-from-harmless-grasshoppers-to-swarming-pests; Accessed Dec. 1, 2025.

"Accelerating Life Sciences Research with Retro Biosciences." *OpenAI*, Jul. 30, 2024, www.openai.com/index/accelerating-life-sciences-research-with-retro-biosciences. Accessed Sep. 7, 2025.

Acosta, Gina. "How Wegmans Keeps Winning." *Progressive Grocer*, 18 Sept. 2020.

Adams, Frederick, and Kenneth Aizawa. "The Coupling-Constitution Fallacy." *The Bounds of Cognition*, Wiley-Blackwell, 2008.

"Age of Mothers at Childbirth". *OECD Family Database*. Organisation for Economic Co-operation and Development, oecd.org/els/soc/SF_2_3_Age_mothers_childbirth.pdf. Accessed Aug. 25, 2025.

Agre, Philip E., and David Chapman. *What Are Plans For?* AI Memo 1050a, Artificial Intelligence Laboratory, MIT, 1989.

"AI Cracks Superbug Problem in Two Days That Took Scientists Years." *BBC News*, Mar. 30, 2025, http://www.bbc.com/news/articles/clyz6e9edy30. Accessed Mar. 31, 2025.

Aiello, Leslie C., and Peter Wheeler. "The Expensive Tissue Hypothesis: The Brain and the Digestive System in Human and Primate Evolution." *Current Anthropology*, vol. 36, no. 2, 1995, pp. 199–221.

Alderson-Day, Ben, and Charles Fernyhough. "Inner Speech: Development, Cognitive Functions, Phenomenology, and Neurobiology." *Psychological Bulletin* 141.5 (2015), pp. 931.

Alexander, Jon, and Ariane Conrad. *Citizens: Why the Key to Fixing Everything Is All of Us*. Canbury Press, 2022.

Allport, D. Alan, et al. "On the Division of Attention: A Disproof of the Single Channel Hypothesis." *Quarterly Journal of Experimental Psychology*, vol. 24, no. 2, 1972, pp. 225–235.

Allport, Floyd Henry. "Methods in the Study of Collective Action Phenomena." *Journal of Social Psycholology* 15(1) 1942, pp. 165–185.

Alvaro, David. "Reprogramming the Rules: How AI Is Transforming De Novo Protein Design in Drug Development." *Pharma's Almanac*, Apr. 15, 2015. www.pharmasalmanac.com/articles/reprogramming-the-rules-how-ai-is-transforming-de-novo-protein-design-in-drug-development. Accessed Sep. 6, 2025.

"Amazon Braket – Get Started with Quantum Computing." *AWS*, aws.amazon.com/blogs/aws/amazon-braket-get-started-with-quantum-computing/. Accessed Aug. 25, 2025.

American Psychiatric Association. "American Adults Express Increasing Anxiousness in Annual Poll; Stress and Sleep Are Key Factors Imp." *Psychiatry.org*, May 4, 2024. psychiatry.org/news-room/news-releases/annual-poll-adults-express-increasing-anxiousness. Accessed Nov. 17, 2025.

American Psychological Association. "Screen Time and Emotional Problems in Kids: A Vicious Circle?" *APA News Releases*, Jun. 9, 2025, https://www.apa.org/news/press/releases/2025/06/screen-time-problems-children. Accessed Nov 17, 2025.

"Americans' Attitudes and Experiences with Privacy Policies and Laws." *Pew Research Center*, Nov. 15, 2019, pewresearch.org/internet/2019/11/15/americans-attitudes-and-experiences-with-privacy-policies-and-laws/. Accessed Aug. 25, 2025.

"Americans' Trust in Scientists and Views of Science Decline in 2023." *Pew Research Center*, Nov. 14, 2023, http://www.pewresearch.org/science/2023/11/14/americans-trust-in-scientists-positive-views-of-science-continue-to-decline/. Accessed Nov. 3, 2025.

Amodio, David M., and Chris D. Frith. "Meeting of Minds: The Medial Frontal Cortex and Social Cognition." *Nature Reviews Neuroscience*, vol. 7, no. 4, 2006, pp. 268–277.

Amunts, Katrin, and Karl Zilles. "Architectonic Mapping of the Human Brain beyond Brodmann." *Neuron*, vol. 88, no. 6, 2015, pp. 1086–1107.

"Annual Life Expectancy at Birth in the United States, from 1850 to 2023, with Projections until 2100." *Statista*, Statista Inc., Jul. 11, 2024, http://www.statista.com/statistics/1040079/life-expectancy-united-states-all-time/. Accessed Nov. 17, 2025.

Antonakakis, Manos, et al. "Understanding the Mirai Botnet." *26th USENIX Security Symposium (USENIX Security 17)*, USENIX Association, 2017, pp. 1093–1110.

Aral, Sinan, and Dylan Walker. "Creating Social Contagion through Viral Product Design: A Randomized Trial of Peer Influence in Networks." *Management Science* 57, no. 9, 2011, pp. 1623–1639.

Arbib, James, and Seba, Tony, *Stellar: A World Beyond Limits, and How to Get There*, 2024.

Arbib, Michael A. *How the Brain Got Language: The Mirror System Hypothesis*. Vol. 16, Oxford University Press, 2012.

Arendt, Hannah. *The Human Condition*. 2nd ed., University of Chicago Press, 1998.

Arias, Elizabeth, et al. "United States Life Tables, 2021." *National Vital Statistics Reports*, vol. 72, no. 12, National Center for Health Statistics, 7 Nov. 2023.

Aristotle. *The Politics*. Translated by Carnes Lord, 2nd ed., University of Chicago Press, 2013.

Armitage, Hanae. "Smart Toilet Monitors for Signs of Disease." *Stanford Medicine News Center*, Apr. 1, 2020, med.stanford.edu/news/all-news/2020/04/smart-toilet-monitors-for-signs-of-disease.html. Accessed Aug. 25, 2025.

Arora, Siddharth, et al. "Detecting and Monitoring the Symptoms of Parkinson's Disease Using Smartphones: A Pilot Study." *Parkinsonism & Related Disorders*, vol. 21, no. 6, 2015, pp. 650–655.

Arute, Frank, et al. "Quantum Supremacy Using a Programmable Superconducting Processor." *Nature*, vol. 574, no. 7779, 2019, pp. 505–510.

Asch, Solomon E. "Effects of Group Pressure on the Modification and Distortion of Judgments." *Groups, Leadership and Men*, edited by H. Guetzkow, Carnegie Press, 1951, pp. 177–190.

Ashby, Ross W. *An Introduction to Cybernetics*, Methuen and Company, 1964.

Athey, Susan, and Joshua S. Gans. "The Impact of Targeting Technology on Advertising Markets and Media Competition." *American Economic Review*, vol. 100, no. 2, 2010, pp. 608–613.

Aur, Dorian. "Where Is the 'Jennifer Aniston neuron'?." *Nature Precedings*, 2010, pp. 1–1.

Aurelius, Marcus. *Meditations*. Penguin, 2015.

"Autonomous Vehicle Market Demand to Reach 4.2 Million Units by 2030." *PR Newswire*, 6 May 2020, prnewswire.com/news-releases/autonomous-vehicle-market-demand-to-reach-4–2-million-units-by-2030-grand-view-research-inc-301035671.html. Accessed Aug. 25, 2025.

"Average Cost of College Over Time: Yearly Tuition Since 1970." *EducationData.org*, 9 Sept. 2024, educationdata.org/average-cost-of-college-by-year. Accessed Nov. 17, 2025.

Ayan, Steve. "There Is No Such Thing as Conscious Thought", *Scientific American*, December 20, 2018. https://www.scientificamerican.com/article/there-is-no-such-thing-as-conscious-thought/ Accessed Nov. 17, 2025.

Ayers, Michael S., and Lynne M. Reder. "A Theoretical Review of the Misinformation Effect: Predictions from an Activation-Based Memory Model." *Psychonomic Bulletin & Review*, vol. 5, no. 1, 1998, pp. 1–21.

Baars, Bernard J. *A Cognitive Theory of Consciousness.* Cambridge University Press, 1988.

Baker, Jean H. *Votes for Women: The Struggle for Suffrage Revisited.* Oxford University Press, 2002.

Ball, Carlos A., ed. *After Marriage Equality: The Future of Lgbt Rights.* NYU Press, 2019.

Barkow, Jerome H., Leda Cosmides, and John Tooby, editors. *The Adapted Mind: Evolutionary Psychology and the Generation of Culture.* Oxford University Press, 1992.

Barks, Sarah K., et al. "Brain Organization of Gorillas Reflects Species Differences In Ecology." *American Journal of Physical Anthropology* 156.2, 2015, pp. 252–262.

Barrett, Lisa Feldman. *How Emotions Are Made: the Secret Life of the Brain.* Houghton Mifflin Harcourt, 2017.

Bartlett, Lesley, and Daniel Schugurensky. "Deschooling Society 50 Years Later: Revisiting Ivan Illich in the Era of COVID-19." *Otras Voces en Educación* 1, no. 1, 2020, pp. 1–19.

Bartol Jr, Thomas M., et al. "Nanoconnectomic Upper Bound on the Variability of Synaptic Plasticity." *Elife* 4, 2015, e10778.

Baudrillard, Jean. *Simulacra and Simulation.* Translated by Sheila Faria Glaser, University of Michigan Press, 1994.

Baumgartner, Frank R., Bryan D. Jones, and John Wilkerson. "Studying Policy Dynamics." In *Policy Dynamics*, ed. Frank R. Baumgartner, and Bryan D. Jones. University of Chicago Press, 2002, pp. 29–46.

Bay, Noel Si-Yang, and Boon-Huat Bay. "Greek Anatomist Herophilus: the Father of Anatomy." *Anatomy & Cell Biology*, vol. 43, no. 4, Dec. 2010, pp. 280–283.

Bazzi, Samuel. "The Roots and Persistence of 'Rugged Individualism' in the United States." *Frontier Culture*, 2017, https://www.bu.edu/econ/files/2017/03/Frontier-Culture-The-Roots-and-Persistence-of-%E2%80%9CRugged-Individualism%E2%80%9D-in-the-United-States-2.pdf. Accessed Nov. 13, 2025.

Bednarik, Robert G. "An Aetiology of Hominin Behaviour." *Homo,* vol. 63.5, 2012, pp. 319–335.

Bednarik, Robert G. "Doing with Less: Hominin Brain Atrophy." HOMO*: Journal of Comparative Human Biology*, vol. 65, no. 6, 2014, pp. 433–49.

Bednarik, Robert G. "The Domestication of Humans." *Anthropologie*, vol. 46, 2008, pp. 1–17.

Bell, Daniel. *The Coming of Post-Industrial Society: A Venture in Social Forecasting.* Basic Books, 1999.

Bell, Gordon, et al. "The Effect of Technology on Near-Term Computer Structures." *Computer*, vol. 5, no. 2, Mar.-Apr. 1972, pp. 29–38.

Bell, Vaughan. "A History of Media Technology Scares, from the Printing Press to Facebook." *Slate*, Feb. 15, 2010, https://slate.com/technology/2010/02/a-history-of-media-technology-scares-from-the-printing-press-to-facebook.html. Accessed Aug. 25, 2025.

Benke, Kurt, and Bruce Tomkins. "Future Food Production Systems: Vertical Farming and Controlled-Environment Agriculture." *Sustainability: Science, Practice and Policy*, vol. 13, no. 1, 2017, pp. 13–26.

Benkler, Yochai. *The Wealth of Networks: How Social Production Transforms Markets and Freedom*. Yale University Press, 2006.

Berger, Jonah. *Contagious: Why Things Catch On*. Simon & Schuster, 2013.

Berreman, Gerald Duane, and Kathleen M. Zaretsky. *Social Inequality: Comparative and Developmental Approaches*. Academic Press, 1981.

Besson, Mireille, et al. "Brain Waves Associated with Musical Incongruities Differ for Musicians and Nonmusicians." *Neuroscience Letters*, vol. 168, 1994, pp. 101–1.

Bharade, Aditi. "Tourists in Hawaii Followed the GPS and Drove Their Car Into Water." *Business Insider*, May 3, 2023, https://www.businessinsider.com/tourists-hawaii-gps-drove-car-into-water-2023–5. Accessed Aug. 24, 2025.

Bibri, Simon E., et al. "Environmentally Sustainable Smart Cities and Their Converging Models: Mapping the Evolution of AI, IoT, and Big Data Solutions." *Energy Informatics*, no. 6, 9, 2023.

Bijker, Wiebe E., Thomas P. Hughes, and Trevor Pinch, editors. *The Social Construction of Technological Systems: New Directions in the Sociology and History of Technology*. MIT Press, 1987.

Biswal, Bharat, et al. "Functional Connectivity in the Motor Cortex of Resting Human Brain Using Echo-Planar MRI." *Magnetic Resonance in Medicine*, vol. 34, no. 4, 1995, pp. 537–541.

Bloom, Harold. Shakespeare: *The Invention of the Human*. Riverhead Books, 1998.

Boesch, Christophe, et al. "Chimpanzee Ethnography Reveals Unexpected Cultural Diversity." *Nature Human Behaviour*, 2020, pp. 1–7.

Bogin, Barry. "Evolutionary Hypotheses for Human Childhood." *American Journal of Physical Anthropology*, vol. 104, no. S25, 1997, pp. 63–89.

Bohler, Sébastien. *Le Bug Humain: Pourquoi Notre Cerveau Nous Pousse à Détruire la Planète et Comment l'en Empêcher* (*Bohler, Sébastien. Le Bug Humain: Pourquoi Notre Cerveau Nous Pousse à Détruire la Planète et Comment l'en Empêcher. Robert Laffont, 2019*). Robert Laffont, 2019.

Bonaventura, Davide, Sergio Esposito, and Giampaolo Bella. "The Iot Breaches Your Household Again." *arXiv preprint arXiv:2407.12159*, 2024.

Borexino Collaboration. "Experimental Evidence of Neutrinos Produced in the CNO Fusion Cycle in the Sun." *Nature*, vol. 587, no. 7835, 2020, pp. 577–582.

Boserup, Ester. *Population and Technological Change: A Study of Long-Term Trends*. Chicago: University of Chicago Press, 1981.

Bostrom, Nick, and Milan M. Ćirković, editors. *Global Catastrophic Risks*. Oxford University Press, 2011.

Bostrom, Nick. "Are We Living in a Computer Simulation?" *The Philosophical Quarterly,* vol. 53.211, 2003, pp. 243–255.

Bowlby, John. *Attachment and Loss*. Basic Books, 1969.

Boyd, Richard. "Richard Boyd, Co-Founder & CEO of Tanjo Inc – Interview Series." *Interview by Antoine Tardif, Unite.AI,* 16 Apr. 2020, updated Dec. 9, 2022, https://www.unite.ai/richard-boyd-the-co-founder-ceo-of-tanjo-inc-interview-series/. Accessed Sep. 6, 2025.

Boyd, Robert, and Peter J. Richerson. *Culture and the Evolutionary Process*. University of Chicago Press, 1985.

Brady, Scott, et al., eds. *Basic Neurochemistry: Molecular, Cellular and Medical Aspects*. Elsevier, 2005.

Branson, Richard. "Every Success Story Is a Tale of Constant Adaptation, Revision, and Change." *Virgin.com,* 22 Jul. 2015. Accessed Aug. 24, 2025.

Bratsberg, Bernt, and Ole Rogeberg. "Flynn Effect and Its Reversal Are Both Environmentally Caused." *Proceedings of the National Academy of Sciences*, vol. 115, no. 26, 2018, pp. 6674–6678.

Bregman, Rutger. "The Real Lord of the Flies: What Happened When Six Boys Were Shipwrecked for 15 Months." *The Guardian*, May 9, 2020, https://www.theguardian.com/books/2020/may/09/the-real-lord-of-the-flies-what-happened-when-six-boys-were-shipwrecked-for-15-months. Accessed Nov. 19, 2025.

Bresnahan, Timothy F., and Manuel Trajtenberg. "General Purpose Technologies 'Engines of Growth?" *Journal of Econometrics*. 65.1. 1995, pp. 83–108.

Breuer, Pablo, and S.J. Terp. "DISARM: A Framework for Analysis of Disinformation Campaigns." *Proceedings of the 2022 IEEE International Conference on Cognitive Security and Information Management* (*COGSIMA*), IEEE, 2022, pp. 1–8.

Brin, David. *Earth*. Bantam Books, 1990.

Brin, David. *Kiln People*. Tor Books, 2002.

Brin, David. "Stones of Significance." *Analog Science Fiction and Fact*, vol. 120, no. 1, Jan. 2000.

Brod, Craig. *Technostress: The Human Cost of the Computer Revolution*. Addison-Wesley, 1984.

Brodmann, Korbinian. *Brodmann's Localisation in the Cerebral Cortex*. Translated by Laurence J. Garey, Springer, 2006.

Brown, Curtis. "Behaviorism: Skinner and Dennett." *Philosophy of Mind*, Trinity University, 2001.

Bruner, Jerome. "Two Modes of Thought." *Actual Minds, Possible Worlds*, Harvard University Press, 1986, pp. 11–43.

Bruner, Robert F. "Feynman's Delight." *Bruner Blog*, Darden School of Business, University of Virginia, Sep. 27, 2009, blogs.darden.virginia.edu/brunerblog/2009/09/feynmans-delight/. Accessed Nov. 19, 2025.

Brynjolfsson, Erik, and Andrew McAfee. *Race Against the Machine: How the Digital Revolution Is Accelerating Innovation, Driving Productivity, and Irreversibly Transforming Employment and the Economy*. Digital Frontier Press, 2011.

Brynjolfsson, Erik, and Andrew McAfee. *The Second Machine Age: Work, Progress, and Prosperity in a Time of Brilliant Technologies*. W. W. Norton & Company, 2014.

Btl. "The Discontinuity Thesis: Why This Time Really Is Different." *Medium*, Jun. 2025, medium.com/@btl101/the-discontinuity-thesis-why-this-time-really-is-different-3-a63fc5bc018. Accessed Sep. 7, 2025.

Bud, Robert. *Penicillin: Triumph and Tragedy*. Oxford University Press, 2007.

Budge, E.A. Wallis. *Thoth, the Hermes of Egypt: A Study of Some Aspects of Theological Thought in Ancient Egypt*. Kegan Paul, Trench, Trubner & Co., 1904.

Buechler, Steven M. *Women's Movements in the United States: Women's Suffrage, Equal Rights, and Beyond*. Rutgers University Press, 1990.

Burch, Noel. *Conscious Competence Learning Model*. Gordon Training International. 1970.

Burgess, Matt. "GPS Signals Are Being Disrupted in Russian Cities", *Wired*, Dec 15, 2022. https://www.wired.com/story/gps-jamming-interference-russia-ukraine/. Ac cessed Aug. 24, 2025.

Burt, Ronald S. *Structural Holes: The Social Structure of Competition*. Harvard University Press, 1992.

Bush, Vannevar. "As We May Think." *The Atlantic Monthly*, vol. 176, no. 1, Jul. 1945, pp. 101–108.

Byrne, Richard W., and Andrew Whiten. *Machiavellian Intelligence: Social Expertise and the Evolution of Intellect in Monkeys, Apes and Humans*. Clarendon Press, 1988.

Byrne, Richard. W., Nadia Corp, and Jeannette M. Byrne. "Manual Dexterity in the Gorilla: Bimanual and Digit Role Differentiation in a Natural Task." *Animal Cognition*, vol. 4, no. 3–4, Nov. 2001, pp. 347–61.

Camlin, Thaddeus. "Addiction and Society: Snowflakes and a Culture of Outrage." *Practical Recovery*, Jun. 23, 2017.

Cannon, Hugh M. "Dealing with the Complexity Paradox in Business Simulation Games." *Developments in Business Simulations and Experiential Exercises*, vol. 22, 1995, pp. 96–102.

Cannon, Hugh M., and Alvin C. Burns. "A Framework for Assessing the Competencies Reflected in Simulation Performance." *Developments in Business Simulation and Experiential Learning* 26, 1999.

Cannon, Hugh M., Andrew H. Feinstein, and Daniel P. Friesen. "Managing Complexity: Applying the Conscious-Competence Model to Experiential Learning." *Devel-*

opments in Business Simulation and Experiential Learning: Proceedings of the Annual ABSEL Conference, vol. 37, 2010.

Cannon, Hugh M., Daniel P. Friesen, Steven J. Lawrence, and Andrew H. Feinstein. "The Simplicity Paradox: Another Look at Complexity in Design of Simulations and Experiential Exercises." *Developments in Business Simulations and Experiential Exercises*, vol. 36, 2009, pp. 243–250.

Carr, Nicholas. "Is Google Making Us Stupid." The Atlantic, 2008, Jul./Aug. version, https://www.theatlantic.com/magazine/archive/2008/07/is-google-making-us-stupid/306868/. Accessed Nov. 19. 2025.

Carr, Nicholas. *The Shallows: What the Internet Is Doing to Our Brains*. W. W. Norton & Company, 2010.

Carroll, Lewis. *Through the Looking-Glass, and What Alice Found There*. Edited by Chris Riddell, HarperCollins, 2015.

Carruthers, Mary. *The Book of Memory*. Cambridge University Press, 1990.

Carruthers, Mary. *The Craft of Thought*. Cambridge University Press, 1998.

Carruthers, Peter. *The Architecture of the Mind: Massive Modularity and the Flexibility of Thought*. Oxford University Press, 2006.

Carruthers, Peter. *The Centered Mind: What the Science of Working Memory Shows Us about the Nature of Human Thought*. Oxford University Press, 2015.

Carruthers, Peter. "The Illusion of Conscious Thought." *Journal of Consciousness Studies*, no. 24.9–10, 2017, pp. 228–252.

Carson, Rachel. *Silent Spring*. Houghton Mifflin, 1962.

Carter, Susan B., et al., editors. *Historical Statistics of the United States: Millennial Edition*. Cambridge University Press, 2006.

Cartwright, Mark. "Pax Romana." *World History Encyclopedia*, Dec. 8, 2015, www.worldhistory.org/Pax_Romana/. Accessed Nov. 19, 2025.

Case, Amber "We Are All Cyborgs Now". *TED*. https://www.ted.com/talks/amber_case_we_are_all_cyborgs_now. Accessed Aug. 23, 2025.

Case, Anne, and Angus Deaton. *Deaths of Despair and the Future of Capitalism*. Princeton University Press, 2020.

Case, Anne, and Angus Deaton. "Rising Morbidity and Mortality in Midlife among White Non-Hispanic Americans in the 21st Century." *Proceedings of the National Academy of Sciences* 112, no. 49, Dec. 8, 2015, pp. 15078–15083.

Cavalli-Sforza, Luigi Luca, and Marcus W. Feldman. *Cultural Transmission and Evolution: A Quantitative Approach*. Princeton University Press, 1981.

Centola, Damon, et al. "Experimental Evidence for Tipping Points in Social Convention." *Science* 360, no. 6393. 2018, pp. 1116–1119.

Cesario, Joseph, et al. "Your Brain Is Not an Onion with a Tiny Reptile Inside." *Current Directions in Psychological Science*, vol. 29, no. 3, 2020, pp. 255–260.

Chalmers, David. "Foreword to Andy Clark's *Supersizing the Mind." Supersizing the Mind: Embodiment, Action, and Cognitive Extension*, by Andy Clark, Oxford University Press, 2008.

Chan, Justin, et al. "Contactless Cardiac Arrest Detection Using Smart Devices." *NPJ Digital Medicine*, vol. 2, no. 1, 2019, pp. 1–8.

Chan, Shirley, et al. "Is Menopause Still Evolving? Evidence from a Longitudinal Study of Multiethnic Populations and Its Relevance to Women's Health." *BMC Women's Health*, vol. 20, 2020, pp. 1–15.

Chandler, Daniel. *The Sapir-Whorf Hypothesis*. 1994.

Chanson-Rolle, Aurelie, et al. "Nutritional Composition of Orange Juice: A Comparative Study between French Commercial and Home-Made Juices." *Food and Nutrition Sciences* no. 7.04, 2016, p. 252.

Chantler, Abigail. "The "Sturm und Drang" Style Revisited." *International Review of the Aesthetics and Sociology of Music*, 2003, pp. 17–31.

"Charting the Future of Biotechnology – An Action Plan for American Security and Prosperity". *National Security Commission on Emerging Biotechnology*. https://www.biotech.senate.gov/final-report/chapters/. Accessed Nov. 17, 2025.

Chartrand, Tanya L., and John A. Bargh. "The Chameleon Effect: The Perception-Behavior Link and Social Interaction." *Journal of Personality and Social Psychology*, vol. 76, no. 6, 1999, pp. 893–910.

Chayka, Kyle. "The Year Creators Took Over." *The New Yorker*, Dec. 11, 2024.

Chen, Wei-Guang, et al. "The Emerging Science of Interoception: Sensing, Integrating, Interpreting, and Regulating Signals within the Self." *Trends in Neurosciences*, vol. 44.1, 2021, pp. 3–16.

Cherlin, Andrew J. "The Deinstitutionalization of American Marriage." Journal of Marriage and Family, vol. 66, no. 4, 2004, pp. 848–861.

Cheung, Albert S. Y. "From Datafication to the Data State: Making Sense of China's Social Credit System and Its Implications." *Law & Social Inquiry*, vol. 47, no. 2, Spring 2022, pp. 230–252.

Cho, Adrian. "Ordinary Computers Can Beat Google's Quantum Computer After All." *Science*, Aug. 2, 2022. science.org/content/article/ordinary-computers-can-beat-google-s-quantum-computer-after-all. Accessed Aug. 25, 2025.

Cho, Adrian. "Yes, We Have Free Will. No, We Absolutely Do Not." *Nautilus*, 7 Feb. 2024, nautil.us/yes-we-have-free-will-no-we-absolutely-do-not-4319. Accessed Aug. 25, 2025.

Chorost, Michael. *Rebuilt: How Becoming Part Computer Made Me More Human*. Houghton Mifflin Harcourt, 2005.

Christensen, Clayton M. *The Innovator's Dilemma: When New Technologies Cause Great Firms To Fail*. Harvard Business Review Press, 2013.

Christensen, Clayton M., Michael E. Raynor, and Rory McDonald. "What Is Disruptive Innovation." *Harvard Business Review*, vol. 93.12, 2015, pp. 44–53.

Christian, Kimberly M., and Richard F. Thompson. "Long-Term Storage of an Associative Memory Trace in the Cerebellum." *Behavioral Neuroscience*, vol. 119, 2005, pp. 256–537.

Chudek, Maciej, and Joseph Henrich. "Culture-Gene Coevolution, Norm-Psychology and the Emergence of Human Prosociality." *Trends in Cognitive Sciences*, vol. 15, no. 5, 2011, pp. 218–226.

Churchill, Winston. *Speech in the House of Commons*, 11 Nov. 1947. *The Churchill Centre*, https://winstonchurchill.org/resources/speeches/1947-1951-the-united-nations/democracy-is-the-worst-form-of-government-except-for-all-those-others-that-have-been-tried/. Accessed Aug. 26, 2025.

Churchill, Winston. "The Sinews of Peace." In *Post-War Speeches*, Houghton Mifflin Co., 1949.

Ciechanowski, Bartosz. "Tesseract." Bartosz Ciechanowski, 2024, ciechanow.ski/tesseract/. Accessed Jul. 13, 2025.

Claidière, Nicolas, et al. "How Darwinian Is Cultural Evolution?" *Philosophical Transactions of the Royal Society B*, vol. 369, 2014, article 20130368.

Clancey, William J. "Becoming a Rover." *Simulation and Its Discontents*, edited by Sherry Turkle, Cambridge University Press, 2009.

Clark, Andy, and David J. Chalmers. "The Extended Mind." *Analysis*, vol. 58, no. 1, 1998, pp. 7–19.

Clark, Andy. *Being There: Putting Brain, Body, and World Together Again.* MIT Press, 1997.

Clark, Andy. *Microcognition.* MIT Press, 1989.

Clark, Andy. *Natural-Born Cyborgs: Minds, Technologies, and the Future of Human Intelligence.* Oxford University Press, 2003.

Clark, Andy. *Supersizing the Mind: Embodiment, Action, and Cognitive Extension.* Oxford University Press, 2008.

Claidière, Nicolas, and Dan Sperber. "The Natural Selection of Fidelity in Social Learning." *Communicative & Integrative Biology*, vol. 3, no. 4, 2010, pp. 1–2.

Cline, Ernest. *Ready Player One.* Crown Publishers, 2011.

Clynes, Manfred E., and Nathan S. Kline. "Cyborgs and Space". *Astronautics*, Sep. 1960.

Cochran, Gregory, and Henry Harpending. *The 10,000-Year Explosion: How Civilization Accelerated Human Evolution.* Basic Books, 2009.

Cohen, Lizabeth. *A Consumers' Republic: the Politics of Mass Consumption in Postwar America.* Knopf, 2003.

Cole, Steven W. "The Social Life of Genes." *Pacific Standard*, in *Stories – The Social Life of Genes – Showcase*, Apr. 6, 2025, https://psmag.com/social-justice/the-social-life-of-genes-64616/. Accessed Nov. 27, 2025.

"Committee Opinion no. 615: Access to Contraception". *American College of Obstetricians and Gynecologists*, vol. 125, no. 1, Jan. 2015, pp. 250–255. https://pubmed.ncbi.nlm.nih.gov/25560140/. Accessed Aug. 25, 2025.

Comstock, Beth. "How to Make a Big Company More Innovative." *Vox*, Nov. 30, 2018.

Conkling, Gary. "'Monetizing Fear' by Bulletproofing Schools." *Oregon Capital Chronicle*, Sep. 7, 2024.

"Connecting America: The National Broadband Plan." *Federal Communications Commission*. 2010, https://transition.fcc.gov/national-broadband-plan/national-broadband-plan.pdf. Accessed Sep. 6, 2025.

Cook, Laurence M., et al. "Selective Bird Predation on the Peppered Moth: The Last Experiment of Michael Majerus." *Biology Letters*, vol. 8, no. 4, 2012, pp. 609–612.

Coolidge, Frederick L., Thomas Wynn, and Karenleigh A. Overmann. "The Evolution of Working Memory." *Working Memory: The Connected Intelligence*, edited by Tracy Packiam Alloway and Ross G. Alloway, Psychology Press, 2013.

Cosmides, Leda, and John Tooby. "From Evolution to Behavior: Evolutionary Psychology as the Missing Link." *The Latest on the Best: Essays on Evolution and Optimality*, edited by J. Dupré, MIT Press, 1987, pp. 227–306.

Cosmides, Leda, and John Tooby. "The Psychological Foundations of Culture." *The Adapted Mind: Evolutionary Psychology and the Generation of Culture*, edited by Jerome H. Barkow, Leda Cosmides, and John Tooby, Oxford University Press, 1992.

Costanza, Robert, John Cumberland, Herman Daly, Robert Goodland, and Richard Norgaard. *An Introduction to Ecological Economics*. St. Lucie Press, 1997.

Covey, Stephen R. *The 7 Habits of Highly Effective People*. Free Press, 1989.

"Criminal Justice Fact Sheet." *NAACP*, 4 Nov. 2022, naacp.org/resources/criminal-justice-fact-sheet. Accessed Jun. 3, 2025.

Crisp, Roger, ed. *Aristotle: Nicomachean Ethics*. Cambridge University Press, 2014.

Cronin, Thomas M. "Speciation and Stasis in Marine Ostracoda: Climatic Modulation of Evolution." *Science*, vol.227.4682, 1985, pp. 60–63.

Csikszentmihalyi, Mihaly. *Flow: The Psychology of Optimal Experience*. Harper & Row, 1990.

Czarnik, Anthony W. "A Sense for Landmines." *Nature, vol.* 394.6692, 1998, pp. 417–418.

Dalai Lama. *The Art of Happiness: A Handbook for Living*. Edited by Howard C. Cutler, Riverhead Books, 1998.

Damasio, Antonio R. *Looking for Spinoza: Joy, Sorrow, and the Feeling Brain*. Houghton Mifflin Harcourt, 2003.

Damasio, Antonio R., and Hanna Damasio. "Brain and Language." *Scientific American*, vol. 267, no. 3, 1992, pp. 88–109.

Damasio, Antonio. "Neural Basis of Emotions." *Scholarpedia*, vol. 6, no. 3, 2011, p. 1804.

Danneels, Erwin. "Disruptive Technology Reconsidered: a Critique and Research Agenda." *Journal of Product Innovation Management*, vol. 21.4, 2004, pp. 246–258.

Dart, Raymond A. "Australopithecus africanus: The Man-Ape of South Africa." *Nature*, vol. 115, no. 2884, Feb. 7, 1925.

Darwin, Charles. *The Descent of Man and Selection in Relation to Sex*. 2nd ed., John Murray, 1874.

Darwin, Charles. *The Origin of Species*. PF Collier & Son, 1909.

Davenport, Christian. "Meet the Most Fascinating Part of the F-35: The $400,000 Helmet." *The Washington Post*, 1 Apr. 2015, www.washingtonpost.com/news/checkpoint/wp/2015/04/01/meet-the-most-fascinating-part-of-the-f-35-the-400000-helmet/. Accessed Aug. 23, 2025.

Davenport, Frances V., et al. "Escalating Costs of Billion-Dollar Disasters in the US: Climate Change Necessitates Disaster Risk Reduction." *Global Environmental Change*, vol. 78, Jan. 2023, http://www.sciencedirect.com/science/article/pii/S2667278222000906. Accessed Feb.19, 2025.

David, Daniela, et al. "Text Neck Syndrome in Children and Adolescents." *International Journal of Environmental Research and Public Health*, vol. 18, no. 4, 2021, 1565.

Davis-Floyd, Robbie, and Joseph Dumit, editors. *Cyborg Babies: From Techno-Sex to Techno-Tots*. Routledge, 1998.

Dawkins, Richard. *The Selfish Gene*. Oxford University Press, 2016.

DeCasien, Alex R., Scott A. Williams, and James P. Higham. "Primate Brain Size Is Predicted by Diet but Not Sociality." *Nature Ecology & Evolution* vol. 1.5, 2017, p. 0112.

Dehaene, Stanislas, and Laurent Cohen. "The Unique Role of the Visual Word Form Area in Reading." *Trends in Cognitive Sciences*, vol. 15, no. 6, 2011, pp. 254–262.

Dehaene, Stanislas. *Consciousness and the Brain: Deciphering How the Brain Codes Our Thoughts*. Viking, 2014.

Dehaene, Stanislas. *Reading in the Brain: the New Science of How We Read*. Penguin Group USA, 2010.

Dehaene, Stanislas. *The Number Sense: How the Mind Creates Mathematics*. Rev. and updated ed., Oxford University Press, 2011.

Dennett, Daniel C. "Beyond Belief". In Woodfield, Andrew, *Thought and Object: Essays On Intentionality*. Clarendon Press, 1982.

Dennett, Daniel C. *Consciousness Explained*. Little, Brown and Company, 1991.

Dennett, Daniel C. *Kinds of Minds*. Basic Books, 1996.

Dennett, Daniel C. "Learning and Labeling." *Mind and Language*, vol. 8, no. 4, 1993, pp. 540–547.

Denver Area Educational Telecommunications Consortium, Inc. v. Federal Communications Commission. U.S. Supreme Court, 1996.

DeSilva, Jeremy M., et al. "When and Why Did Human Brains Decrease in Size? A New Change-Point Analysis and Insights from Brain Evolution in Ants." *Frontiers in Eco logy and Evolution*, vol. 9, 2021, 742639.

DeSilver, Drew. "The Polarization in Today's Congress Has Roots That Go Back Decades." *Pew Research Center*, Mar. 10, 2022.

Dhority, Lynn. *The Act Approach: The Use of Suggestion for Integrative Learning*. Psychology Press, 1991.

Diamond, Jared. *Guns, Germs, and Steel: The Fates of Human Societies*. W. W. Norton, 1997.

Dickens, Charles. *A Tale of Two Cities*. 1859. Penguin Classics, 2003, p. 1.

Diener, Ed, Richard E. Lucas, and Christie Napa Scollon. "Beyond the Hedonic Treadmill: Revising the Adaptation Theory of Well-Being." *American Psychologist* 61, no. 4, 2006, 305.

Dijksterhuis, Ap, and John A. Bargh. "The Perception-Behavior Expressway: Automatic Effects of Social Perception on Social Behavior." *Advances in Experimental Social Psychology*, vol. 33, Academic Press, 2001, pp. 1–40.

Dinan, Timothy G., et al. "Psychobiotics: a Novel Class of Psychotropic." *Biological Psychiatry*, vol. 74, no. 10, 15 Nov. 2013, pp. 720–726.

Dixson, Alan F. *Primate Sexuality: Comparative Studies of the Prosimians, Monkeys, Apes, and Humans*. 2nd ed., Oxford University Press, 2012.

Dobbs, David. "The Social Life of Genes". *Pacific Standard Magazine*. Sep. 3, 2013. https://psmag.com/social-justice/the-social-life-of-genes-64616/. Accessed Nov. 17, 2025.

"Don't Judge a Fund by the Book". *Bloomberg*, Jun. 25, 2001, https://www.bloomberg.com/news/articles/2001-06-25/dont-judge-a-fund-by-the-book. Accessed Dec 13, 2025.

Donald, Merlin. *A Mind So Rare: the Evolution of Human Consciousness*. W. W. Norton & Company, 2001.

Donald, Merlin. On the Evolution of Representational Capacities. *Behavioral and Brain Sciences*, vol. *16, 1993*, pp. 775–785.

Donald, M. *Origins of the Modern Mind: Three Stages in the Evolution of Culture and Cognition*. Harvard University Press. 1991.

Doran, Diane M., et al. "Western Lowland Gorilla Diet and Resource Availability: New Evidence, Cross-Site Comparisons, and Reflections on Indirect Sampling Methods." *American Journal of Primatology: Official Journal of the American Society of Primatologists*, vol. 58.3, 2002, pp. 91–116.

Dosi, Giovanni. "Technological Paradigms and Technological Trajectories: a Suggested Interpretation of the Determinants and Directions of Technical Change." *Research Policy*, vol. 11.3, 1982, pp. 147–162.

Downes, Larry. *The Laws of Disruption: Harnessing the new forces that govern life and business in the digital age*. Basic Books, 2009.

Drexler, K. Eric. *Reframing Superintelligence*. 2019.

Druzhinin, Nikolai M. "Gosudarstvennye Krest'yane I Reforma Pd Kiseleva". *State Peasants and the Reform of Pd Kiselev*, Vols I-II. Moscow, 1946.

Dubner, Stephen J. "If We're All in It for Ourselves, Who Are We?" *Freakonomics Radio*, Aug. 7, 2024, https://freakonomics.com/podcast/if-were-all-in-it-for-ourselves-who-are-we/. Accessed Jun. 13, 2025.

Dunbar, Robin I. M. "The Social Brain Hypothesis." *Evolutionary Anthropology*, vol. 6, 1998, pp. 178–190.

Dunbar, Robin I. M. "The Social Brain: Psychological Underpinnings and Implications for the Structure of Organizations." *Current Directions in Psychological Science*, vol. 23, no. 2, 2014, pp. 109–114.

Dunlap, Thomas R. *DDT: Scientists, Citizens, and Public Policy*. Princeton University Press, 1981.

Durkheim, Émile. *The Division of Labor in Society*. Translated by Lewis A. Coser, Free Press, 1984.

Dutton, Edward, et al. "The Negative Flynn Effect: a Systematic Literature Review." *Intelligence*, vol. 59, 2016, pp. 163–169.

EdgeDelta Blog. "In 2023, Facebook Produced Around 4 PB on Average per Day." *EdgeDelta*, Mar. 11, 2024. https://edgedelta.com/company/blog/how-much-data-is-created-per-day. Accessed Nov. 18, 2025.

Edwards, Jim. "The Russians Are Screwing with the GPS System to Send Bogus Navigation Data to Thousands of Ships" *Business Insider*, Apr. 14, 2019. https://www.businessinsider.com/gnss-hacking-spoofing-jamming-russians-screwing-with-gps-2019-4. Accessed Nov. 17, 2025.

Einstein, Albert. *The World as I See It*. Citadel Press, 1993.

El Saddik, Abdulmotaleb, et al. "Metaverse and Extended Universe (XV, pp. Opportunities and Challenges for Consumer Technologies," *IEEE Consumer Electronics Magazine*, vol. 13, no.3, May 2024, pp. 25–26.

Eldredge, Niles and Stephen J. Gould "Punctuated Equilibria: an Alternative to Phyletic Gradualism", 1972, In T.J.M. Schopf, ed., *Models in Paleobiology*. Freeman Cooper. pp. 82–115. Reprinted in N. Eldredge Time Frames. Princeton Univ. Press, 1985, pp. 193–223.

Ellul, Jacques, "The Autonomy of Technique", *Controlling Technology: Contemporary Is sues*, Katz, Eric, Light, Andrew, and Thompson, William B., Eds. Prometheus Books, 2003.

Ellul, Jacques. *The Technological Society*. Translated by John Wilkinson, Vintage Books, 1964.

Endsley, Mica R. "Design and Evaluation for Situation Awareness Enhancement." *Proceedings of the Human Factors Society Annual Meeting*, vol. 32, no. 2, Sage Publications, 1988.

Engelbart, Douglas C. "A Conceptual Framework." *Douglas C. Engelbart Papers*, Stanford University Libraries, web.stanford.edu/dept/SUL/library/extra4/sloan/mouse-site/EngelbartPapers/B5_F18_ConceptFrameworkPt1.html. Accessed Aug. 23, 2025.

Engelbart, Douglas C. "Augmenting Human Intellect: a Conceptual Framework." *Stanford Research Institute*, Oct. 1962, https://www.dougengelbart.org/pubs/papers/scanned/Doug_Engelbart-AugmentingHumanIntellect.pdf. Accessed Jul. 7, 2025.

Engelbart, Douglas C. "The Better We Get at Getting Better, the Faster We Will Get Better." *QuoteFancy*, https://quotefancy.com/quote/1177039/Douglas-Engelbart-The-better-we-get-at-getting-better-the-faster-we-will-get-better. Accessed Sep. 2, 2025.

Enquist, Magnus, Stefano Ghirlanda, and Johan Lind. *The Human Evolutionary Transition: from Animal Intelligence to Culture*. Princeton University Press, 2023.

Enquist, Magnus, Stefano Ghirlanda, Andreas Eriksson, and Kimmo Eriksson. "The Human Evolutionary Transition: From Ecological Dominance to Directional Selection." *Bulletin of Mathematical Biology*, vol. 72, no. 5, 2010, pp. 1006–1029.

"Epistemology." *Encyclopædia Britannica*, Encyclopædia Britannica, Inc., Aug. 2, 2025, https://www.britannica.com/topic/epistemology. Accessed Sep. 17, 2025.

Epstein, David. *Range: Why Generalists Triumph in a Specialized World*. Penguin, 2021.

Ericsson, K. Anders, and Neil Charness. "Expert Performance: Its Structure and Acquisition." *American Psychologist*, vol. 49, no. 8, 1994, pp. 725–747.

Ericsson, K. Anders. "Guest Post: The Danger of Delegating Education to Journalists." *Radical Scholarship*, 3 Nov. 2014. Archived by Wayback Machine, https://web.archive.org/web/20190320062202/https://radicalscholarship.wordpress.com/2014/11/03/guest-post-the-danger-of-delegating-education-to-journalists-k-anders-ericsson/. Accessed Aug. 24. 2025.

Ericsson, K. Anders. "Training History, Deliberate Practice and Elite Sports Performance: An Analysis in Response to Tucker and Collins Review–What Makes Champions?" *British Journal of Sports Medicine*, vol. 47, no. 9, 2013, pp. 533–535.

Evenson, Robert E., and Douglas Gollin. "Assessing the Impact of the Green Revolution, 1960 to 2000." *Science*, vol. 300, no. 5620, 2003, pp. 758–762.

"F-35 Gen III Helmet-Mounted Display System." *Collins Aerospace*, https://www.collinsaerospace.com/what-we-do/industries/military-and-defense/displays-and-controls/airborne/helmet-mounted-displays/f-35-gen-iii-helmet-mounted-display-system. Accessed Nov. 19, 2025.

"Factbox: Quotes from Trump Supporters at U.S. Republican Convention." *Reuters*, Aug. 25, 2020. reuters.com/article/usa-election-convention-quotes/factbox-quotes-from-trump-supporters-at-u-s-republican-convention-idINKBN25L08K. Accessed Aug. 25, 2025.

Fairservice Jr., Walter A., *The Threshold of Civilization*. Scribner, 1975.

Farenga, Patrick. "Homeschooling: Creating Alternatives to Education." *Bulletin of Science, Technology & Society* 18, no. 2. May, 1998, pp. 127–133.

Fazio, Russell H., and Michael A. Olson. "Implicit Measures in Social Cognition Research: Their Meaning and Use." *Annual Review of Psychology*, vol. 54, no. 1, 2003, pp. 297–327.

Feffer, John. "You Can't Just Vote Out Trumpism." *The Nation*, Jul. 2, 2020, http://www.thenation.com/article/politics/trump-election-2020/. Accessed Aug. 26, 2025.

"First Births to Older Mothers, 1972–2006." *CDC National Center for Health Statistics*, Apr. 2010, cdc.gov/nchs/data/databriefs/db68.pdf. Accessed Aug. 25, 2025.

Firth, Joseph, et al. "The 'Online Brain': How the Internet May Be Changing Our Cognition." *World Psychiatry*, vol. 18, no. 2, 2019, pp. 119–129.

Fischer, Claude S., and Michael Hout. *Century of Difference: How America Changed in the Last One Hundred Years*. Russell Sage Foundation, 2006.

Fiske, Alan Page. "Relational Models Theory: a Contemporary Overview." *Relational Models Theory: a Contemporary Overview*, edited by Nick Haslam, Lawrence Erlbaum Associates Publishers, 2004, pp. 3–25.

Fiske, Susan T. Tufts, and Shelley E. Taylor. *Social Cognition: from Brains to Culture*. Sage, 2020.

Fiske, Susan T., and Shelley E. Taylor. *Social Cognition: from Brains to Culture*. Sage, 2013.

Flor, Nick V., and Edwin Hutchins. "Analyzing Distributed Cognition in Software Teams: a Case Study of Team Programming during Perfective Software Maintenance." *Proceedings of the Fourth Annual Workshop on Empirical Studies of Programmers*, edited by Jurgen Koenemann-Belliveau, Thomas G. Moher, and Scott P. Robertson, 1991, pp. 36–59.

Flynn, James R. "The Mean IQ of Americans: Massive Gains 1932 to 1978." *Psychological Bulletin*, vol. 95, no. 1, 1984, p. 29.

Flynn, James R. *What Is Intelligence? Beyond the Flynn Effect*. Cambridge University Press, 2007.

Fodor, Jerry A. "Modules, Frames, Fridgeons, Sleeping Dogs, and the Music of the Spheres." *The Robot's Dilemma*, Ablex, 1987, pp. 139–149.

Fodor, Jerry A. *The Language of Thought*. Harvester Press, 1975.

Fodor, Jerry A. *The Mind Doesn't Work That Way: the Scope and Limits of Computational Psychology*. MIT Press, 2001.

Fodor, Jerry A. *The Modularity of Mind: an Essay on Faculty Psychology*. MIT Press, 1983.

Fox, Eva. "Tesla Autopilot Is Over 9 Times Safer than Human Drivers in Cars of Other Manufacturers." *Tesmanian*, Jan. 10, 2023, https://www.tesmanian.com/blogs/tesmanian-blog/tesla-autopilot-is-x9-safer-tesla-vehicles-are-x2-safer-than-cars-of-ot her-brands?srsltid=AfmBOopq91L7HcPnPCk_FuAeGagEffihykIlfhzyJt9TFvE7nWhCexpu. Accessed Nov. 19, 2025.

Fox, Justin. "The Disruption Myth" *The Atlantic*, Oct. 2014, https://www.theatlantic.com/magazine/archive/2014/10/the-disruption-myth/379348/. Accessed Dec. 13, 2025.

Frankl, Viktor E. *Man's Search for Meaning*. Translated by Ilse Lasch, Beacon Press, 2006.

Freud, Sigmund. "The Unconscious." *The Standard Edition of the Complete Psychological Works of Sigmund Freud*, translated by James Strachey, vol. 14, Hogarth Press, 1957, pp. 166–215.

Friedman, Milton. *Capitalism and Freedom*. University of Chicago Press, 1962.

Fry, Richard. "Millennials Are the Largest Generation in the U.S. Labor Force." *Pew Research Center*, Apr. 11, 2018, https://www.pewresearch.org/short-reads/2018/04/11/millennials-largest-generation-us-labor-force/. Accessed Nov. 19, 2025.

Fujii, Jessica A., et al. "Ecological Drivers of Variation in Tool-Use Frequency across Sea Otter Populations." *Behavioral Ecology*, vol. 26, no. 2, 2015, pp. 519–526.

Fuller, R. Buckminster. *Critical Path*. St. Martin's Press, 1981.

Fuller, R. Buckminster. *Operating Manual for Spaceship Earth*. Southern Illinois University Press, 1969.

Gabor, Dennis. *Inventing the Future*. Alfred A. Knopf, 1964.

Gaines, Jeff, and Jim Sorensen. *Fat, Broke, and Busy: Escaping America's "Never Enough" Culture."* Jul. 25, 2013.

Galam, Serge, and Taksu Cheon. "Tipping Points in Opinion Dynamics: A Universal Formula in Five Dimensions." *Frontiers in Physics* 8, no. 566580, 2020.

Gallagher, Shaun, and Dan Zahavi. *The Phenomenological Mind*. 2nd ed., Routledge, 2020.

Gallup. "American Pride Slips to New Low." *Gallup Poll*, Jun. 30, 2025, https://news.gallup.com/poll/692150/american-pride-slips-new-low.aspx. Accessed Nov. 19, 2025.

Gallup. "U.S. Depression Rates Reach New Highs." *Gallup News*, Mar. 26, 2025, news.gallup.com/poll/505745/depression-rates-reach-new-highs.aspx. Accessed Nov. 5, 2025.

Gamanut, Radu, et al. "The Mouse Cortical Connectome, Characterized by an Ultra-Dense Cortical Graph, Maintains Specificity by Distinct Connectivity Profiles." *Neuron*, vol. 97, no. 3, 2017, pp. 698–715.

Gazit, Irit, and Joseph Terkel. "Domination of Olfaction Over Vision in Explosives Detection by Dogs." *Applied Animal Behaviour Science*, vol. 82, no. 1, 2003, pp. 65–73.

Gazzaley, Adam, and Larry D. Rosen. *The Distracted Mind: Ancient Brains in a High-Tech World*. MIT Press, 2016.

Gazzaniga, Michael S. *The Mind's Past*. University of California Press, 2000.

Gazzaniga, Michael S. "The Split Brain in Man." *Scientific American*, vol. 217, no. 2, 1967, pp. 24–29.

"GDP per Capita (Current US\$)." *World Bank*, data.worldbank.org/indicator/NY.GDP.PCAP.CD. Accessed Aug. 25, 2025.

Geary, David C. *The Origin of Mind: Evolution of Brain, Cognition, and General Intelligence*. American Psychological Association, 2005.

Geary, James. "Brain-Machine Interface (BMI)." *Edge.org*, Feb. 21, 2025, https://www.edge.org/response-detail/11756. Accessed Aug. 23, 2025.

Gee, Henry. *The Decline and Fall of the Human Empire: Why Our Species Is on the Edge of Extinction*. First U.S. edition, St. Martin's Press, 2025.

Geels, Frank W. "Micro-Foundations of the Multi-Level Perspective on Socio-Technical Transitions: Developing a Multi-Dimensional Model of Agency through Crossovers between Social Constructivism, Evolutionary Economics and Neo-Institutional Theory." *Technological Forecasting and Social Change*, no. 152. 2020, pp. 119–894.

Geels, Frank W. "Technological Transitions as Evolutionary Reconfiguration Processes: a Multi-Level Perspective and a Case-Study." *Research Policy*, vol. 31, no. 8–9, 2002, pp. 1257–1274.

Geels, Frank W. "The Multi-Level Perspective on Sustainability Transitions: Responses to Seven Criticisms." *Environmental Innovation and Societal Transitions*, vol. 1, no. 1, 2011, pp. 24–40.

Geels, Frank W., and Johan Schot. "Typology of Transition Pathways." *Research Policy* no. 36.3. 2007, pp. 399–417.

Geertz, Clifford. *The Interpretation of Cultures: Selected Essays*. Basic Books, 1973.

Gell-Mann, Murray. *The Quark and the Jaguar: Adventures in the Simple and the Complex*. W.H. Freeman, 1994.

Gelman, Susan A. "Learning from Others: Children's Construction of Concepts." *Annual Review of Psychology*, vol. 60, 2009, pp. 115–140.

"Generation Z and Mental Health." *Annie E. Casey Foundation*, Mar. 3, 2021, https://www.aecf.org/blog/generation-z-and-mental-health. Accessed Nov. 9, 2025.

Georgescu-Roegen, Nicholas. *The Entropy Law and the Economic Process*. Harvard University Press, 1971.

Gerstel, Naomi. "The Third Shift: Gender and Care Work outside the Home." *Qualitative Sociology*, vol. 23, no. 4, 2000, pp. 467–483.

Gesner, Conrad. *Bibliotheca Universalis*. Zurich, 1545.

Gibson, William. *Neuromancer*. Ace Books, 1984.

Giedion, Sigfried. *Mechanization Takes Command: a Contribution to Anonymous History*. Oxford University Press, 1948.

Gilbert, Daniel T., et al. "On Cognitive Busyness: When Person Perceivers Meet Persons Perceived." *Journal of Personality and Social Psychology*, vol. 54, 1988, pp. 733–740.

Gillian, Duncan. "How Abu Dhabi's New Quantum Computer Could Help Solve the Mysteries of Science." *The National*, Aug. 18, 2021, https://www.thenationalnews.com/uae/science/2021/08/18/how-abu-dhabis-new-quantum-computer-could-help-solve-the-mysteries-of-science/. Accessed Nov. 19, 2025.

Gilliland, Mike, and Euvie Ivanova, hosts. "FTP133: Jim Rutt – How to Transition to a GameB Society." *Future Thinkers*, Dec. 8, 2020, https://futurethinkers.org/jim-rutt-gameb/. Accessed Aug. 26, 2025.

Gladwell, Malcolm. *Outliers: The Story of Success*. Little, Brown, 2008.

Gladwell, Malcolm. *Talking to Strangers – What We Should Know about the People We Don't Know*. Little Brown and Company, 2019.

Gladwell, Malcolm. *The Tipping Point: How Little Things Can Make a Big Difference*. Little, Brown and Company, 2000.

Glanz, James. "Power, Pollution and the Internet." *The New York Times*, Sep. 22, 2012.

Gleick, James. *Chaos: Making a New Science*. Viking Press, 1987.

Glenn, Siedah Garrett, and Michael Jackson. "Man in the Mirror." *Bad*, performed by Michael Jackson, Epic Records, 1987.

"Global GDP Over the Long Run." *Our World in Data*, ourworldindata.org/grapher/global-gdp-over-the-long-run. Accessed Aug. 25, 2025.

Gobble, MaryAnn The Case Against Disruptive Innovation. Research–Technology Management, vol. 58(1), 2015, pp. 59–63.

Goethe, Johann Wolfgang von. "The Sorcerer's Apprentice." Translated by Katrin Gygax, *Gygatext*, 2013, https://www.gygatext.ch/english_translations_zurich_sorcerers_apprentice.html. Accessed Nov. 12, 2025.

Golding, William. *Lord of the Flies*. Faber & Faber, 1954.

Goody, Jack. *The Domestication of the Savage Mind*. Cambridge University Press, 1977.

Gould, Stephen J. *The Mismeasure of Man*. W. W. Norton, 1981.

Gould, Stephen J. "The Spice of Life". *Leader to Leader*. 2000, no.15, pp.14–19.

Gould, Stephen J., and Elisabeth S. Vrba. "Exaptation – a Missing Term in the Science of Form." *Paleobiology*. 1982, pp. 4–15.

Graham, Billy. *The Holy Spirit: Activating God's Power in Your Life*. Word Books, 1978.

Granovetter, Mark S. "The Strength of Weak Ties." *American Journal of Sociology*, vol. 78, no. 6, May 1973, pp. 1360–1380.

Grant, Peter R., and B. Rosemary Grant. *How and Why Species Multiply: the Radiation of Darwin's Finches*. Princeton University Press, 2007.

Grant, Peter R., and B. Rosemary Grant. "Unpredictable Evolution in a 30-Year Study of Darwin's Finches." *Science*, vol. 296, no. 5568, 2002, pp. 707–711.

"Grasshopper Behavior." *BioExpedition*, Jun. 28, 2024, www.bioexpedition.com/grasshopper-behavior/. Accessed Sep. 23, 2025.

Gray, Chris Hables, (ed.). *The Cyborg Handbook*. Routledge, 1995.

Gray, Richard. "Be A-R-fraid: Augmented Reality Game Turns Your Home into Horror Story as Zombies, Demons Stalk You." *Daily Mail*, May 1, 2015. www.dailymail.co.uk/sciencetech/article-3063974/Be-AR-fraid-afraid-Augmented-reality-game-turns-home-horror-story-zombies-demons-stalk-you.html. Accessed Aug. 23, 2025.

Gray, Terry. "Some of My Favorite Quotations." *University of Washington*, 2000, https://staff.washington.edu/gray/misc/quotes.html. Accessed Nov. 11, 2025.

Graylin, Alvin W., and Louis Rosenberg. *Our Next Reality: How the AI-Powered Metaverse Will Reshape the World*. Foreword by Neal Stephenson, Nicholas Brealey Publishing, 2024.

Graziano, Michael S. A. *Consciousness and the Social Brain*. Oxford University Press, 2013.

Greenwald, Anthony G., et al. "A Unified Theory of Implicit Attitudes, Stereotypes, Self-Esteem, and Self-Concept." *Psychological Review*, vol. 109, no. 1, 2002, p. 3.

Gregory, Richard L., *The Intelligent Eye*. Weidenfeld and Nicolson, 1970.

Greising, Jack H. and John Quinn. "First Hague Peace Conference." *EBSCO Research Starters*, www.ebsco.com/research-starters/history/first-hague-peace-conference, 2023. Accessed Dec 3, 2025.

Griffiths, Paul E., and Karola Stotz. "How the Mind Grows: A Developmental Perspective on the Biology of Cognition." *Synthese*, vol. 122, no. 1–2, 2000, pp. 29–5.

Grimm, Volker, et al. "Pattern-Oriented Modeling of Agent-Based Complex Systems: Lessons from Ecology." *Science*, vol. 310, no. 5750, 2005, pp. 987–991.

"Growth of the World's Urban and Rural Population, 1920–2000." *United Nations*, 1969, population.un.org/wup/Archive/Files/studies/United%20Nations%20(1969)%20-%20Growth%20of%20the%20World's%20Urban%20and%20Rural%20Population,%201920-2000.pdf. Accessed Aug. 25, 2025.

Gu, DF, et al. "The Use of Long PCR to Confirm Three Common Alleles at the CYP2A6 Locus and the Relationship Between Genotype and Smoking Habit." *Annals of Human Genetics*, vol. 64, no. 5, 2000, pp. 383–390.

Guilfoyle, Kimberly. "RNC 2020 Day 1." *CNN Politics*, Aug. 25, 2020, cnn.com/politics/live-news/rnc-2020-day-1/. Accessed Aug. 25, 2025.

Gumplowicz, Ludwig. *The Outlines of Sociology*. American Academy of Political and Social Science, 1899.

Hagel, John, et al. "The Big Shift: Measuring the Forces of Change." *Deloitte Center for the Edge*, 2009, www2.deloitte.com/us/en/pages/center-for-the-edge/articles/big-shift-measuring-forces-of-change.html. Accessed Aug. 19, 2025.

Hagel, John, et al. "*The New Reality: Constant Disruption*." *Harvard Business Review*, Jan. 17, 2009, hbr.org/2009/01/the-new-reality-constant-disru. Accessed Dec. 5, 2025.

"Hague Convention." Encyclopedia Britannica, Jul. 20, 1998, www.britannica.com/event/Hague-Conventions. Accessed Mar 16, 2024.

"Hague Conventions of 1899 and 1907." *Wikipedia, Wikimedia Foundation,* last edited Sep. 25, 2025, en.wikipedia.org/wiki/Hague_Conventions_of_1899_and_1907.

Haidt, Jonathan. *The Happiness Hypothesis: Finding Modern Truth in Ancient Wisdom.* Basic Books, 2006.

Haines, Michael R. "American Fertility In Transition: New Estimates of Birth Rates in the United States, 1900–1910." *Demography, no.* 26.1. 1989, pp. 137–148.

Haines, Michael R. "Fertility and Mortality in the United States, 1790–2000." *A Population History of North America,* edited by Michael R. Haines and Richard H. Steckel, Cambridge University Press, 2000, pp. 139–230.

Hall, Chris. "The More This Rolex Costs, the More You Want It. Here's Why." *Wired,* 6 Sep. 6, 2024, https://www.wired.com/story/the-more-this-rolex-costs-the-more-you-want-it-this-is-why/. Accessed Nov. 19, 2025.

Hall, Edward T. *Beyond Culture.* Doubleday, 1976.

Halpin, Harry. "Does the Web Extend the Mind?" *Proceedings of the 5th Annual ACM Web Science Conference* (*WebSci '13*), ACM, 2013, pp. 139–147.

Halsband, Ulrike. "Bilingual and Multilingual Language Processing." *Journal of Physiology-Paris,* vol. 99, no. 4–6, 2006, pp. 355–369.

Hamid, Iman, et al. "Rapid Adaptation to Malaria Facilitated by Admixture in the Human Population of Cabo Verde." *eLife,* vol. 10, 2021, e631.

Hampton, Simon J. "Can Evolutionary Psychology Learn from the Instinct Debate?" *History of the Human Sciences,* vol. 19, no. 4, 2006, pp. 57–74.

Han, Shihui, et al. "Cultural Influences on Neural Substrates of Attentional Control." *Psychological Science,* vol. 19, no. 1, 2008, pp. 12–17.

Harari, Yuval N. "Yuval Noah Harari on Big Data, Google and the End of Free Will." *Financial Times,* Aug. 26, 2016, www.ft.com/content/50bb4830-6a4c-11e6-ae5b-a7cc5dd5a28c. Accessed Aug. 23, 2025.

Harari, Yuval Noah. *21 Lessons for the 21st Century.* Spiegel & Grau, 2018.

Haraway, Donna J. "Cyborg Manifesto: Science, Technology, and Social-Feminist in the Late 20th Century." *Social Review,* no. 80. 1985, pp. 65–108.

Harford, Tim. "Big Data: a Big Mistake?" *Significance,* vol. 11, no. 5, 2014, pp. 14–19.

Harrington, Anne. "At the Intersection of Knowledge and Values: Fragments of a Dialogue in Woods Hole, Massachusetts, Aug. 1990." *So Human a Brain: Knowledge and Values in the Neurosciences,* Birkhäuser Boston, 1992, pp. 247–324.

Harris, Lasana T., et al. "Attributions on the Brain: Neuro-Imaging Dispositional Inferences, Beyond Theory of Mind." *NeuroImage,* vol. 28, no. 4, 2005, pp. 763–769.

Hawken, Paul. *Blessed Unrest: How the Largest Movement in the World Came into Being, and Why No One Saw It Coming.* Viking, 2007.

Hawken, Paul. *The Ecology of Commerce: A Declaration of Sustainability.* HarperBusiness, 1993.

Hawken, Paul. "You Are Brilliant, and the Earth Is Hiring." *The NAMTA Journal*, vol. 38, no. 1, Winter 2013, pp. 269–72.

Hawkes, Kristen. "The Grandmother Effect." *Nature*, vol. 428, no. 6979, 2004, pp. 128–129.

Hawking, Stephen. Interview. *San Jose Mercury News*, Jan. 23, 2000.

Hawkins, Jeff, with Sandra Blakeslee. *A Thousand Brains: a New Theory of Intelligence*. Basic Books, 2021.

Hawks, John, et al. "Recent Acceleration of Human Adaptive Evolution." *Proceedings of the National Academy of Sciences*, vol. 104, no. 52, 2007, pp. 20753–20758.

Hawthorn Center. "The Deeper Benefits of a Digital Detox." *Hawthorn Center*, May 22, 2025, https://www.hawthorncenter.com/the-deeper-benefits-of-a-digital-detox/. Accessed 19 Mar. 2025.

Hayek, Friedrich A. *The Sensory Order: an Inquiry into the Foundations of Theoretical Psychology*. University of Chicago Press, 1952.

Haynes, John-Dylan, et al. "Reading Hidden Intentions in the Human Brain Using High-Resolution fMRI." *Current Biology*, vol. 17, no. 4, 2007, pp. 323–328.

He, Qinghua, et al. "Brain Anatomy Alterations Associated with Social Networking Site (SNS) Addiction." *Scientific Reports*, vol. 7, no. 1, 2017, 45064.

He, Qinghua, et al. "Excess Social Media Use in Normal Populations Is Associated with Amygdala-Striatal but Not with Prefrontal Morphology." *Psychiatry Research: Neuroimaging*, vol. 269, 2017, pp. 31–35.

Hegel, Georg Wilhelm Friedrich. *The Phenomenology of Mind*. Translated by J. B. Baillie, Allen and Unwin, 1910.

Heidegger, Martin. *Being and Time*. Translated by John Macquarrie and Edward Robinson, Harper & Row, 1962.

Heidegger, Martin. *What Is Called Thinking?* Translated by J. Glenn Gray, Harper & Row, 1968.

Heimans, Jeremy, and Henry Timms. *New Power: How Power Works in Our Hyperconnected World–and How to Make It Work for You*. Doubleday, 2018.

Helpman, Elhanan, editor. *General Purpose Technologies and Economic Growth*. MIT Press, 1998.

Heinemann, William, and Donald J. Munro. *Individualism and Holism: Studies in Confucian and Taoist Values*. Donald J. Munro, editor. Center for Chinese Studies, University of Michigan, 1985.

Hendler, James, and Jennifer Golbeck. "Metcalfe's Law, Web 2.0, and the Semantic Web." *Journal of Web Semantics*, vol. 6, no. 1, 2008, pp. 14–20.

Henneberg, Maciej. "Brain Size/Body Weight Variability in Homo Sapiens: Consequences for Interpreting Hominid Evolution." *HOMO Journal of Comparative Human Biology*, vol. 39, 1990, pp. 121–130.

Henneberg, Maciej. "Decrease of Human Skull Size in the Holocene." *Human Biology*, vol. 60, no. 3, 1988, pp. 395–405.

Henneberg, Maciej. "The Rate of Human Morphological Microevolution and Taxonomic Diversity of Hominids." *Studies in Historical Anthropology*, vol. 4, 2006, pp. 49–59.

Henneberg, Maciej, and Arthur Saniotis. "Evolutionary Origins of Human Brain and Spirituality." *Anthropologischer Anzeiger*, vol. 67, 2009, pp. 427–438.

Henrich, Joseph, and Richard McElreath. "The Evolution of Cultural Evolution." *Evolutionary Anthropology*, vol. 12, no. 3, 2003, pp. 123–135.

Hermer-Vazquez, Linda, et al. "Language, Space, and the Development of Cognitive Flexibility in Humans: The Case of Two Spatial Memory Tasks." *Cognition*, vol. 79, no. 3, 2001, pp. 263–299.

Herr, Hugh. "The Intersection of Man and Machine." *NIH Catalyst*, vol. 29, no. 5, 2012.

Herrel, Anthony, et al. "Rapid Large-Scale Evolutionary Divergence in Morphology and Performance Associated with Exploitation of a Different Dietary Resource." *Pr oceedings of the National Academy of Sciences*, vol. 105, no. 12, 2008, pp. 4792–4795.

Hertz, Erica, James R. Hebert, and Joan Landon. "Social and Environmental Factors and Life Expectancy, Infant Mortality, and Maternal Mortality Rates: Results of a Cross-National Comparison." *Social Science & Medicine*, no. 39.1, 1994, pp. 105–114.

Heston, Charlton. "From My Cold, Dead Hands." *Speech at the NRA Annual Meeting*, Charlotte, NC, May 20, 2000. Transcript available via C-SPAN, www.c-span.org/video/?155368-1/nra-annual-meeting. Accessed Sep. 7, 2025.

Heyes, Cecilia. "Automatic Imitation." *Psychological Bulletin*, vol. 137, no. 3, 2011, pp. 463–483.

Hilbert, Martin. "Digital Technology and Social Change: the Digital Transformation of Society from a Historical Perspective." *PLOS ONE*, Jul. 15, 2020, www.ncbi.nlm.nih.gov/pmc/articles/PMC7366943/. Accessed Nov. 19, 2025.

Hill, Charles WL, and Frank T. Rothaermel. "The Performance of Incumbent Firms in the Face of Radical Technological Innovation." *Academy of Management Review*, vol. 28.2, 2003, pp. 257–274.

"History and Status of the U.S. Road System." *Assessing and Managing the Ecological Impacts of Paved Roads*, National Academies Press, 2005, pp. 49–65.

Hobbes, Hobbes, Thomas. "Third Objections." In René Descartes, *The Philosophical Writings of Descartes*, Vol. 2, translated by John Cottingham, Robert Stoothoff, and Dugald Murdoch, Cambridge University Press, 1984, pp. 122–123 (AT VII, 172–173).

Hockfield, Susan. *The Age of Living Machines: How Biology Will Build the Next Technology Revolution*. WW Norton & Company, 2019.

Hoffman, Reid, and Ben Casnocha. *The Start-up of You: Adapt to the Future, Invest in Yourself, and Transform Your Career*. Random House, 2012.

Hofstadter, Douglas R. *Gödel, Escher, Bach: an Eternal Golden Braid*. 20th anniversary ed., Basic Books, 1999.

Hohmann, Gottfried, and Barbara Fruth. "Culture in Bonobos? Between-Species and Within-Species Variation in Behavior." *Current Anthropology*, no. 44.4, 2003, pp. 563–571.

Holland, John H. *Emergence: from Chaos to Order*. Perseus Publishing, 1998.

Houston, Ben. *Exocortex Definition*. May 2000. http://www.exocortex.org/ben/exocortex. Accessed Aug. 24, 2025.

Howard, Cosmo. "What did a 'fair go' originally mean to Australians?" *Australian Journal of Political Science* 58, no. 2, 2023. pp. 210–229.

Howard, Gardner. *The Mind's New Science: a History of the Cognitive Revolution*. Basic Books, 1985.

Howard, Newton, and Erik Cambria. "Intention Awareness: Improving upon Situation Awareness in Human-Centric Environments." *Human-Centric Computing and Information Sciences*, vol. 3, no. 1, 2013, pp. 1–17.

Howell, William Smiley. *The Empathic Communicator*. Wadsworth Publishing Company, 1982.

Hrdy, Sarah Blaffer. *The Woman That Never Evolved*. Rev. ed., Harvard University Press, 1999.

Hull, Isabel V. *A Scrap of Paper: Breaking and Making International Law during the Great War*. Cornell University Press, 2014.

"Humanity at Its Finest, 20,000 Bikers Respond to 6-Yr-Old Kilian..." YouTube, http://www.youtube.com/watch?v=MZ85aFEivGo. Accessed Dec. 3, 2024.

Humphries, Mark. *The Spike: an Epic Journey through the Brain in 2.1 Seconds*. Princeton University Press, 2021.

Hunt, Gavin R. "Manufacture and Use of Hook-Tools by New Caledonian Crows." *Nature*, vol. 379, no. 6562, 1996, pp. 249–251.

Hutchins, Edwin. *Cognition in the Wild*. MIT Press, 1995.

Hutchins, Edwin. *Culture and Inference: a Trobriand Case Study*. Harvard University Press, 1980.

Hutchins, Edwin. *"How a Cockpit Remembers Its Speeds." Cognitive Science*, vol. 19, no. 3, 1995, pp. 265–288.

Hạnh, Thích Nhất. *Being Peace*. Parallax Press, 1987.

"'I can't afford McDonald's anymore': Customers slam McDonald's prices after paying almost \25 for 3 items." *Reddit*, Aug. 30, 2023, http://www.reddit.com/r/LateStageCapitalism/comments/17wn5n8/i_cant_afford_mcdonalds_anymore_customers_slam/. Accessed Sep. 9, 2025.

Iacoboni, Marco. *Mirroring People: The Science of Empathy and How We Connect with Others*. Picador, 2009.

Iddon, Paul. "What Is GPS Jamming and Why Is Israel Using It?" *The New Arab*, Apr. 22, 2024, https://www.newarab.com/analysis/what-gps-jamming-and-why-israel-using-it. Accessed Aug. 24, 2025.

Illich, Ivan. *Deschooling Society*. Harper & Row, 1971.

Inglehart, Ronald, and Christian Welzel. *Modernization, Cultural Change, and Democracy: The Human Development Sequence*. Cambridge University Press, 2005.

International Renewable Energy Agency (IRENA). "Renewable Power Generation Costs in 2024." *IRENA*, 2024, https://www.irena.org/Publications/2025/Jun/Renewable-Power-Generation-Costs-in-2024. Accessed Nov. 19, 2025.

IoT Analytics. "Number of Connected IoT Devices Growing 13 % to 18.8 Billion Globally." *State of IoT Summer 2024*, Sep. 3, 2024, https://iot-analytics.com/product/state-of-iot-summer-2024/. Accessed Nov 19, 2025.

Isaac, Mike. "Uber Is Said to Be Raising Money at a $62.5 Billion Valuation." *The New York Times*, 4 Dec. 2015.

"It's the Economy, Stupid." *Political Dictionary*, https://politicaldictionary.com/words/its-the-economy-stupid/. Accessed Aug. 26, 2025.

Jablonski, Nina G., and Robin H. Crompton. "Feeding Behavior, Mastication, and Tooth Wear in the Western Tarsier (Tarsius Bancanus)." *International Journal of Primatology*, vol.15.1, 1994, pp. 29–59.

Jack, William, and Tavneet Suri. "Risk Sharing and Transactions Costs: Evidence from Kenya's Mobile Money Revolution." *American Economic Review*, vol. 104, no. 1, 2014, pp. 183–223.

Jacobs, Jane. *The Death and Life of Great American Cities*. Random House, 1961.

Jacoby, Sarah F., et al. "Historic Redlining, Structural Racism, and Firearm Violence: A Structural Equation Modeling Approach." *Preventive Medicine*, vol. 153, 2021.

Jahan, Sarwat, Ahmed Saber Mahmud, and Chris Papageorgiou. "What Is Keynesian Economics?" *Finance & Development*, vol. 51, no. 3, 2014, pp. 53–54.

Jal, Emmanual, *My Life Is Art*, Counterpoint, 2023.

James, Clive. "Global Status of Commercialized Biotech/GM Crops: 2011." *ISAAA Briefs*, no. 43, International Service for the Acquisition of Agri-biotech Applications, 2011, https://www.isaaa.org/resources/publications/briefs/43/. Accessed Nov. 19, 2025.

James, Samuel D. *Digital Liturgies: Rediscovering Christian Wisdom in an Online Age*. Crossway, 2025.

James, William. "Great Men and Their Environment." *The Will to Believe and Other Essays in Popular Philosophy*, Harvard University Press, 1979, pp. 216–254.

James, William. "Is Life Worth Living?" *International Journal of Ethics*, vol. 6, no. 1, 1895, pp. 1–29.

James, William. *The Principles of Psychology*. Henry Holt and Company, 1890.

Jeffries, Vincent, and H. Edward Ransford. *Social Stratification: a Multiple Hierarchy Approach*. Allyn and Bacon, 1980.

Jerison, Harry J. *Evolution of the Brain and Intelligence*. Academic Press, 1973.

Jobs, Steve. "'You've Got to Find What You Love,' Jobs Says." Stanford University Commencement Address, Jun. 12, 2005. *Stanford News*, https://news.stanford.edu/stories/2005/06/steve-jobs-2005-graduates-stay-hungry-stay-foolish. Accessed Nov. 19, 2025.

Johansson, Roland S., and Åke Bernhard Vallbo. "Tactile Sensibility in the Human Hand: Relative and Absolute Densities of Four Types of Mechanoreceptive Units in Glabrous Skin." *The Journal of Physiology*, vol. 286, no. 1, 1979, pp. 283–300.

Johnson, Steven. *Extra Life: a Short History of Living Longer*. Riverhead, 2021.

Johnson, Steven. *Where Good Ideas Come from: the Natural History of Innovation*. Penguin, 2011.

Jokinen, Eeva, et al. "The Fourth Shift: on the Borders of Home, Labour and Affects." *Research plan submitted to the Academy of Finland*, Jan. 31, 2007, 4kaanne.wordpress.com/2009/02/14/neljas-kaanne-tutkimusprojekti/. Accessed Aug. 26, 2025.

Jonas, Gilbert. *Freedom's Sword: The NAACP and the Struggle Against Racism in America, 1909–1969*. Routledge, 2005.

Jones, Bryan D., and Frank R. Baumgartner. "From There to Here: Punctuated Equilibrium to the General Punctuation Thesis to a Theory of Government Information Processing." *Policy Studies Journal*, vol. 40.1, 2012, pp. 1–20.

Jones, Jeffrey M. "Confidence in U.S. Institutions Down; Average at New Low." *Gallup*, 5 Jul. 2022, news.gallup.com/poll/394283/confidence-institutions-down-average-new-low.aspx. Accessed Aug. 23, 2025.

Joseph, Rhawn. *The Right Brain and the Unconscious: Discovering the Stranger Within*. Springer, 2013.

June, Carl H., et al. "CAR T Cell Immunotherapy for Human Cancer." *Science*, vol. 359, no. 6382, 2018, pp. 1361–1365.

Kahane, Henry, and Renée Kahane. "Humanistic Linguistics." *The Journal of Aesthetic Education*, vol. 17, no. 4, 1983, pp. 65–89.

Kahneman, Daniel, et al. "Experimental Tests of the Endowment Effect and the Coase Theorem." *Journal of Political Economy*, vol. 98, 1990, pp. 1325–1348.

Kahneman, Daniel. *Attention and Effort*. Prentice-Hall, 1973.

Kahneman, Daniel. "Thinking, Fast and Slow | Daniel Kahneman | Talks at Google." *YouTube*, uploaded by Talks at Google, https://www.youtube.com/watch?v=CjVQJdIrDJo. Accessed Aug. 25, 2025.

Kan, Michael. "Artist Manipulates Google Maps Traffic View with Cart Full of Smartphones" *PC Mag*, February 2, 2020, https://www.pcmag.com/news/artist-manipulates-google-maps-traffic-view-with-cart-full-of-smartphones. Accessed Aug.24,2025.

Kanazawa, Satoshi. "The Savanna Principle." *Managerial and Decision Economics*, vol. 25.1, 2004, pp. 41–54.

Kanherkar, Riya R., et al. "Epigenetics across the Human Lifespan." *Frontiers in Cell and Developmental Biology*, vol. 2, 2014, 49.

"Kasa Smart Plugs." Amazon. https://www.amazon.com/Kasa-Smart-Required-Certified-. Accessed Dec. 2, 2025.

Katz, Eric, et al., editors. *Controlling Technology: Contemporary Issues.* Prometheus Books, 2003.

Katz, Lawrence F., and Alan B. Krueger. "The Rise and Nature of Alternative Work Arrangements in the United States, 1995–2015." *ILR Review*, vol. 72, no. 2, 2019, pp. 382–416.

Kay, Alan. "The Best Way to Predict the Future Is to Invent It." *EDUCOM Bulletin*, 1983.

Keen, Andrew. "Keen On... Ray Kurzweil: How Computers Will Reverse Engineer the Human Mind by 2029." *TechCrunch*, Dec. 3, 2012, techcrunch.com/2012/12/03/keen-on-ray-kurzweil-how-computers-will-reverse-engineer-the-human-mind-by-2029/ . Accessed Jul 27, 2024.

Keesara, Sirina, Andrea Jonas, and Kevin Schulman. "Covid-19 and Health Care's Digital Revolution." *New England Journal of Medicine*, vol. 382, no. 23, 2020, p. e82.

Keil, Frank C. "Words, Moms, and Things: Language as a Road Map to Reality." *Monographs of the Society for Research in Child Development*, vol. 63, no. 1, 1998, pp. 152–158.

Kelly, Kevin. *Out of Control: The New Biology of Machines, Social Systems, and the Economic World.* Addison-Wesley, 1994.

Kelly, Kevin. *The Inevitable: Understanding the 12 Technological Forces That Will Shape Our Future.* Viking, 2016.

Kelly, Kevin. "The Technium." *Edge*, edge.org/conversation/kevin_kelly-the-technium. Accessed 2 Dec. 2020.

Kelly, Kevin. *What Technology Wants.* Penguin, 2010.

Kendra, James M., and Tricia Wachtendorf. *American Dunkirk: The Waterborne Evacuation of Manhattan on 9/11.* Temple University Press, 2016.

Kennedy, Anthony M. *Packingham v. North Carolina.* 582 U.S. 98. Supreme Court of the United States, 2017.

Kennedy, John F. "Commencement Address at American University, Jun. 10, 1963." *John F. Kennedy Presidential Library and Museum*, https://www.jfklibrary.org/archives/other-resources/john-f-kennedy-speeches/american-university-19630610. Accessed on Nov. 19, 2025.

Keynes, John Maynard. *Economic Possibilities for our Grandchildren*, 1930. https://roar-assets-auto.rbl.ms/documents/43407/Intro_Session1.pdf. Accessed Nov. 12, 2025.

Keynes, John Maynard. *The General Theory of Employment, Interest and Money.* Macmillan, 1936.

Kingdon, Jonathan. *Lowly Origin: Where, When, and Why Our Ancestors First Stood Up.* Princeton University Press, 2003.

Kirsh, David. "The Intelligent Use of Space." *Artificial Intelligence*, vol. 73, 1995, pp. 31–68.

Kissinger, Henry A., et al. *The Age of AI: and Our Human Future*. Little, Brown and Company, 2021.

Knierim, James. "Cerebellum." *Neuroscience Online*, University of Texas Health Science Center at Houston, 2020, https://nba.uth.tmc.edu/neuroscience/m/s3/chapter05.html. Accessed on Nov. 19, 2025.

Knight, Rob. *Follow Your Gut: The Enormous Impact of Tiny Microbes.* TED Books, 2015.

Knight, Will. "Alpha Zero's 'Alien' Chess Shows the Power, and the Peculiarity, of AI." *MIT Technology Review*, Dec. 8, 2017, http://www.technologyreview.com/2017/12/08/147199/alpha-zeros-alien-chess-shows-the-power-and-the-peculiarity-of-ai/. Accessed Aug. 23, 2025.

Knill, David C., and Alexandre Pouget. "The Bayesian Brain: the Role of Uncertainty in Neural Coding and Computation." *Trends in Neurosciences*, vol. 27, no. 12, 2004, pp. 712–19.

Kondepudi, Dilip, et al. "Dissipative Structures and Irreversibility in Nature: Celebrating 100th Birth Anniversary of Ilya Prigogine (1917–2003)." *Chaos: an Interdisciplinary Journal of Nonlinear Science*, vol. 27, no. 10, 2017.

Korcok, M. "Blood Donations Dwindle in US after Post-Sept. 11 Wastage." *PMC*, 2002, http://www.ncbi.nlm.nih.gov/pmc/articles/PMC128413/. Accessed Jul. 16, 2024.

Korte, Martin. "The Impact of the Digital Revolution on Human Brain and Behavior: Where Do We Stand?" *Dialogues in Clinical Neuroscience*, vol. 22, no. 2, 2020, pp. 101–111.

Kosslyn, Stephen M. *Image and Mind.* Harvard University Press, 1980.

Krigsman, Michael, host. "When AI Goes Off the Rails: an Executive Action Plan." *CXOTalk*, episode 888, featuring Dr. David A. Bray, 25 Jul. 2025, www.cxotalk.com/episode/when-ai-goes-off-the-rails-an-executive-action-plan. Accessed Sep. 1, 2025.

Kuhn, Thomas S. "Second Thoughts on Paradigms". In: Suppe, Frederick (ed.) *The Essential Tension.* University of Chicago Press, 1977, pp. 293–319.

Kuhn, Thomas S. *The Structure of Scientific Revolutions.* University of Chicago Press, 1962.

Kumar, Naveen. "Big Data Statistics 2025 (Growth & Market Data)." *DemandSage*, Jun. 24, 2025, demandsage.com/big-data-statistics/. Accessed Nov. 19, 2025.

Kurzweil, Ray. "The Law of Accelerating Returns." *Alan Turing: Life and Legacy of a Great Thinker*, Springer, 2004.

Kurzweil, Ray. *The Singularity Is Near: When Humans Transcend Biology.* Penguin, 2006.

Kuznets, Simon. *Modern Economic Growth: Rate, Structure and Spread.* Yale University Press, 1966.

Kögel, Tomas, and Alexia Prskawetz. "Agricultural Productivity Growth and Escape from the Malthusian Trap." *Journal of Economic Growth*, vol. 6, 2001, pp. 337–357.

Kühn, Simone, et al. "Playing Super Mario Induces Structural Brain Plasticity: Gray Matter Changes Resulting from Training with a Commercial Video Game." *Molecular Psychiatry*, vol. 19, no. 2, 2014, pp. 265–271.

Laland, Kevin N., and Gillian R. Brown. *Sense and Nonsense: Evolutionary Perspectives on Human Behaviour.* 2nd ed., Oxford University Press, 2011.

Laland, Kevin N., and Marcus W. Feldman. *Niche Construction: the Neglected Process in Evolution.* Princeton University Press, 2003.

Laland, Kevin N., et al. "Niche Construction, Biological Evolution and Cultural Change." *The Behavioral and Brain Sciences*, vol. 23, 2000, pp. 131–175.

"Land Rover Debuts 'See-Through' Virtual Imaging Technology in Discovery Vision Concept." *Media.LandRover.com*, 8 Apr. 2014, media.landrover.com/en-us/news/2014/04/land-rover-debuts-see-through-virtual-imaging-technology-discovery-vision-concept. Accessed Aug. 23, 2025.

Lang, Elmar Wolfgang, et al. "Brain Connectivity Analysis: a Short Survey." *Computational Intelligence and Neuroscience*, 2012.

Lanier, Jaron. *Who Owns the Future?* Simon & Schuster, 2013.

Lashley, Karl S. "In Search of the Engram." *Society of Experimental Biology*, Symposium, vol. 4, 1950, pp. 454–482.

Lashley, Karl S. "The Behavioristic Interpretation of Ionsciousness." *Psychological Review*, vol. 30, 1923, pp. 237–272 (Part I) & pp. 329–353 (Part II).

Lasswell, Harold D. *Propaganda Technique in the World War.* Ravenio Books, 1927.

Lau, Hakwan, and Richard E. Passingham. "Relative Blindsight in Normal Observers and the Neural Correlate of Visual Consciousness." *Proceedings of the National Academy of Sciences*, vol. 103, no. 49, 2006, pp. 18763–68.

Le Bon, Gustave. *The Crowd: a Study of the Popular Mind.* Dover Publications, 2002.

Leach, Helen M. "Human Domestication Reconsidered." *Current Anthropology*, vol. 44, no. 3, 2003, pp. 349–368.

Lee, Richard B., and Irven DeVore, editors. *Man the Hunter.* Aldine Publishing Company, 1968.

Lee, Timothy L. "Are Self-Driving Cars Already Safer Than Human Drivers?" *Hacker News*, 9 Sep. 2023.

Lenski, Gerhard Emmanuel. *Power and Privilege: a Theory of Social Stratification.* UNC Press Books, 1984.

Leonardi, Michela, et al. "The Evolution of Lactase Persistence in Europe: A Synthesis of Archaeological and Genetic Evidence." *International Dairy Journal*, vol. 22, no. 2, 2012, pp. 88–97.

Leopold, Aldo. *A Sand County Almanac: and Sketches Here and There.* Oxford University Press, 1949.

Lepore, Jill. "The Disruption Machine". The New Yorker, Jun. 16, 2014. https://www.newyorker.com/magazine/2014/06/23/the-disruption-machine. Accessed Dec. 12. 2025.

Levine, Rick, et al. *The Cluetrain Manifesto*. Basic Books, 2009.

Levy, Daniel, et al. "Evidence for a Gene Influencing Blood Pressure on Chromosome 17: Genome Scan Linkage Results for Longitudinal Blood Pressure Phenotypes in Subjects from the Framingham Heart Study." *Hypertension*, vol. 36, no. 4, 2000, pp. 477–483.

Lewens, Tim. *Darwin*. Routledge, 2007.

Lewis, David Levering. *W.E.B. Du Bois: Biography of a Race, 1868–1919*. Henry Holt, 1993.

Lewontin, Richard C., Steven Rose, and Leon J. Kamin. *Not in Our Genes: Biology, Ideology and Human Nature*. Pantheon Books, 1984.

Li, Qing. *Forest Bathing: How Trees Can Help You Find Health and Happiness*. Viking, 2018.

Li, Shuangjun, et al. "Critical review on mobile direct air capture: Concept expansion, characteristic description, and performance evaluation." *Matter*, vol. 7.3, 2024, pp. 889–933.

Li, Ting-Kai, et al. "Genetic and Environmental Influences on Alcohol Metabolism in Humans." *Alcoholism: Clinical and Experimental Research*, vol. 25, no. 1, 2001, pp. 136–144.

Libet, Benjamin, et al. "Time of Conscious Intention to Act in Relation to Onset of Cerebral Activities (Readiness-Potential, pp. The Unconscious Initiation of a Freely Voluntary Act." *Brain*, vol. 106, 1983, pp. 623–642.

Libet, Benjamin. "Unconscious Cerebral Initiative and the Role of Conscious Will in Voluntary Action." *Behavioral and Brain Sciences*, vol. 8, no. 4, 1985, pp. 529–56.

Licklider, Joseph C.R. "Man–Computer Symbiosis." *IRE Transactions on Human Factors in Electronics*, vol. 1, 2008, pp. 4–11.

Licklider, Joseph C.R. "Memorandum for Members and Affiliates of the Intergalactic Computer Network." *Advanced Research Projects Agency*, Apr. 23, 1963, https://worrydream.com/refs/Licklider_1963_-_Members_and_Affiliates_of_the_Intergalactic_Computer_Network.pdf. Accessed Sep. 23, 2024.

Licklider, Joseph Carl Robnett. "In Memoriam, J.C.R. Licklider, 1915–1990." *IEEE Annals of the History of Computing*, vol. 12, no. 1, 1990, pp. 3–4.

Liebherr, Magnus, et al. "Smartphones and Attention, Curse or Blessing?–A Review on the Effects of Smartphone Usage on Attention, Inhibition, and Working Memory." *Computers in Human Behavior Reports*, vol. 1, 2020, 100005.

"Life Expectancy." *Our World in Data*, ourworldindata.org/life-expectancy. Accessed Aug. 25, 2025.

"Lifestyle Diseases Pose New Burden on Africa." *UN Africa Renewal*, Dec. 2016, un.org/africarenewal/magazine/december-2016-march-2017/lifestyle-diseases-pose-new--burden-africa. Accessed Aug. 25, 2025.

Lin, Yutang. *My Country and My People*. London, 1936.

Linda. "Robert Macfarlane's Landmarks." *Under the Gables*, 4 Mar. 2017, underthegables.blogspot.com/2017/03/robert-macfarlanes-landmarks.html. Accessed Aug. 23, 2025.

Lindblom, Jessica, and Tom Ziemke. "Social Situatedness of Natural and Artificial Intelligence: Vygotsky and Beyond." *Adaptive Behavior*, vol. 11, no. 2, 2003, pp. 79–96.

Lipsitz, George. *The Possessive Investment in Whiteness: How White People Profit from Identity Politics*. Revised and expanded ed., Temple University Press, 2006.

Liu, Cixin. *Death's End*. Vol. 3. Macmillan, 2016.

Loch, Christoph H., and Bernardo A. Huberman. "A Punctuated-Equilibrium Model of Technology Diffusion." *Management Science*, vol.45.2, 1999, pp. 160–177.

Locke, John. *Two Treatises of Government*. Edited by Peter Laslett, Cambridge University Press, 1988.

"Locusts, Facts and Photos." *National Geographic*, May 3, 2021, www.nationalgeographic.com/animals/invertebrates/facts/locusts. Accessed Dec. 2, 2025.

Lonsdorf, Elizabeth V. "Sex Differences in the Development of Termite-Fishing Skills in the Wild Chimpanzees, Pan troglodytes schweinfurthii, of Gombe National Park, Tanzania." *Animal Behaviour*, vol. 70, no. 3, 2005, pp. 673–683.

Loudoun County, Virginia. "Loudoun County, VA – Data Centers and Virtual Infrastructure." *Press release, Loudoun County Government*, https://biz.loudoun.gov/key-business-sectors/data-centers/. Accessed Sep. 7, 2025.

Lumsden, Charles J., and Edward O. Wilson. *Genes, Mind, and Culture*. Harvard University Press, 1981.

Lund, Kristine. "Modelling the Co-elaboration of Knowledge: Connecting Cognitive, Linguistic, Social and Interactional Systems." *Language Is a Complex Adaptive System: Explorations and Evidence*, edited by Kristine Lund et al., Language Science Press, 2022, pp. 35–48.

Lundgaard, Stine Schmieg, and Claus Andreas Foss Rosenstand. *Investigating Disruption: a Literature Review of Core Concepts of Disruptive Innovation Theory*. Aalborg University Press, 2019.

Luthar, Suniya S. "Children of the Affluent: Challenges to Well-Being." *Current Directions in Psychological Science*, vol. 12, no. 2, 2003, pp. 49–53.

Luthar, Suniya S. "The Culture of Affluence: Psychological Costs of Material Wealth." *Child Development*, vol. 74, no. 6, 2003, pp. 1581–1593.

Luthar, Suniya S., and Shawn J. Latendresse. "Children of the Affluent: Challenges to Well-Being." *Current Directions in Psychological Science*, vol. 14, no. 1, 2005, pp. 49–53.

Lévy, Pierre. *Collective Intelligence: Mankind's Emerging World in Cyberspace*. Perseus Books, 1997.

Macfarlane, Robert, and Jackie Morris. *The Lost Words*. Hamish Hamilton, 2017.

Machlup, Fritz. *The Production and Distribution of Knowledge in the United States*. Princeton University Press, 1962.

MacLean, Paul D. *A Triune Concept of the Brain and Behaviour*. University of Toronto Press, 1973.

MacLean, Paul. D. "Man and His Animal Brains." *Modern Medicine*, vol. 32, 1964, pp. 95–106.

Macrae, C. Neil, and Galen V. Bodenhausen. "Social Cognition: Categorical Person Perception." *British Journal of Psychology*, vol. 92, no. 1, 2001, pp. 239–255.

Madison, Paige. "Why the 'Missing Link' Fossil Was Almost Missed." *National Geographic*, Oct. 18, 2024.

Madrigal, Alexis C. "The Hut Where the Internet Began." *The Atlantic*, Jul. 8, 2013, http://www.theatlantic.com/technology/archive/2013/07/the-hut-where-the-internet-began/277551/. Accessed Aug. 23, 2025.

Maguire, Eleanor A., Richard S. J. Frackowiak, and Christopher D. Frith. "Recalling Routes around London: Activation of the Right Hippocampus in Taxi Drivers." *Journal of Neuroscience*, vol. 17, no. 18, 1997, pp. 7103–7110.

Malafouris, Lambros. "The Cognitive Basis of Material Engagement: Where Brain, Body and Culture Conflate." *Rethinking Materiality: the Engagement of Mind with the Material World*, edited by Elizabeth DeMarrais, Colin Gosden, and Chris Renfrew, McDonald Institute for Archaeological Research, 2004, pp. 53–62.

Malafouris, Lambros. "Thinking as 'Thinging': Psychology with Things." *Current Directions in Psychological Science*, vol. 29, no. 1, 2020, pp. 3–8.

Manusov, Valerie. "Mimicry or Synchrony: the Effects of Intentionality Attributions for Nonverbal Mirroring Behavior." *Communication Quarterly*, vol. 40, no. 1, 1992, pp. 69–83.

Maor, Elad, et al. "Voice Signal Characteristics Are Independently Associated with Coronary Artery Disease." *Mayo Clinic Proceedings*, vol. 93, no. 7, Elsevier, 2018.

Marcelle, Enitan T., et al. "The Role of the Hippocampus in Working Memory and Word Reading: Novel Neural Correlates of Reading Among Youth Living in the Context of Economic Disadvantage." *Developmental Cognitive Neuroscience*, vol. 71, 2025, pp. 101491.

Marcus, Lilit."Raw Dogging: the New Travel Trend." *CNN Travel*, Jun. 28, 2024, https://www.cnn.com/2024/06/28/travel/raw-dogging-travel-trend-explainer-intl-hnk/index.html. Accessed Dec 9, 2025.

Marcuse, Herbert. *Repressive Tolerance*. 1965. Reprinted in *A Critique of Pure Tolerance*, by Robert Paul Wolff, Barrington, Moore Jr., and Herbert Marcuse, Beacon Press, 1969, pp. 81–123.

Marlowe, Frank W. “Hunter-Gatherers and Human Evolution.” *Evolutionary Anthropology: Issues, News, and Reviews*, vol. 14, no. 2, 2005, pp. 54–67.

Masnick, Mike. “Can Google Be Held Liable for Man Who Died Following Google Maps Over a Collapsed Bridge?” *Techdirt*, 25 Sept. 2023, https://www.techdirt.com/2023/09/25/can-google-be-held-liable-for-man-who-died-following-google-maps-over-a-collapsed-bridge/. Accessed Aug. 24, 2025.

Masri-Zada, Tariq, et al. ”The Impact of Social Media and Technology on Child and Adolescent Mental Health.” *National Institutes of Health*, 2025.

“Mass Murder Leaves 6 Dead, Including Unborn Child, in Indianapolis.” *WANE*, Jan. 24, 2021, www.wane.com/news/mass-murder-leaves-6-dead-including-unborn-child-in-indianapolis/. Accessed Aug. 23, 2025.

Maxwell, John C. *Developing The Leader within You.* Thomas Nelson Inc, 1993.

Mayer, Emeran. *The Mind-Gut Connection: How the Hidden Conversation within Our Bodies Impacts Our Mood, Our Choices, and Our Overall Health.* Harper Wave, 2016.

Mayr, Ernst “Change of Genetic Environment and Evolution” In J. Huxley, A. C. Hardy and E. B. Ford. Evolution as a Process. Allen and Unwin, 1954, pp. 157–180.

Mayr, Ernst. *Populations, Species, and Evolution: an Abridgment of Animal Species and Evolution.* Vol. 19. Harvard University Press, 1970.

Mayr, Ernst. “Typological versus Population Thinking.” *Conceptual Issues in Evolutionary Biology*, edited by Elliott Sober, 3rd ed., MIT Press, 2006, pp. 325–328.

Mazmanian, Sarkis K., et al. ”An Immunomodulatory Molecule of Symbiotic Bacteria Directs Maturation of the Host Immune System.” *Cell*, vol. 122, no. 1, Jul. 15, 2005, pp. 107–118.

McArthur, Leslie Z. “What Grabs You? The Role of Attention in Impression Formation and Causal Attribution.” *Social Cognition: The Ontario Symposium*, vol. 1, 1981.

McCarthy, John, and Patrick J. Hayes. “Some Philosophical Problems from the Standpoint of Artificial Intelligence.” *Stanford University*, 1969, www-formal.stanford.edu/jmc/mcchay69.pdf. Accessed Sep. 1, 2025.

McCarthy-Jones, Simon. “The Autonomous Mind: the Right to Freedom of Thought in the Twenty-First Century.” *Frontiers in Artificial Intelligence*, vol. 2, 2019, article 19.

McClelland, James L., et al. ”The Appeal of Parallel Distributed Processing.” *Parallel Distributed Processing*, Volume 2, edited by James L. McClelland and David E. Rumelhart, MIT Press, 1986.

McCrann, Saoirse, et al. ”Smartphone Use as a Possible Risk Factor for Myopia.” *Clinical and Experimental Optometry*, vol. 104, no. 1, 2021, pp. 35–41.

McCullough, David. *The American Spirit: Who We Are and What We Stand for.* Simon and Schuster, 2017.

McCusker, James. “Should We Be Alarmed about Changing American Values?” *HeraldNet*, Aug. 31, 2019, heraldnet.com/business/should-we-be-alarmed-about-changing-american-values/. Accessed Aug. 25, 2025.

McDonald, Kerry, and Peter Gray. *Unschooled: Raising Curious, Well-Educated Children Outside the Conventional Classroom.* Chicago Review Press, 2019.

McDougall, William. *The Group Mind: a Sketch of the Principles of Collective Psychology.* Cambridge University Press, 1920.

McGilchrist, Iain. *The Master and His Emissary: the Divided Brain and the Making of the Western World.* Yale University Press, 2019.

McGilchrist, Iain. *The Matter with Things: Our Brains, Our Delusions, and the Unmaking of the World.* Perspectiva Press, 2021.

Mcgregor, Jena. "Clayton Chistenson's Innovation Brain". *Bloomberg*, Jun. 18, 2007. https://www.bloomberg.com/news/articles/2007-06-15/clayton-christensens-innovation-brainbusinessweek-business-news-stock-market-and-financial-advice. Accessed Dec. 13, 2025.

McKay, Ryan T., and Daniel C. Dennett. "The Evolution of Misbelief." *Behavioral and Brain Sciences*, vol. 32, 2009, pp. 493–561.

McKinsey & Company. "McKinsey on Finance" *McKinsey & Company*, no. 84, Dec. 2023, www.mckinsey.com/~/media/mckinsey/business%20functions/strategy%20and%20corporate%20finance/our%20insights/mckinsey%20on-finance%20number84/mckinsey-on-finance-number84.pdf. Accessed Sep. 7, 2025.

McLuhan, Marshall. *Understanding Media: the Extensions of Man.* The MIT Press, 1994.

Mead, Margaret. *Coming of Age in Samoa: a Psychological Study of Primitive Youth for Western Civilisation.* William Morrow, 1928.

Mead, Margaret. *Toward a Human Science: the Relevance of Anthropology to Contemporary Problems*, edited by Frank G. Sommers, Transaction Publishers, 1991.

"Measuring America: 10 Years and Counting." *Measure of America*, Social Science Research Council, 2018, measureofamerica.org/10yearsandcounting/. Accessed Aug. 26, 2025.

Megginson, Leon C. "Lessons from Europe for American Business." *Southwestern Social Science Quarterly*, vol. 44, no. 1, Jun. 1963, pp. 3–13.

Melvill, Henry. *The Golden Lectures, Sermons Delivered at St. Margaret's church, Lothbury, 1853–56. Selected from the Penny pulpit.* United Kingdom, n.p, 1854.

Menary, Richard. "Attacking the Bounds of Cognition." *Philosophical Psychology*, vol. 19, no. 3, 2006, pp. 329–344.

Menary, Richard. *Cognitive Integration: Mind and Cognition Unbounded.* Palgrave Macmillan, 2007.

Menary, Richard, editor. *The Extended Mind.* MIT Press, 2010.

Mensah, Julia, et al. Combating Noncommunicable Diseases in Kenya: an Investment Case. World *Bank*, https://documents1.worldbank.org/curated/en/428881586197529642/pdf/Combating-Noncommunicable-Diseases-in-Kenya-An-Investment-Case.pdf, 2020. Accessed Dec. 2, 2025.

Merriam-Webster. "Artificial Intelligence". https://www.merriam-webster.com/dictionary/artificial%20intelligence. Accessed Aug. 27, 2021.

Meshi, Dar, et al. "The Emerging Neuroscience of Social Media." *Trends in Cognitive Sciences*, vol. 19, no. 12, 2015, pp. 771–782.

Metcalfe, Tom. "For First Time, Scientists Detect 'Ghostly' Signal That Reveals Engine of Universe." *NBC News*, 25 Nov. 2020, nbcnews.com/science/space/first-time-scientists-detect-ghostly-signal-reveals-engine-universe-n1248982. Accessed Aug. 25, 2025.

Metzinger, Thomas K. "The Myth of Cognitive Agency: Subpersonal Thinking As a Cyclically Recurring Loss of Mental Autonomy." *Frontiers in Psychology*, vol. 4, 2013, 68140.

Michael, Katina. "Uberveillance: Microchipping People and the Assault on Privacy." *Quadrant*, vol. LIII, no. 3, 2009, pp. 85–89.

Michael, Katina, Roba Abbas, and Savvas Papagiannidis, "The Social Implications of XR: Promises, Perils, and Potential" *IEEE Technology and Society Magazine*, vol. 43, Mar. 2024 no. 1, pp. 91–108.

Mihai, Andrei. "New Finnish Invention Allows Cars to Easily Detect Black Ice in Real Time." *ZME Science*, Nov. 27, 2017, https://www.zmescience.com/science/news-science/black-ice-detection-car-27112017/. Accessed Nov.19, 2025.

Miller, George Armitage. *The Science of Mental Life*. Hutchinson, 1964.

Milne, Alan Alexander *Winnie-the-Pooh*. Illustrated by E. H. Shepard, Methuen & Co. Ltd., 1926.

Minkel, J. R. "The 2003 Northeast Blackout-Five Years Later." *Scientific American*, Aug. 13, 2008.

Minkin, Rachel, and Kiley Hurst. "The Experiences of U.S. Adults Who Don't Have Children." *Pew Research Center*, 25 Jul. 2024.

Mitchell, Jason P., et al. "The Link between Social Cognition and Self-Referential Thought in the Medial Prefrontal Cortex." *Journal of Cognitive Neuroscience*, vol. 17, no. 8, 2005, pp. 1306–1315.

Mitchell, Jason P., Todd F. Heatherton, and C. Neil Macrae. "Distinct Neural Systems Subserve Person and Object Knowledge." *Proceedings of the National Academy of Sciences*, vol. 99.23. 2002, pp. 15238–15243.

Mithen, Steven J. *The Prehistory of the Mind: a Search for the Origins of Art, Religion and Science*. Thames and Hudson, 1996.

Moazed, Alex and Nicholas L. Johnson. "Why Clayton Christensen Is Wrong about Uber and Disruptive Innovation", *Techcrunch*, Feb. 27, 2016, https://techcrunch.com/2016/02/27/why-clayton-christensen-is-wrong-about-uber-and-disruptive-innovation/. Accessed Dec. 13, 2025.

Moisala, Mona. *Brain Activations Related to Attention and Working Memory and Their Association with Technology-Mediated Activities*. Unigrafia, 2017.

Moller, Aage R., editor. *Reprogramming the Brain*. Vol. 157, Elsevier, 2006.

Moneyball. Directed by Bennett Miller, performances by Brad Pitt and Jonah Hill, Columbia Pictures, 2011.

Moon, Mariella "Here's What We Know about Magic Leap So Far." *Engadget*, Feb. 25, 2015, https://www.engadget.com/2015-02-25-magic-leap-explainer.html. Accessed Aug. 23, 2025.

Moore, Lorna G. "Human Genetic Adaptation to High Altitude." *High Altitude Medicine & Biology*, vol. 2, no. 2, 2001, pp. 257–279.

Moore, Lorna G. "Human Genetic Adaptation to High Altitudes: Current Status and Future Prospects." *Quaternary International*, vol. 461, 2017, pp. 4–13.

Morrison, Ryan. "ChatGPT Will Have PhD-Level Intelligence in the Next 2 Years." *Tom's Guide*, Jun. 24, 2024, https://www.tomsguide.com/ai/chatgpt/openai-cto-says-chatgpt-will-have-phd-level-intelligence-in-the-next-2-years. Accessed Nov. 19 , 2025.

Moser, Emily, and Michael McCulloch. "Canine Scent Detection of Human Cancers: a Review of Methods and Accuracy." *Journal of Veterinary Behavior*, vol. 5, no. 3, 2010, pp. 145–152.

Mostafavi, Hakhamanesh, et al. "Identifying Genetic Variants That Affect Viability in Large Cohorts." *PLOS Biology*, vol. 15, no. 9, 2017, e2002458.

Mucci, Tim. "What Is the Technological Singularity?" *IBM*, Jun. 7, 2024, https://www.ibm.com/think/topics/technological-singularity. Accessed Nov. 19, 2025.

Mukherjee, Sahana, and Jens Manuel Krogstad. "Most U.S. Voters Say Immigrants–No Matter Their Legal Status–Mostly Take Jobs Citizens Don't Want." *Pew Research Center*, Oct. 21, 2024.

Mullen, Sean. "When Grasshoppers Go Biblical: Serotonin Causes Locusts to Swarm." *Scientific American*, Feb. 20, 2024, www.scientificamerican.com/article/when-grasshoppers-go-bibl/. Accessed Sep. 23, 2025.

Mumford, Lewis. *Technics and Civilization*. Harcourt, Brace and Company, 1934.

Mumford, Lewis. *The Myth of the Machine: Technics and Human Development*. Vol. 1. Harcourt Brace Jovanovich, 1967.

Musgrave, Stephanie, et al. "The Ontogeny of Termite Gathering Among Chimpanzees in the Goualougo Triangle, Republic of Congo." *American Journal of Physical Anthropology*, vol. 174, no. 2, 2021, pp. 187–200.

Myhrvold, Nathan. "The Next Fifty Years of Software." *ACM97 Conference*, 1997.

Nakanishi, Masaki, et al. "Online Voluntary Eye Blink Detection Using Electrooculogram." *Proceedings of the 2012 International Symposium on Nonlinear Theory and Its Applications (NOLTA 2012)*, Palma, Mallorca, Spain, October 22–26, 2012. IEICE Pr oceeding Series 1, 2012.

National Center for Education Statistics (NCES). "Table 318.10. Degrees Conferred by Postsecondary Institutions, by Level of Degree and Sex of Student: Selected Years,

1869–70 through 2031–32." *Digest of Education Statistics 2023*, U.S. Dept. of Education, Oct. 2023.

National Center for Education Statistics. "Undergraduate Enrollment." *Condition of Education 2023, U.S. Dept. of Education*, 2023, https://nces.ed.gov/programs/coe/indicator/cha. Accessed Sep. 6, 2025.

National Conference of State Legislatures. "Cannabis Overview." *NCSL*, updated 2025, https://www.ncsl.org/civil-and-criminal-justice/cannabis-overview. Accessed Sep. 6, 2025.

"National Household Travel Survey Daily Travel Quick Facts." *U.S. Bureau of Transportation Statistics*, bts.gov/statistical-products/surveys/national-household-travel-survey-daily-travel-quick-facts. Accessed Aug. 25, 2025.

Naughton, John. "John Brockman: The Man Who Runs the World's Smartest Website." *The Observer*, Jan. 8, 2012, www.theguardian.com/technology/2012/jan/08/john-brockman-edge-interview-john-naughton. Accessed Aug. 23, 2025.

Naím, Moisés. *The End of Power: from Boardrooms to Battlefields and Churches to States, Why Being In Charge Isn't What It Used to Be*. Basic Books, 2013.

NCES Condition of Education. "Undergraduate Degree Fields." 2024, https://nces.ed.gov/programs/coe/indicator/cta. Accessed Sep. 6, 2025.

Neisser, Ulric. *Cognitive Psychology*. Appleton-Century-Crofts, 1967.

Neurath, Otto. "Museums of the Future (1933)." *Empiricism and Sociology*, edited by Marie Neurath and Robert Cohen, Reidel Publishing Company, 1973, pp. 218–223.

"New Census Data Shows the Nation Is Diversifying Even Faster than Predicted." *Brookings*, brookings.edu/research/new-census-data-shows-the-nation-is-diversifying-even-faster-than-predicted/. Accessed Aug. 25, 2025.

Newcomb, Doug "Eyes On With Heads-Up Display Car Tech". *PCMag*, Jul. 18, 2014. www.pcmag.com/article2/0,2817,2461037,00.asp. Accessed Aug. 23, 2025.

Newman, Lily Hay. "Hacker Lexicon: What Is an Attack Surface?" *Wired*, 30 Mar. 2017, https://www.wired.com/2017/03/hacker-lexicon-attack-surface/. Accessed Nov. 19, 2025.

Newman, Mark E. J. "Modularity and Community Structure in Networks." *Proceedings of the National Academy of Sciences*, vol. 103, no. 23, 2006, pp. 8577–8582.

Newport, Cal. *Deep Work: Rules for Focused Success in a Distracted World*. Hachette UK, 2016.

Newport, Cal. *Digital Minimalism: Choosing a Focused Life in a Noisy World*. Penguin, 2019.

Newport, Frank. "Church Attendance Has Declined in Most U.S. Religious Groups." *Gallup*, Mar. 26, 2024, news.gallup.com/poll/642548/church-attendance-declined-religious-groups.aspx. Accessed Aug. 27, 2025.

"News Platform Fact Sheet, 2024." *Pew Research Center*, Sep. 17, 2024.

"Nest Labs. Energy Savings from the Nest Learning Thermostat: Energy Bill Analysis Results." *Nest*, Feb. 2015, https://storage.googleapis.com/nest-public-downloads/press/documents/energy-savings-white-paper.pdf. Accessed Sep. 5, 2025.

Nisbett, Richard E. *The Geography of Thought: How Asians and Westerners Think Differently ... and Why*. Free Press, 2003.

Nisbett, Richard E., and Timothy D. Wilson. "Telling More than We Can Know: Verbal Reports on Mental Processes." *Psychological Review*, vol. 84, no. 3, 1977, p. 231.

North, Douglass C. *Structure and Change in Economic History*. W. W. Norton, 1981.

Noto La Diega, Guido, and Ian Walden. "Contracting for the 'Internet of Things': Looking into the Nest." *European Journal of Law and Technology*, vol. 7.2, 2016.

Nuyens, Filip M., et al. "The Empirical Analysis of Non-Problematic Video Gaming and Cognitive Skills: a Systematic Review." *International Journal of Mental Health and Addiction*, vol. 17.2, 2019, pp. 389–414.

Obama, Barack. "Remarks by the President at the Acceptance of the Nobel Peace Prize. Oslo City Hall, Oslo, Norway, Dec. 10, 2009." *The White House*, https://obama whitehouse.archives.gov/the-press-office/remarks-president-acceptance-nobel-pe ace-prize. Accessed Aug. 3, 2025.

Obergefell v. Hodges, 576 U.S. 644. Supreme Court of the United States, 2015. *Legal Information Institute, Cornell Law School*, www.law.cornell.edu/supremecourt/text/14-556. Accessed Jun. 12, 2025.

Odling-Smee, F. John, et al. "Niche construction: the neglected process in evolution. 37." *Princeton University Press Olson M (1992) Intuition and the logistic equation. Trends Ecol Evol*, vol 7, 2003, p. 314.

OECD. "Economic Policy Reforms 2021." *OECD Publishing*, 2021. https://www.oecd.org/en/publications/economic-policy-reforms-2021_3c796721-en.html. Accessed Aug. 26, 2025.

Ogawa, Masahiro. "The Theories of Ivan Illich and Paulo Freire and Environmental Education: Re-thinking from the Standpoint of Adult and Community Education." *Japanese Journal of Environmental Education* 19, no. 3, 2010, pp. 3–29.

O'Leary, Dennis DM. "Do Cortical Areas Emerge from a Protocortex?" *Trends in Neurosciences* 12.10, 1989, pp. 400–406.

Olson, David R. *The World on Paper: the Conceptual and Cognitive Implications of Writing and Reading*. Cambridge University Press, 1996.

Olsson, Sam. "App Downloads Statistics." *Kurve*, on Oct. 2, 2023, kurve.co.uk/blog/app-downloads-statistics. Accessed Aug. 25, 2025.

"One in Five: Disparities in Crime and Policing." *The Sentencing Project*, 16 Jan. 2024, www.sentencingproject.org/reports/one-in-five-disparities-in-crime-and-policing/. Accessed Jul. 17, 2025.

"Online Dating Now Dominates Relationship Building, Study Shows." *Global Dating Insights*, 15 Oct. 2024, www.globaldatinginsights.com/featured/online-dating-now-dominates-relationship-building-study-shows/. Accessed Jun. 3, 2025.

Osorio, Daniel, and Misha Vorobyev. "Photoreceptor Spectral Sensitivities in Terrestrial Animals: Adaptations for Luminance and Colour Vision." *Proceedings of the Royal Society B: Biological Sciences*, vol. 272, no. 1574, 2005, pp. 1745–1752.

Ostrom, Elinor. *Governing the Commons: The Evolution of Institutions for Collective Action*. Cambridge University Press, 1990.

Ottery, Christine "Cyborg anthropologist Amber Case's memo to her future self". *WIRED UK*, May 23, 2016, https://www.wired.com/story/memo-to-future-self/. Accessed Aug. 23, 2025.

Oura Health Ltd. *Oura Ring*. https://ouraring.com. Accessed Aug. 18, 2025.

Outram, Dorinda. *The Enlightenment*. 3rd ed., Cambridge University Press, 2019.

Oxford English Dictionary. "Thought, n.," *Oxford English Dictionary*, https://www.oed.com/dictionary/thought_n? Accessed Jul. 1, 2020.

Pagel, Mark, Chris Venditti, and Andrew Meade. "Large Punctuational Contribution of Speciation to Evolutionary Divergence at the Molecular Level." *Science*, vol. 3-14.5796, 2006, pp. 119–121.

Pandey, Erica. "Sean Parker: Facebook Was Designed to Exploit Human 'Vulnerability.'" *Axios*, Nov. 9, 2017, axios.com/sean-parker-facebook-was-designed-to-exploit-human-vulnerability-1513306782-6d18fa32-5438-4e60-af71-13d126b58e41.html. Accessed Aug. 25, 2025.

Panksepp, Jaak, and Jules B. Panksepp. "The Seven Sins of Evolutionary Psychology." *Evolution and Cognition*, vol. 6.2, 2000, pp. 108–131.

Pantev, Christo, et al. "Increased Auditory Cortical Representation in Musicians." *Nature*, vol. 392, 1998, pp. 811–814.

Pardi, Norbert, et al. "mRNA Vaccines – a New Era in Vaccinology." *Nature Reviews Drug Discovery*, vol. 17, no. 4, 2018, pp. 261–279.

Pardo, Bryce, et al. "The Future of Fentanyl and Other Synthetic Opioids". *RAND Corporation*, 2019, http://www.rand.org/pubs/research_reports/RR3117.html. Accessed Jul. 16, 2025.

Park, Hae-Jeong, and Karl Friston. "Structural and Functional Brain Networks: from Connections to Cognition." *Science*, vol. 342, no. 6158, 2013, p. 1238411.

Parker, Charlen. "Raleigh NC Trash Driver Shocks Council with a Powerful Speech". *YouTube*, https://www.youtube.com/watch?v=Xqy73nmKovs. Accessed Aug. 18, 20 25.

Patel, Shiwali. "With Fewer Than 10 Transgender NCAA Athletes, the Outcry Is Out of Proportion." *The Washington Post*, Jul. 5, 2025.

Pearce-Duvet, Jessica MC. "The Origin of Human Pathogens: Evaluating the Role of Agriculture and Domestic Animals in the Evolution of Human Disease." *Biological Reviews*, vol. 81, no. 3, 2006, pp. 369–382.

Perez, Carlota. *Technological Revolutions and Financial Capital*. Edward Elgar Publishing, 2003.

Perrow, Charles. *Normal Accidents: Living with High-Risk Technologies*. Updated ed., Princeton University Press, 1999.

"Persistent Forecasting of Disruptive Technologies¬Report 2". *National Research Council*. National Academies Press, 2010.

Pevsner, Jonathan. "Leonardo Da Vinci's Contributions to Neuroscience." *Trends in Neurosciences*, vol. 25.4, 2002, pp. 217–220.

Pew Research Center. "A Majority of U.S. Teens Fear a Shooting Could Happen at Their School–and Most Parents Share Their Concern." *Pew Research Center*, Apr. 18, 2018.

Pew Research Center. "How Americans View Data Privacy." *Pew Research Center*, Oct. 18, 2023, www.pewresearch.org/internet/2023/10/18/how-americans-view-data-privacy/. Accessed Nov. 4, 2025.

Pew Research Center. "How COVID-19 Changed U.S. Workplaces." *Pew Research Center*, Feb. 12, 2025, https://www.pewresearch.org/politics/2025/02/12/how-covid-19-changed-u-s-workplaces/. Accessed Apr. 23, 2025.

Pew Research Center. "Share of 40-year-olds in US who have never married reaches new high." *Pew Research Center*, Jun. 28, 2023, pewresearch.org/short-reads/2023/06/28/a-record-high-share-of-40-year-olds-in-the-us-have-never-been-married/. Accessed Apr. 13, 2025.

Piaget, Jean, and Eleanor Duckworth. "Genetic Epistemology." *American Behavioral Scientist*, vol. 13.3, 1970, pp. 459–480.

Pianka, Eric R. "On r- and K-Selection." *The American Naturalist*, vol. 104, no. 940, 1970, pp. 592–597.

Picard, Joe. "They're in the Top 10% of Earners. They Still Don't Feel Rich." *The Wall Street Journal*, Jun. 30, 2025.

Piercy, Katie. "How do Grasshoppers Reproduce?" *Meadowia*, April 2, 2021, meadowia.com/how-do-grasshoppers-reproduce/. Accessed on Dec. 2, 2025.

Pinch, Trevor, and Wiebe Bijker. "The Social Construction of Facts and Artifacts: Or How the Sociology of Science and the Sociology of Technology Might Benefit Each Other." *The Social Construction of Technological Systems: New Directions in the Sociology and History of Technology*, edited by Wiebe Bijker et al., MIT Press, 1987, pp. 17–50.

Pinch, Trevor. "The Social Construction of Technology (SCOT, The Old, the New, and the Nonhuman)" *Material Culture and Technology in Everyday Life: Ethnographic Approaches*, 2009, pp. 45–58.

Pinker, Steven, "Is the World Getting Better or Worse? A Look at the Numbers". *YouTube.* https://www.youtube.com/watch?v=yCm9Ng0bbEQ. Accessed Nov. 17, 2025.

Pinker, Steven. *The Language Instinct: How the Mind Creates Language.* Penguin UK, 2003.

Pinker, Steven, and Michael Shermer. "Progress and Its Discontents: the Edge Annual Question 2016." *Edge*, edge.org/3rd_culture/pinker_rose/part2/part2_p2.html. Accessed Aug. 25, 2025.

Pinker, Steven. *How the Mind Works.* Penguin UK, 2003.

Pinker, Steven. *The Blank Slate: the Modern Denial of Human Nature.* Viking, 2002.

Pitt, Joseph C. "The Autonomy of Technology." *Controlling Technology: Contemporary Issues*, edited by Eric Katz et al., Prometheus Books, 2003.

Plato. *Theaetetus.* Translated by Harold North Fowler, in *Plato in Twelve Volumes*, Vol. 12, Harvard University Press / Heinemann, 1921. 189e.

Popkin, Jeremy D. *News and Politics in the Age of Revolution: Jean Luzac's" Gazette de Leyde".* Cornell University Press, 2015.

Porter, Mason Alexander, Jukka-Pekka Onnela, and Peter J. Mucha. "Communities in Networks." *Notices of the American Mathematical Society*, Vol. 56, No. 9, 2009, pp. 1082–1097.

Posner, Michael I., and Mary K. Rothbart. *Attention, Genes, and Development: How Attention Shapes the Brain, Behavior, and Learning.* Oxford University Press, 2007.

Postman, Neil. *Amusing Ourselves to Death: Public Discourse in the Age of Show Business.* Penguin Books, 1985.

"Privacy Policy." *DoorDash*, help.doordash.com/consumers/s/privacy-policy-us. Accessed Dec. 9, 2020.

"Public Sees America's Future in Decline on Many Fronts." *Pew Research Center Social & Demographic Trends Project*, Mar. 21, 2019, pewresearch.org/social-trends/2019/03/21/public-sees-americas-future-in-decline-on-many-fronts/. Accessed Feb. 19, 2024.

Pundir, Meenakshi, et al. "Emerging Biotechnologies for Evaluating Disruption of Stress, Sleep, and Circadian Rhythm Mechanism Using Aptamer-Based Detection of Salivary Biomarkers." *Biotechnology Advances*, vol. 59, 2022, 107961.

"Purple Community Service." *Purple.ai*, purple.ai/blogs/purple-community-service/. Accessed Aug. 25, 2025.

Putnam, Robert D. *Bowling Alone: The Collapse and Revival of American Community.* Simon & Schuster, 2000.

Putnam, Robert D., Lewis Feldstein, and Donald J. Cohen. *Better Together: Restoring the American Community.* Simon & Schuster, 2004.

Quartz, Steven R., and Terrence J. Sejnowski. "The Neural Basis of Cognitive Development: A Constructivist Manifesto." *Behavioral and Brain Sciences*, vol. 20, no. 4, 1997, pp. 537–556.

"Racial Inequality in the United States." *U.S. Department of the Treasury*, 21 Jul. 2022, home.treasury.gov/news/featured-stories/racial-inequality-in-the-united-states. A ccessed Mar. 15, 2024.

Raichle, Marcus E. "The Brain's Default Mode Network." *Annual Review of Neuroscience*, vol. 38, 2015, pp. 433–447.

Raichle, Marcus E., and Debra A. Gusnard. "Appraising the Brain's Energy Budget." *Proceedings of the National Academy of Sciences*, vol. 99, no. 16, 2002, pp. 10237–10239.

Rainie, Lee."Cable and Satellite TV Use Has Dropped Dramatically in the U.S. since 2015." *Pew Research Center*, Mar. 17, 2021, http://www.pewresearch.org/short-reads/2021/03/17/cable-and-satellite-tv-use-has-dropped-dramatically-in-the-u-s-since-2015/. Accessed Feb. 22, 2024.

Rajaperumal, T. A., and C. Christopher Columbus. "Transforming the electrical grid: the role of AI in advancing smart, sustainable, and secure energy systems." *Energy Informatics*, vol. 8.1, 2025, 51.

Rao, Pallavi. "A Visual Breakdown of Who Owns America's Wealth." Visual Capitalist, Mar. 4, 2025, www.visualcapitalist.com/a-visual-breakdown-of-who-owns-americas-wealth/. Accessed Nov. 19, 2025.

"Raw Dogging: The New Travel Trend." *CNN Travel*, Jun. 28, 2024, https://www.cnn.com/2024/06/28/travel/raw-dogging-travel-trend-explainer-intl-hnk/index.html. Accessed Nov. 3, 2024.

Raworth, Kate. *Doughnut Economics: Seven Ways to Think Like a 21st-Century Economist.* Chelsea Green Publishing, 2017.

Ready Player One. Directed by Steven Spielberg, Warner Bros. Pictures, 2018.

"Reckoning." *Halo*, created by Kyle Killen and Steven Kane, season 1, episode 9, Paramount+, May 12, 2022.

Richard, Alison F., and Robert E. Dewar. "Lemur ecology." *Annual Review of Ecology and Systematics*, 1991, pp. 145–175.

Richerson, Peter J., and Robert Boyd. *Not by Genes Alone: How Culture Transformed Human Evolution.* University of Chicago Press, 2008.

Riddle, Karyn, et al. "Binge Watching and Serial Viewing: Comparing New Media Viewing Habits in 2015 and 2020." *Psychology of Popular Media*, vol. 11, no. 2, 2022, pp. 187–197.

Ridley, Matt. *Nature via Nurture: Genes, Experience, and What Makes Us Human.* HarperCollins, 2003.

Ridley, Matt. *The Origins of Virtue*. Penguin UK, 1997.

Ridley, Matt. *The Red Queen: Sex and the Evolution of Human Nature*. Macmillan, 1994.

Riemer, Kai, and Robert B. Johnston. "Disruption As Worldview Change: A Kuhnian Analysis of the Digital Music Revolution." *Journal of Information Technology*, vol. 34.4, 2019, pp. 350–370.

Rimfeld, Kaili, et al. "Genetic Influence on Social Outcomes During and After the Soviet Era in Estonia." *Nature Human Behaviour*, vol. 2, no. 4, 2018, pp. 269–275.

Rip, Arie, and René Kemp. "Technological Change." *Human Choice and Climate Change. Vol. 2: Resources and Technology*, edited by Steve Rayner and Elizabeth L. Malone, Battelle Press, 1998, pp. 327–399.

Robbins, Philip, and Murat Aydede, editors. *The Cambridge Handbook of Situated Cognition*. Cambridge University Press, 2009.

Robinson, Scott E., and Floun'say R. Caver. "Punctuated Equilibrium and Congressional Budgeting." *Political Research Quarterly*, vol. 59.1, 2006, pp. 161–166.

Robson, David. "It Only Takes 35% of People to Change the World." *BBC*, May 13, 2019, https://www.bbc.com/future/article/20190513-it-only-takes-35-of-people-to-change-the-world. Accessed Dec. 9, 2025.

Rocher, Luc, et al. " Estimating the Success of Re-identifications in Incomplete Datasets Using Generative Models." *Nature Communications*, vol. 10, no. 1, 2019, pp. 1–9.

Rogers, Everett M., Arvind Singhal, and Margaret M. Quinlan. "Diffusion of Innovations." *An Integrated Approach to Communication Theory and Research*, Routledge, 2014, pp. 432–448.

Romanelli, Elaine, and Michael L. Tushman. "Organizational Transformation As Punctuated Equilibrium: an Empirical Test." *Academy of Management Journal*, vol. 37.5, 1994, pp. 1141–1166.

Romer, Paul M. "Endogenous Technological Change." *Journal of Political Economy*, vol. 98, no. 5, 1990, pp. S71–S102.

Roosevelt, Theodore. *The Autobiography of Theodore Roosevelt: with Illustrations*. Charles Scribner's Sons, 1913.

Rosenthal, David M. *Consciousness and Mind*. Oxford University Press, 2005.

Rosenthal, Sara Brin, et al. "Revealing the Hidden Networks of Interaction in Mobile Animal Groups Allows Prediction of Complex Behavioral Contagion." *Proceedings of the National Academy of Sciences*, vol. 112.15, 2015, pp. 4690–4695.

Rosling, Hans, Ola Rosling, and Anna Rosling Rönnlund. *Factfulness: Ten Reasons We're Wrong About the World -and Why Things Are Better Than You Think*. Flatiron Books, 2018.

Ross, Edward Alsworth. *Social Psychology: an Outline and Source Book*. Macmillan, 1908.

Rousseau, Jean-Jacques. *The Social Contract*. Translated by Maurice Cranston, Penguin Classics, 2004.

Rowlands, Mark. *The Body in Mind: Understanding Cognitive Processes*. Cambridge University Press, 1999.

Rubin, David C. *Memory in Oral Traditions: the Cognitive Psychology of Epic, Ballads, and Counting-Out Rhymes*. Oxford University Press, 1995.

Ruhlman, Michael. *Grocery: the Buying and Selling of Food in America*. Abrams, 2017.

Rupert, Robert D. "Challenges to the Hypothesis of Extended Cognition." *The Journal of Philosophy*, vol. 101, no. 8, 2004, pp. 389–428.

Russell, Bertrand. *Power: a New Social Analysis*. Psychology Press, 2004.

Russell, Stuart, and Peter Norvig. *Artificial Intelligence: a Modern Approach, 3rd edition*. Pearson, 2009.

Rutt, Jim. "EP 190 Peter Turchin on Cliodynamics and End Times." *YouTube, uploaded by IBHA International Big History Association*, https://www.youtube.com/watch?v=-UKSm39YJGw. Accessed Nov.18, 2025.

Sabatini, Fabio, and Francesco Sarracino. "Keeping Up With the E-Joneses: Do Online Social Networks Raise Social Comparisons?" *Economics Discussion Papers, No 2018–43*. Kiel Institute for the World Economy, *2018*. http://www.economics-ejournal.org/economics/discussionpapers/2018–43. Accessed Nov. 5, 2024.

Sacasas, Michael. "Borg Complex: a Primer." *The Frailest Thing*, Mar. 1, 2013, thefrailestthing.com/2013/03/01/borg-complex-a-primer/. Accessed Dec. 21, 2023.

Sacks, Oliver. *The Man Who Mistook His Wife for a Hat: And Other Clinical Tales*. Picador, 1986.

Sagan, Carl. *The Demon-Haunted World: Science as a Candle in the Dark*. Random House, 1995.

Sakai, Kuniyoshi L. "Language Acquisition and Brain Development." *Science*, vol. 310, no. 5749, 2005, pp. 815 –819.

Sandberg, Anders, and Nick Bostrom. *Whole Brain Emulation: a Roadmap. Technical Report #2008-3, Future of Humanity Institute*, University of Oxford, 2008.

Santos, Laurie R., et al. "Rhesus Monkeys (Macaca mulatta) Know What Others Can and Cannot Hear." *Animal Behaviour*, vol. 71, no. 5, 2006, pp. 1175–1181.

Sanz, Crickette, et al. "Design Complexity in Termite-Fishing Tools of Chimpanzees (Pan troglodytes)." *Biology Letters*, vol. 5, no. 3, 2009, pp. 293–296.

Sapolsky, Robert M. *Determined: a Science of Life without Free Will*. Penguin Press, 2023.

Sarter, Martin, et al. "The Cognitive Neuroscience of Sustained Attention: Where Top-Down Meets Bottom-Up." *Brain Research Reviews*, vol. 35, no. 2, 2001, pp. 146–160.

Scannell, Patrick. *The Great Irony of Technology*. Brill Press. Forthcoming.

Scannell, Patrick. "The Great Irony of Technology: Why Technology Leaves Us Feeling Worse and What To Do About It." YouTube, uploaded by Public Interest Technology (PIT) Colloquium, Date unknown, www.youtube.com/watch?v=gf7yGBSG1aU. Accessed Aug. 23, 2025.

Scannell, Patrick. *The History of Thought, Book 2: The Co-evolution of Technology and Cognition in the Human* Species. Forthcoming.

Schaake, Marietje. *The Tech Coup: How to Save Democracy from Silicon Valley*. Stanford University Press, 2024.

Scheler, Max (ed.). *Versuche zu einer Soziologie des Wissens. München und Leipzig: Duncker & Humblot, 1924*. Karl Mannheim. *Ideology and Utopia: An Introduction to the Sociology of Knowledge*. Translated by Louis Wirth and Edward Shils. Harcourt, Brace and Company, 1936.

Schenkel, Wolfgang. "Color Terms in Ancient Egyptian and Coptic." *Anthropology of Color: Interdisciplinary Multilevel Modeling*, edited by Robert E. MacLaury et al., John Benjamins, 2007, pp. 211–228.

Schilling, David. R. "Knowledge Doubling Every 12 Months, Soon to be Every 12 Hours." http://www.industrytap.com/knowledge-doubling-every-12-months-soon-to-be-every-12-hours/3950. Accessed Dec. 1, 2020.

Schlaggar, Bradley L., and Bruce D. McCandliss. "Development of Neural Systems for Reading." *Annual Review Neuroscience*, vol. 30.1, 2007, pp. 475–503.

Schlaug, Gottfried, et al. "Effects of Music Training on the Child's Brain and Cognitive Development." *Annals of the New York Academy of Sciences*, vol. 1060, 2005, pp. 219–230.

Schmidt, Eric. Interview by Reid Hoffman. "You should be running toward AI, with Eric Schmidt." *Masters of Scale*, Apr. 5, 2023, mastersofscale.com/you-should-be-running-toward-ai-eric-schmidt/. Accessed Jan. 17, 2024.

Schor, Juliet B. *After the Gig: How the Sharing Economy Got Hijacked and How to Win It Back*. University of California Press, 2020.

Schumacher, Ernst F. *Small Is Beautiful: Economics as if People Mattered*. Harper & Row, 1973.

Schumpeter, Joseph A. *Capitalism, Socialism, and Democracy*. Harper and Brothers, 1942.

Schumpeter, Joseph A. "Carl Menger." *Ten Great Economists: from Marx to Keynes*. Routledge, 1969.

Schumpeter, Joseph A. *The Theory of Economic Development: an Inquiry into Profits, Capital, Credit, Interest, and the Business Cycle*. Harvard University Press, 1934.

Schwab, Klaus. *The Fourth Industrial Revolution*. Crown Business, 2017.

Schweizer, Valerie J. "Marriage: More than a Century of Change, 1900–2022." *National Center for Family & Marriage Research*, bgsu.edu/ncfmr/resources/data/family-profiles/FP-24-10.html. Accessed Aug. 27, 2024.

Scott-Phillips, Thom, and Dan Sperber. "The Mutual Relevance of Teaching and Cultural Attraction." *Behavioral and Brain Sciences*, vol. 38, 2015, pp. 44–45.

"Search for High Energy 5.5 Mev Solar Axions with the Complete Borexino Dataset." *INFN Laboratori Nazionali del Gran Sasso*, https://borex.lngs.infn.it/papers/articles/search-for-high-energy-solar-axions-with-complete-borexino-dataset/ Accessed Aug. 25, 2025.

Selfridge, Oliver G. "Pandemonium: a Paradigm for Learning." *Symposium on the Mechanization of Thought Processes*, HM Stationery Office, 1959.

Selous, Edmund. "Thought-Transference (or What?) in Birds." 1931.

Semendeferi, Katerina, et al. "Prefrontal Cortex in Humans and Apes: A Comparative Study of Area 10." *American Journal of Physical Anthropology*, vol. 114, 2001, pp. 224–241.

Semon, Richard. "Engraphic Action of Stimuli on the Individual." *The Mneme*, translated by Louis Simon, George Allen & Unwin, 1921.

Sen, Amartya. *Development as Freedom*. Knopf, 1999.

Senator Bernie Sanders: "American Dream or American Nightmare? Depends Whom You Ask." *KWTX*, kwtx.com/content/news/American-Dream-or-American-nightmare\-Depends-whom-you-ask-558040921.html. Accessed Aug. 25, 2025.

Seth, Anil. "Your Brain Hallucinates Your Conscious Reality." *TED*, Jun. 2017,www.ted.com/talks/anil_seth_your_brain_hallucinates_your_conscious_reality. Accessed Nov. 18, 2025.

"Sewage Spill During the Blackout Exposed a Lingering City Problem." *The New York Times*, Aug. 28, 2003.

Shahani, Bhagwan. "The Human Blink Reflex." *Journal of Neurology, Neurosurgery & Psychiatry*, vol. 33, no. 6, Dec. 1970, pp. 792–800.

Shakespeare, William. *Hamlet*. Act 3, Scene 2. Edited by Barbara A. Mowat and Paul Werstine, Folger Shakespeare Library, 1992.

Shankland, Stephen. "IBM's Quantum Computing Ambitions Get Exponential, Like Moore's Law." *CNET*, Mar. 3, 2019, cnet.com/news/ibms-quantum-computing-ambitions-get-exponential-like-moores-law/. Accessed Aug. 25, 2025.

Shapiro, Carl, and Hal R. Varian. *Information Rules*. Harvard Business Press, 1999.

Shapiro, Lawrence. *Embodied Cognition*. 2nd ed., Routledge, 2019.

Sherwood, Chet C., et al. "Invariant Synapse Density and Neuronal Connectivity Scaling in Primate Neocortical Evolution." *Cerebral Cortex*, vol. 30.10, 2020, pp. 5604–5615.

Shulevitz, Judith. "Alexa, Should We Trust You." *The Atlantic*, Nov. 2018, https://www.theatlantic.com/magazine/archive/2018/11/alexa-how-will-you-change-us/570844/. Accessed Dec. 2, 2025.

Simon, Herbert A. *The Sciences of the Artificial*. 3rd ed., MIT Press, 1996.

Singer, Jenny. "What to Know About Daunte Wright, Who Was Fatally Shot by Police at a Traffic Stop." *Glamour*, Apr. 13, 2021.

"Singles Around Me." Google Play Store, play.google.com/store/apps/details?id=com.singlesaroundme.android&hl=en. Accessed Aug. 23, 2025.

Sire, Tom. "How Many Servers Are There in the World?" *Tech Pulsion*, techpulsion.com/how-many-servers-are-there-in-the-world/. Jun. 30, 2025. Accessed Aug. 25, 2025.

Skinner, B. F. *Science and Human Behavior*. Macmillan, 1953.

Skinner, B. F. "The Real Question Is Not Whether Machines Think But Whether Men Do." *Contingencies of Reinforcement: A Theoretical Analysis*, 1969.

Skinner, B. F. *Verbal Behavior*. Appleton-Century-Crofts, 1957.

Slater, Stanley F., Jakki J. Mohr, and Sanjit Sengupta. "Radical Product Innovation Capability: Literature Review, Synthesis, and Illustrative Research Propositions." *Journal of Product Innovation Management*, vol. 31.3, 2014, pp. 552–566.

Smart, Paul R. "Human-Extended Machine Cognition." *Cognitive Systems Research*, vol. 49, 2018, pp. 9–23.

Smart, Paul R. "The Web-Extended Mind." *Metaphilosophy*, vol. 43, no. 4, 2012, pp. 426–445.

SmartFeed USA. https://smartfeedusa.com/ *SmartFeed USA*, Accessed Aug. 18, 2025.

Smil, Vaclav. *Growth: From Microorganisms to Megacities*. MIT Press, 2019.

Snyder, Cynthia K., and Anne-Marie Chang. "Mobile Technology, Sleep, and Circadian Disruption." *Sleep and Health*, Academic Press, 2019, pp. 159–170.

"Social." *Encyclopedia.com*, http://www.encyclopedia.com/humanities/dictionaries-thesauruses-pictures-and-press-releases/social-0. Accessed Aug. 3, 2024.

Solow, Robert M. "A Contribution to the Theory of Economic Growth." *Quarterly Journal of Economics* 70, no. 1. Feb. 1956, pp. 65–94.

Son, Masayoshi. "SoftBank World 2024". *The Sun*, Oct. 3, 2024.

Soon, Chun Siong, et al. "Unconscious Determinants of Free Decisions in the Human Brain." *Nature Neuroscience*, vol. 11, no. 5, 2008, pp. 543–545.

Sparrow, Betsy, et al. "Google Effects on Memory: Cognitive Consequences of Having Information at Our Fingertips." *Science*, vol. 333, no. 6043, Jun. 14, 2011, pp. 776–778.

Sparrow, Betsy, Jenny Liu, and Daniel M. Wegner. "Google Effects on Memory: Cognitive Consequences of Having Information at Our Fingertips." *Science*, vol. 333, no. 6040, 2011, pp. 776–778.

Sperber, Dan. "Anthropology and Psychology: towards an Epidemiology of Representations." *Man*, vol. 20, no. 1, 1985, pp. 73–89.

Sperber, Dan. "Culturally Transmitted Misbeliefs." *Behavioral and Brain Sciences*, vol. 32, 2009, pp. 534–535.

Sporns, Olaf, et al. "The Human Connectome: A Structural Description of the Human Brain." *PLoS Computational Biology*, vol. 1, no. 4, 2005, p. e42.

Sporns, Olaf. "Connectome Networks: from Cells to Systems." *Micro-, Meso-and Macro-Connectomics of the Brain*, Springer, 2016, pp. 107–127.

Sprevak, Mark. "Extended Cognition and Functionalism." *Journal of Philosophy*, vol. 106, no. 9, 2009, pp. 503–527.

Squire, Larry R., and Stuart M. Zola. "Episodic Memory, Semantic Memory, and Amnesia." *Hippocampus*, vol. 8, no. 3, 1998, pp. 205–211.

Sreenivasan, Sameet, et al. "The Role of Network Topology on the Initial Spreading of Social Behaviors." *Physical Review E* vol. 84, no. 3, 2011, pp. 036108.

Stallworth, Charles. "The Elites Have Stopped Hiding Their Hatred of the Working Class." *Newsweek*, Oct. 12, 2022.

Stanley, Deborah A. "Getting Implicit about the Unconscious: Do We Really Know What Social Neuroscience Has to Offer Implicit Social Cognition Research?" *Journal of Social Issues*, vol. 72, no. 1, 2016, pp. 107–118.

Stanton, Andrew. "Plurality of Americans Believe 'Civil War' Likely in Next Decade–Poll." *Newsweek*, Jun. 17, 2025, http://www.newsweek.com/plurality-americans-believe-civil-war-likely-poll-2086771. Accessed Aug. 26, 2025.

Starr, Michelle. "Facial Recognition App Matches Strangers to Online Profiles." *CNET*, Jan. 7, 2014, www.cnet.com/news/facial-recognition-app-matches-strangers-to-online-profiles. Accessed Aug. 23. 2025.

Steinmetz, Joseph E., et al. 1987. "Initial Localization of the Acoustic Conditioned Stimulus Projection System to the Cerebellum Essential for Classical Eyelid Conditioning." Proceedings of the National Academy of Science, vol. 84, 1987, pp. 3531–3535.

Steiner, Pierre. "CS Peirce and Artificial Intelligence: Historical Heritage and (New) Theoretical Stakes." *Philosophy and Theory of Artificial Intelligence*, Springer, 2013, pp. 265–276.

Stephan, Heinz, et al. "New and Revised Data on Volumes of Brain Structures in Insectivores and Primates." *Folia Primatologica*, vol. 35(1), 1981, pp. 1–29.

Sterelny, Kim. "Externalism, Epistemic Artefacts and the Extended Mind." *The Externalist Challenge*, 2004, pp. 239–254.

Sterelny, Kim. "From Hominins to Humans: How Sapiens Became Behaviourally Modern." *Philosophical Transactions of the Royal Society B: Biological Sciences*, vol. 366, no. 1566, 2011, pp. 809–822.

Sterelny, Kim. "Minds: Extended or Scaffolded?." *Phenomenology and the Cognitive Sciences* vol. 9.4, 2010, pp. 465–481.

Stern, Nicholas. "Green and Intelligent: the Role of AI in the Climate Transition." *npj Climate Action*, 2025.

Stoeger, Angela S., and Paul Manger. "Vocal Learning in Elephants: Neural Bases and Adaptive Context." *Current Opinion in Neurobiology*, vol. 28, 2014, pp. 101–107.

Stone, Linda. "Beyond Simple Multi-Tasking: Continuous Partial Attention." *Linda Stone*, 8 Mar. 2014, lindastone.net/2009/11/30/beyond-simple-multi-tasking-continuous-partial-attention/. Accessed Sep. 13, 2022.

Stonier, Tom. "The Evolution of Machine Intelligence". *Beyond Information*. Springer, 1992.

Suchman, Lucy A. *Plans and Situated Actions: the Problem of Human–Machine Communication*. Cambridge University Press, 1987.

Suddendorf, Thomas, D. R. Addis, and M. C. Corballis. "Mental Time Travel and Shaping of the Human Mind". Bar M, ed. *Predictions in the Brain: Using Our Past to Generate a Future*, 2011, pp. 344–354.

Sundararajan, Arun. *The Sharing Economy: the End of Employment and the Rise of Crowd-Based Capitalism*. MIT Press, 2016.

SUNY American Government Courses. "The Evolution of the Media." *SUNY American Government*, courses.lumenlearning.com/suny-amgovernment/chapter/the-evolution-of-the-media/. Accessed Oct 5, 2025.

Suri, Tavneet, and William Jack. "The Long-Run Poverty and Gender Impacts of Mobile Money." *Science*, vol. 354, no. 6317, 2016, pp. 1288–1292.

Suri, Tavneet. "Mobile Money." *Annual Review of Economics*, vol. 9, 2017, pp. 497–520, www.annualreviews.org/doi/10.1146/annurev-economics-063016-103638. Accessed Aug. 19, 2025.

Sutton, John. "Exograms and Interdisciplinarity: History, the Extended Mind, and the Civilizing Process." *The Extended Mind*, edited by Richard Menary, MIT Press, 2010, pp. 189–225.

Sutton, John. "Remembering." *Cambridge Handbook of Situated Cognition*, edited by Philip Robbins and Murat Aydede, Cambridge University Press, 2009, pp. 217–235.

Svendsen, Lars Fr. H. *A Philosophy of Boredom*. Translated by John Irons, Reaktion Books, 2005.

Swanson, Bret. "Moore's Law at 50" *American Enterprise Institute*, Nov. 10, 2015, https://www.aei.org/research-products/report/moores-law-at-50-the-performance-and-prospects-of-the-exponential-economy/. Accessed on Nov. 19, 2025.

Takahashi, Kazutoshi, and Shinya Yamanaka. "Induction of Pluripotent Stem Cells from Mouse Embryonic and Adult Fibroblast Cultures by Defined Factors." *Cell*, vol. 126, no. 4, 2006, pp. 663–676.

Takemoto, Hiroyuki, Yoshi Kawamoto, and Takeshi Furuichi. "How Did Bonobos Come to Range South of The Congo River? Reconsideration of the Divergence of Pan Paniscus from Other Pan Populations." *Evolutionary Anthropology: Issues, News, and Reviews*, vol. 24.5, 2015, pp. 170–184.

Tarafdar, Monideepa, et al. "Impact of Technostress on End-User Satisfaction and Performance." *Journal of Management Information Systems*, vol. 27, no. 3, 2011, pp. 303–334.

Tarde, Gabriel. *The Laws of Imitation*. Translated by Elsie Clews Parsons, Henry Holt, 1903.

"Tarsier." *Wikipedia*, en.wikipedia.org/wiki/Tarsier. Accessed Jul. 7, 2025.

Taylor, Timothy. *The Artificial Ape: How Technology Changed the Course of Human Evolution*. St. Martin's Press, 2010.

Taylor, Shelley E. "Social Cognition and Health." *APA Handbook of Personality and Social Psychology, Volume 1: Attitudes and Social Cognition*. American Psychological Association, 2015, pp. 339–361.

Taylor, Shelley E. "Social Cognition and Health." *Personality and Social Psychology Bulletin*, vol. 8, no. 3, 1982, pp. 549–562.

"Tech Life Predictions for 2030." *World Economic Forum*, 2017, weforum.org/agenda/2017/10/tech-life-predictions-for-2030/. Accessed Aug. 25, 2025.

Teixeira, Thales S., and Greg Piechota. *Unlocking The Customer Value Chain: How Decoupling Drives Consumer Disruption*. Currency, 2019.

Tenhunen, Marja-Liisa. "Why Finland Is the Happiest Country in the World?" *Cogito-Multidisciplinary Research Journal* 3, 2020, pp. 18–28.

Tennie, Claudio, et al. "Ratcheting Up the Ratchet: on the Evolution of Cumulative Culture." *Philosophical Transactions of the Royal Society B: Biological Sciences*, vol. 364, no. 1528, 2009, pp. 2405–2415.

The Matrix. Directed by the Wachowskis, Warner Bros., 1999.

The Social Dilemma. Directed by Jeff Orlowski, performances by Tristan Harris, Shoshana Zuboff, Jaron Lanier, and others, Exposure Labs, 2020.

Theis, Thomas N., and H-S. Philip Wong. "The End of Moore's Law: a New Beginning for Information Technology." *Computing in Science & Engineering*, vol. 19, no. 2, 2017, pp. 41–50.

"Thinking, Fast and Slow, Daniel Kahneman, Talks at Google". *YouTube*, uploaded by Talks at Google, https://www.youtube.com/watch?v=CjVQJdIrDJo. Accessed Aug. 25, 2025.

Thinley, Jigmi Y., and Janette Hartz-Karp. "National Progress, Sustainability and Higher Goals: the Case of Bhutan's Gross National Happiness." *Sustainable Earth* 2, no. 1, 2019, p. 11.

"This Is Water by David Foster Wallace (Full Transcript and Audio)" *FarnamStreet*, https://fs.blog/2012/04/david-foster-wallace-this-is-water/. Accessed Aug. 19, 2025.

Thompson, Clive. *Smarter Than You Think: How Technology Is Changing Our Minds for the Better*. Penguin Press, 2013.

Thompson, Evan. *Mind in Life: Biology, Phenomenology, and the Sciences of Mind*. Harvard University Press, 2010.

Thompson, Richard F. *Foundations of Physiological Psychology*. Harper & Row, New York, 1967.

Thoreau, Henry David. *Civil Disobedience*. 1849. Project Gutenberg, https://www.gutenberg.org/ebooks/71. Accessed Nov. 19, 2025.

"Time Poverty." Wikipedia, https://en.wikipedia.org/wiki/Time_poverty. Accessed Sep. 16, 2025.

Tocqueville, Alexis de. *Democracy in America*. Translated by Harvey C. Mansfield and Delba Winthrop, University of Chicago Press, 2000.

Toffler, Alvin. *Future Shock*. Random House, 1970.

Tolkien, John Ronald Reuel. *The Hobbit*. Houghton Mifflin Harcourt, 2012.

Tomasello, Michael, et al. "Cultural Learning." *Behavioral and Brain Sciences*, vol. 16, 1993, pp. 495–552.

Tomasello, Michael. *The Cultural Origins of Human Cognition*. Harvard University Press, 1999.

"Tongan Castaways;." *Wikipedia*, last edited Jun. 17, 2025, en.wikipedia.org/wiki/Tongan_castaways. Accessed Aug. 23, 2025.

Tooby, John, and Leda Cosmides. "Evolutionary Psychology and the Generation of Culture, Part I: Theoretical Considerations." *Ethology and Sociobiology* vol. 10.1–3, 1989, pp. 29–49.

"TOP500." *Wikipedia*, Wikimedia Foundation, en.wikipedia.org/wiki/TOP500. Accessed Apr. 13, 2023.

Topol, Eric J. *Deep Medicine: How Artificial Intelligence Can Make Healthcare Human Again*. Basic Books, 2019.

Toward a Human Science: The Relevance of Anthropology to Contemporary Problems, edited by Frank G. Sommers, Transaction Publishers, 1991.

Toynbee, Arnold Joseph, and David Churchill Somervell. *Civilization on Trial*. Oxford University Press, 1948.

Trerotola, Marco, et al. "Epigenetic Inheritance and the Missing Heritability." *Human Genomics*, vol. 9, no. 1, 2015, 17.

Trevino, Joshua. "The Overton Window." *Swords Crossed*, Apr. 29, 2006, *SwordsCrossed.org (blog)*, archived at the Wayback Machine, Jul. 16, 2006, web.archive.org/web/20060716000532/ http://www.swordscrossed.org/?p=50. Accessed Aug. 23, 2025.

Triplett, Norman. "The Dynamogenic Factors in Pacemaking and Competition." *American Journal of Psychology*, vol. 9, 1898, pp. 507–533.

Tulving, Endel. *Elements of Episodic Memory*. Oxford University Press, 1983.

Turchin, Peter. *Ultrasociety: How 10,000 Years of War Made Humans the Greatest Cooperators on Earth*. Beresta Books, 2016.

Turkle, Sherry. *Alone Together: Why We Expect More from Technology and Less from Each Other*. 3rd ed., Basic Books, 2017.

Tversky, Amos, and Daniel Kahneman. "Judgment under Uncertainty: Heuristics and Biases." *Science*, vol. 185, no. 4157, 1974, pp. 1124–1131.

Twenge, Jean M. *iGen: Why Today's Super-Connected Kids Are Growing Up Less Rebellious, More Tolerant, Less Happy–and Completely Unprepared for Adulthood–and What That Means for the Rest of Us*. Atria Books, 2017.

Twenge, Jean M., and W. Keith Campbell. "Associations Between Screen Time and Lower Psychological Well-Being Among Children and Adolescents: Evidence from a Longitudinal Study." *Preventive Medicine*, vol. 116, 2018, pp. 135–142.

Tylor, Edward B. *Primitive Culture: Researches into the Development of Mythology, Philosophy, Religion, Language, Art, and Custom.* J. Murray, 1871.

"The Unfinished Revolution, Strategy and Means for Coping with Complex Problems", *Doug Engelbart's Colloquium at Stanford*, Apr. 2000, https://www.dougengelbart.org/colloquium/colloquium.html. Accessed Nov 20, 2025.

"UBS Chief Investment Office. How the Data Universe Could Grow More Than 10 Times from 2020 to 2030." *UBS*, Mar. 21, 2024, https://www.ubs.com/us/en/wealth-management/insights/market-news/article/_jcr_content.0116445243.file/PS9jb25 0ZW50L2RhbS9pbXBvcnRlZC9jaW9yZXNlYXJjaC9wZGYvMTUvOTYvMzIvOS8x NTk2MzI5L2VuL3VzLzE1OTYzMjkucGRm/1596329.pdf. Accessed Sep. 5, 2025.

U.S. Bureau of Economic Analysis. "Real Gross Domestic Product (*GDPC1*)." *Federal Reserve Bank of St. Louis*, https://fred.stlouisfed.org/series/GDPC1. Accessed Aug. 26, 2025.

U.S. Bureau of Labor Statistics. "May 2023 National Occupational Employment and Wage Estimates." *OEWS, U.S. Dept. of Labor*, Apr. 3, 2024, https://www.bls.gov/oes/2023/may/oes_nat.html. Accessed Sep. 6, 2025.

U.S. Census Bureau. "Annual Retail Trade Survey Shows Impact of Online Shopping on Retail Sales During COVID-19 Pandemic." *Census.gov*, Apr. 2022, http://www.census.gov/library/stories/2022/04/ecommerce-sales-surged-during-pandemic.html. Accessed Jul. 2, 2025.

U.S. Census Bureau. "Table A. Resident Population–Selected Years 1790 to 2020." *Historical National Population Estimates*, https://www.census.gov/data/tables/time-series/dec/popchange-data-text.html. Accessed Jul. 2, 2025.

U.S. Census Bureau. "The Number of People Primarily Working from Home Tripled Between 2019 and 2021." *US Census Bureau*, 15 Sept. 2022, https://www.census.gov/newsroom/press-releases/2022/people-working-from-home.html. Accessed Sep. 6, 2025.

U.S. Department of Health and Human Services. "Protecting Youth Mental Health: the U.S. Surgeon General's Advisory." *Office of the Surgeon General, 2021*, www.hhs.gov/sites/default/files/surgeon-general-youth-mental-health-advisory.pdf. Accessed Sep. 6, 2025.

U.S. Senate Select Committee on Intelligence. "Senate Intel Committee Releases Bipartisan Report on Russia's Use of Social Media." *Senate Select Committee on Intelligence*, Oct. 8, 2019, http://www.intelligence.senate.gov/2019/10/08/press-senate-intel-committee-releases-bipartisan-report-russia-e2–80–99s-use-social-media/. Accessed Oct. 22, 2025.

Upload. Created by Greg Daniels, Amazon Prime Video, 2020–2023.

"US Night Fighter Radars of WWII." *HyperWar: U.S. Navy Reference*, ibiblio.org/hyperwar/USN/ref/NightFighterRadars/index.html. Accessed Aug. 23, 2025.

Uslu, Handan. *Understanding the Relationship Between Media Use and Maladaptive Daydreaming*. Diss. Georgetown University, 2015.

Vaisey, Stephen. "Socrates, Skinner, and Aristotle: Three Ways of Thinking About Culture in Action 1." *Sociological Forum*. Vol. 23. No. 3. Blackwell Publishing Ltd, 2008.

Valkhof, Bart, et al. "Data Volume Is Soaring. Here's How the ICT Sector Can Sustainably Handle the Surge." *World Economic Forum*, May 22, 2024, https://www.weforum.org/stories/2024/05/data-growth-drives-ict-energy-innovation/. Accessed Sep. 5, 2025.

Van Valen, Leigh. "A New Evolutionary Law (1973)." *Foundations of Macroecology; University of Chicago Press: Chicago, IL, USA*, 2014, pp. 284–314.

Veblen, Thorstein. *The Theory of the Leisure Class*. Oxford University Press, 2007.

Verhoef, Peter C., et al. "Digital Transformation: a Multidisciplinary Reflection and Research Agenda." *Journal of Business Research*, vol. 101, Nov. 2019, pp. 889–913.

Verne Harnish, *Scaling Up: How a Few Companies Make It...and Why the Rest Don't*. Gazelles, Inc., 2014.

Vesoulis, Abby. "Police Looking for Kidnapped Girl Tipped Off after Suspect Bought Diapers: Reports." *Newsweek*, Nov. 12, 2021, https://www.newsweek.com/police-looking-kidnapped-girl-tipped-off-after-suspect-bought-diapers-reports-1645999. Accessed Jul. 29, 2025.

Vico, Giambattista. *The New Science of Giambattista Vico*. Translated by Thomas Goddard Bergin and Max Harold Fisch, 3rd ed., Cornell University Press, 1984.

Villmoare, Brian, and Mark Grabowski. "Did the Transition to Complex Societies in the Holocene Drive a Reduction in Brain Size? A Reassessment of the DeSilva et al. (2021) Hypothesis." *Frontiers in Ecology and Evolution*, vol. 10, 2022, 963568.

Vinge, Vernor. "True Names." *Binary Star No. 5*, Dell, 1981.

Visual Capitalist, 8 Feb. 2025, http://www.visualcapitalist.com/visualized-the-1s-share-of-u-s-wealth-over-time-1989–2024/. Accessed Sep. 2, 2025.

Vogt, Kathleen. "Ancient Skepticism." *Stanford Encyclopedia of Philosophy*, Stanford University, Nov. 4, 1997, https://plato.stanford.edu/archives/fall2007/entries/skepticism-ancient/. Accessed Jun. 8, 2025.

Vollset, Stein Emil, et al. "Fertility, Mortality, Migration, and Population Scenarios for 195 Countries and Territories from 2017 to 2100: a Forecasting Analysis for the Global Burden of Disease Study." *The Lancet*, vol. 396, no. 10258, 2020, pp. 1285–1306.

"Volunteers Remember 9/11." *American Red Cross Greater New York*, American Red Cross, http://www.redcross.org/local/new-york/greater-new-york/volunteer/volunteers-remember-9–11.html. Accessed Sep. 27, 2025.

Vygotsky, Lev S. *Mind in Society: the Development of Higher Psychological Processes*. Harvard University Press, 1978.

Wall-Scheffler, Cara, et al. "The Myth of Man the Hunter: Women's Contribution to the Hunt across Ethnographic Contexts." *PLOS One*, vol. 18, no. 6, 2023, e028710.

Wallace, Alfred Russel. *Darwinism: an Exposition of the Theory of Natural Selection with Some of Its Applications*. Macmillan and Co., 1889.

Wallace, Lewis. "Don't Bogart That E-Mail." *Wired*, May 5, 2005.

Walter, Jens, and Ruth Ley. "The Human Gut Microbiome: Ecology and Recent Evolutionary Changes." *Annual Review of Microbiology*, vol. 65, no. 1, 2011, pp. 411–429.

Washington Post Staff. "Where Did the COVID Aid Money Go?" *The Washington Post*, Sep. 8, 2022, https://www.washingtonpost.com/business/interactive/2022/covid-money-trail-investigation-explained/. Accessed Aug. 26, 2025.

Watson, James D. "Discovering the Brain." *National Academy of Sciences*, Washington, DC, 1992.

Watson, John B. "Psychology as the Behaviorist Views It." *Psychological Review*, vol. 20, no. 2, 1913, pp. 158–177.

Watts, Duncan J. *Six Degrees: The Science of a Connected Age*. W. W. Norton, 2003.

Waxman, Olivia B. "The Electricity Lobby Was Behind the First National Christmas Tree Lighting." *Time*, Dec. 1, 2016. *Time.com*.

"We Read 150 Privacy Policies. They Were an Incomprehensible Disaster." *The New York Times*, Jun. 12, 2019, nytimes.com/interactive/2019/06/12/opinion/facebook-google-privacy-policies.html. Accessed Aug. 25, 2025.

Wegner, Daniel M. *The Illusion of Conscious Will*. MIT Press, 2002.

Wegner, Daniel M. "Transactive Memory: A Contemporary Analysis of the Group Mind." *Theories of Group Behavior*, edited by Brian Mullen and George R.Goethals, Springer, 1987, pp. 185–208.

Weinberger, David. *Everyday Chaos: Technology, Complexity, and How We're Thriving in a New World of Possibility*. Harvard Business Press, 2019.

Weinberger, David. *Too Big to Know: Rethinking Knowledge Now That the Facts Aren't the Facts, Experts Are Everywhere, and the Smartest Person in the Room Is the Room*. Basic Books, 2011.

Weisbord, Marvin R., and Sandra Janoff. *Don't Just Do Something, Stand There!: Ten Principles for Leading Meetings That Matter*. Berrett-Koehler Publishers, 2007.

Weise, Elizabeth, and Santucci, Jeanine ."AT&T Cell Service Outage Disasters." *USA Today*, Feb. 22, 2024, https://www.usatoday.com/story/news/nation/2024/02/22/att-cell-service-outage-disasters/72697379007/. Accessed Aug. 24, 2025.

Welman, Santino, et al. "Searching for the Haplorrhine Heterotherm: Field and Laboratory Evidence." *Frontiers in Physiology*, vol. 8, 2017, article 745.

Wharton, Edith. *The Ghost Stories of Edith Wharton*. Dover Publications, 2000.

"What Is 10G?", *Cable Labs*, https://www.cablelabs.com/10g. Accessed Aug. 19, 2025.

"What's Really Driving Disruption (It's Not Technology)", Wharton, Mar. 28, 2019, https://knowledge.wharton.upenn.edu/article/what-drives-disruption/ Accessed Dec. 23, 2025.

Wheeler, Michael. *Reconstructing the Cognitive World: The Next Step.* MIT Press, 2005.

White, Frances J., and Richard W. Wrangham. "Feeding Competition and Patch Size in the Chimpanzee Species *Pan paniscus* and *Pan troglodytes.*" *Behaviour*, vol. 105, nos. 1–2, 1988, pp. 148–64.

Whitelocks, Sadie. "Computer Games Leave Children with 'Dementia', Warns Top Neurologist." *Daily Mail*, Oct. 14, 2011, www.dailymail.co.uk/health/article-2049040/Computer-games-leave-children-dementia-warns-neurologist.html. Accessed Aug. 23, 2025.

Whiten, Andrew, et al. "Conformity to Cultural Norms of Tool Use in Chimpanzees." *Nature*, vol. 437, 2005, pp. 737–740.

Whiten, Andrew, et al. "Culture Evolves." *Philosophical Transactions of the Royal Society B: Biological Sciences* vol. 366.1567, 2011, pp. 938–948.

Whybrow, Peter C. *American Mania: When More Is Not Enough.* W.W. Norton & Company, 2005.

Wickens, Christopher D., and Daniel Gopher. "Control Theory Measures of Tracking as Indices of Attention Allocation Strategies." *Human Factors*, vol. 19, no. 4, 1977, pp. 349–365.

Wiener, Norbert. *Cybernetics: Or Control and Communication in the Animal and the Machine.* MIT Press, 1948.

Wilber, Ken. *A Brief History of Everything.* Shambhala, 1996.

Wilmer, Henry H., and Jason M. Chein. "Mobile Technology Habits: Patterns of Association Among Device Usage, Intertemporal Preference, Impulse Control, and Reward Sensitivity." *Psychonomic Bulletin & Review*, vol. 23, no. 5, 2016, pp. 1607–1614.

Wilson, Edward O. *The Social Conquest of Earth*, Liveright Pub. Corporation, 2012.

Wilson, Robert A. *Boundaries of the Mind: the Individual in the Fragile Sciences.* Cambridge University Press, 2004.

Wilson, Timothy D. *Strangers to Ourselves: Discovering the Adaptive Unconscious.* Harvard University Press, 2002.

Winner, Langdon. *The Whale and the Reactor: a Search for Limits in an Age of High Technology.* University of Chicago Press, 1986.

Wittman, Anna Blackburn, and L. Lewis Wall. "The Evolutionary Origins of Obstructed Labor: Bipedalism, Encephalization, and the Human Obstetric Dilemma." *Obstetrical & Gynecological Survey*, vol. 62, no. 11, 2007, pp. 739–748.

"Women in the Labor Force: a Databook." *U.S. Bureau of Labor Statistics*, Mar. 2006, bls.gov/opub/mlr/2006/03/art3full.pdf. Accessed Aug. 25, 2025.

Woollett, Katherine, and Eleanor A. Maguire. "Acquiring 'The Knowledge' of London's Layout Drives Structural Brain Changes.", *Current Biology*, vol. 21, no. 24, 2011, pp. 2109–2114.

"Word Lens." Wikipedia, updated 5 months ago, en.wikipedia.org/wiki/Word_Lens. Accessed Aug. 23, 2025.

World Economic Forum. "The Global Risks Report 2020. 15th ed.," *World Economic Fo rum*, Jan. 15, 2020, www3.weforum.org/docs/WEF_Global_Risk_Report_2020.pdf. Accessed Aug. 23, 2025.

World Economic Forum. "The Global Risks Report 2025. 20th ed.", *World Economic Forum*, Jan. 15, 2025, reports.weforum.org/docs/WEF_Global_Risks_Report_2025.pdf. Accessed Aug. 23, 2025.

"The World's Cities in 2018: Data Booklet." *United Nations*, 2018, un.org/en/events/citiesday/assets/pdf/the_worlds_cities_in_2018_data_booklet.pdf. Accessed Aug. 25, 2025.

Worrall, John. "Structural Realism: the Best of Both Worlds?" *Dialectica*, vol. 43, no. 1–2, 1989, pp. 99–124.

Wrangham, Richard. *Catching Fire: How Cooking Made Us Human*. Basic Books, 2009.

Wrensch, Margaret R., et al. "CYP1A1 Variants and Smoking-Related Lung Cancer in San Francisco Bay Area Latinos and African Americans." *International Journal of Cancer*, vol. 113, no. 1, 2005, pp. 141–147.

Wundt, Wilhelm. *The Language of Gestures*. Translated by J. S. Thayer, Mouton, 1973.

Wylie, Alison. "Archaeological Cables and Tacking: the Implications of Practice for Bernstein's 'Options beyond Objectivism and Relativism.'" *Philosophy of the Social Sciences*, vol. 19, no. 1, 1989, pp. 1–18.

Wynn, Thomas, and Frederick L. Coolidge. "Implications of a Strict Standard for Recognizing Modern Cognition in Prehistory." *Cognitive Archaeology and Human Evolution*, 2009, pp. 117–127.

"xAI." *xAI*, x.ai/. Accessed Aug. 25, 2025.

Yamanaka, Shinya. "Induced Pluripotent Stem Cells: Past, Present, and Future." *Cell Stem Cell*, vol. 10, no. 6, 2012, pp. 678–684.

Yang, Andrew. *The War on Normal People: The Truth About America's Disappearing Jobs and Why Universal Basic Income Is Our Future*. Hachette Books, 2018.

Yang, Lin, et al. "Trends in Sedentary Behavior among the US Population, 2001–2016." *JAMA*, vol. 321, no. 16, 2019, pp. 1587–1597.

Yeatman, Jason D., et al. "Development of White Matter and Reading Skills." *Proceedings of the National Academy of Sciences*, vol. 109.44, 2012, pp. E3045–E3053.

Yeats, William Butler. "The Second Coming." *The Poetry Foundation*, 2010, http://www.poetryfoundation.org/poems/43290/the-second-coming. Accessed Apr. 17, 2025.

"Your GPS Thinks You Are in Cairo, but This Is Not a Cyberattack." *Ynetnews*, Dec. 3, 2023, https://www.ynetnews.com/article/sygirj5bp. Accessed Aug. 24, 2025.

Zaino, Rami, Vian Ahmed, Ahmed M. Alhammadi, and Mohamad Alghoush. "Electric Vehicle Adoption: a Comprehensive Systematic Review of Technological, Environmental, Organizational and Policy Impacts." *World Electric Vehicle Journal*, vol. 15, no. 8, 2024, art. 375.

Zhang, Haobo, et al. "Grey Matter Correlates of Three Language Tests in Non-Demented Older Adults." *PLOS ONE*, vol. 8, no. 11, 2013, e80215.

Zheng, Yongnian. "Harmonious Society and Harmonious World: China's Policy Discourse under Hu Jintao." *China Policy Institute Briefing*, no. 26, 2007, https://www.nottingham.ac.uk/iaps/documents/cpi/briefings/briefing-26-harmonious-society-and-harmonious-world.pdf. Accessed Jul. 8, 2025.

Zimmerman, Robert A., and Larissa T. Bilaniuk. "Pediatric Head Trauma." *Neuroimaging Clinics of North America*, vol. 4, no. 2, 1994, pp. 349–366.

Zipes, Jack. *Fairy Tales and the Art of Subversion: the Classical Genre for Children and the Process of Civilization*. Routledge, 1983.

Zuboff, Shoshana. *The Age of Surveillance Capitalism: the Fight for a Human Future at the New Frontier of Power*. Profile Books, 2019.

Zuckerman, Ethan. "The Internet's Original Sin." *The Atlantic*, Aug. 14, 2014.

Index

www.ingramcontent.com/pod-product-compliance
Lightning Source LLC
LaVergne TN
LVHW020717290926
849104LV00003B/25
9789004737143